Table of Contents

Contents of Reference Volume 1

intro . . . . Introduction to base reference manual 1

about . . . . Display information about your Stata 6
adjust . . . . Tables of adjusted means and proportions 7
adoupdate . . . . Update user-written ado-files 24
alpha . . . . Compute interitem correlations (covariances) and Cronbach's alpha 28
ameans . . . . Arithmetic, geometric, and harmonic means 35
anova . . . . Analysis of variance and covariance 38
anova postestimation . . . . Postestimation tools for anova 80
areg . . . . Linear regression with a large dummy-variable set 97
areg postestimation . . . . Postestimation tools for areg 103
asclogit . . . . Alternative-specific conditional logit (McFadden's choice) model 105
asclogit postestimation . . . . Postestimation tools for asclogit 114
asmprobit . . . . Alternative-specific multinomial probit regression 121
asmprobit postestimation . . . . Postestimation tools for asmprobit 142
asroprobit . . . . Alternative-specific rank-ordered probit regression 151
asroprobit postestimation . . . . Postestimation tools for asroprobit 162

BIC note . . . . Calculating and interpreting BIC 169
binreg . . . . Generalized linear models: Extensions to the binomial family 174
binreg postestimation . . . . Postestimation tools for binreg 183
biprobit . . . . Bivariate probit regression 186
biprobit postestimation . . . . Postestimation tools for biprobit 193
bitest . . . . Binomial probability test 196
bootstrap . . . . Bootstrap sampling and estimation 200
bootstrap postestimation . . . . Postestimation tools for bootstrap 220
boxcox . . . . Box–Cox regression models 224
boxcox postestimation . . . . Postestimation tools for boxcox 234
brier . . . . Brier score decomposition 236
bsample . . . . Sampling with replacement 242
bstat . . . . Report bootstrap results 249

centile . . . . Report centile and confidence interval 256
ci . . . . Confidence intervals for means, proportions, and counts 262
clogit . . . . Conditional (fixed-effects) logistic regression 274
clogit postestimation . . . . Postestimation tools for clogit 288
cloglog . . . . Complementary log-log regression 292
cloglog postestimation . . . . Postestimation tools for cloglog 300
cnreg . . . . Censored-normal regression 302
cnreg postestimation . . . . Postestimation tools for cnreg 307
cnsreg . . . . Constrained linear regression 310
cnsreg postestimation . . . . Postestimation tools for cnsreg 315
constraint . . . . Define and list constraints 317
copyright . . . . Display copyright information 320
correlate . . . . Correlations (covariances) of variables or coefficients 321
cumul . . . . Cumulative distribution 329
cusum . . . . Graph cumulative spectral distribution 333

db . . . . Launch dialog 337

diagnostic plots . . . . Distributional diagnostic plots 339

display . . . . Substitute for a hand calculator 350

do . . . . Execute commands from a file 351

doedit . . . . Edit do-files and other text files 352

dotplot . . . . Comparative scatterplots 353

dstdize . . . . Direct and indirect standardization 360

dydx . . . . Calculate numeric derivatives and integrals 375

eivreg . . . . Errors-in-variables regression 380

eivreg postestimation . . . . Postestimation tools for eivreg 384

error messages . . . . Error messages and return codes 386

estat . . . . Postestimation statistics 387

estimates . . . . Save and manipulate estimation results 395

estimates describe . . . . Describe estimation results 399

estimates for . . . . Repeat postestimation command across models 401

estimates notes . . . . Add notes to estimation results 403

estimates replay . . . . Redisplay estimation results 405

estimates save . . . . Save and use estimation results 407

estimates stats . . . . Model statistics 410

estimates store . . . . Store and restore estimation results 412

estimates table . . . . Compare estimation results 414

estimates title . . . . Set title for estimation results 419

estimation options . . . . Estimation options 420

exit . . . . Exit Stata 422

exlogistic . . . . Exact logistic regression 423

exlogistic postestimation . . . . Postestimation tools for exlogistic 440

expoisson . . . . Exact Poisson regression 444

expoisson postestimation . . . . Postestimation tools for expoisson 456

fracpoly . . . . Fractional polynomial regression 458

fracpoly postestimation . . . . Postestimation tools for fracpoly 472

frontier . . . . Stochastic frontier models 476

frontier postestimation . . . . Postestimation tools for frontier 490

gllamm . . . . Generalized linear and latent mixed models 492

glm . . . . Generalized linear models 493

glm postestimation . . . . Postestimation tools for glm 523

glogit . . . . Logit and probit regression for grouped data 529

glogit postestimation . . . . Postestimation tools for glogit, gprobit, blogit, and bprobit 538

grmeanby . . . . Graph means and medians by categorical variables 540

hausman . . . . Hausman specification test 542

heckman . . . . Heckman selection model 551

heckman postestimation . . . . Postestimation tools for heckman 564

heckprob . . . . Probit model with sample selection 570

heckprob postestimation . . . . Postestimation tools for heckprob 577

help . . . . Display online help 582

hetprob . . . . Heteroskedastic probit model 584

hetprob postestimation . . . . Postestimation tools for hetprob 591

histogram . . . . Histograms for continuous and categorical variables 594

hsearch . . . . Search help files 605

STATA BASE REFERENCE MANUAL
VOLUME 1
A–H
RELEASE 10

A Stata Press Publication
StataCorp LP
College Station, Texas

Stata Press, 4905 Lakeway Drive, College Station, Texas 77845

The suggested citation for this software is

StataCorp. 2007. *Stata Statistical Software: Release 10*. College Station, TX: StataCorp LP.

Contents of Reference Volume 2

inequality .... Inequality measures   1
intreg .... Interval regression   4
intreg postestimation .... Postestimation tools for intreg 12
ivprobit .... Probit model with endogenous regressors 15
ivprobit postestimation .... Postestimation tools for ivprobit 26
ivregress .... Single-equation instrumental-variables regression 30
ivregress postestimation .... Postestimation tools for ivregress 45
ivtobit .... Tobit model with endogenous regressors 57
ivtobit postestimation .... Postestimation tools for ivtobit 66

jackknife .... Jackknife estimation 70
jackknife postestimation .... Postestimation tools for jackknife 82

kappa .... Interrater agreement 83
kdensity .... Univariate kernel density estimation 97
ksmirnov .... Kolmogorov–Smirnov equality-of-distributions test 107
kwallis .... Kruskal–Wallis equality-of-populations rank test 111

ladder .... Ladder of powers 114
level .... Set default confidence level 121
lincom .... Linear combinations of estimators 123
linktest .... Specification link test for single-equation models 130
lnskew0 .... Find zero-skewness log or Box–Cox transform 136
log .... Echo copy of session to file or device 140
logistic .... Logistic regression, reporting odds ratios 144
logistic postestimation .... Postestimation tools for logistic 153
logit .... Logistic regression, reporting coefficients 178
logit postestimation .... Postestimation tools for logit 190
loneway .... Large one-way ANOVA, random effects, and reliability 194
lowess .... Lowess smoothing 200
lpoly .... Kernel-weighted local polynomial smoothing 206
lrtest .... Likelihood-ratio test after estimation 216
lv .... Letter-value displays 226

matsize .... Set the maximum number of variables in a model 232
maximize .... Details of iterative maximization 234
mean .... Estimate means 241
mean postestimation .... Postestimation tools for mean 251
meta .... Meta-analysis 253
mfp .... Multivariable fractional polynomial models 257
mfp postestimation .... Postestimation tools for mfp 268
mfx .... Obtain marginal effects or elasticities after estimation 269
mkspline .... Linear and restricted cubic spline construction 285
ml .... Maximum likelihood estimation 291
mlogit .... Multinomial (polytomous) logistic regression 315
mlogit postestimation .... Postestimation tools for mlogit 327
more .... The —more— message 335
mprobit .... Multinomial probit regression 337
mprobit postestimation .... Postestimation tools for mprobit 343
mvreg .... Multivariate regression 346
mvreg postestimation .... Postestimation tools for mvreg 352

nbreg . . . . Negative binomial regression 354
nbreg postestimation . . . . Postestimation tools for nbreg and gnbreg 365
nestreg . . . . Nested model statistics 369
net . . . . Install and manage user-written additions from the Internet 375
net search . . . . Search the Internet for installable packages 393
netio . . . . Control Internet connections 397
news . . . . Report Stata news 400
nl . . . . Nonlinear least-squares estimation 401
nl postestimation . . . . Postestimation tools for nl 420
nlcom . . . . Nonlinear combinations of estimators 422
nlogit . . . . Nested logit regression 433
nlogit postestimation . . . . Postestimation tools for nlogit 454
nlsur . . . . Estimation of nonlinear systems of equations 459
nlsur postestimation . . . . Postestimation tools for nlsur 479
nptrend . . . . Test for trend across ordered groups 481

ologit . . . . Ordered logistic regression 485
ologit postestimation . . . . Postestimation tools for ologit 493
oneway . . . . One-way analysis of variance 497
oprobit . . . . Ordered probit regression 507
oprobit postestimation . . . . Postestimation tools for oprobit 511
orthog . . . . Orthogonalize variables and compute orthogonal polynomials 514

pcorr . . . . Partial correlation coefficients 520
permute . . . . Monte Carlo permutation tests 522
pk . . . . Pharmacokinetic (biopharmaceutical) data 532
pkcollapse . . . . Generate pharmacokinetic measurement dataset 540
pkcross . . . . Analyze crossover experiments 543
pkequiv . . . . Perform bioequivalence tests 551
pkexamine . . . . Calculate pharmacokinetic measures 557
pkshape . . . . Reshape (pharmacokinetic) Latin-square data 563
pksumm . . . . Summarize pharmacokinetic data 571
poisson . . . . Poisson regression 576
poisson postestimation . . . . Postestimation tools for poisson 585
predict . . . . Obtain predictions, residuals, etc., after estimation 590
predictnl . . . . Obtain nonlinear predictions, standard errors, etc., after estimation 601
probit . . . . Probit regression 613
probit postestimation . . . . Postestimation tools for probit and dprobit 629
proportion . . . . Estimate proportions 633
proportion postestimation . . . . Postestimation tools for proportion 638
prtest . . . . One- and two-sample tests of proportions 639

Contents of Reference Volume 3

qc . . . . Quality control charts 1
qreg . . . . Quantile regression 14
qreg postestimation . . . . Postestimation tools for qreg, iqreg, sqreg, and bsqreg 33
query . . . . Display system parameters 35

ranksum . . . . Equality tests on unmatched data 41
ratio . . . . Estimate ratios 47
ratio postestimation . . . . Postestimation tools for ratio 55
reg3 . . . . Three-stage estimation for systems of simultaneous equations 56

reg3 postestimation . . . . Postestimation tools for reg3   76
regress . . . . Linear regression   79
regress postestimation . . . . Postestimation tools for regress 102
regress postestimation time series . . . . Postestimation tools for regress with time series 146
#review . . . . Review previous commands 156
roc . . . . Receiver operating characteristic (ROC) analysis 157
rocfit . . . . Fit ROC models 178
rocfit postestimation . . . . Postestimation tools for rocfit 184
rologit . . . . Rank-ordered logistic regression 188
rologit postestimation . . . . Postestimation tools for rologit 203
rreg . . . . Robust regression 205
rreg postestimation . . . . Postestimation tools for rreg 211
runtest . . . . Test for random order 213

sampsi . . . . Sample size and power determination 219
saved results . . . . Saved results 230
scobit . . . . Skewed logistic regression 234
scobit postestimation . . . . Postestimation tools for scobit 242
sdtest . . . . Variance-comparison tests 245
search . . . . Search Stata documentation 251
serrbar . . . . Graph standard error bar chart 257
set . . . . Overview of system parameters 260
set_defaults . . . . Reset system parameters to original Stata defaults 271
signrank . . . . Equality tests on matched data 273
simulate . . . . Monte Carlo simulations 279
sj . . . . Stata Journal and STB installation instructions 286
sktest . . . . Skewness and kurtosis test for normality 289
slogit . . . . Stereotype logistic regression 293
slogit postestimation . . . . Postestimation tools for slogit 305
smooth . . . . Robust nonlinear smoother 308
spearman . . . . Spearman's and Kendall's correlations 315
spikeplot . . . . Spike plots and rootograms 324
ssc . . . . Install and uninstall packages from SSC 328
stem . . . . Stem-and-leaf displays 334
stepwise . . . . Stepwise estimation 338
suest . . . . Seemingly unrelated estimation 347
summarize . . . . Summary statistics 365
sunflower . . . . Density-distribution sunflower plots 373
sureg . . . . Zellner's seemingly unrelated regression 379
sureg postestimation . . . . Postestimation tools for sureg 387
swilk . . . . Shapiro–Wilk and Shapiro–Francia tests for normality 390
symmetry . . . . Symmetry and marginal homogeneity tests 394

table . . . . Tables of summary statistics 402
tabstat . . . . Display table of summary statistics 412
tabulate oneway . . . . One-way tables of frequencies 417
tabulate twoway . . . . Two-way tables of frequencies 425
tabulate, summarize() . . . . One- and two-way tables of summary statistics 441
test . . . . Test linear hypotheses after estimation 446
testnl . . . . Test nonlinear hypotheses after estimation 464
tetrachoric . . . . Tetrachoric correlations for binary variables 472
tobit . . . . Tobit regression 482

tobit postestimation . . . . Postestimation tools for tobit   488
total . . . . Estimate totals   492
total postestimation . . . . Postestimation tools for total   498
translate . . . . Print and translate logs   500
treatreg . . . . Treatment-effects model   509
treatreg postestimation . . . . Postestimation tools for treatreg   520
truncreg . . . . Truncated regression   523
truncreg postestimation . . . . Postestimation tools for truncreg   529
ttest . . . . Mean-comparison tests   532

update . . . . Update Stata   541

vce_option . . . . Variance estimators   546
view . . . . View files and logs   551
vwls . . . . Variance-weighted least squares   554
vwls postestimation . . . . Postestimation tools for vwls   560

which . . . . Display location and version for an ado-file   561

xi . . . . Interaction expansion   563

zinb . . . . Zero-inflated negative binomial regression   573
zinb postestimation . . . . Postestimation tools for zinb   579
zip . . . . Zero-inflated Poisson regression   581
zip postestimation . . . . Postestimation tools for zip   587
ztnb . . . . Zero-truncated negative binomial regression   589
ztnb postestimation . . . . Postestimation tools for ztnb   596
ztp . . . . Zero-truncated Poisson regression   599
ztp postestimation . . . . Postestimation tools for ztp   603

Cross-Referencing the Documentation

When reading this manual, you will find references to other Stata manuals. For example,

[U] **26 Overview of Stata estimation commands**
[XT] **xtabond**
[D] **reshape**

The first is a reference to chapter 26, *Overview of Stata estimation commands* in the *Stata User's Guide*, the second is a reference to the xtabond entry in the *Longitudinal/Panel-Data Reference Manual*, and the third is a reference to the reshape entry in the *Data Management Reference Manual*.

All the manuals in the Stata Documentation have a shorthand notation:

[GSM]   *Getting Started with Stata for Macintosh*
[GSU]   *Getting Started with Stata for Unix*
[GSW]   *Getting Started with Stata for Windows*

[U]   *Stata User's Guide*

[R]   *Stata Base Reference Manual*

[D]   *Stata Data Management Reference Manual*

[G]   *Stata Graphics Reference Manual*

[P]   *Stata Programming Reference Manual*

[XT]   *Stata Longitudinal/Panel-Data Reference Manual*

[MV]   *Stata Multivariate Statistics Reference Manual*

[SVY]   *Stata Survey Data Reference Manual*

[ST]   *Stata Survival Analysis and Epidemiological Tables Reference Manual*

[TS]   *Stata Time-Series Reference Manual*

[I]   *Stata Quick Reference and Index*

[M]   *Mata Reference Manual*

Detailed information about each of these manuals may be found online at

http://www.stata-press.com/manuals/

Title

intro — Introduction to base reference manual

Description

This entry describes the organization of the reference manuals.

Remarks

The complete list of reference manuals is as follows:

[R]	*Stata Base Reference Manual*
	Volume 1, A–H
	Volume 2, I–P
	Volume 3, Q–Z
[D]	*Stata Data Management Reference Manual*
[G]	*Stata Graphics Reference Manual*
[P]	*Stata Programming Reference Manual*
[XT]	*Stata Longitudinal/Panel-Data Reference Manual*
[MV]	*Stata Multivariate Statistics Reference Manual*
[SVY]	*Stata Survey Data Reference Manual*
[ST]	*Stata Survival Analysis and Epidemiological Tables Reference Manual*
[TS]	*Stata Time-Series Reference Manual*
[I]	*Stata Quick Reference and Index*
[M]	*Mata Reference Manual*

When we refer to "reference manuals", we mean all manuals listed above.

When we refer to the *Stata* reference manuals, we mean all manuals listed above except the *Mata Reference Manual*.

When we refer to the *Base Reference Manual*, we mean just the three-volume *Base Reference Manual*, known as [R].

When we refer to the specialty manuals, we mean all the manuals listed above, except [R] and [I], the *Quick Reference and Index*.

Detailed information about each of these manuals can be found online at

http://www.stata-press.com/manuals/

1

Arrangement of the Stata Reference manuals

Each manual contains the following sections:

- Table of Contents.
  At the beginning of volume 1 of [R], the *Stata Base Reference Manual*, is a table of contents for the three volumes.

- Cross-Referencing the Documentation.
  This section lists all the manuals and explains how they are cross-referenced.

- Introduction.
  This entry—usually called intro—provides an overview of the manual. In the specialty manuals, this introduction suggests entries that you might want to read first and provides information about new features.

  Each specialty manual contains an overview of the commands described in it.

- Entries.
  Entries are arranged in alphabetical order. Most entries describe Stata commands, but some entries discuss concepts, and others provide overviews.

  Entries that describe estimation commands are followed by an entry discussing postestimation commands that are available for use after the estimation command. For example, the **xtlogit** entry in the [XT] manual is followed by the **xtlogit postestimation** entry.

- Index.
  At the end of each manual is an index. The index for the entire three-volume *Stata Base Reference Manual* is found at the end of the third volume.

The *Stata Quick Reference and Index*, [I], contains a combined index for all the manuals and a subject table of contents for all the manuals and the *User's Guide*. It also contains quick-reference information on many subjects, such as the estimation commands.

To find information and commands quickly, use Stata's search command; see [R] **search** (see the entry search in the [R] manual). You can broaden your search to the Internet by using search, all to find commands and extensions written by Stata users.

Arrangement of each entry

Entries in the *Stata Reference* manuals generally contain the following sections:

> *Syntax*
> *Description*
> *Options*
> *Remarks*
> *Saved Results*
> *Methods and Formulas*
> *References*
> *Also See*

Syntax

A command's syntax diagram shows how to type the command, indicates all possible options, and gives the minimal allowed abbreviations for all the items in the command. For instance, the syntax diagram for the summarize command is

<u>su</u>mmarize [*varlist*] [*if*] [*in*] [*weight*] [, *options*]

options	description
Main	
<u>d</u>etail	display additional statistics
<u>mean</u>only	suppress the display; calculate only the mean; programmer's option
<u>f</u>ormat	use variable's display format
<u>separator</u>(#)	draw separator line after every # variables; default is separator(5)

varlist may contain time-series operators; see [U] **11.4.3 Time-series varlists**.

by is allowed; see [D] **by**.

aweights, fweights, and iweights are allowed. However, iweights may not be used with the detail option; see [U] **11.1.6 weight**.

Items in the typewriter-style font should be typed exactly as they appear in the diagram, although they may be abbreviated. <u>Underlining</u> indicates the shortest abbreviations where abbreviations are allowed. For instance, <u>su</u>mmarize may be abbreviated su, sum, summ, etc., or it may be spelled out completely. Items in the typewriter font that are not underlined may not be abbreviated.

Square brackets denote optional items. In the syntax diagram above, *varlist*, *if*, *in*, *weight*, and the *options* are optional.

The *options* are listed in a table immediately following the diagram, along with a brief description of each.

Items typed in *italics* represent arguments for which you are to substitute variable names, observation numbers, and the like.

The diagrams use the following symbols:

#	Indicates a literal number, e.g., 5; see [U] **12.2 Numbers**.
[ ]	Anything enclosed in brackets is optional.
{ }	At least one of the items enclosed in braces must appear.
|	The vertical bar separates alternatives.
%fmt	Any Stata format, e.g., %8.2f; see [U] **12.5 Formats: controlling how data are displayed**.
depvar	The dependent variable in an estimation command; see [U] **20 Estimation and postestimation commands**.
exp	Any algebraic expression, e.g., (5+myvar)/2; see [U] **13 Functions and expressions**.
filename	Any filename; see [U] **11.6 File-naming conventions**.
indepvars	The independent variables in an estimation command; see [U] **20 Estimation and postestimation commands**.
newvar	A variable that will be created by the current command; see [U] **11.4.2 Lists of new variables**.
numlist	A list of numbers; see [U] **11.1.8 numlist**.
oldvar	A previously created variable; see [U] **11.4.1 Lists of existing variables**.
options	A list of options; see [U] **11.1.7 options**.
range	An observation range, e.g., 5/20; see [U] **11.1.4 in range**.
"string"	Any string of characters enclosed in double quotes; see [U] **12.4 Strings**.

varlist	A list of variable names; see [U] **11.4 varlists**. If *varlist* allows time-series operators, a note to that effect will be shown below the syntax diagram; see [U] **11.4.3 Time-series varlists**.
varname	A variable name; see [U] **11.3 Naming conventions**.
weight	A [*wgttype=exp*] modifier; see [U] **11.1.6 weight** and [U] **20.17 Weighted estimation**.
xvar	The variable to be displayed on the horizontal axis.
yvar	The variable to be displayed on the vertical axis.

The *Syntax* section will indicate whether time-series operators may be used with a command. `summarize` allows time-series operators.

If a command allows prefix commands, this will be indicated immediately following the table of options. The `summarize` command allows `by`.

If a command allows weights, the types of weights allowed will be specified, with the default weight listed first. `summarize` allows `aweights`, `fweights`, and `iweights`, and if the type of weight is not specified, the default is `aweights`.

Description

Following the syntax diagram is a brief description of the purpose of the command.

Options

If the command allows any options, they are explained here, and for dialog users the location of the options in the dialog is indicated. For instance, in the **logistic** entry in this manual, the *Options* section looks like this:

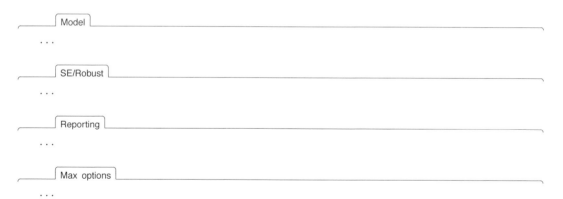

Remarks

The explanations under *Description* and *Options* are exceedingly brief and technical; they are designed to provide a quick summary. The remarks explain in English what the preceding technical jargon means. Examples are used to illustrate the command.

Saved Results

Commands are classified as e-class, r-class, s-class, or n-class, according to whether they save calculated results in e(), r(), s(), or not at all. These results can then be used in subroutines by other programs (ado-files). Such saved results are documented here; see [U] **18.8 Accessing results calculated by other programs** and [U] **18.9 Accessing results calculated by estimation commands**.

Methods and Formulas

The techniques and formulas used in obtaining the results are described here as tersely and technically as possible. If a command is implemented as an ado-file, that is indicated here.

References

Published sources are listed that either were directly referenced in the preceding text or might be of interest.

Also See

Other manual entries are listed that might also interest you.

Also See

[U] **1.1 Getting Started with Stata**

Title

> **about** — Display information about your Stata

Syntax

```
about
```

Description

about displays information about your version of Stata.

Remarks

about displays information about the release number, flavor, serial number, and license for your Stata. If you are running Stata for Windows, information about memory is also displayed:

```
. about

Stata/SE 10.0 for Windows
Born 28 Jul 2007
Copyright (C) 1985-2007

Total physical memory:       506600 KB
Available physical memory:    83040 KB

10-user Stata for Windows (network) perpetual license:
        Serial number:  8199040000
          Licensed to:  Alan R. Riley
                        StataCorp
```

or

```
. about

Stata/IC 10.0 for Macintosh
Born 28 Jul 2007
Copyright (C) 1985-2007

Single-user Stata for Macintosh perpetual license:
        Serial number:  49040001
          Licensed to:  Chinh Nguyen
                        StataCorp
```

Also See

[R] **which** — Display location and version for an ado-file

[U] **5 Flavors of Stata**

Title

adjust — Tables of adjusted means and proportions

Syntax

adjust [*var* [= #] ...] [*if*] [*in*] [, *options*]

options	description
Main	
by(*varlist*)	compute and display predictions for each level of variables
Options	
xb	linear prediction; the default
pr	predicted probabilities
exp	exponentiated linear prediction
se	display standard error of the prediction; default is none
stdf	display standard error of the forecast; default is none
ci	display confidence or prediction intervals
level(#)	set confidence level; default is level(95)
vertical	stack confidence intervals
equation(*eqno*)	use *eqno* equation in a multiple-equation system
nooffset	ignore offset or exposure variable (if any) when making predictions
generate(*newvar1* [*newvar2*])	generate prediction variable, error variable
More options	
replace	replace data in memory with table
label(*text*)	prediction label
selabel(*text*)	error-term label
cilabel(*text*)	confidence-interval label
format(%*fmt*)	display format for numbers in table
nokey	suppress table key
noheader	suppress table header
center	center numbers in cells; default is to right-align
left	left-align column labels; default is to right-align
cellwidth(#)	cell width
csepwidth(#)	column separation
scsepwidth(#)	supercolumn separation
stubwidth(#)	left stub width

Description

After an estimation command, adjust provides adjusted predictions of $\mathbf{x}\beta$ (the means in a linear-regression setting), probabilities (available after some estimation commands), or exponentiated linear predictions. The estimate is computed for each level of the by() variables, setting the variables

7

specified in [*var* [= #] . . .] to their mean or to the specified number if the = # part is specified. If by() is not specified, adjust produces results as if by() defined one group. Variables used in the estimation command but not included in either the by() variable list or the adjust variable list are left at their current values, observation by observation. Here adjust displays the average estimated prediction (or the corresponding probability or exponentiated prediction), substituting the mean of these unspecified variables within each group defined by the variables in the by() option.

Options

⌐ Main ⌐

by(*varlist*) specifies the variables whose levels determine the subsets of the data for which adjusted predictions are to be computed. The variables in by() need not have been involved in the original estimation command. A maximum of seven variables may be specified in the by() option. If by() is not specified, the results are computed treating all the data as one group.

⌐ Options ⌐

xb, the default, specifies that the linear prediction from the estimation command be displayed. This produces predicted value (a mean in the linear-regression setting) and is equivalent to the xb option of predict. Depending on the estimation command, the xb values may not be in the original units of the dependent variable.

pr is an alternative to xb that specifies that predicted probability be displayed. The pr option is not available after all commands.

exp is an alternative to xb that specifies that the exponentiated linear prediction, $\exp(\mathbf{x}\beta)$, be displayed. Depending on the estimation command, the resulting quantity might be called an "incidence rate", a "hazard ratio", etc.

se specifies that the standard error of the linear prediction be displayed. This is equivalent to the stdp option of predict.

stdf specifies that the standard error of the forecast of the linear prediction be displayed. This is equivalent to the stdf option of predict and is available only after estimation commands that support the stdf predict option.

ci specifies that a confidence interval be displayed. The confidence interval is for the displayed estimate as determined by the xb, pr, or exp option—producing an interval for the adjusted linear prediction, probability, or exponentiated linear prediction. When stdf is specified, a prediction interval is produced, which is, by definition, wider than the corresponding confidence interval.

level(#) specifies the confidence level, as a percentage, for confidence or prediction intervals. The default is level(95) or as set by set level; see [R] **level**.

vertical requests that the endpoints of confidence or prediction intervals be stacked vertically on display. You must specify more than two variables in the by() option for vertical results to be produced.

equation(*eqno*) specifies the equation in a multiple-equation system that is to be used in the adjust command. This option is allowed only after multiple-equation estimation commands.

nooffset is relevant only if you specified offset(*varname*) or exposure(*varname*) when you fitted your model. It modifies the calculations made by adjust so that they ignore the offset or exposure variable.

generate(*newvar1* [*newvar2*]) generates one or two new variables. If one variable is specified, the adjusted linear prediction for each observation is generated in *newvar1* (holding the appropriate variables to their means or to other specified values). If pr is specified, the adjusted linear prediction is transformed to a probability. If exp is specified, the exponentiated prediction is returned. If *newvar2* is specified, the standard error from either the se option or the stdf option is placed in the second variable.

─────| More options |──

replace specifies that the data in memory be replaced with data containing 1 observation per cell corresponding to the table produced by the adjust command.

label(*text*), selabel(*text*), and cilabel(*text*) allow you to change the labels for the displayed predictions (from the xb, pr, or exp option), error terms (from the se or stdf option), and confidence intervals (from the ci option). label() and selabel() also change the variable labels for the variables created by the generate() option.

format(*%fmt*) specifies the display format for presenting the numbers in the table; see [U] **12.5 Formats: controlling how data are displayed**. format(%8.0g) is the default. Standard errors and confidence intervals are further formatted for output by automatic enclosure within parentheses or square brackets.

nokey and noheader suppress the display of the table key and header information.

center specifies that results be centered in the table's cells. The default is to right-align the results. For centering to work well, specify a display format, too, such as format(%9.2f).

left specifies that column labels be left-aligned. The default is to right-align them to distinguish them from supercolumn labels, which are left-aligned.

cellwidth(*#*) specifies the width of the cell in units of digit widths; 10 means the space occupied by 10 digits, which is 0123456789. The default is not a fixed number but a number chosen to spread the table out while presenting a reasonable number of columns across the page.

csepwidth(*#*) specifies the separation between columns in units of digit widths. The default is not a fixed number but a number chosen according to what Stata thinks looks best.

scsepwidth(*#*) specifies the separation between supercolumns in units of digit widths. The default is not a fixed number but a number chosen according to what Stata thinks looks best.

stubwidth(*#*) specifies the width of the left stub of the table in units of digit widths. The default is not a fixed number but a number chosen according to what Stata thinks looks best.

Remarks

adjust is a postestimation command; see [U] **20 Estimation and postestimation commands**. adjust is really a front-end process for predict; see [R] **predict**. It sets up the values at which predictions are desired and then displays the predictions in tabular form; the data remain unchanged. adjust's options control the labeling of the predictions, errors, and confidence intervals. tabdisp is used to produce the final table; see [P] **tabdisp**.

If you have restricted your estimation command to a portion of the data by using if or in, then you will generally want to use the same conditions with adjust. This task is easily done by including if e(sample) with the adjust command. However, there may be legitimate reasons for using different data to perform the estimation and to obtain adjusted predictions (i.e., out-of-sample adjusted predictions).

If you perform an estimation command using weights and follow that with the `adjust` command, letting some of the variables vary observation by observation, the weights will not be used in constructing the adjusted table. This fact is not a concern when all variables of the estimation command are specified in `adjust`'s variable list and `by()` option variable list.

An interesting way to use `adjust` is after using `xi` with the estimation command; see example 5 below. Another interesting use of `adjust` is to specify a categorical variable both in the variable list being set to a specific value and as a `by` variable. This method is helpful in examining the predictions for several groups as if they were set at a particular group's value (possibly also holding some other variables to certain values).

▷ Example 1

Using the automobile dataset, let's understand automobile price as a function of whether the car was manufactured domestically. We first look at the average price for domestic and foreign cars.

```
. use http://www.stata-press.com/data/r10/auto
(1978 Automobile Data)
. table foreign, c(mean price sd price) format(%8.3f)
```

Car type	mean(price)	sd(price)
Domestic	6072.423	3097.104
Foreign	6384.682	2621.915

These average prices do not seem far apart. However, we do not think that these averages are telling us the full story since there are other variables that affect the price of automobiles. As an (admittedly fake) first attempt at modeling automobile price, we decide to perform a regression.

```
. regress price mpg weight turn foreign
```

Source	SS	df	MS
Model	334771309	4	83692827.3
Residual	300294087	69	4352088.22
Total	635065396	73	8699525.97

Number of obs =	74
F(4, 69) =	19.23
Prob > F =	0.0000
R-squared =	0.5271
Adj R-squared =	0.4997
Root MSE =	2086.2

price	Coef.	Std. Err.	t	P>|t|	[95% Conf. Interval]
mpg	-.4660076	73.51407	-0.01	0.995	-147.1226	146.1905
weight	4.284532	.7404967	5.79	0.000	2.807282	5.761782
turn	-229.2059	114.2423	-2.01	0.049	-457.1131	-1.298679
foreign	3221.415	706.4847	4.56	0.000	1812.017	4630.813
_cons	1368.197	4887.597	0.28	0.780	-8382.292	11118.69

We now ask ourselves what the prices of domestic and foreign cars would be if they were compared at identical values of the explanatory variables `mpg`, `weight`, and `turn` under this regression model. We compare the prices predicted from the model with these three variables set at their respective mean values. The `adjust` command provides the answer.

```
. adjust mpg weight turn, by(foreign) se ci center
```

Dependent variable: price   Command: regress
Covariates set to mean: mpg = 21.297297, weight = 3019.4595, turn = 39.648649

Car type	xb	stdp	lb	ub
Domestic	5207.54	(320.823)	[4567.52	5847.56]
Foreign	8428.95	(552.515)	[7326.72	9531.19]

Key: xb   = Linear Prediction
     stdp   = Standard Error
     [lb , ub] = [95% Confidence Interval]

We see that, under this regression model, foreign cars are predicted to be more expensive than domestic cars when compared at the average values of mpg, weight, and turn. The standard errors and confidence intervals are also presented. What if, instead of obtaining prediction standard errors and confidence intervals, we wanted to know the forecast standard errors and prediction intervals?

```
. adjust mpg weight turn, by(foreign) stdf ci center
```

Dependent variable: price   Command: regress
Covariates set to mean: mpg = 21.297297, weight = 3019.4595, turn = 39.648647

Car type	xb	stdf	lb	ub
Domestic	5207.54	(2110.69)	[996.826	9418.25]
Foreign	8428.95	(2158.09)	[4123.68	12734.2]

Key: xb   = Linear Prediction
     stdf   = Standard Error (forecast)
     [lb , ub] = [95% Prediction Interval]

In these two tables, the reported means are the same, but their reported standard errors differ. The first table uses the se option (stdp option of predict), which reports the standard error of the linear prediction. This standard error measures uncertainty of the mean originating from the uncertainty of the estimated model coefficients. If our model had been fitted on an infinite population, we would be certain of the model's coefficients and hence the mean of each population. These standard errors would then be zero.

Even if we knew the mean with certainty—even if the stdp standard errors were zero—we would be uncertain about the price of each car. Individual cars have individual prices drawn from a distribution whose mean we would know: $\text{price}_j = \mu_j + \epsilon_j$.

In the second table, the reported standard errors are noticeably larger than those in the first. These standard errors reflect our total uncertainty of the price of individual cars, $\sqrt{\sigma_\mu^2 + \sigma_\epsilon^2}$. This uncertainty is based on our uncertainty of the mean itself, σ_μ—the prediction's standard error—and the inherent uncertainty because of the unmeasured characteristics of the individual cars themselves, σ_ϵ, the residual standard error. These two components, appropriately combined, are called the forecast standard error and are obtained with the stdf option.

The first table presents confidence intervals based on the stdp standard error, and the second table presents what are commonly called prediction intervals based on the larger stdf forecast standard errors. Confidence intervals are intervals on our prediction for the mean. Prediction intervals are intervals on our prediction for individual observations.

◁

▷ Example 2

We also want to know the predicted cost for domestic and foreign cars when some of the variables are set at particular values. For instance, if we want to compare the predicted price when mpg is 25, turn is 35.2, and weight is at its mean value, we type the following:

. adjust mpg=25 turn=35.2 weight, by(foreign)

```
        Dependent variable: price      Command: regress
     Covariate set to mean: weight = 3019.4595
   Covariates set to value: mpg = 25, turn = 35.2
```

Car type	xb
Domestic	6225.47
Foreign	9446.88

Key: xb = Linear Prediction

Imagine that we now want to find the predicted average price of foreign and domestic cars under this regression model when mpg and weight are set to their overall means but the turn variable is left alone, meaning that it takes on the values observed car by car.

. adjust mpg weight, by(foreign) se gen(pred err)

```
        Dependent variable: price      Command: regress
         Created variables: pred, err
       Variable left as is: turn
     Covariates set to mean: mpg = 21.297297, weight = 3019.4595
```

Car type	xb	stdp
Domestic	4796.42	(342.766)
Foreign	9400.69	(621.791)

Key: xb   = Linear Prediction
      stdp = Standard Error

Specifying gen(pred err) generates prediction and error variables that we can use. Let's look at a few observations and compare them with the actual price and the predicted price by using predict with no constraints.

```
. predict pred2, xb
. list foreign turn price pred2 pred err in 47/58, sepby(foreign)
```

	foreign	turn	price	pred2	pred	err
47.	Domestic	42	5,798	7585.93	4668.595	373.002
48.	Domestic	42	4,934	6600.487	4668.595	373.002
49.	Domestic	45	5,222	4798.425	3980.978	628.3688
50.	Domestic	40	4,723	5901.609	5127.007	314.9057
51.	Domestic	43	4,424	6156.588	4439.389	445.2716
52.	Domestic	41	4,172	3484.962	4897.801	325.7249
53.	Foreign	37	9,690	8226.297	9036.04	548.7275
54.	Foreign	36	6,295	5196.463	9265.246	589.1915
55.	Foreign	34	9,735	8138.971	9723.657	719.4052
56.	Foreign	35	6,229	6711.028	9494.451	647.5291
57.	Foreign	32	4,589	5893.468	10182.07	890.1414
58.	Foreign	34	5,079	6554.16	9723.657	719.4052

Note the differences in the predictions from the `predict` command and those from `adjust`. `predict` uses each observation's values for the variables in the regression model. `adjust` substitutes certain values for some or all of the variables in the regression model depending on what was specified in the `adjust` command. The first produces predicted values for the cars. The second produces predicted values for the cars with certain characteristics changed.

◁

▷ Example 3

We wish to look at the predicted cost of domestic and foreign cars by repair record (`rep78`) under the current regression model, holding `mpg` and `weight` to their means and allowing `turn` to vary. `adjust` lets us do this, even though we did not include `rep78` in the regression model.

(Continued on next page)

```
. adjust mpg weight, by(rep78 foreign) se ci center format(%9.2f)
```

```
        Dependent variable: price     Command: regress
          Variable left as is: turn
     Covariates set to mean: mpg = 21.289856, weight = 3032.029
```

Repair Record 1978	Car type	
	Domestic	Foreign
1	4951.66 (322.08) [4309.13,5594.19]	
2	4407.30 (471.15) [3467.38,5347.21]	
3	4790.37 (351.31) [4089.51,5491.22]	9471.91 (632.09) [8210.93,10732.89]
4	4722.45 (368.21) [3987.90,5457.01]	9548.31 (653.36) [8244.89,10851.73]
5	6097.69 (579.71) [4941.19,7254.19]	9420.97 (618.80) [8186.50,10655.45]

```
  Key:  Linear Prediction
        (Standard Error)
        [95% Confidence Interval]
```

This is an example of a two-way table produced by adjust. Up to seven-way tables are possible.

The means of mpg and weight in this run of adjust are not the same as in the previous runs. The rep78 variable (which was not a part of the regression) has five missing values, and adjust does casewise deletion when confronted with missing values.

◁

▷ Example 4

What if we wanted to find the predicted cost of foreign and domestic cars under this regression model while setting weight to its mean, letting mpg and turn vary, and pretending that all the cars are domestic? adjust handles this analysis since it allows variables to be set to a specific value for the prediction, and yet these variables may still differentiate subsets of the data in the by() option.

```
. adjust weight foreign=0, by(foreign) se ci center
```

```
     Dependent variable: price      Command: regress
   Variables left as is: mpg, turn
 Covariate set to mean: weight = 3019.4595
Covariate set to value: foreign = 0
```

Car type	xb	stdp	lb	ub
Domestic	4797.11	(363.698)	[4071.55	5522.66]
Foreign	6177.65	(637.952)	[4904.97	7450.33]

```
     Key:  xb        =  Linear Prediction
           stdp      =  Standard Error
           [lb , ub] =  [95% Confidence Interval]
```

In this table, we obtain the predicted prices of all cars as if they were domestic. The $6,178 prediction, for instance, is the average predicted price of our sample of foreign cars if they instead had been domestic and had average weight. The foreign-car sample has a different prediction from that of the domestic-car sample because the cars in the two samples have different mpg and turn, and we left these differences just as we observed them.

We now do the same, except that we treat all cars as if they were foreign by typing

```
. adjust weight foreign=1, by(foreign) se ci center
```

```
     Dependent variable: price      Command: regress
   Variables left as is: mpg, turn
 Covariate set to mean: weight = 3019.4595
Covariate set to value: foreign = 1
```

Car type	xb	stdp	lb	ub
Domestic	8018.52	(627.607)	[6766.48	9270.56]
Foreign	9399.07	(685.01)	[8032.51	10765.6]

```
     Key:  xb        =  Linear Prediction
           stdp      =  Standard Error
           [lb , ub] =  [95% Confidence Interval]
```

Put either way, the tables report the same difference in prices due to mpg and turn between the domestic and foreign car samples: 9399.07 − 8018.52 or 6177.65 − 4797.11.

◁

▷ Example 5

What if we decide to include the **rep78** variable in the regression model of our earlier examples? An easy way to do that is with **xi**; see [R] **xi**.

(Continued on next page)

```
. xi: regress price mpg weight turn I.rep78
I.rep78          _Irep78_1-5          (naturally coded; _Irep78_1 omitted)
```

Source	SS	df	MS		
Model	242819042	7	34688434.6		
Residual	333977917	61	5475047.82		
Total	576796959	68	8482308.22		

```
                                          Number of obs =      69
                                          F(  7,     61) =    6.34
                                          Prob > F      =  0.0000
                                          R-squared     =  0.4210
                                          Adj R-squared =  0.3545
                                          Root MSE      =  2339.9
```

price	Coef.	Std. Err.	t	P>|t|	[95% Conf. Interval]	
mpg	-86.24749	84.98694	-1.01	0.314	-256.1894	83.69441
weight	3.39851	.8279604	4.10	0.000	1.742901	5.05412
turn	-321.7209	136.6736	-2.35	0.022	-595.0167	-48.42515
_Irep78_2	1143.126	1860.47	0.61	0.541	-2577.113	4863.365
_Irep78_3	1074.757	1715.121	0.63	0.533	-2354.84	4504.354
_Irep78_4	1541.853	1759.521	0.88	0.384	-1976.527	5060.234
_Irep78_5	2813.323	1849.747	1.52	0.133	-885.4749	6512.121
_cons	9030.873	5599.464	1.61	0.112	-2165.946	20227.69

Now we wish to return to our two-way adjusted table presented in example 3 and examine the adjusted predictions under this new regression model. We will set mpg and weight to their mean values, allow turn to vary, and obtain the predictions for domestic and foreign cars by repair record.

```
. adjust mpg weight, by(foreign rep78) se
```

```
    Dependent variable: price     Command: regress
    Variables left as is: turn, _Irep78_2, _Irep78_3, _Irep78_4, _Irep78_5
Covariates set to mean: mpg = 21.289856, weight = 3032.0291
```

Car type	Repair Record 1978				
	1	2	3	4	5
Domestic	4308.5	4687.54	5156.86	5528.63	8730.43
	(1655.28)	(861.486)	(509.245)	(677.039)	(880.691)
Foreign			7206.34	7780.68	8873.42
			(759.055)	(798.379)	(910.117)

```
Key:  Linear Prediction
      (Standard Error)
```

adjust can use the original rep78 variable in the by() option. The output says that the variables created by xi are left as is along with the turn variable. This statement is true, but with the rep78 variable in the by() option, adjust still produces the desired results.

If you have used xi in your estimation command, you can freely use in the by() option the original variables on which xi operated. However, when setting these variables to specific values, you must use the names produced by xi. Say that we wish to create the same adjusted prediction table as before, but we now want to treat all the data as if the data had a repair record of 3. Here is how we would do it:

```
. adjust mpg weight _Irep78_2=0 _Irep78_3=1 _Irep78_4=0 _Irep78_5=0, by(for rep78) se
```

```
   Dependent variable: price     Command: regress
   Variable left as is: turn
Covariates set to mean: mpg = 21.289856, weight = 3032.0291
Covariates set to value: _Irep78_2 = 0, _Irep78_3 = 1, _Irep78_4 = 0,
                         _Irep78_5 = 0
```

Car type	Repair Record 1978				
	1	2	3	4	5
Domestic	5383.26	4619.17	5156.86	5061.54	6991.86
	(468.873)	(653.654)	(509.245)	(530.558)	(687.006)
Foreign			7206.34	7313.58	7134.85
			(759.055)	(796.544)	(734.568)

```
Key:  Linear Prediction
      (Standard Error)
```

If we wanted to do the same thing but set the repair record to 1 (the level dropped by xi), then in the adjust command we would set all the variables that xi created to zero.

```
. adjust mpg weight _Irep78_2=0 _Irep78_3=0 _Irep78_4=0 _Irep78_5=0, by(for rep78) se
```
◁

▷ Example 6

The adjust command also works after multiple-equation models. Let us take our first example regression and replace it with a corresponding multivariate regression using mvreg:

```
. mvreg gear_ratio price displ = mpg weight turn foreign
```

Equation	Obs	Parms	RMSE	"R-sq"	F	P
gear_ratio	74	5	.2632419	0.6854	37.58139	0.0000
price	74	5	2086.166	0.5271	19.2305	0.0000
displacement	74	5	40.98463	0.8118	74.38435	0.0000

	Coef.	Std. Err.	t	P>|t|	[95% Conf. Interval]	
gear_ratio						
mpg	.0098519	.0092763	1.06	0.292	-.0086538	.0283577
weight	-.0002951	.0000934	-3.16	0.002	-.0004815	-.0001087
turn	.012322	.0144156	0.85	0.396	-.0164364	.0410804
foreign	.4308748	.0891474	4.83	0.000	.2530305	.6087191
_cons	3.079496	.6167391	4.99	0.000	1.849135	4.309857
price						
mpg	-.4660076	73.51407	-0.01	0.995	-147.1226	146.1905
weight	4.284532	.7404967	5.79	0.000	2.807282	5.761782
turn	-229.2059	114.2423	-2.01	0.049	-457.1131	-1.298679
foreign	3221.415	706.4847	4.56	0.000	1812.017	4630.813
_cons	1368.197	4887.597	0.28	0.780	-8382.292	11118.69
displacement						
mpg	.2235286	1.444251	0.15	0.877	-2.657673	3.10473
weight	.1003079	.0145477	6.90	0.000	.071286	.1293298
turn	-.4931961	2.244395	-0.22	0.827	-4.970641	3.984249
foreign	-26.13646	13.87953	-1.88	0.064	-53.82537	1.552458
_cons	-83.01403	96.02129	-0.86	0.390	-274.5713	108.5432

We simply indicate which equation from the multiple-equation model is to be used by adjust. The equation may be specified by number or by name. We use equation(price) below, but we could have specified equation(#2) instead.

```
. adjust mpg weight turn, by(foreign) se ci center equation(price)
```

```
              Equation: price      Command: mvreg
Covariates set to mean: mpg = 21.297297, weight = 3019.4595, turn = 39.648647
```

Car type	xb	stdp	lb	ub
Domestic	5207.54	(320.823)	[4567.52	5847.56]
Foreign	8428.95	(552.515)	[7326.72	9531.19]

```
      Key:  xb         =  Linear Prediction
            stdp       =  Standard Error
            [lb , ub]  =  [95% Confidence Interval]
```

As expected, this table is the same as that produced in example 1.

◁

▷ Example 7

adjust following anova helps us explore the underlying cause for significant terms in the analysis-of-variance (ANOVA) or analysis-of-covariance (ANCOVA) table. The sysage.dta dataset illustrating ANCOVA in [R] **anova** provides an example.

```
. use http://www.stata-press.com/data/r10/sysage, clear
(Systolic Blood Pressure Data)
. anova systolic drug disease drug*disease age, continuous(age)
```

		Number of obs =	58	R-squared	= 0.6826	
		Root MSE	= 8.1164	Adj R-squared =	0.5980	

Source	Partial SS	df	MS	F	Prob > F
Model	6375.73983	12	531.311652	8.07	0.0000
drug	2762.01519	3	920.671731	13.98	0.0000
disease	605.323109	2	302.661555	4.59	0.0153
drug*disease	608.728232	6	101.454705	1.54	0.1871
age	2116.40132	1	2116.40132	32.13	0.0000
Residual	2964.41534	45	65.8758965		
Total	9340.15517	57	163.862371		

age is a significant covariate and drug and disease are significant factors, while the interaction is not. However, the interaction does not look as if it can be completely ignored. To further explore this interaction, we use adjust:

. adjust age, by(disease drug) se f(%6.3f) replace label(Predicted Systolic)

Dependent variable: systolic   Command: anova
Covariate set to mean: age = 45.155174

Patient's Disease	Drug Used 1	2	3	4
1	28.641 (3.316)	26.972 (3.634)	15.184 (4.690)	19.164 (3.760)
2	29.809 (4.068)	32.427 (4.063)	4.288 (3.630)	12.294 (3.315)
3	17.999 (3.654)	19.153 (3.318)	7.427 (4.063)	13.630 (3.631)

Key: Predicted Systolic
       (Standard Error)

From this table, we see that drug 3 gives the lowest predicted systolic measure and that the systolic measure for disease 2 varies more widely across the different drugs (a low of 4.288 for drug 3 and a high of 32.427 for drug 2).

We specified the `replace` option so that, at this point, the data from the adjusted table are now in memory. Remember to specify `replace` only if you have previously saved your original data.

. list drug disease xb stdp, sepby(disease)

	drug	disease	xb	stdp
1.	1	1	28.64146	3.315753
2.	2	1	26.97245	3.634287
3.	3	1	15.18372	4.690391
4.	4	1	19.16392	3.760155
5.	1	2	29.80869	4.067505
6.	2	2	32.42668	4.062614
7.	3	2	4.287933	3.629817
8.	4	2	12.29404	3.314871
9.	1	3	17.99923	3.654392
10.	2	3	19.15318	3.318073
11.	3	3	7.426678	4.062614
12.	4	3	13.63019	3.631155

A graph of the interaction can be produced from these data.

(Continued on next page)

```
. scatter xb drug, sort(disease drug) c(L) m(i) mlabel(disease) mlabp(0) mlabsiz(medium)
```

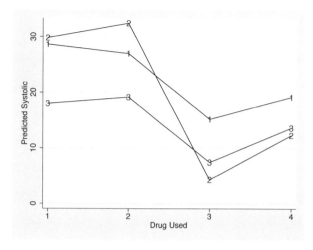

This simple example shows one of the many uses of the data after using the `replace` option.

◁

▷ Example 8

To illustrate the `pr` option of `adjust`, we turn to the low-birthweight dataset discussed in [R] **logistic** that was obtained from Hosmer and Lemeshow (2000). Below we show a `logistic` model (`probit`, `logit`, and several other commands are also allowed with the `pr` option of `adjust`).

```
. use http://www.stata-press.com/data/r10/lbw, clear
(Hosmer & Lemeshow data)

. xi: logistic low age lwt i.race smoke, nolog
i.race          _Irace_1-3          (naturally coded; _Irace_1 omitted)
Logistic regression                      Number of obs   =        189
                                         LR chi2(5)      =      20.08
                                         Prob > chi2     =     0.0012
Log likelihood = -107.29639              Pseudo R2       =     0.0856
```

low	Odds Ratio	Std. Err.	z	P>|z|	[95% Conf. Interval]	
age	.9777443	.0334083	-0.66	0.510	.9144097	1.045466
lwt	.9875761	.006305	-1.96	0.050	.9752956	1.000011
_Irace_2	3.425372	1.771281	2.38	0.017	1.243215	9.437768
_Irace_3	2.5692	1.069301	2.27	0.023	1.136391	5.808555
smoke	2.870346	1.09067	2.77	0.006	1.363	6.044672

Here we will ask for adjusted probability estimates under this model, setting `lwt` to its mean and `age` to 30. We will ask for these adjusted probability estimates by `race` and `smoke`.

```
. adjust age = 30 lwt, by(race smoke) pr ci format(%7.4f)
```

```
    Dependent variable: low      Command: logistic
    Variables left as is: _Irace_2, _Irace_3
   Covariate set to mean: lwt = 129.8201
  Covariate set to value: age = 30
```

race	smoked during pregnancy 0	1
white	0.1226 [0.0572,0.2434]	0.2863 [0.1670,0.4452]
black	0.3237 [0.1412,0.5822]	0.5787 [0.3110,0.8070]
other	0.2642 [0.1464,0.4291]	0.5075 [0.2838,0.7283]

```
      Key:  Probability
            [95% Confidence Interval]
```

The confidence interval is for the adjusted estimated probabilities. We obtain both the probabilities and the confidence intervals by first computing the linear predictions and associated confidence intervals and then transforming them to probabilities. If we had not specified the pr option, we would have obtained only the adjusted linear predictions.

```
. adjust age = 30 lwt, by(race smoke) ci format(%7.4f) noheader
```

race	smoked during pregnancy 0	1
white	-1.9681 [-2.8020,-1.1341]	-0.9136 [-1.6070,-0.2202]
black	-0.7368 [-1.8053,0.3317]	0.3176 [-0.7954,1.4305]
other	-1.0245 [-1.7633,-0.2856]	0.0300 [-0.9259,0.9859]

```
      Key:  Linear Prediction
            [95% Confidence Interval]
```

We can easily confirm that the entries in this table can be transformed to the probabilities in the previous table. For instance, $.1226 = \exp(-1.9681)/\{1 + \exp(-1.9681)\}$.

◁

▷ Example 9

In this example, we specify the exp option for adjust following an stcox regression on the drugtr.dta data introduced in the section *Cox regression with censored data* in [ST] **stcox**. To make the adjust table more meaningful, we generate a new age variable that has the lowest age in the data (47 in this case) subtracted from the age. This step will produce hazard ratios in the adjust table relative to someone aged 47 years instead of to a newborn.

```
. use http://www.stata-press.com/data/r10/drugtr
(Patient Survival in Drug Trial)

. generate age0 = age-47

. stcox drug age0, nolog

        failure _d:  died
   analysis time _t:  studytime

Cox regression -- Breslow method for ties

No. of subjects =          48                   Number of obs    =          48
No. of failures =          31
Time at risk    =         744
                                                LR chi2(2)       =       33.18
Log likelihood  =   -83.323546                  Prob > chi2      =      0.0000
```

_t	Haz. Ratio	Std. Err.	z	P>|z|	[95% Conf. Interval]	
drug	.1048772	.0477017	-4.96	0.000	.0430057	.2557622
age0	1.120325	.0417711	3.05	0.002	1.041375	1.20526

Using either age0 or age with stcox produces identical results—the reported hazard ratios are relative to a one-unit change in the variable. If we wanted to see the hazard ratios for the placebo and treatment drug when age is 50 (age0 is 3) with age 47 (age0 is 0) as the base, we could use adjust:

```
. adjust age0=3, by(drug) exp stubwidth(11) label(Hazard Ratio)
```

```
      Dependent variable: _t     Command: stcox
   Covariate set to value: age0 = 3
```

Drug type (0=placebo)	exp(xb)
0	1.40615
1	.147473

```
   Key:  exp(xb)  =  Hazard Ratio
```

These hazard ratios are for 50-year-olds with respect to 47-year-olds on placebo. Without the exp option, we would have obtained a table of the linear predictions instead of the hazard ratios.

◁

Methods and Formulas

adjust is implemented as an ado-file.

Acknowledgment

The adjust command is based in part on the adjmean and adjprop commands (Garrett 1995, 1998) written by Joanne Garrett of the University of North Carolina at Chapel Hill.

References

Conroy, R. M. 2002. Choosing an appropriate real-life measure of effect size: The case of a continuous predictor and a binary outcome. *Stata Journal* 2: 290–295.

Garrett, J. M. 1995. sg33: Calculation of adjusted means and adjusted proportions. *Stata Technical Bulletin* 24: 22–25. Reprinted in *Stata Technical Bulletin Reprints*, vol. 4, pp. 161–165.

——. 1998. sg33.1: Enhancements for calculation of adjusted means and adjusted proportions. *Stata Technical Bulletin* 43: 16–24. Reprinted in *Stata Technical Bulletin Reprints*, vol. 8, pp. 111–123.

Higbee, K. T. 1998. sg89: Adjusted predictions and probabilities after estimation. *Stata Technical Bulletin* 44: 30–37. Reprinted in *Stata Technical Bulletin Reprints*, vol. 8, pp. 165–173.

Hosmer, D. W., Jr., and S. Lemeshow. 2000. *Applied Logistic Regression.* 2nd ed. New York: Wiley.

Also See

[R] **table** — Tables of summary statistics

[U] **20 Estimation and postestimation commands**

Title

> **adoupdate** — Update user-written ado-files

Syntax

> adoupdate $\left[pkglist\right]$ $\left[,\ options\right]$

options	description
update	perform update; default is to list packages that have updates, but not to update them
all	include packages that might have updates; default is to list or update only packages that are known to have updates
ssconly	check only packages obtained from SSC; default is to check all installed packages
dir(*dir*)	check packages installed in *dir*; default is to check those installed in PLUS
verbose	provide output to assist in debugging network problems

Description

User-written additions to Stata are called packages. These packages can add remarkable abilities to Stata. Packages are found and installed by using ssc, findit, and net.

User-written packages are updated by their developers, just as official Stata software is updated by StataCorp.

To determine whether your official Stata software is up to date, and to update it if not, you use update.

To determine whether your user-written additions are up to date, and to update them if not, you use adoupdate.

Options

update specifies that packages with updates be updated. The default is simply to list the packages that could be updated without actually performing the update.

The first time you adoupdate, do not specify this option. Once you see adoupdate work, you will be more comfortable with it. Then type

    . adoupdate, update

The packages that can be updated will be listed and updated.

all is rarely specified. Sometimes, adoupdate cannot determine whether a package you previously installed has been updated. adoupdate can determine that the package is still available over the web but is unsure whether the package has changed. Usually, the package has not changed, but if you want to be certain that you are using the latest version, reinstall from the source.

Specifying all does this. Typing

    . adoupdate, all

24

adds such packages to the displayed list as needing updating but does not update them. Typing

```
. adoupdate, update all
```

lists such packages and updates them.

ssconly is a popular option. Many packages are available from the SSC (Statistical Software Components) archive—often called the Boston College Archive—which is provided by http://www.repec.org. Many users find most of what they want there. See [R] **ssc** for more information on the SSC.

ssconly specifies that adoupdate check only packages obtained from that source. Specifying this option is popular because SSC always provides distribution dates, and so adoupdate can be certain whether an update exists.

dir(*dir*) specifies which installed packages be checked. The default is dir(PLUS), and that is probably correct. If you are responsible for maintaining a large system, however, you may have previously installed packages in dir(SITE), where they are shared across users. See [P] **sysdir** for an explanation of these directory codewords. You may also specify an actual directory name, such as C:\mydir.

verbose is specified when you suspect network problems. It provides more detailed output that may help you diagnose the problem.

Remarks

Do not confuse adoupdate with update. Use adoupdate to update user-written files. Use update to update the components (including ado-files) of the official Stata software. To use either command, you must be connected to the Internet.

Remarks are presented under the following headings:

> *Using adoupdate*
> *Possible problem the first time you run adoupdate and the solution*
> *Notes for developers*

Using adoupdate

The first time you try adoupdate, type

```
. adoupdate
```

That is, do not specify the update option. adoupdate without update produces a report but does not update any files. The first time you run adoupdate, you may see messages such as

```
. adoupdate
(note:  package utx was installed more than once; older copy removed)
(remaining output omitted)
```

Having the same packages installed multiple times is common; adoupdate cleans that up.

The second time you run adoupdate, pick one package to update. Suppose that the report indicates that package st0008 has an update available. Type

```
. adoupdate st0008, update
```

You can specify one or many packages after the adoupdate command. You can even use wildcards such as st* to mean all packages that start with st or st*8 to mean all packages that start with st and end with 8. You can do that with or without the update option.

Finally, you can let `adoupdate` update all your user-written additions:

```
. adoupdate, update
```

Possible problem the first time you run adoupdate and the solution

The first time you run `adoupdate`, you might get many duplicate messages:

```
. adoupdate
(note: package ___ installed more than once; older copy removed)
(note: package ___ installed more than once; older copy removed)
(note: package ___ installed more than once; older copy removed)
...
(note: package ___ installed more than once; older copy removed)
(remaining output omitted)
```

Some users have hundreds of duplicates. You might even see the same package name repeated more than once:

```
(note: package stylus installed more than once; older copy removed)
(note: package stylus installed more than once; older copy removed)
```

That means that the package was duplicated twice.

Stata tolerates duplicates, and you did nothing wrong when you previously installed and updated packages. `adoupdate`, however, needs the duplicates removed, mainly so that it does not keep checking the same files.

Solution: Just let `adoupdate` run. `adoupdate` will run faster next time, when there are no (or just a few) duplicates.

Notes for developers

`adoupdate` reports whether an installed package is up to date by comparing its distribution date with that of the package available over the web.

If you are distributing software, include the line

```
d Distribution-Date: date
```

somewhere in your `.pkg` file. The capitalization of `Distribution-Date` does not matter, but include the hyphen and the colon as shown. Code the date in either of two formats:

all numeric:	*yyyymmdd*, for example, `20050508`
Stata standard:	*ddMONyyyy*, for example, `08may2005`

Saved Results

`adoupdate` saves the following in `r()`:

Macros

r(pkglist)   a space-separated list of package names that need updating (update not specified)
             or that were updated (update specified)

Also See

[R] **ssc** — Install and uninstall packages from SSC

[R] **search** — Search Stata documentation

[R] **net** — Install and manage user-written additions from the Internet

[R] **update** — Update Stata

Title

> **alpha** — Compute interitem correlations (covariances) and Cronbach's alpha

Syntax

> alpha *varlist* [*if*] [*in*] [, *options*]

options	description
Options	
<u>a</u>sis	take sign of each item as is
<u>c</u>asewise	delete cases with missing values
<u>d</u>etail	list individual interitem correlations and covariances
<u>g</u>enerate(*newvar*)	save the generated scale in *newvar*
<u>i</u>tem	display item-test and item-rest correlations
<u>l</u>abel	include variable labels in output table
<u>m</u>in(*#*)	must have at least *#* observations for inclusion
<u>r</u>everse(*varlist*)	reverse signs of these variables
<u>s</u>td	standardize items in the scale to mean 0, variance 1

by is allowed; see [D] **by**.

Description

alpha computes the interitem correlations or covariances for all pairs of variables in *varlist* and Cronbach's α statistic for the scale formed from them. At least two variables must be specified with alpha.

Options

> Options

asis specifies that the sense (sign) of each item be taken as presented in the data. The default is to determine the sense empirically and reverse the scorings for any that enter negatively.

casewise specifies that cases with missing values be deleted listwise. The default is pairwise computation of covariances/correlations.

detail lists the individual interitem correlations and covariances.

generate(*newvar*) specifies that the scale constructed from *varlist* be stored in *newvar*. Unless asis is specified, the sense of items entering negatively is automatically reversed. If std is also specified, the scale is constructed by using standardized (mean 0, variance 1) values of the individual items. Unlike most Stata commands, generate() does not use casewise deletion. A score is created for every observation for which there is a response to at least one item (one variable in *varlist* is not missing). The summative score is divided by the number of items over which the sum is calculated.

item specifies that item-test and item-rest correlations and the effects of removing an item from the scale be displayed. item is valid only when more than two variables are specified in *varlist*.

`label` requests that the detailed output table be displayed in a compact format that enables the inclusion of variable labels.

`min(#)` specifies that only cases with at least # observations be included in the computations. `casewise` is a shorthand for `min(k)`, where k is the number of variables in *varlist*.

`reverse(varlist)` specifies that the signs (directions) of the variables (items) in *varlist* be reversed. Any variables specified in `reverse()` that are not also included in `alpha`'s *varlist* are ignored.

`std` specifies that the items in the scale be standardized (mean 0, variance 1) before summing.

Remarks

Cronbach's alpha (Cronbach 1951) assesses the reliability of a summative rating (Likert 1932) scale composed of the variables (called *items*) specified. The set of items is often called a *test* or *battery*. A scale is simply the sum of the individual item scores, reversing the scoring for statements that have negative correlations with the factor (e.g., attitude) being measured. Scales can be formed by using the raw item scores or standardized item scores.

The reliability α is defined as the square of the correlation between the measured scale and the underlying factor. If you think of a test as being composed of a random sample of items from a hypothetical domain of items designed to measure the same thing, α represents the expected correlation of one test with an alternative form containing the same number of items. The square root of α is the estimated correlation of a test with errorless true scores (Nunnally and Bernstein 1994, 235).

In addition to reporting α, `alpha` generates the summative scale from the items (variables) specified and automatically reverses the sense of any when necessary. Stata's decision can be overridden by specifying the `reverse(varlist)` option.

Because it concerns reliability in measuring an unobserved factor, α is related to factor analysis. The test should be designed to measure one factor, and, since the scale will be composed of an unweighted sum, the factor loadings should all contribute roughly equal information to the score. Both of these assumptions can be verified with `factor`; see [MV] **factor**. Equality of factor loadings can also be assessed by using the `item` option.

▷ Example 1

To illustrate `alpha`, we apply it, first without and then with the `item` option, to the automobile dataset after randomly introducing missing values:

```
. use http://www.stata-press.com/data/r10/automiss
(1978 Automobile Data)
. alpha price headroom rep78 trunk weight length turn displ, std
Test scale = mean(standardized items)
Reversed item: rep78
Average interitem correlation:      0.5251
Number of items in the scale:            8
Scale reliability coefficient:      0.8984
```

The scale derived from our somewhat arbitrarily chosen automobile items (variables) appears to be reasonable since the estimated correlation between it and the underlying factor it measures is $\sqrt{.8984} \approx .9478$ and the estimated correlation between this battery of eight items and all other eight-item batteries from the same domain is .8984. Since the "items" are not on the same scale, it is important that `std` was specified so that the scale and its reliability were based on the sum of standardized variables. We could obtain the scale in a new variable called `sc` with the option `gen(sc)`.

Though the scale appears reasonable, we include the `item` option to determine if all the items fit the scale:

```
. alpha price headroom rep78 trunk weight length turn displ, std item
Test scale = mean(standardized items)
```

Item	Obs	Sign	item-test correlation	item-rest correlation	average inter-item correlation	alpha
price	70	+	0.5260	0.3719	0.5993	0.9128
headroom	66	+	0.6716	0.5497	0.5542	0.8969
rep78	61	−	0.4874	0.3398	0.6040	0.9143
trunk	69	+	0.7979	0.7144	0.5159	0.8818
weight	64	+	0.9404	0.9096	0.4747	0.8635
length	69	+	0.9382	0.9076	0.4725	0.8625
turn	66	+	0.8678	0.8071	0.4948	0.8727
displacement	63	+	0.8992	0.8496	0.4852	0.8684
Test scale					0.5251	0.8984

"Test" denotes the additive scale; here .5251 is the average interitem correlation, and .8984 is the alpha coefficient for a test scale based on all items.

"Obs" shows the number of nonmissing values of the items; "Sign" indicates the direction in which an item variable entered the scale; "−" denotes that the item was reversed. The remaining four columns in the table provide information on the effect of one item on the scale.

Column four gives the item-test correlations. Apart from the sign of the correlation for items that entered the scale in reversed order, these correlations are the same numbers as those computed by the commands

```
. alpha price headroom rep78 trunk weight length turn displ, std gen(sc)
. pwcorr sc price headroom rep78 trunk weight length turn displ
```

Typically, the item-test correlations should be roughly the same for all items. Item-test correlations may not be adequate to detect items that fit poorly because the poorly fitting items may distort the scale. Accordingly, it may be more useful to consider item-rest correlations (Nunnally and Bernstein 1994), i.e., the correlation between an item and the scale that is formed by all other items. The average interitem correlations (covariances if `std` is omitted) of all items, excluding one, are shown in column six. Finally, column seven gives Cronbach's α for the test scale, which consists of all but the one item.

Here neither the `price` item nor the `rep78` item seems to fit well in the scale in all respects. The item-test and item-rest correlations of `price` and `rep78` are much lower than those of the other items. The average interitem correlation increases substantially by removing either `price` or `rep78`; apparently they do not correlate strongly with the other items. Finally, we see that Cronbach's α coefficient will increase from .8984 to .9128 if the item `price` is dropped, and it will increase from .8984 to .9143 if `rep78` is dropped. For well-fitting items, we would of course expect that α decreases by shortening the test.

◁

▷ Example 2

The variable names for the automobile data are reasonably informative. This may not always be true; items in batteries commonly used to measure personality traits, attitudes, or values, etc., are usually named with indexed names such as `item12a`, `item12b`, etc. The `label` option forces `alpha` to produce the same statistical information in a more compact format that leaves room to include variable (item) labels. In this compact format, `alpha` excludes the number of nonmissing values of the items, displays the statistics using fewer digits, and uses somewhat cryptic headers:

```
. alpha price headroom rep78 trunk weight length turn displ, std item label detail
Test scale = mean(standardized items)
```

Items	S	it-cor	ir-cor	ii-cor	alpha	label
price	+	0.526	0.372	0.599	0.913	Price
headroom	+	0.672	0.550	0.554	0.897	Headroom (in.)
rep78	−	0.487	0.340	0.604	0.914	Repair Record 1978
trunk	+	0.798	0.714	0.516	0.882	Trunk space (cu. ft.)
weight	+	0.940	0.910	0.475	0.863	Weight (lbs.)
length	+	0.938	0.908	0.473	0.862	Length (in.)
turn	+	0.868	0.807	0.495	0.873	Turn Circle (ft.)
displacement	+	0.899	0.850	0.485	0.868	Displacement (cu. in.)
Test scale				0.525	0.898	mean(standardized items)

Interitem correlations (reverse applied) (obs=pairwise, see below)

	price	headroom	rep78	trunk
price	1.0000			
headroom	0.1174	1.0000		
rep78	-0.0479	0.1955	1.0000	
trunk	0.2748	0.6841	0.2777	1.0000
weight	0.5093	0.5464	0.3624	0.6486
length	0.4511	0.5823	0.3162	0.7404
turn	0.3528	0.4067	0.4715	0.5900
displacement	0.5537	0.5166	0.3391	0.6471

	weight	length	turn	displacement
weight	1.0000			
length	0.9425	1.0000		
turn	0.8712	0.8589	1.0000	
displacement	0.8753	0.8422	0.7723	1.0000

Pairwise number of observations

	price	headroom	rep78	trunk
price	70			
headroom	62	66		
rep78	59	54	61	
trunk	65	61	59	69
weight	60	56	52	60
length	66	61	58	64
turn	62	58	56	62
displacement	59	58	51	58

	weight	length	turn	displacement
weight	64			
length	60	69		
turn	57	61	66	
displacement	54	58	56	63

Since the `detail` option was also specified, the interitem correlation matrix was printed, together with the number of observations used for each entry (since these varied across the matrix). Note the negative sign attached to `rep78` in the output, indicating the sense in which it entered the scale.

Better-looking output with less-cryptic headers is produced if the linesize is set to a value of at least 100:

```
. set linesize 100
. alpha price headroom rep78 trunk weight length turn displ, std item label
Test scale = mean(standardized items)
```

Item	Obs	Sign	item-test corr.	item-rest corr.	inter-item corr.	alpha	Label
price	70	+	0.5260	0.3719	0.5993	0.9128	Price
headroom	62	+	0.6716	0.5497	0.5542	0.8969	Headroom (in.)
rep78	59	−	0.4874	0.3398	0.6040	0.9143	Repair Record 1978
trunk	65	+	0.7979	0.7144	0.5159	0.8818	Trunk space (cu. ft.)
weight	60	+	0.9404	0.9096	0.4747	0.8635	Weight (lbs.)
length	66	+	0.9382	0.9076	0.4725	0.8625	Length (in.)
turn	62	+	0.8678	0.8071	0.4948	0.8727	Turn Circle (ft.)
displacement	59	+	0.8992	0.8496	0.4852	0.8684	Displacement (cu. in.)
Test scale					0.5251	0.8984	mean(standardized items)

◁

Users of `alpha` require some standard for judging values of α. We paraphrase Nunnally and Bernstein (1994, 265): In the early stages of research, modest reliability of 0.70 or higher will suffice; values in excess of 0.80 often waste time and funds. In contrast, where measurements on individuals are of interest, a reliability of 0.80 may not be nearly high enough. Even with a reliability of 0.90, the standard error of measurement is almost one-third as large as the standard deviation of test scores; a reliability of 0.90 is the minimum that should be tolerated, and a reliability of 0.95 should be considered the desirable standard.

Saved Results

`alpha` saves the following in `r()`:

Scalars
r(alpha)	scale reliability coefficient
r(k)	number of items in the scale
r(cov)	average interitem covariance
r(rho)	average interitem correlation if `std` is specified

Matrices
r(Alpha)	scale reliability coefficient
r(ItemTestCorr)	item-test correlation
r(ItemRestCorr)	item-rest correlation
r(MeanInterItemCov)	average interitem covariance
r(MeanInterItemCorr)	average interitem correlation if `std` is specified

If the option `item` is specified, results are saved as row matrices for the `k` subscales when one variable is removed.

Methods and Formulas

alpha is implemented as an ado-file.

Let x_i, $i = 1, \ldots, k$, be the variables over which α is to be calculated. Let s_i be the sign with which x_i enters the scale. If asis is specified, $s_i = 1$ for all i. Otherwise, principal-factor analysis is performed on x_i, and the first factor's score is predicted; see [MV] **factor**. s_i is -1 if correlation of the x_i and the predicted score is negative and $+1$ otherwise.

Let r_{ij} be the correlation between x_i and x_j, c_{ij} be the covariance, and n_{ij} be the number of observations used in calculating the correlation or covariance. The average correlation is

$$\overline{r} = \frac{\sum\limits_{i=2}^{k} \sum\limits_{j=1}^{i-1} s_i s_j n_{ij} r_{ij}}{\sum\limits_{i=2}^{k} \sum\limits_{j=1}^{i-1} n_{ij}}$$

and the average covariance similarly is

$$\overline{c} = \frac{\sum\limits_{i=2}^{k} \sum\limits_{j=1}^{i-1} s_i s_j n_{ij} c_{ij}}{\sum\limits_{i=2}^{k} \sum\limits_{j=1}^{i-1} n_{ij}}$$

Let c_{ii} denote the variance of x_i, and define the average variance as

$$\overline{v} = \frac{\sum\limits_{i=1}^{k} n_{ii} c_{ii}}{\sum\limits_{i=1}^{k} n_{ii}}$$

If std is specified, the scale reliability α is calculated as defined by the general form of the Spearman–Brown Prophecy Formula (Nunnally and Bernstein 1994, 232; Allen and Yen 1979, 85–88):

$$\alpha = \frac{k\overline{r}}{1 + (k-1)\overline{r}}$$

This expression corresponds to α under the assumption that the summative rating is the sum of the standardized variables (Nunnally and Bernstein 1994, 234). If std is not specified, α is defined (Nunnally and Bernstein 1994, 232, 234) as

$$\alpha = \frac{k\overline{c}}{\overline{v} + (k-1)\overline{c}}$$

Let x_{ij} reflect the value of item i in the jth observation. If std is specified, the jth value of the scale computed from the k x_{ij} items is

$$S_j = \frac{1}{k_j} \sum\limits_{i=1}^{k} s_i S(x_{ij})$$

where $S()$ is the function that returns the standardized (mean 0, variance 1) value if x_{ij} is not missing and returns zero if x_{ij} is missing. k_j is the number of nonmissing values in x_{ij}, $i = 1, \ldots, k$. If std is not specified, $S()$ is the function that returns x_{ij} or returns missing if x_{ij} is missing.

Lee Joseph Cronbach (1916–2001) was an American psychometrician and educational psychologist who worked principally on measurement theory, program evaluation, and instruction. He taught and researched at the State College of Washington, the University of Chicago, the University of Illinois, and Stanford. Cronbach's initial paper on alpha led to a theory of test reliability, generalizability theory, a comprehensive statistical model for identifying sources of measurement error.

Acknowledgment

This improved version of alpha was written by Jeroen Weesie of the Department of Sociology, Utrecht University, The Netherlands.

References

Acock, A. C. 2006. *A Gentle Introduction to Stata*. College Station, TX: Stata Press.

Allen, M. J., and W. M. Yen. 1979. *Introduction to Measurement Theory*. Monterey, CA: Brooks/Cole.

Bleda, M. J., and A. Tobias. 2000. sg143: Cronbach's alpha one-sided confidence interval. *Stata Technical Bulletin* 56: 26–27. Reprinted in *Stata Technical Bulletin Reprints*, vol. 10, pp. 187–189.

Cronbach, L. J. 1951. Coefficient alpha and the internal structure of tests. *Psychometrika* 16: 297–334.

Kuder, G. F., and M. W. Richardson. 1937. The theory of the estimation of test reliability. *Psychometrika* 2: 151–160.

Likert, R. A. 1932. A technique for the measurement of attitudes. *Archives of Psychology* 140: 5–55.

Nunnally, J. C., and I. H. Bernstein. 1994. *Psychometric Theory*. 3rd ed. New York: McGraw–Hill.

Shavelson, R. J., and G. Gleser. 2002. Lee J. Cronbach (1916–2001). *American Psychologist* 57: 60–361.

Tarlov, A. R., J. E. Ware, Jr., S. Greenfield, E. C. Nelson, E. Perrin, and M. Zubkoff. 1989. The medical outcomes study. An application of methods for monitoring the results of medical care. *Journal of the American Medical Association* 262: 925–930.

Weesie, J. 1997. sg66: Enhancements to the alpha command. *Stata Technical Bulletin* 35: 32–34. Reprinted in *Stata Technical Bulletin Reprints*, vol. 6, pp. 176–179.

Also See

[MV] **factor** — Factor analysis

Title

---

ameans — Arithmetic, geometric, and harmonic means

Syntax

ameans [*varlist*] [*if*] [*in*] [*weight*] [, *options*]

options	description
Main	
<u>a</u>dd(*#*)	add *#* to each variable in *varlist*
<u>o</u>nly	add *#* only to variables with nonpositive values
<u>l</u>evel(*#*)	set confidence level; default is level(95)

by is allowed; see [D] **by**.
aweights and fweights are allowed; see [U] **11.1.6 weight**.

Description

ameans computes the arithmetic, geometric, and harmonic means, with their corresponding confidence intervals, for each variable in *varlist* or for all the variables in the data if *varlist* is not specified. gmeans and hmeans are synonyms for ameans.

If you simply want arithmetic means and corresponding confidence intervals, see [R] **ci**.

Options

> Main

add(*#*) adds the value *#* to each variable in *varlist* before computing the means and confidence intervals. This option is useful when analyzing variables with nonpositive values.

only modifies the action of the add(*#*) option so that it adds *#* only to variables with at least one nonpositive value.

level(*#*) specifies the confidence level, as a percentage, for confidence intervals. The default is level(95) or as set by set level; see [U] **20.7 Specifying the width of confidence intervals**.

Remarks

▷ Example 1

We have a dataset containing 8 observations on a variable named x. The eight values are 5, 4, −4, −5, 0, 0, *missing*, and 7.

```
. ameans x
```

Variable	Type	Obs	Mean	[95% Conf. Interval]	
x	Arithmetic	7	1	-3.204405	5.204405
	Geometric	3	5.192494	2.57899	10.45448
	Harmonic	3	5.060241	3.023008	15.5179

35

```
. ameans x, add(5)
```

Variable	Type	Obs	Mean	[95% Conf. Interval]	
x	Arithmetic	7	6	1.795595	10.2044 *
	Geometric	6	5.477226	2.1096	14.22071 *
	Harmonic	6	3.540984	.	. *

(*) 5 was added to the variables prior to calculating the results.
Missing values in confidence intervals for harmonic mean indicate
that confidence interval is undefined for corresponding variables.
Consult Reference Manual for details.

The number of observations displayed for the arithmetic mean is the number of nonmissing observations. The number of observations displayed for the geometric and harmonic means is the number of nonmissing, positive observations. Specifying the add(5) option produces 3 more positive observations. The confidence interval for the harmonic mean is not reported; see *Methods and Formulas* below.

◁

Saved Results

ameans saves the following in r():

Scalars

r(N)	number of nonmissing observations; used for arithmetic mean
r(N_pos)	number of nonmissing positive observations; used for geometric and harmonic means
r(mean)	arithmetic mean
r(lb)	lower bound of confidence interval for arithmetic mean
r(ub)	upper bound of confidence interval for arithmetic mean
r(Var)	variance of untransformed data
r(mean_g)	geometric mean
r(lb_g)	lower bound of confidence interval for geometric mean
r(ub_g)	upper bound of confidence interval for geometric mean
r(Var_g)	variance of $\ln x_i$
r(mean_h)	harmonic mean
r(lb_h)	lower bound of confidence interval for harmonic mean
r(ub_h)	upper bound of confidence interval for harmonic mean
r(Var_h)	variance of $1/x_i$

Methods and Formulas

ameans is implemented as an ado-file.

See Armitage and Berry (2002) or Snedecor and Cochran (1989). For a history of the concept of the mean, see Plackett (1958).

When restricted to the same set of values (i.e., to positive values), the arithmetic mean ($\overline{x}$) is greater than or equal to the geometric mean, which in turn is greater than or equal to the harmonic mean. Equality holds only if all values within a sample are equal to a positive constant.

The arithmetic mean and its confidence interval are identical to those provided by ci; see [R] **ci**.

To compute the geometric mean, `ameans` first creates $u_j = \ln x_j$ for all positive x_j. The arithmetic mean of the u_j and its confidence interval are then computed as in `ci`. Let $\overline{u}$ be the resulting mean, and let $[L, U]$ be the corresponding confidence interval. The geometric mean is then $\exp(\overline{u})$, and its confidence interval is $[\exp(L), \exp(U)]$.

The same procedure is followed for the harmonic mean, except that then, $u_j = 1/x_j$. The harmonic mean is then $1/\overline{u}$, and its confidence interval is $[1/U, 1/L]$ if L is greater than zero. If L is not greater than zero, this confidence interval is not defined, and missing values are reported.

When weights are specified, `ameans` applies the weights to the transformed values, $u_j = \ln x_j$ and $u_j = 1/x_j$, respectively, when computing the geometric and harmonic means. For details on how the weights are used to compute the mean and variance of the u_j, see [R] **summarize**. Without weights, the formula for the geometric mean reduces to

$$\exp\left\{\frac{1}{n}\sum_j \ln(x_j)\right\}$$

Without weights, the formula for the harmonic mean is

$$\frac{n}{\sum_j \dfrac{1}{x_j}}$$

Acknowledgments

This improved version of `ameans` is based on the `gmci` command (Carlin, Vidmar, and Ramalheira 1998) and was written by John Carlin, University of Melbourne, Australia; Suzanna Vidmar, University of Melbourne, Australia; and Carlos Ramalheira, Coimbra University Hospital, Portugal.

References

Armitage, P., and G. Berry. 2002. *Statistical Methods in Medical Research*. 4th ed. Oxford: Blackwell.

Carlin, J., S. Vidmar, and C. Ramalheira. 1998. sg75: Geometric means and confidence intervals. *Stata Technical Bulletin* 41: 23–25. Reprinted in *Stata Technical Bulletin Reprints*, vol. 7, pp. 197–199.

Keynes, J. M. 1911. The principal averages and the laws of error which lead to them. *Journal of the Royal Statistical Society* 74: 322–331.

Kotz, S., and N. L. Johnson, ed. 1985. *Encyclopedia of Statistical Sciences*, vol. 1, vol. 3. New York: Wiley.

Plackett, R. L. 1958. The principle of the arithmetic mean. *Biometrika* 45: 130–135.

Snedecor, G. W., and W. G. Cochran. 1989. *Statistical Methods*. 8th ed. Ames, IA: Iowa State University Press.

Also See

[R] **ci** — Confidence intervals for means, proportions, and counts

[R] **mean** — Estimate means

[R] **summarize** — Summary statistics

[SVY] **svy estimation** — Estimation commands for survey data

Title

> **anova** — Analysis of variance and covariance

Syntax

$$\underline{\text{an}}\text{ova } \textit{varname } \big[\textit{term }[\,/\,]\,\big[\textit{term }[\,/\,]\,\ldots\,\big]\,\big]\ \big[\textit{if}\,\big]\ \big[\textit{in}\,\big]\ \big[\textit{weight}\,\big]\ \big[\,,\ \textit{options}\,\big]$$

where *term* is of the form $\textit{varname}\big[\,\{\ *\ |\ |\ \}\textit{varname}\big[\,\ldots\,\big]\,\big]$

options	description
Model	
<u>cat</u>egory(*varlist*)	variables in *terms* that are categorical or class
<u>cl</u>ass(*varlist*)	synonym for category(*varlist*)
<u>cont</u>inuous(*varlist*)	variables in *terms* that are continuous
<u>rep</u>eated(*varlist*)	variables in *terms* that are repeated-measures variables
partial	use partial (or marginal) sums of squares
sequential	use sequential sums of squares
<u>nocons</u>tant	suppress constant term
Adv. model	
bse(*term*)	between-subjects error term in repeated-measures ANOVA
bseunit(*varname*)	variable representing lowest unit in the between-subjects error term
grouping(*varname*)	grouping variable for computing pooled covariance matrix
Reporting	
<u>reg</u>ress	display the regression table
[no]<u>an</u>ova	display or suppress the ANOVA table
<u>detail</u>	show mapping from values to level numbers for categorical variables

by is allowed; see [U] **11.1.10 Prefix commands**.
aweights and fweights are allowed; see [U] **11.1.6 weight**.
See [U] **20 Estimation and postestimation commands** for more capabilities of estimation commands.

Description

The anova command fits analysis-of-variance (ANOVA) and analysis-of-covariance (ANCOVA) models for balanced and unbalanced designs, including designs with missing cells; for repeated-measures ANOVA; and for factorial, nested, or mixed designs. anova can also be used to produce regression estimates by those who have no interest in ANOVA and ANCOVA output.

If you want to fit one-way ANOVA models, you may find the oneway or loneway command more convenient; see [R] **oneway** and [R] **loneway**. If you are interested in MANOVA or MANCOVA, see [MV] **manova**.

38

Options

Model

category(*varlist*) indicates the names of the variables in the *terms* that are categorical or class variables. Stata ordinarily assumes that all variables are categorical variables, so this option need not be specified. If you specify this option, however, the variables referred to in the *terms* that are not listed in the category() option are assumed to be continuous. Also see the class() and continuous() options.

class(*varlist*) is a synonym for category(*varlist*).

continuous(*varlist*) indicates the names of the variables in the *terms* that are continuous. Stata ordinarily assumes that all variables are categorical. Also see the category() and class() options.

repeated(*varlist*) indicates the names of the categorical variables in the *terms* that are to be treated as repeated-measures variables in a repeated-measures ANOVA or ANCOVA.

partial presents the ANOVA table using partial (or marginal) sums of squares. This setting is the default. Also see the sequential option.

sequential presents the ANOVA table using sequential sums of squares.

noconstant suppresses the constant term (intercept) from the ANOVA or regression model.

Adv. model

bse(*term*) indicates the between-subjects error term in a repeated-measures ANOVA. This option is needed only in the rare case when the anova command cannot automatically determine the between-subjects error term.

bseunit(*varname*) indicates the variable representing the lowest unit in the between-subjects error term in a repeated-measures ANOVA. This option is rarely needed since the anova command automatically selects the first variable listed in the between-subjects error term as the default for this option.

grouping(*varname*) indicates a variable that determines which observations are grouped together in computing the covariance matrices that will be pooled and used in a repeated-measures ANOVA. This option is rarely needed since the anova command automatically selects the combination of all variables except the first (or as specified in the bseunit() option) in the between-subjects error term as the default for grouping observations.

Reporting

regress presents the regression output corresponding to the specified model. Specifying regress implies the noanova option, so if you want both the regression output and the ANOVA table, you must also specify the anova option. You need not specify the regress option at the time of estimation. You can obtain the underlying regression estimates at any time by typing anova, regress.

[no]anova indicates that the ANOVA table be or not be displayed. The anova command typically displays the ANOVA table, and in those cases, the noanova option suppresses the display. For instance, typing anova, detail noanova would show the detail output for the last ANOVA model while suppressing the ANOVA table itself.

If you specify the regress option, the ANOVA table is automatically suppressed. Then also specifying the anova option would show both the regression output and the ANOVA table.

detail presents a table showing the actual values of the categorical variables along with their mapping into level numbers. You do not have to specify this option at the time of estimation. You can obtain the output at any time by typing anova, detail.

Remarks

Remarks are presented under the following headings:

> *Introduction*
> *One-way ANOVA*
> *Two-way ANOVA*
> *N-way ANOVA*
> *ANCOVA*
> *Nested designs*
> *Mixed designs*
> *Latin-square designs*
> *Repeated-measures ANOVA*
> *Weighted data*
> *How Stata numbers levels*
> *anova as an alternative to regress*

Introduction

anova uses least squares to fit linear models and can report both regression and ANOVA output. The command is designed to be most useful for fitting ANOVA or ANCOVA models (henceforth referred to simply as ANOVA models) but is also useful for fitting regression models with many categorical variables.

If your interest is in formal ANOVA, read this entire entry. If your interest is solely in regression, you may want to skip to *anova as an alternative to regress*; also see [R] **regress**.

If your interest is in one-way ANOVA, you may find the oneway command more convenient; see [R] **oneway**.

ANOVA was pioneered by Fisher. It features prominently in his texts on statistical methods and the design of experiments (1925, 1935). Many books discuss ANOVA; see, for instance, Altman (1991); van Belle et al. (2004); Cobb (1998); Snedecor and Cochran (1989); or Winer, Brown, and Michels (1991). For a classic source, see Scheffé (1959). Kennedy and Gentle (1980) discuss ANOVA's computing problems. Edwards (1985) is concerned primarily with the relationship between multiple regression and ANOVA. Acock (2006, chap. 9) and Rabe-Hesketh and Everitt (2004, chap. 4 and 5) illustrate their discussion with Stata output. Repeated-measures ANOVA is discussed in Winer, Brown, and Michels (1991); Kuehl (2000); and Milliken and Johnson (1984). Pioneering work in repeated-measures ANOVA can be found in Box (1954); Geisser and Greenhouse (1958); Huynh and Feldt (1976); and Huynh (1978).

One-way ANOVA

anova, entered without options, performs and reports standard ANOVA. For instance, to perform a one-way layout of a variable called endog on exog, you would type anova endog exog.

▷ Example 1

We run an experiment varying the amount of fertilizer used in growing apple trees. We test four concentrations, using each concentration in three groves of 12 trees each. Later in the year, we measure the average weight of the fruit.

If all had gone well, we would have had 3 observations on the average weight for each of the four concentrations. Instead, two of the groves were mistakenly leveled by a confused man on a large bulldozer. We are left with the following data:

```
. use http://www.stata-press.com/data/r10/apple
(Apple trees)

. list, abbrev(10) sepby(treatment)
```

	treatment	weight
1.	1	117.5
2.	1	113.8
3.	1	104.4
4.	2	48.9
5.	2	50.4
6.	2	58.9
7.	3	70.4
8.	3	86.9
9.	4	87.7
10.	4	67.3

To obtain one-way ANOVA results, we type

```
. anova weight treatment
```

	Number of obs = 10	R-squared = 0.9147
	Root MSE = 9.07002	Adj R-squared = 0.8721

Source	Partial SS	df	MS	F	Prob > F
Model	5295.54433	3	1765.18144	21.46	0.0013
treatment	5295.54433	3	1765.18144	21.46	0.0013
Residual	493.591667	6	82.2652778		
Total	5789.136	9	643.237333		

We find significant (at better than the 1% level) differences among the four concentrations.

Although the output is a usual ANOVA table, let's run through it anyway. Above the table is a summary of the underlying regression. The model was fitted on 10 observations, and the root mean squared error (Root MSE) is 9.07. The R^2 for the model is 0.9147, and the adjusted R^2 is 0.8721.

The first line of the table summarizes the model. The sum of squares (Partial SS) for the model is 5295.5 with 3 degrees of freedom (df). This line results in a mean square (MS) of $5295.5/3 \approx 1765.2$. The corresponding F statistic is 21.46 and has a significance level of 0.0013. Thus the model appears to be significant at the 0.13% level.

The next line summarizes the first (and only) term in the model, treatment. Since there is only one term, the line is identical to that for the overall model.

The third line summarizes the residual. The residual sum of squares is 493.59 with 6 degrees of freedom, resulting in a mean squared error of 82.27. The square root of this latter number is reported as the Root MSE.

The model plus the residual sum of squares equals the total sum of squares, which is reported as 5789.1 in the last line of the table. This is the total sum of squares of weight after removal of the mean. Similarly, the model plus the residual degrees of freedom sum to the total degrees of freedom, 9. Remember that there are 10 observations. Subtracting 1 for the mean, we are left with 9 total degrees of freedom.

◁

❏ Technical Note

Rather than using the anova command, we could have performed this analysis by using the oneway command. Example 1 in [R] **oneway** repeats this same analysis. You may wish to compare the output.

❏

To see the underlying regression corresponding to a model after fitting an ANOVA model using the anova command, you can obtain the regression output by typing anova, regress.

▷ Example 2

Returning to the apple tree experiment, we found that the fertilizer concentration appears to significantly affect the average weight of the fruit. Although that finding is interesting, we next want to know which concentration appears to grow the heaviest fruit. To find out, we examine the underlying regression coefficients.

```
. anova, regress
```

Source	SS	df	MS
Model	5295.54433	3	1765.18144
Residual	493.591667	6	82.2652778
Total	5789.136	9	643.237333

Number of obs =	10
F(3, 6) =	21.46
Prob > F =	0.0013
R-squared =	0.9147
Adj R-squared =	0.8721
Root MSE =	9.07

weight	Coef.	Std. Err.	t	P>|t|	[95% Conf. Interval]
_cons	77.5	6.413473	12.08	0.000	61.8068 93.1932
treatment					
1	34.4	8.279758	4.15	0.006	14.14016 54.65984
2	-24.76667	8.279758	-2.99	0.024	-45.02651 -4.506828
3	1.15	9.070021	0.13	0.903	-21.04354 23.34354
4	(dropped)				

See [R] **regress** for an explanation of how to read this table. In summary, we find that concentration 1 produces significantly heavier fruits than concentration 4, the omitted (dropped) group; concentration 2 produces significantly lighter fruits; and concentrations 3 and 4 appear to be roughly equivalent.

◁

❏ Technical Note

After fitting the model using `anova`, we could have simply typed `regress` rather than `anova, regress` to obtain the same output. `regress` typed without arguments displays the last regression. Since `anova` defines a regression, typing `regress` after `anova` is another way of displaying the regression.

❏

▷ Example 3

We previously typed `anova weight treatment` to produce and display the ANOVA table for our apple tree experiment. We typed `anova, regress` to obtain the regression coefficients. If we want to redisplay the ANOVA table, we can type `anova` without arguments:

```
. anova
```

	Number of obs =		10	R-squared	=	0.9147
	Root MSE	= 9.07002		Adj R-squared	=	0.8721
Source	Partial SS	df	MS		F	Prob > F
Model	5295.54433	3	1765.18144		21.46	0.0013
treatment	5295.54433	3	1765.18144		21.46	0.0013
Residual	493.591667	6	82.2652778			
Total	5789.136	9	643.237333			

◁

Two-way ANOVA

You can include multiple explanatory variables with the `anova` command, and you can specify interactions by placing '`*`' between the variable names. For instance, typing `anova y a b` performs a two-way layout of `y` on `a` and `b`. Typing `anova y a b a*b` performs a full two-way factorial layout.

With the default partial sums of squares, when you specify interacted terms, the order of the terms does not matter. Typing `anova y a b a*b` is the same as typing `anova y b a b*a`.

▷ Example 4

The classic two-way factorial ANOVA problem, at least as far as computer manuals are concerned, is a two-way ANOVA design from Afifi and Azen (1979).

Fifty-eight patients, each suffering from one of three different diseases, were randomly assigned to one of four different drug treatments, and the change in their systolic blood pressure was recorded. Here are the data:

	Disease 1	Disease 2	Disease 3
Drug 1	42, 44, 36 13, 19, 22	33, 26, 33 21	31, –3, 25 25, 24
Drug 2	28, 23, 34 42, 13	34, 33, 31 36	3, 26, 28 32, 4, 16
Drug 3	1, 29, 19	11, 9, 7 1, –6	21, 1, 9 3
Drug 4	24, 9, 22 –2, 15	27, 12, 12 –5, 16, 15	22, 7, 25 5, 12

Let's assume that we have entered these data into Stata and stored the data as systolic.dta. Below we use the data, list the first 10 observations, summarize the variables, and tabulate the control variables:

```
. use http://www.stata-press.com/data/r10/systolic
(Systolic Blood Pressure Data)
. list in 1/10
```

	drug	disease	systolic
1.	1	1	42
2.	1	1	44
3.	1	1	36
4.	1	1	13
5.	1	1	19
6.	1	1	22
7.	1	2	33
8.	1	2	26
9.	1	2	33
10.	1	2	21

```
. summarize
```

Variable	Obs	Mean	Std. Dev.	Min	Max
drug	58	2.5	1.158493	1	4
disease	58	2.017241	.8269873	1	3
systolic	58	18.87931	12.80087	-6	44

```
. tabulate drug disease
```

Drug Used	Patient's Disease			Total
	1	2	3	
1	6	4	5	15
2	5	4	6	15
3	3	5	4	12
4	5	6	5	16
Total	19	19	20	58

Each observation in our data corresponds to one patient, and for each patient we record drug, disease, and the increase in the systolic blood pressure, systolic. The tabulation reveals that the data are not balanced—there are not equal numbers of patients in each drug–disease cell. Stata does not require that the data be balanced. We can perform a two-way factorial ANOVA by typing

```
. anova systolic drug disease drug*disease
```

```
                           Number of obs =      58     R-squared     =  0.4560
                           Root MSE      = 10.5096     Adj R-squared =  0.3259
```

Source	Partial SS	df	MS	F	Prob > F
Model	4259.33851	11	387.212591	3.51	0.0013
drug	2997.47186	3	999.157287	9.05	0.0001
disease	415.873046	2	207.936523	1.88	0.1637
drug*disease	707.266259	6	117.87771	1.07	0.3958
Residual	5080.81667	46	110.452536		
Total	9340.15517	57	163.862371		

Although Stata's `table` command does not perform ANOVA, it can produce useful summary tables of your data (see [R] **table**):

```
. table drug disease, c(mean systolic) row col f(%8.2f)
```

Drug Used	Patient's Disease 1	2	3	Total
1	29.33	28.25	20.40	26.07
2	28.00	33.50	18.17	25.53
3	16.33	4.40	8.50	8.75
4	13.60	12.83	14.20	13.50
Total	22.79	18.21	15.80	18.88

◁

❑ Technical Note

How do you interpret the significance of terms like `drug` and `disease` in unbalanced data? If you are familiar with SAS, the sums of squares and the F statistic reported by Stata correspond to SAS type III sums of squares. (Stata can also calculate sequential sums of squares, but we will postpone that topic for now.)

Let's think in terms of the following table:

	Disease 1	Disease 2	Disease 3	
Drug 1	μ_{11}	μ_{12}	μ_{13}	$\mu_{1\cdot}$
Drug 2	μ_{21}	μ_{22}	μ_{23}	$\mu_{2\cdot}$
Drug 3	μ_{31}	μ_{32}	μ_{33}	$\mu_{3\cdot}$
Drug 4	μ_{41}	μ_{42}	μ_{43}	$\mu_{4\cdot}$
	$\mu_{\cdot1}$	$\mu_{\cdot2}$	$\mu_{\cdot3}$	$\mu_{\cdot\cdot}$

In this diagram, μ_{ij} is the mean increase in systolic blood pressure associated with drug i and disease j, while $\mu_{i\cdot}$ is the mean for drug i, $\mu_{\cdot j}$ is the mean for disease j, and $\mu_{\cdot\cdot}$ is the overall mean.

If the data are balanced, meaning that there are an equal number of observations going into the calculation of each mean μ_{ij}, the row means, $\mu_{i\cdot}$, are given by

$$\mu_{i\cdot} = \frac{\mu_{i1} + \mu_{i2} + \mu_{i3}}{3}$$

In our case, the data are not balanced, but we define the $\mu_{i\cdot}$ according to that formula anyway. The test for the main effect of drug is the test that

$$\mu_{1\cdot} = \mu_{2\cdot} = \mu_{3\cdot} = \mu_{4\cdot}$$

To be absolutely clear, the F test of the term `drug`, called the *main effect* of drug, is formally equivalent to the test of the three constraints:

$$\frac{\mu_{11} + \mu_{12} + \mu_{13}}{3} = \frac{\mu_{21} + \mu_{22} + \mu_{23}}{3}$$

$$\frac{\mu_{11} + \mu_{12} + \mu_{13}}{3} = \frac{\mu_{31} + \mu_{32} + \mu_{33}}{3}$$

$$\frac{\mu_{11} + \mu_{12} + \mu_{13}}{3} = \frac{\mu_{41} + \mu_{42} + \mu_{43}}{3}$$

In our data, we obtain a significant F statistic of 9.05 and thus reject those constraints. ❑

❑ Technical Note

Stata can display the symbolic form underlying the test statistics it presents, as well as display other test statistics and their symbolic forms; see *Obtaining symbolic forms* in [R] **anova postestimation**. Here is the result of requesting the symbolic form for the main effect of drug in our data:

```
. test drug, symbolic
_cons         0
drug
        1       r1
        2       r2
        3       r3
        4       -(r1+r2+r3)
disease
        1       0
        2       0
        3       0
drug*disease
    1   1       1/3 r1
    1   2       1/3 r1
    1   3       1/3 r1
    2   1       1/3 r2
    2   2       1/3 r2
    2   3       1/3 r2
    3   1       1/3 r3
    3   2       1/3 r3
    3   3       1/3 r3
    4   1       -1/3 (r1+r2+r3)
    4   2       -1/3 (r1+r2+r3)
    4   3       -1/3 (r1+r2+r3)
```

This says exactly what we said in the previous technical note. ❑

❑ Technical Note

Saying that there is no main effect of a variable is not the same as saying that it has no effect at all. Stata's ability to perform ANOVA on unbalanced data can easily be put to ill use.

For example, consider the following table of the probability of surviving a bout with one of two diseases according to the drug administered to you:

	Disease 1	Disease 2
Drug 1	1	0
Drug 2	0	1

If you have disease 1 and are administered drug 1, you live. If you have disease 2 and are administered drug 2, you live. In all other cases, you die.

This table has no main effects of either drug or disease, although there is a large interaction effect. You might now be tempted to reason that since there is only an interaction effect, you would be indifferent between the two drugs in the absence of knowledge about which disease infects you. Given an equal chance of having either disease, you reason that it does not matter which drug is administered to you—either way, your chances of surviving are 0.5.

You may not, however, have an equal chance of having either disease. If you knew that disease 1 was 100 times more likely to occur in the population, and if you knew that you had one of the two diseases, you would express a strong preference for receiving drug 1.

When you calculate the significance of main effects on unbalanced data, you must ask yourself why the data are unbalanced. If the data are unbalanced for random reasons and you are making predictions for a balanced population, the test of the main effect makes perfect sense. If, however, the data are unbalanced because the underlying populations are unbalanced and you are making predictions for such unbalanced populations, the test of the main effect may be practically—if not statistically—meaningless.

❑

▷ Example 5

Stata can perform ANOVA not only on unbalanced populations, but also on populations that are so unbalanced that entire cells are missing. For instance, using our systolic blood pressure data, let's refit the model eliminating the drug 1–disease 1 cell. Since anova follows the same syntax as all other Stata commands, we can explicitly specify the data to be used by typing the if qualifier at the end of the anova command. Here we want to use the data that are not for drug 1 and disease 1:

```
. anova systolic drug disease drug*disease if !(drug==1 & disease==1)
```

| | Number of obs = | 52 | R-squared | = 0.4545 |
| | Root MSE | = 10.1615 | Adj R-squared = | 0.3215 |

Source	Partial SS	df	MS	F	Prob > F
Model	3527.95897	10	352.795897	3.42	0.0025
drug	2686.57832	3	895.526107	8.67	0.0001
disease	327.792598	2	163.896299	1.59	0.2168
drug*disease	703.007602	5	140.60152	1.36	0.2586
Residual	4233.48333	41	103.255691		
Total	7761.44231	51	152.185143		

◁

(Continued on next page)

❑ Technical Note

The test of the main effect of drug in the presence of missing cells is more complicated than that for unbalanced data. Our underlying tableau now has the following form:

	Disease 1	Disease 2	Disease 3	
Drug 1		μ_{12}	μ_{13}	
Drug 2	μ_{21}	μ_{22}	μ_{23}	$\mu_{2\cdot}$
Drug 3	μ_{31}	μ_{32}	μ_{33}	$\mu_{3\cdot}$
Drug 4	μ_{41}	μ_{42}	μ_{43}	$\mu_{4\cdot}$
		$\mu_{\cdot2}$	$\mu_{\cdot3}$	

The hole in the drug 1–disease 1 cell indicates that the mean is unobserved. Considering the main effect of drug, the test is unchanged for the rows in which all the cells are defined:

$$\mu_{2\cdot} = \mu_{3\cdot} = \mu_{4\cdot}$$

The first row, however, requires special attention. Here we want the average outcome for drug 1, which is averaged only over diseases 2 and 3, to be equal to the average values of all other drugs averaged over those same two diseases:

$$\frac{\mu_{12} + \mu_{13}}{2} = \frac{\left(\mu_{22} + \mu_{23}\right)/2 + \left(\mu_{32} + \mu_{33}\right)/2 + \left(\mu_{42} + \mu_{43}\right)/2}{3}$$

Thus the test contains three constraints:

$$\frac{\mu_{21} + \mu_{22} + \mu_{23}}{3} = \frac{\mu_{31} + \mu_{32} + \mu_{33}}{3}$$

$$\frac{\mu_{21} + \mu_{22} + \mu_{23}}{3} = \frac{\mu_{41} + \mu_{42} + \mu_{43}}{3}$$

$$\frac{\mu_{12} + \mu_{13}}{2} = \frac{\mu_{22} + \mu_{23} + \mu_{32} + \mu_{33} + \mu_{42} + \mu_{43}}{6}$$

❑

Stata can calculate two types of sums of squares, *partial* and *sequential*. If you do not specify which sums of squares to calculate, Stata calculates partial sums of squares. The technical notes above have gone into great detail about the definition and use of partial sums of squares. Use the `sequential` option to obtain sequential sums of squares.

❑ Technical Note

Before we illustrate sequential sums of squares, consider one more feature of the partial sums. If you know how such things are calculated, you may worry that the terms must be specified in some particular order, that Stata would balk or, even worse, produce different results if you typed, say, `anova drug*disease drug disease` rather than `anova drug disease drug*disease`. We assure you that is not the case.

When you type a model, Stata internally reorganizes the terms, forms the cross-product matrix, inverts it, converts the result to an upper-Hermite form, and then performs the hypothesis tests. As a final touch, Stata reports the results in the same order that you typed the terms.

❑

▷ **Example 6**

We wish to estimate the effects on systolic blood pressure of drug and disease using sequential sums of squares. We want to introduce disease first, then drug, and finally, the interaction of drug and disease:

```
. anova systolic disease drug disease*drug, sequential
                          Number of obs =      58    R-squared     =  0.4560
                          Root MSE      = 10.5096    Adj R-squared =  0.3259

        Source |    Seq. SS     df       MS            F     Prob > F

         Model |  4259.33851    11   387.212591       3.51    0.0013

       disease |   488.639383    2   244.319691       2.21    0.1210
          drug |  3063.43286     3  1021.14429        9.25    0.0001
  disease*drug |   707.266259    6   117.87771        1.07    0.3958

      Residual |  5080.81667    46   110.452536

         Total |  9340.15517    57   163.862371
```

The F statistic on disease is now 2.21. When we fitted this same model by using partial sums of squares, the statistic was 1.88.

◁

N-way ANOVA

You may include high-order interaction terms, such as a third-order interaction between the variables A, B, and C, by typing A*B*C.

▷ **Example 7**

We wish to determine the operating conditions that maximize yield for a manufacturing process. There are three temperature settings, two chemical supply companies, and two mixing methods under investigation. Three observations are obtained for each combination of these three factors.

```
. use http://www.stata-press.com/data/r10/manuf
(manufacturing process data)

. describe

Contains data from http://www.stata-press.com/data/r10/manuf.dta
  obs:           36                          manufacturing process data
 vars:            4                          2 Jan 2007 13:28
 size:          288 (93.5% of memory free)

              storage  display    value
variable name   type   format     label      variable label

temperature    byte    %9.0g      temp       machine temperature setting
chemical       byte    %9.0g      supplier   chemical supplier
method         byte    %9.0g      meth       mixing method
yield          byte    %9.0g                 product yield

Sorted by:
```

We wish to perform a three-way factorial ANOVA.

```
. anova yield temp chem temp*chem meth temp*meth chem*meth temp*chem*meth
```

	Number of obs =	36	R-squared	=	0.5474
	Root MSE	= 2.62996	Adj R-squared =		0.3399

Source	Partial SS	df	MS	F	Prob > F
Model	200.75	11	18.25	2.64	0.0227
temperature	30.50	2	15.25	2.20	0.1321
chemical	12.25	1	12.25	1.77	0.1958
temperature*chemical	24.50	2	12.25	1.77	0.1917
method	42.25	1	42.25	6.11	0.0209
temperature*method	87.50	2	43.75	6.33	0.0062
chemical*method	.25	1	.25	0.04	0.8508
temperature*chemical*method	3.50	2	1.75	0.25	0.7785
Residual	166.00	24	6.91666667		
Total	366.75	35	10.4785714		

The interaction between temperature and method appears to be the important story in these data. A table of means for this interaction is given below.

```
. table method temp, c(mean yield) row col f(%8.2f)
```

mixing method	machine temperature setting			
	low	medium	high	Total
stir	7.50	6.00	6.00	6.50
fold	5.50	9.00	11.50	8.67
Total	6.50	7.50	8.75	7.58

We decide to use the folding method of mixing and a high temperature in our manufacturing process.

◁

ANCOVA

You can include multiple explanatory variables with the anova command, but unless you explicitly state otherwise, all the variables are interpreted as *categorical variables*. Using the continuous(*varlist*) option, you can designate variables as *continuous* and thus perform ANCOVA.

▷ Example 8

Let's now return to example 4, the systolic blood pressure example. Remember that the data contained three variables: systolic, the increase in systolic blood pressure; disease, which of four diseases the patient endured; and drug, which of three drugs were administered. We have another variant of this dataset that includes one more variable, the age of the patient, age. (This example, which was based on real data, now enters the fictional world. We concocted the age data based on nothing more than our own fanciful imagination.)

```
. use http://www.stata-press.com/data/r10/sysage
(Systolic Blood Pressure Data)
. summarize age
```

Variable	Obs	Mean	Std. Dev.	Min	Max
age	58	45.15517	13.9985	20	73

Let's now estimate `systolic` as a function of drug, disease, age, and disease interacted with age:

```
. anova systolic drug disease age disease*age, continuous(age)
```

	Number of obs =	58		R-squared	= 0.6221
	Root MSE	= 8.48737		Adj R-squared =	0.5604

Source	Partial SS	df	MS	F	Prob > F
Model	5810.41855	8	726.302318	10.08	0.0000
drug	2791.94475	3	930.648251	12.92	0.0000
disease	129.092158	2	64.5460789	0.90	0.4148
age	1817.80067	1	1817.80067	25.23	0.0000
disease*age	43.4069507	2	21.7034754	0.30	0.7412
Residual	3529.73663	49	72.0354414		
Total	9340.15517	57	163.862371		

We typed `anova systolic drug disease age disease*age, continuous(age)` to obtain the above estimates. Rather than typing `continuous(age)`, we could have typed `category(drug disease)`.

In general, we can specify either the continuous variables or the categorical variables. If we do not specify either option, Stata assumes that all the variables are categorical. If we specify the `continuous(`*varlist*`)` option, Stata assumes that the remaining variables are categorical. If we specify the `category(`*varlist*`)` option, Stata assumes that the remaining variables are continuous. We suggest typing whichever list is shorter.

We can type `class(`*varlist*`)` rather than `category(`*varlist*`)`, as they are synonyms.

Remember that we can specify interactions by typing *varname**varname. We have seen examples of interacting categorical variables with categorical variables and, directly above, a categorical variable (`disease`) with a continuous variable (`age`).

We can also interact continuous variables with continuous variables. To include an age^2 term in our model, we could type `age*age`. If we also wanted to interact the categorical variable `disease` with the age^2 term, we could type `disease*age*age` (or even `age*disease*age`).

◁

Nested designs

In addition to specifying interaction terms, nested terms can also be specified in an ANOVA. A vertical bar is used to indicate nesting: A|B is read as A nested within B. A|B|C is read as A nested within B, which is nested within C. A|B*C is read as A is nested within the interaction of B and C. A*B|C is read as the interaction of A and B, which is nested within C.

Different error terms can be specified for different parts of the model. The forward slash is used to indicate that the next term in the model is the error term for what precedes it. For instance, `anova y A / B|A` indicates that the F test for A is to be tested by using the mean square from B|A in the denominator. Error terms (terms following the slash) are generally not tested unless they are themselves followed by a slash. Residual error is the default error term.

For example, consider $A / B / C$, where A, B, and C may be arbitrarily complex terms. Then anova will report A tested by B and B tested by C. If we add one more slash on the end to form $A / B / C /$, anova will also report C tested by the residual error.

▷ Example 9

We have collected data from a manufacturer that is evaluating which of five different brands of machinery to buy to perform a particular function in an assembly line. Twenty assembly-line employees were selected at random for training on these machines, with four employees assigned to learn a particular machine. The output from each employee (operator) on the brand of machine for which he trained was measured during four trial periods. In this example, the operator is nested within machine. Because of sickness and employee resignations, the final data are not balanced. The following table gives the mean output and sample size for each machine and operator combination.

```
. use http://www.stata-press.com/data/r10/machine
(machine data)

. table machine operator, c(mean output n output) col f(%8.2f)
```

five brands of machine	operator nested in machine 1	2	3	4	Total
1	9.15	9.48	8.27	8.20	8.75
	2	4	3	4	13
2	15.03	11.55	11.45	11.52	12.47
	3	2	2	4	11
3	11.27	10.13	11.13		10.84
	3	3	3		9
4	16.10	18.97	15.35	16.60	16.65
	3	3	4	3	13
5	15.30	14.35	10.43		13.63
	4	4	3		11

Assuming that operator is random (i.e., we wish to infer to the larger population of possible operators) and machine is fixed (i.e., only these five machines are of interest), the typical test for machine uses operator nested within machine as the error term. operator nested within machine can be tested by residual error. Our earlier warning concerning designs with unplanned missing cells and/or unbalanced cell sizes also applies to interpreting the ANOVA results from this unbalanced nested example.

```
. anova output machine / operator|machine /
                          Number of obs =      57   R-squared     =  0.8661
                          Root MSE      = 1.47089   Adj R-squared =  0.8077

              Source | Partial SS    df       MS            F     Prob > F

               Model | 545.822288    17   32.1071934      14.84    0.0000

             machine | 430.980792     4   107.745198      13.82    0.0001
    operator|machine | 101.353804    13   7.79644648

    operator|machine | 101.353804    13   7.79644648       3.60    0.0009

            Residual |  84.3766582   39   2.16350406

               Total | 630.198947    56   11.2535526
```

operator|machine is preceded by a slash, indicating that it is the error term for the terms before it (here, machine). operator|machine is also followed by a slash that indicates it should be tested with residual error. The output lists the operator|machine term twice, once as the error term for machine, and again as a term tested by residual error. A line is placed in the ANOVA table to separate the two. In general, a dividing line is placed in the output to separate the terms into groups that are tested with the same error term. The overall model is tested by residual error and is separated from the rest of the table by a blank line at the top of the table.

The results indicate that the machines are not all equal and that there are significant differences between operators.

◁

▷ Example 10

A medical researcher comes to you for help in analyzing some data he has collected. Two skin rash treatment protocols were tested at eight clinics (four clinics for each protocol). Three doctors were selected at random from each of the clinics to administer the particular protocol to four of their patients. Each patient was treated for four separate rash patches, and the response was measured. The data are described below.

```
. use http://www.stata-press.com/data/r10/rash
(skin rash data)

. describe

Contains data from http://www.stata-press.com/data/r10/rash.dta
  obs:          384                          skin rash data
  vars:           5                          19 Dec 2006 02:19
  size:       3,456 (93.0% of memory free)   (_dta has notes)

              storage  display    value
variable name  type    format     label      variable label

response       byte     %9.0g
treatment      byte     %9.0g                 2 treatment protocols
clinic         byte     %9.0g                 4 clinics per treatment
doctor         byte     %9.0g                 3 doctors per clinic
patient        byte     %9.0g                 4 patients per doctor

Sorted by:
```

The researcher wants to determine if the treatment protocols are significantly different. He would also like to know if clinic, doctor, and patient are significant effects. In this analysis, treatment is a fixed factor, whereas clinic, doctor, and patient are random factors. Our first thought is that there are more powerful experimental designs that could have been used. For instance, a better design would be to test both treatments on each patient. However, the data have already been collected, so we proceed with the analysis of this fully nested ANOVA design.

In the following anova command, we use abbreviations for the variable names, which can sometimes make long ANOVA model statements easier to read.

```
. anova response t / c|t / d|c|t / p|d|c|t /
                          Number of obs =      384    R-squared     =  0.5040
                          Root MSE      = 11.6192    Adj R-squared =  0.3404

                 Source │  Partial SS   df      MS            F     Prob > F
               ─────────┼──────────────────────────────────────────────────
                  Model │  39505.9896   95  415.852522        3.08    0.0000

              treatment │  4240.04167    1  4240.04167        9.79    0.0204
        clinic|treatment │  2599.48958    6  433.248264
               ─────────┼──────────────────────────────────────────────────
        clinic|treatment │  2599.48958    6  433.248264        0.93    0.4982
  doctor|clinic|treatment │  7429.58333   16  464.348958
               ─────────┼──────────────────────────────────────────────────
  doctor|clinic|treatment │  7429.58333   16  464.348958        1.32    0.2065
   patient|doctor|clinic| │
              treatment │  25236.875    72  350.512153
               ─────────┼──────────────────────────────────────────────────
   patient|doctor|clinic| │
              treatment │  25236.875    72  350.512153        2.60    0.0000

               Residual │  38882.00    288  135.006944
               ─────────┼──────────────────────────────────────────────────
                  Total │  78387.9896  383  204.66838
```

We conclude that the two treatment protocols are significantly different. We also see that clinic and doctor are not significant, whereas patient is significant. The mean response for each of the two treatment protocols is shown below.

```
. table treatment, c(mean response) f(%8.2f) stubwidth(11)
```

2 treatment protocols	mean(response)
1	39.18
2	45.83

Treatment protocol 2 gives the larger response.

◁

Mixed designs

An ANOVA can consist of both nested and crossed terms. A split-plot ANOVA design provides an example.

▷ Example 11

Two reading programs and three skill-enhancement techniques are under investigation. Ten classes of first-grade students were randomly assigned so that five classes were taught with one reading program and another five classes were taught with the other. The 30 students in each class were divided into six groups with five students each. Within each class, the six groups were divided randomly so that each of the three skill-enhancement techniques was taught to two of the groups within each class. At the end of the school year, a reading assessment test was administered to all the students. In this split-plot ANOVA, the whole-plot treatment is the two reading programs, and the split-plot treatment is the three skill-enhancement techniques.

```
. use http://www.stata-press.com/data/r10/reading
(Reading experiment data)

. describe
Contains data from http://www.stata-press.com/data/r10/reading.dta
  obs:           300                        Reading experiment data
  vars:            5                        2 Jan 2007 05:15
  size:         2,700 (86.2% of memory free)  (_dta has notes)

              storage  display    value
variable name   type   format     label    variable label

score          byte    %9.0g               reading score
program        byte    %9.0g               reading program
class          byte    %9.0g               class nested in program
skill          byte    %9.0g               skill enhancement technique
group          byte    %9.0g               group nested in class and skill

Sorted by:
```

In this split-plot ANOVA, the error term for `program` is `class` nested within `program`. The error term for `skill` and the `program` by `skill` interaction is the `class` by `skill` interaction nested within `program`. Other terms are also involved in the model and can be seen below.

Our `anova` command is too long to fit on one line of this manual. Where we have chosen to break the command into multiple lines is arbitrary. If we were typing this command into Stata, we would just type along and let Stata automatically wrap across lines, as necessary.

(Continued on next page)

```
. anova score prog / class|prog skill prog*skill / class*skill|prog
> / group|class*skill|prog /
```

	Number of obs =	300	R-squared = 0.3738		
	Root MSE = 14.6268		Adj R-squared = 0.2199		
Source	Partial SS	df	MS	F	Prob > F
Model	30656.5167	59	519.601977	2.43	0.0000
program	4493.07	1	4493.07	8.73	0.0183
class|program	4116.61333	8	514.576667		
skill	1122.64667	2	561.323333	1.54	0.2450
program*skill	5694.62	2	2847.31	7.80	0.0043
class*skill|program	5841.46667	16	365.091667		
class*skill|program	5841.46667	16	365.091667	1.17	0.3463
group|class*skill|program	9388.1	30	312.936667		
group|class*skill|program	9388.1	30	312.936667	1.46	0.0636
Residual	51346.4	240	213.943333		
Total	82002.9167	299	274.257246		

The `program*skill` term is significant, as is the `program` term for these particular data. Let us look at a table of mean scores by `program` and `skill`.

```
. table prog skill, c(mean score) row col f(%8.2f)
```

reading program	skill enhancement technique			
	1	2	3	Total
1	68.16	52.86	61.54	60.85
2	50.70	56.54	52.10	53.11
Total	59.43	54.70	56.82	56.98

The first reading program and the first skill-enhancement technique perform best when combined. Clearly, there is an interaction between the reading program and the skill-enhancement technique. ◁

❑ Technical Note

There are several valid ways to write complicated anova terms. In the reading experiment example (example 11), we had a term `group|class*skill|program`. This term can be read as `group` nested within both `class` and `skill` and further nested within `program`. You can also write this term as `group|class*skill*program` or `group|program*class*skill` or `group|skill*class|program`, etc. All variations will produce the same result. Some people prefer having only one '|' in a term and would use `group|class*skill*program`, which is read as `group` nested within `class`, `skill`, and `program`.

❑

Latin-square designs

You can use anova to analyze a Latin-square design. Consider the following example, published in Snedecor and Cochran (1989).

▷ Example 12

Data from a Latin-square design are as follows:

Row	Column 1	Column 2	Column 3	Column 4	Column 5
1	257(B)	230(E)	279(A)	287(C)	202(D)
2	245(D)	283(A)	245(E)	280(B)	260(C)
3	182(E)	252(B)	280(C)	246(D)	250(A)
4	203(A)	204(C)	227(D)	193(E)	259(B)
5	231(C)	271(D)	266(B)	334(A)	338(E)

In Stata, the data might appear as follows:

```
. use http://www.stata-press.com/data/r10/latinsq
. list
```

	row	c1	c2	c3	c4	c5
1.	1	257	230	279	287	202
2.	2	245	283	245	280	260
3.	3	182	252	280	246	250
4.	4	203	204	227	193	259
5.	5	231	271	266	334	338

Before anova can be used on these data, the data must be organized so that the outcome measurement is in one column. reshape is inadequate for this task because there is information about the treatments in the sequence of these observations. pkshape is designed to reshape this type of data; see [R] **pkshape**.

(*Continued on next page*)

```
. pkshape row row c1-c5, order(beacd daebc ebcda acdeb cdbae)
. list
```

	sequence	outcome	treat	carry	period
1.	1	257	1	0	1
2.	2	245	5	0	1
3.	3	182	2	0	1
4.	4	203	3	0	1
5.	5	231	4	0	1
6.	1	230	2	1	2
7.	2	283	3	5	2
8.	3	252	1	2	2
9.	4	204	4	3	2
10.	5	271	5	4	2
11.	1	279	3	2	3
12.	2	245	2	3	3
13.	3	280	4	1	3
14.	4	227	5	4	3
15.	5	266	1	5	3
16.	1	287	4	3	4
17.	2	280	1	2	4
18.	3	246	5	4	4
19.	4	193	2	5	4
20.	5	334	3	1	4
21.	1	202	5	4	5
22.	2	260	4	1	5
23.	3	250	3	5	5
24.	4	259	1	2	5
25.	5	338	2	3	5

```
. anova outcome sequence period treat
```

```
                        Number of obs =       25    R-squared     =  0.6536
                        Root MSE      = 32.4901    Adj R-squared =  0.3073

          Source |   Partial SS    df       MS            F     Prob > F

           Model |    23904.08     12   1992.00667       1.89     0.1426

        sequence |    13601.36      4   3400.34          3.22     0.0516
          period |     6146.16      4   1536.54          1.46     0.2758
           treat |     4156.56      4   1039.14          0.98     0.4523

        Residual |    12667.28     12   1055.60667

           Total |    36571.36     24   1523.80667
```

◁

These methods will work with any type of Latin-square design, including those with replicated measurements. For more information, see [R] **pk**, [R] **pkcross**, and [R] **pkshape**.

Repeated-measures ANOVA

One approach for analyzing repeated-measures data is to use multivariate ANOVA (MANOVA); see [MV] **manova**. In this approach, the data are placed in wide form (see [D] **reshape**), and the repeated measures enter the MANOVA as dependent variables.

A second approach for analyzing repeated measures is to use `anova`. However, one of the underlying assumptions for the F tests in ANOVA is independence of observations. In a repeated-measures design, this assumption is almost certainly violated or is at least suspect. In a repeated-measures ANOVA, the subjects (or whatever the experimental units are called) are observed for each level of one or more of the other categorical variables in the model. These variables are called the repeated-measure variables. Observations from the same subject are likely to be correlated.

The approach used in repeated-measures ANOVA to correct for this lack of independence is to apply a correction to the degrees of freedom of the F test for terms in the model that involve repeated measures. This correction factor, ϵ, lies between the reciprocal of the degrees of freedom for the repeated term and 1. Box (1954) provided the pioneering work in this area. Milliken and Johnson (1984) refer to the lower bound of this correction factor as Box's conservative correction factor. Winer, Brown, and Michels (1991) call it simply the conservative correction factor.

Geisser and Greenhouse (1958) provide an estimate for the correction factor called the Greenhouse–Geisser ϵ. This value is estimated from the data. Huynh and Feldt (1976) show that the Greenhouse–Geisser ϵ tends to be conservatively biased. They provide a revised correction factor called the Huynh–Feldt ϵ. When the Huynh–Feldt ϵ exceeds 1, it is set to 1. Thus there is a natural ordering for these correction factors:

$$\text{Box's conservative } \epsilon \leq \text{Greenhouse–Geisser } \epsilon \leq \text{Huynh–Feldt } \epsilon \leq 1$$

A correction factor of 1 is the same as no correction.

`anova` with the `repeated()` option computes these correction factors and displays the revised test results in a table that follows the standard ANOVA table. In the resulting table, H-F stands for Huynh–Feldt, G-G stands for Greenhouse–Geisser, and Box stands for Box's conservative ϵ.

▷ Example 13

This example is taken from table 4.3 of Winer, Brown, and Michels (1991). The reaction time for five subjects each tested with four drugs was recorded in the variable `score`. Here is a table of the data (see [P] **tabdisp** if you are unfamiliar with `tabdisp`):

```
. use http://www.stata-press.com/data/r10/t43
(T4.3 -- Winer, Brown, Michels)

. tabdisp person drug, cellvar(score)
```

person	drug 1	2	3	4
1	30	28	16	34
2	14	18	10	22
3	24	20	18	30
4	38	34	20	44
5	26	28	14	30

`drug` is the repeated variable in this simple repeated-measures ANOVA example. The ANOVA is specified as follows:

```
. anova score person drug, repeated(drug)
                             Number of obs =      20     R-squared     =  0.9244
                             Root MSE      = 3.06594     Adj R-squared =  0.8803

          Source | Partial SS     df       MS              F     Prob > F

           Model |      1379       7       197          20.96     0.0000

          person |     680.8       4       170.2        18.11     0.0001
            drug |     698.2       3   232.733333        24.76     0.0000

        Residual |     112.8      12       9.4

           Total |    1491.8      19   78.5157895
```

```
Between-subjects error term:  person
                   Levels:  5            (4 df)
      Lowest b.s.e. variable:  person

Repeated variable: drug
                                   Huynh-Feldt epsilon        =  1.0789
                                  *Huynh-Feldt epsilon reset to 1.0000
                                   Greenhouse-Geisser epsilon =  0.6049
                                   Box's conservative epsilon =  0.3333

                                   ───────── Prob > F ─────────
          Source |   df     F    Regular    H-F      G-G      Box

            drug |    3   24.76   0.0000   0.0000   0.0006   0.0076
        Residual |   12
```

Here the Huynh–Feldt ϵ is 1.0789, which is larger than 1. It is reset to 1, which is the same as making no adjustment to the standard test computed in the main ANOVA table. The Greenhouse–Geisser ϵ is 0.6049, and its associated p-value is computed from an F ratio of 24.76 using 1.8147 ($= 3\epsilon$) and 7.2588 ($= 12\epsilon$) degrees of freedom. Box's conservative ϵ is set equal to the reciprocal of the degrees of freedom for the repeated term. Here it is 1/3, so that Box's conservative test is computed using 1 and 4 degrees of freedom for the observed F ratio of 24.76.

Even for Box's conservative ϵ, drug is significant with a p-value of 0.0076. The following table gives the mean score (i.e., response time) for each of the four drugs:

```
. table drug, c(mean score) f(%8.2f)

      drug | mean(score)

         1 |       26.40
         2 |       25.60
         3 |       15.60
         4 |       32.00
```

The ANOVA table for this example provides an F test for person, but you should ignore it. An appropriate test for person would require replication (i.e., multiple measurements for person and drug combinations). Also, without replication there is no test available for investigating the interaction between person and drug.

◁

▷ Example 14

Table 7.7 of Winer, Brown, and Michels (1991) provides another repeated-measures ANOVA example. There are four dial shapes and two methods for calibrating dials. Subjects are nested within calibration method, and an accuracy score is obtained. The data are shown below.

```
. use http://www.stata-press.com/data/r10/t77
(T7.7 -- Winer, Brown, Michels)
. tabdisp shape subject calib, cell(score)
```

4 dial	2 methods for calibrating dials and subject nested in calib					
shapes	———— 1 ————			———— 2 ————		
	1	2	3	1	2	3
1	0	3	4	4	5	7
2	0	1	3	2	4	5
3	5	5	6	7	6	8
4	3	4	2	8	6	9

The calibration method and dial shapes are fixed factors, whereas subjects are random. The appropriate test for calibration method uses the nested subject term as the error term. Both the dial shape and the interaction between dial shape and calibration method are tested with the dial shape by subject interaction nested within calibration method. Here we drop this term from the `anova` command, and it becomes residual error. The dial shape is the repeated variable since each subject is tested with all four dial shapes. Here is the `anova` command that produces the desired results:

```
. anova score calib / subject|calib shape calib*shape, repeated(shape)
```

| | | Number of obs = | 24 | R-squared | = 0.8925 |
| | | Root MSE = 1.11181 | | Adj R-squared = 0.7939 |

Source	Partial SS	df	MS	F	Prob > F
Model	123.125	11	11.1931818	9.06	0.0003
calib	51.0416667	1	51.0416667	11.89	0.0261
subject|calib	17.1666667	4	4.29166667		
shape	47.4583333	3	15.8194444	12.80	0.0005
calib*shape	7.45833333	3	2.48611111	2.01	0.1662
Residual	14.8333333	12	1.23611111		
Total	137.958333	23	5.99818841		

```
Between-subjects error term:  subject|calib
                    Levels:  6          (4 df)
      Lowest b.s.e. variable:  subject
      Covariance pooled over:  calib      (for repeated variable)
Repeated variable: shape
```

		Huynh-Feldt epsilon	= 0.8483
		Greenhouse-Geisser epsilon = 0.4751
		Box's conservative epsilon = 0.3333

| | | | ———— Prob > F ———— | | |
Source	df	F	Regular	H-F	G-G	Box
shape	3	12.80	0.0005	0.0011	0.0099	0.0232
calib*shape	3	2.01	0.1662	0.1791	0.2152	0.2291
Residual	12					

The repeated-measure ϵ corrections are applied to any terms that are tested in the main ANOVA table and have the repeated variable in the term. These ϵ corrections are given in a table below the main ANOVA table. Here the repeated-measures tests for `shape` and `calib*shape` are presented.

Calibration method is significant, as is dial shape. The interaction between calibration method and dial shape is not significant. The repeated-measure ϵ corrections do not change these conclusions, but they do change the significance level for the tests on `shape` and `calib*shape`. Here, though, unlike in the previous example, the Huynh–Feldt ϵ is less than 1.

The following table gives the mean score for dial shapes and calibration method. Since the interaction was not significant, we focus our attention on the row and the column labeled "Total".

```
. table shape calib, c(mean score) row col f(%8.2f)
```

4 dial shapes	2 methods for calibrating dials		
	1	2	Total
1	2.33	5.33	3.83
2	1.33	3.67	2.50
3	5.33	7.00	6.17
4	3.00	7.67	5.33
Total	3.00	5.92	4.46

◁

❏ Technical Note

The computation of the Greenhouse–Geisser and Huynh–Feldt epsilons in a repeated-measures ANOVA requires the number of levels and degrees of freedom for the between-subjects error term, as well as a value computed from a pooled covariance matrix. The observations are grouped based on all but the lowest-level variable in the between-subjects error term. The covariance over the repeated variables is computed for each resulting group, and then these covariance matrices are pooled. The dimension of the pooled covariance matrix is the number of levels of the repeated variable (or combination of levels for multiple repeated variables). In example 14, there are four levels of the repeated variable (`shape`), so the resulting covariance matrix is 4×4.

The `anova` command automatically attempts to determine the between-subjects error term and the lowest-level variable in the between-subjects error term to group the observations for computation of the pooled covariance matrix. `anova` issues an error message indicating that the `bse()` or `bseunit()` option is required when `anova` cannot determine them. You may override the default selections of `anova` by specifying the `bse()`, `bseunit()`, or `grouping()` option. The term specified in the `bse()` option must be a term in the ANOVA model.

The default selection for the between-subjects error term (the `bse()` option) is the interaction of the nonrepeated categorical variables in the ANOVA model. The first variable listed in the between-subjects error term is automatically selected as the lowest-level variable in the between-subjects error term but can be overridden with the `bseunit(`*varname*`)` option. *varname* is often a term, such as subject or subsample within subject, and is most often listed first in the term because of the nesting notation of ANOVA. This term makes sense in most repeated-measures ANOVA designs when the terms of the model are written in standard form. For instance, in example 14, there were three categorical variables (`subject`, `calib`, and `shape`), with `shape` the repeated variable. Here `anova` looked for a term involving only `subject` and `calib` to determine the between-subjects error term. It found `subject|calib` as the term with six levels and 4 degrees of freedom. `anova` then picked `subject` as the default for the `bseunit()` option (the lowest variable in the between-subjects error term) since it was listed first in the term.

The grouping of observations proceeds, based on the different combinations of values of the variables in the between-subjects error term, excluding the lowest level variable (as found by default or as specified with the `bseunit()` option). You may specify the `grouping()` option to change the default grouping used in computing the pooled covariance matrix.

The between-subjects error term, number of levels, degrees of freedom, lowest variable in the term, and grouping information are presented after the main ANOVA table and before the rest of the repeated-measures output.

❏

▷ Example 15

Data with two repeated variables are given in table 7.13 of Winer, Brown, and Michels (1991). The accuracy scores of subjects making adjustments to three dials during three different periods are recorded. Three subjects are exposed to a certain noise background level, whereas a different set of three subjects is exposed to a different noise background level. Here is a table of accuracy scores for the `noise`, `subject`, `period`, and `dial` variables:

```
. use http://www.stata-press.com/data/r10/t713
(T7.13 -- Winer, Brown, Michels)
. tabdisp subject dial period, by(noise) cell(score) stubwidth(11)
```

noise background and subject nested in noise	10 minute time periods and dial								
	1			2			3		
	1	2	3	1	2	3	1	2	3
1									
1	45	53	60	40	52	57	28	37	46
2	35	41	50	30	37	47	25	32	41
3	60	65	75	58	54	70	40	47	50
2									
1	50	48	61	25	34	51	16	23	35
2	42	45	55	30	37	43	22	27	37
3	56	60	77	40	39	57	31	29	46

`noise`, `period`, and `dial` are fixed, whereas `subject` is random. Both `period` and `dial` are repeated variables. The ANOVA for this example is specified below.

(Continued on next page)

```
. anova score noise / subject|noise period noise*period
> / period*subject|noise dial noise*dial
> / dial*subject|noise period*dial noise*period*dial, repeated(period dial)
```

	Number of obs =	54	R-squared	= 0.9872
	Root MSE = 2.81859		Adj R-squared =	0.9576

Source	Partial SS	df	MS	F	Prob > F
Model	9797.72222	37	264.803303	33.33	0.0000
noise	468.166667	1	468.166667	0.75	0.4348
subject|noise	2491.11111	4	622.777778		
period	3722.33333	2	1861.16667	63.39	0.0000
noise*period	333.00	2	166.50	5.67	0.0293
period*subject|noise	234.888889	8	29.3611111		
dial	2370.33333	2	1185.16667	89.82	0.0000
noise*dial	50.3333333	2	25.1666667	1.91	0.2102
dial*subject|noise	105.555556	8	13.1944444		
period*dial	10.6666667	4	2.66666667	0.34	0.8499
noise*period*dial	11.3333333	4	2.83333333	0.36	0.8357
Residual	127.111111	16	7.94444444		
Total	9924.83333	53	187.261006		

```
Between-subjects error term:  subject|noise
                    Levels:  6         (4 df)
      Lowest b.s.e. variable:  subject
      Covariance pooled over:  noise     (for repeated variables)
```

Repeated variable: period

	Huynh-Feldt epsilon	= 1.0668
	*Huynh-Feldt epsilon reset to 1.0000	
	Greenhouse-Geisser epsilon =	0.6476
	Box's conservative epsilon =	0.5000

| | | | | Prob > F | | |
Source	df	F	Regular	H-F	G-G	Box
period	2	63.39	0.0000	0.0000	0.0003	0.0013
noise*period	2	5.67	0.0293	0.0293	0.0569	0.0759
period*subject|noise	8					

Repeated variable: dial

	Huynh-Feldt epsilon	= 2.0788
	*Huynh-Feldt epsilon reset to 1.0000	
	Greenhouse-Geisser epsilon =	0.9171
	Box's conservative epsilon =	0.5000

| | | | | Prob > F | | |
Source	df	F	Regular	H-F	G-G	Box
dial	2	89.82	0.0000	0.0000	0.0000	0.0007
noise*dial	2	1.91	0.2102	0.2102	0.2152	0.2394
dial*subject|noise	8					

```
Repeated variables: period*dial
```

```
                                    Huynh-Feldt epsilon         =  1.3258
                                   *Huynh-Feldt epsilon reset to 1.0000
                                    Greenhouse-Geisser epsilon =  0.5134
                                    Box's conservative epsilon =  0.2500
```

| | | | ───── Prob > F ───── | | | |
Source	df	F	Regular	H-F	G-G	Box
period*dial	4	0.34	0.8499	0.8499	0.7295	0.5934
noise*period*dial	4	0.36	0.8357	0.8357	0.7156	0.5825
Residual	16					

For each repeated variable and for each combination of interactions of repeated variables, there are different ϵ correction values. The `anova` command produces tables for each applicable combination.

The two most significant factors in this model appear to be `dial` and `period`. The `noise` by `period` interaction also may be significant, depending on the correction factor you use. Below are tables of the mean accuracy score for dial and for period by noise.

```
. table dial, c(mean score) f(%8.2f)
```

dial	mean(score)
1	37.39
2	42.22
3	53.22

```
. table noise period, c(mean score) f(%8.2f) row col stubwidth(10)
```

| noise background | 10 minute time periods | | | |
	1	2	3	Total
1	53.78	49.44	38.44	47.22
2	54.89	39.56	29.56	41.33
Total	54.33	44.50	34.00	44.28

We see that dial shape 3 produces the highest score and that the scores tend to decrease over the periods.

◁

Example 15 had two repeated-measurement variables. Up to four repeated-measurement variables may be specified in the `anova` command.

Weighted data

Like all estimation commands, `anova` can produce estimates on weighted data. See [U] **11.1.6 weight** for details on specifying the weight.

▷ Example 16

We wish to investigate the prevalence of byssinosis, a form of pneumoconiosis that can afflict workers exposed to cotton dust. We have data on 5,419 workers in a large cotton mill. We know whether each worker smokes, his or her race, and the dustiness of the work area. The variables are

smokes	Smoker or nonsmoker in the last five years.
race	White or other.
workplace	1 (most dusty), 2 (less dusty), 3 (least dusty).

We wish to fit an ANOVA model explaining the prevalence of byssinosis according to a full factorial model of smokes, race, and workplace.

The data are unbalanced. Moreover, although we have data on 5,419 workers, the data are grouped according to the explanatory variables, along with some other variables, resulting in 72 observations. For each observation, we know the number of workers in the group (pop), the prevalence of byssinosis (prob), and the values of the three explanatory variables. Thus we wish to fit a three-way factorial model on grouped data.

We begin by showing a bit of the data, which are from Higgins and Koch (1977).

```
. use http://www.stata-press.com/data/r10/byssin
(Byssinosis incidence)

. describe

Contains data from http://www.stata-press.com/data/r10/byssin.dta
  obs:            72                          Byssinosis incidence
  vars:            5                          19 Dec 2006 07:04
  size:         1,152 (89.7% of memory free)
```

variable name	storage type	display format	value label	variable label
smokes	int	%8.0g	smokes	Smokes
race	int	%8.0g	race	Race
workplace	int	%8.0g	workplace	Dustiness of workplace
pop	int	%8.0g		Population size
prob	float	%9.0g		Prevalence of byssinosis

```
Sorted by:
. list in 1/5, abbrev(10) divider
```

	smokes	race	workplace	pop	prob
1.	yes	white	most	40	.075
2.	yes	white	less	74	0
3.	yes	white	least	260	.0076923
4.	yes	other	most	164	.152439
5.	yes	other	less	88	0

The first observation in the data represents a group of 40 white workers who smoke and work in a "most" dusty work area. Of those 40 workers, 7.5% have byssinosis. The second observation represents a group of 74 white workers who also smoke but who work in a "less" dusty environment. None of those workers has byssinosis.

Almost every Stata command allows weights. Here we want to weight the data by pop. We can, for instance, make a table of the number of workers by their smoking status and race:

```
. tabulate smokes race [freq=pop]
```

Smokes	Race other	white	Total
no	799	1,431	2,230
yes	1,104	2,085	3,189
Total	1,903	3,516	5,419

The [freq=pop] at the end of the tabulate command tells Stata to count each observation as representing pop persons. When making the tally, tabulate treats the first observation as representing 40 workers, the second as representing 74 workers, and so on.

Similarly, we can make a table of the dustiness of the workplace:

```
. tabulate workplace [freq=pop]
```

Dustiness of workplace	Freq.	Percent	Cum.
least	3,450	63.66	63.66
less	1,300	23.99	87.65
most	669	12.35	100.00
Total	5,419	100.00	

We can discover the average incidence of byssinosis among these workers by typing

```
. summarize prob [freq=pop]
```

Variable	Obs	Mean	Std. Dev.	Min	Max
prob	5419	.0304484	.0567373	0	.287037

We discover that 3.04% of these workers have byssinosis. Across all cells, the byssinosis rates vary from 0 to 28.7%. Just to prove that there might be something here, let's obtain the average incidence rates according to the dustiness of the workplace:

```
. table workplace smokes race [freq=pop], c(mean prob)
```

Dustiness of workplace	Race and Smokes other		white	
	no	yes	no	yes
least	.0107527	.0101523	.0081549	.0162774
less	.02	.0081633	.0136612	.0143149
most	.0820896	.1679105	.0833333	.2295082

Let's now fit the ANOVA model.

(Continued on next page)

```
. anova prob workplace smokes race workplace*smokes workplace*race
> smokes*race workplace*smokes*race  [aweight=pop]
(sum of wgt is   5.4190e+03)
```

| | Number of obs = | | 65 | R-squared | = 0.8300 |
| | Root MSE | = .025902 | | Adj R-squared = | 0.7948 |

Source	Partial SS	df	MS	F	Prob > F
Model	.173646538	11	.015786049	23.53	0.0000
workplace	.097625175	2	.048812588	72.76	0.0000
smokes	.013030812	1	.013030812	19.42	0.0001
race	.001094723	1	.001094723	1.63	0.2070
workplace*smokes	.019690342	2	.009845171	14.67	0.0000
workplace*race	.001352516	2	.000676258	1.01	0.3718
smokes*race	.001662874	1	.001662874	2.48	0.1214
workplace*smokes*race	.000950841	2	.00047542	0.71	0.4969
Residual	.035557766	53	.000670901		
Total	.209204304	64	.003268817		

Of course, if we want to see the underlying regression, we could type `anova, regress` or simply `regress`.   ◁

How Stata numbers levels

We have covered all aspects of the `anova` command except one mechanical detail: the definition of a categorical variable. Throughout the above text, we have made statements such as, "Stata assumes that all the variables are categorical unless you explicitly state otherwise". Although the statement makes perfect sense, anyone familiar with statistical programs knows that it leaves something unsaid. Most programs place additional requirements on the definition of categorical variables, namely, that they be sequential integers.

Stata does not care whether your variables are integers, whether they are positive, or whether the numbering scheme is sequential or jumps around helter-skelter. Stata will figure it out.

▷ Example 17

Let's prove our claim. Below we have a small dataset containing three variables: `outcome`, `scheme1`, and `scheme2`.

```
. use http://www.stata-press.com/data/r10/levels
. list
```

	outcome	scheme1	scheme2
1.	28	2	-1.1
2.	23	2	-1.1
3.	34	2	-1.1
4.	42	2	-1.1
5.	13	2	-1.1
6.	29	3	2
7.	19	3	2
8.	9	5	2.1
9.	-2	5	2.1
10.	15	5	2.1

Let's think of an ANOVA model of outcome on scheme1 and outcome on scheme2. If we look closely, we will see that scheme1 and scheme2 are two different ways of labeling the same thing. In scheme1, the levels are labeled 2, 3, and 5; in scheme2, those same levels are labeled −1.1, 2, and 2.1.

anova produces the same output, no matter which labeling scheme we use:

```
. anova outcome scheme1
                          Number of obs =       10    R-squared     =  0.5474
                          Root MSE      = 9.86094    Adj R-squared =  0.4181

        Source |  Partial SS    df        MS            F     Prob > F

         Model |  823.333333     2   411.666667        4.23    0.0624

       scheme1 |  823.333333     2   411.666667        4.23    0.0624

      Residual |  680.666667     7   97.2380952

         Total |        1504     9   167.111111
```

```
. anova outcome scheme2
                          Number of obs =       10    R-squared     =  0.5474
                          Root MSE      = 9.86094    Adj R-squared =  0.4181

        Source |  Partial SS    df        MS            F     Prob > F

         Model |  823.333333     2   411.666667        4.23    0.0624

       scheme2 |  823.333333     2   411.666667        4.23    0.0624

      Residual |  680.666667     7   97.2380952

         Total |        1504     9   167.111111
```

◁

❑ Technical Note

Although statistical results are unaffected by the numbering scheme you use, there is a minor difference in how Stata presents regression results (as well as the results of test with the symbolic option). Below we use anova with the regress option to refit our model based on scheme1. The regress option tells Stata to present the regression table rather than the ANOVA output.

```
. anova outcome scheme1, regress
        Source |     SS      df       MS              Number of obs =      10
                                                      F(  2,     7) =    4.23
         Model |  823.333333   2   411.666667         Prob > F      =  0.0624
      Residual |  680.666667   7   97.2380952         R-squared     =  0.5474
                                                      Adj R-squared =  0.4181
         Total |        1504   9   167.111111         Root MSE      =  9.8609

       outcome |     Coef.   Std. Err.      t    P>|t|     [95% Conf. Interval]

         _cons |   7.333333   5.693215    1.29   0.239    -6.128981    20.79565
       scheme1 |
             2 |   20.66667   7.201411    2.87   0.024     3.638036     37.6953
             3 |   16.66667   9.001763    1.85   0.107    -4.619122    37.95245
             5 |  (dropped)
```

In labeling the regression coefficients, Stata used our numbering scheme. The first level of `scheme1` is labeled 2, the second level 3, and the third level 5, just as we numbered them.

Let's now repeat the experiment with the `scheme2` numbering scheme:

```
. anova outcome scheme2, regress
```

Source	SS	df	MS		
Model	823.333333	2	411.666667	Number of obs =	10
Residual	680.666667	7	97.2380952	F(2, 7) =	4.23
				Prob > F =	0.0624
				R-squared =	0.5474
Total	1504	9	167.111111	Adj R-squared =	0.4181
				Root MSE =	9.8609

outcome	Coef.	Std. Err.	t	P>|t|	[95% Conf. Interval]	
_cons	7.333333	5.693215	1.29	0.239	-6.128981	20.79565
scheme2						
1	20.66667	7.201411	2.87	0.024	3.638036	37.6953
2	16.66667	9.001763	1.85	0.107	-4.619122	37.95245
3	(dropped)					

Remember that `scheme2` labeled the first level −1.1, the second level 2, and the third level 2.1. This time, however, our regression results are labeled 1, 2, and 3.

Stata uses our numbering scheme to label results whenever all our numbers are positive integers less than 99. The first numbering scheme meets that criterion; the second numbering scheme does not. In the second case, Stata created a mapping of our values onto the set of sequential integers. Thus level −1.1 is labeled 1, level 2 is labeled 2, and level 2.1 is labeled 3.

Stata will show us the mapping if we specify the `detail` option. We request that Stata show us only the mapping and not present the ANOVA table:

```
. anova, detail noanova
```

Factor	Value	Value	Value	Value
scheme2	1 -1.1	2 2	3 2.1	

Stata makes a small table. In each entry, the number to the left is how Stata labels the level in its regression output. The number to the right is the true value of the underlying variable.

❏

❏ Technical Note

Since the labeling affects only the output of regression and `test` with the `symbolic` option, it may not seem important. There is, however, a case when you need to know how Stata labels the levels. If you use the `test` command to construct your own tests that explicitly refer to individual coefficients of the underlying regression model, you need to refer to those levels as Stata refers to them. Before you panic, realize that you are unlikely to be performing tests of that type. Most tests can be performed without explicitly referring to the coefficients, so the labeling is irrelevant.

Nevertheless, the coefficient on the "third" level of `scheme2`, which you crazily labeled 2.1, is referred to as `_coef[scheme2[3]]`; see [U] **13.5 Accessing coefficients and standard errors**. On the other hand, the "third" level of `scheme1`, which you more reasonably labeled 5, is referred to as `_coef[scheme1[5]]`. You can find out how Stata labeled the levels from the regression output or from the `detail` output.

If this labeling scheme seems confusing and you plan to perform tests directly on the coefficients of the underlying regression, label your levels sequentially with integers starting at 1, as most other programs require. Or, if you decide to opt for the freedom of having arbitrary labeling schemes, adopt a labeling scheme that uses only positive integers less than 99. This way, your labeling scheme will be Stata's labeling scheme. If you use a truly arbitrary labeling scheme that contains floating-point numbers, however, Stata will create a mapping, and you will have to know that mapping to refer to the coefficients.

❏

Even with nested designs, the scheme you use in numbering the nested levels does not matter, except as noted above. If you have B|A (meaning B nested within A), you may choose to number the levels of B within each level of A

```
A  1  1  1  2  2  2
B  1  2  3  1  2  3
```

or you may number the levels of B disregarding A:

```
A  1  1  1  2  2  2
B  1  2  3  4  5  6
```

The numbering of the levels does not have to be sequential or composed of integer values.

▷ Example 18

The machine example (example 9) provides an illustration. In that example, each operator was numbered sequentially, starting with 1, within each machine. We create op2, which numbers the operators in a strange way, and show the result of running anova, substituting op2 for operator.

```
. use http://www.stata-press.com/data/r10/machine
(machine data)
. generate op2 = 100.4*machine + 10.3*operator
```

Ronald Aylmer Fisher (1890–1962) (Sir Ronald from 1952) studied mathematics at Cambridge. Even before he finished his studies, he had published on statistics. He worked as a statistician at Rothamsted Experimental Station (1919–1933), as professor of eugenics at University College London (1933–1943), as professor of genetics at Cambridge (1943–1957), and in retirement at the SIRO Division of Mathematical Statistics in Adelaide. His many fundamental and applied contributions to statistics and genetics mark him as one of the greatest statisticians of all time, including original work on tests of significance, distribution theory, theory of estimation, fiducial inference, and design of experiments.

(Continued on next page)

```
. anova output machine / op2|machine /, detail
```

Factor	Value	Value	Value	Value
machine	1 1 5 5	2 2	3 3	4 4
op2	1 110.7 5 211.1 9 311.5 13 422.2 17 522.6	2 121 6 221.4 10 321.8 14 432.5 18 532.9	3 131.3 7 231.7 11 332.1 15 442.8	4 141.6 8 242 12 411.9 16 512.3

```
                          Number of obs =       57    R-squared     =  0.8661
                          Root MSE      = 1.47089    Adj R-squared =  0.8077
```

Source	Partial SS	df	MS	F	Prob > F
Model	545.822288	17	32.1071934	14.84	0.0000
machine	430.980792	4	107.745198	13.82	0.0001
op2|machine	101.353804	13	7.79644648		
op2|machine	101.353804	13	7.79644648	3.60	0.0009
Residual	84.3766582	39	2.16350406		
Total	630.198947	56	11.2535526		

This output agrees with our previous result.

◁

anova as an alternative to regress

Even if you have no interest in formal ANOVA, you may still find the anova command useful. It can construct dummy variables on the fly, and when you are fitting regressions containing many dummy variables, anova is quicker than regress in producing the estimates.

Read the first part of this entry just to become familiar with the anova syntax. Then remember the following rules of thumb:

1. Specify the regress option on the end of the anova command. This way, you will see only the regression table and not be bothered with the ANOVA table.

2. Specify the continuous(*varlist*) option so that Stata will know which variables in your model are continuous, or, if it is more convenient, specify the category(*varlist*) option so that Stata will know which variables are categorical.

3. Use the ANOVA table to obtain the overall significance of categorical variables (dummy sets) in certain circumstances, namely, when the categorical variable is not interacted with any other categorical variable in the model. For instance, if you type

anova drate region age, continuous(age)

the F statistic reported in the ANOVA table for region is the overall significance of the dummy set for region. The same is true if you type

anova drate region age region*age, continuous(age)

The same is *not* true if you type

anova drate region age region*sex, continuous(age)

Here the F statistic on `region` is the effect of `region` after orthogonalization with the `region*sex` dummies. If you do not understand this rule, avoid the ANOVA table.

▷ Example 19

We have census data recording the death rate (`drate`) and median age (`age`) for each state. The dataset also includes the region of the country in which each state is located (`region`):

```
. use http://www.stata-press.com/data/r10/census2
(1980 Census data by state)
. summarize drate age region
```

Variable	Obs	Mean	Std. Dev.	Min	Max
drate	50	84.3	13.07318	40	107
age	50	29.5	1.752549	24	35
region	50	2.66	1.061574	1	4

`age` is coded in integral years from 24 to 35, and `region` is coded from 1 to 4, with 1 standing for the Northeast, 2 for the North Central, 3 for the South, and 4 for the West.

When we examine the data more closely, we discover large differences in the death rate across regions of the country:

```
. tabulate region, summarize(drate)
```

Census region	Summary of Death Rate Mean	Std. Dev.	Freq.
NE	93.444444	7.0553368	9
N Cntrl	88.916667	5.5833899	12
South	88.3125	8.5457104	16
West	68.769231	13.342625	13
Total	84.3	13.073185	50

Naturally, we wonder if these differences might not be explained by differences in the median ages of the populations. To find out, we fit a regression model of `drate` on `region` and `age`. Here we decide to start by looking at the ANOVA table because there are no interacted variables. The table will report exactly what we expect it to report—the significance of dummy sets.

```
. anova drate region age
```

```
                       Number of obs =      50    R-squared     = 0.7927
                       Root MSE      = 6.7583    Adj R-squared = 0.7328
```

Source	Partial SS	df	MS	F	Prob > F
Model	6638.86529	11	603.533208	13.21	0.0000
region	1320.00973	3	440.003244	9.63	0.0001
age	2237.24937	8	279.656171	6.12	0.0000
Residual	1735.63471	38	45.6745977		
Total	8374.5	49	170.908163		

We have the answer to our question: differences in median ages do not eliminate the differences in death rates across the four regions. The ANOVA table summarizes the two terms in the model, `region` and `age`. The `region` term contains 3 degrees of freedom, and the `age` term contains 8 degrees of freedom. Both are significant at better than the 1% level.

The age term contains 8 degrees of freedom. Since we did not explicitly indicate that age was to be treated as a continuous variable, it was treated as *categorical*, meaning that unique coefficients were estimated for each level of age. The only clue of this labeling is that the number of degrees of freedom associated with the age term exceeds 1. The labeling becomes more obvious if we review the regression coefficients:

```
. anova, regress
```

Source	SS	df	MS		
Model	6638.86529	11	603.533208		
Residual	1735.63471	38	45.6745977		
Total	8374.5	49	170.908163		

Number of obs = 50
F(11, 38) = 13.21
Prob > F = 0.0000
R-squared = 0.7927
Adj R-squared = 0.7328
Root MSE = 6.7583

drate	Coef.	Std. Err.	t	P>|t|	[95% Conf. Interval]	
_cons	93.925	7.295544	12.87	0.000	79.15594	108.6941
region						
1	13.37147	4.195344	3.19	0.003	4.878439	21.8645
2	13.81431	2.974005	4.65	0.000	7.793748	19.83486
3	13.075	2.747792	4.76	0.000	7.512391	18.63762
4	(dropped)					
age						
24	-38.925	9.944825	-3.91	0.000	-59.05724	-18.79275
26	-53.925	9.944825	-5.42	0.000	-74.05724	-33.79275
27	-24.61666	8.015933	-3.07	0.004	-40.84407	-8.389257
28	-26.26488	7.393098	-3.55	0.001	-41.23143	-11.29834
29	-20.064	7.23115	-2.77	0.009	-34.7027	-5.425299
30	-18.05497	7.074489	-2.55	0.015	-32.37652	-3.733414
31	-9.011922	8.011676	-1.12	0.268	-25.23071	7.206869
32	-11.89646	8.384027	-1.42	0.164	-28.86904	5.07611
35	(dropped)					

If we want to treat age as a continuous variable, we must include the continuous(*varlist*) option. Below, only one variable, age, is continuous.

```
. anova drate region age, continuous(age)
```

Number of obs = 50   R-squared = 0.7203
Root MSE = 7.21483   Adj R-squared = 0.6954

Source	Partial SS	df	MS	F	Prob > F
Model	6032.08254	4	1508.02064	28.97	0.0000
region	1645.66228	3	548.554092	10.54	0.0000
age	1630.46662	1	1630.46662	31.32	0.0000
Residual	2342.41746	45	52.0537213		
Total	8374.5	49	170.908163		

The age term now has 1 degree of freedom. The regression coefficients are

```
. anova, regress
```

Source	SS	df	MS
Model	6032.08254	4	1508.02064
Residual	2342.41746	45	52.0537213
Total	8374.5	49	170.908163

```
Number of obs =      50
F(  4,     45) =   28.97
Prob > F      =  0.0000
R-squared     =  0.7203
Adj R-squared =  0.6954
Root MSE      =  7.2148
```

drate	Coef.	Std. Err.	t	P>|t|	[95% Conf. Interval]	
_cons	-41.97859	19.88906	-2.11	0.040	-82.03722	-1.919959
region						
1	13.37578	3.723447	3.59	0.001	5.876378	20.87519
2	15.16831	3.022157	5.02	0.000	9.081374	21.25525
3	14.07378	2.865756	4.91	0.000	8.301846	19.8457
4	(dropped)					
age	3.922947	.7009425	5.60	0.000	2.511177	5.334718

Although we started analyzing these data to explain the regional differences in death rate, let's focus on the effect of age for a moment. In our first model, each level of age had a unique death rate associated with it. For instance, the predicted death rate in a northeastern state with a median age of 24 was

$$93.93 + 13.37 - 38.93 \approx 68.37$$

whereas the predicted death rate from our current model is

$$-41.98 + 13.38 + 3.92 \cdot 24 \approx 65.48$$

Our previous model had an R^2 of 0.7927, whereas our current model has an R^2 of 0.7203. This "small" loss of predictive power accompanies a gain of 7 degrees of freedom, so we suspect that the continuous-age model is as good as the discrete-age model.

◁

❏ Technical Note

There is enough information in the two ANOVA tables to attach a statistical significance to our suspicion that the loss of predictive power is offset by the savings in degrees of freedom. Since the continuous-age model is nested within the discrete-age model, we can perform a standard Chow test. For those of us who know such formulas off the top of our heads, the F statistic is

$$\frac{(2342.41746 - 1735.63471)/7}{45.6745977} = 1.90$$

There is, however, a better way.

We can find out whether our continuous model is as good as our discrete model by putting age in the model twice: once as a continuous variable and once as a categorical variable. The categorical variable will then measure deviations around the straight line implied by the continuous variable, and the F test for the significance of the categorical variable will test whether those deviations are jointly zero.

There is no way to tell Stata that the variable age in a model is continuous in one instance and categorical in another. As far as Stata is concerned, every variable in the model is either continuous or categorical, so we must fool Stata by creating the same variable with two different names:

```
. generate cage=age
. anova drate region cage age, continuous(cage)
```

	Number of obs =		50	R-squared	= 0.7927
	Root MSE	=	6.7583	Adj R-squared =	0.7328

Source	Partial SS	df	MS	F	Prob > F
Model	6638.86529	11	603.533208	13.21	0.0000
region	1320.00973	3	440.003244	9.63	0.0001
cage	91.9613243	1	91.9613243	2.01	0.1641
age	606.782747	7	86.6832496	1.90	0.0970
Residual	1735.63471	38	45.6745977		
Total	8374.5	49	170.908163		

We first created a new variable called `cage` that is identical to the `age` variable. We then told Stata to fit an ANOVA model of `drate` on the variables `region`, `cage`, and `age`, where the variable `cage` is continuous. Since we did not specify otherwise, Stata assumes that the other two variables, `region` and, most importantly, `age`, are categorical.

We find that the F test for the significance of the (categorical) `age` variable is 1.90, just as we calculated above. It is significant at the 9.7% level. If we hold to a 5% significance level, we cannot reject the hypothesis that the effect of `age` is linear.

❑

▷ Example 20

In our census data, we still find significant differences across the regions after controlling for the median age of the population. We might now wonder whether the regional differences are differences in level—independent of age—or are instead differences in the regional effects of age. Just as we can interact categorical variables with other categorical variables, we can interact categorical variables with continuous variables.

```
. anova drate region age region*age, continuous(age)
```

	Number of obs =		50	R-squared	= 0.7365
	Root MSE	=	7.24852	Adj R-squared =	0.6926

Source	Partial SS	df	MS	F	Prob > F
Model	6167.7737	7	881.110529	16.77	0.0000
region	188.713602	3	62.9045339	1.20	0.3225
age	873.425599	1	873.425599	16.62	0.0002
region*age	135.691162	3	45.2303874	0.86	0.4689
Residual	2206.7263	42	52.5411023		
Total	8374.5	49	170.908163		

The `region*age` term in our model measures the differences in slopes across the regions. We cannot reject the hypothesis that there are no such differences. The `region` effect is now "insignificant". This status does not mean that there are no regional differences in death rates since each test is a *marginal* or *partial* test. Here, with `region*age` included in the model, `region` is being tested at the point where `age` is zero. Apart from this value not existing in the dataset, it is also a long way from the mean value of `age`, so the test of `region` at this point is meaningless (although it is valid if you acknowledge what is being tested).

To obtain a more sensible test of `region`, we can subtract the mean from the `age` variable and use this in the model.

```
. quietly summarize age
. generate mage = age - r(mean)
. anova drate region mage region*mage, continuous(mage)
```

		Number of obs =	50	R-squared	= 0.7365
		Root MSE = 7.24852		Adj R-squared =	0.6926

Source	Partial SS	df	MS	F	Prob > F
Model	6167.7737	7	881.110529	16.77	0.0000
region	1166.14735	3	388.715783	7.40	0.0004
mage	873.425599	1	873.425599	16.62	0.0002
region*mage	135.691162	3	45.2303874	0.86	0.4689
Residual	2206.7263	42	52.5411023		
Total	8374.5	49	170.908163		

`region` is significant when tested at the mean of the `age` variable.

◁

Saved Results

`anova` saves the following in `e()`:

Scalars

e(N)	number of observations
e(mss)	model sum of squares
e(df_m)	model degrees of freedom
e(rss)	residual sum of squares
e(df_r)	residual degrees of freedom
e(r2)	R-squared
e(r2_a)	adjusted R-squared
e(F)	F statistic
e(rmse)	root mean squared error
e(ll)	log likelihood
e(ll_0)	log likelihood, constant-only model
e(ss_#)	sum of squares for term #
e(df_#)	numerator degrees of freedom for term #
e(ssdenom_#)	denominator sum of squares for term # (when using nonresidual error)
e(dfdenom_#)	denominator degrees of freedom for term # (when using nonresidual error)
e(F_#)	F statistic for term # (if computed)
e(N_bse)	number of levels of the between-subjects error term
e(df_bse)	degrees of freedom for the between-subjects error term
e(box#)	Box's conservative epsilon for a particular combination of repeated variables (repeated() only)
e(gg#)	Greenhouse–Geisser epsilon for a particular combination of repeated variables (repeated() only)
e(hf#)	Huynh–Feldt epsilon for a particular combination of repeated variables (repeated() only)

Macros

e(cmd)	anova
e(cmdline)	command as typed
e(depvar)	name of dependent variable
e(varnames)	names of the right-hand-side variables
e(term_#)	term #
e(errorterm_#)	error term for term # (when using nonresidual error)
e(sstype)	type of sum of squares; sequential or partial
e(repvars)	names of repeated variables (repeated() only)
e(repvar#)	names of repeated variables for a particular combination (repeated() only)
e(model)	ols
e(wtype)	weight type
e(wexp)	weight expression
e(properties)	b_nonames V_nonames
e(estat_cmd)	program used to implement estat
e(predict)	program used to implement predict

Matrices

e(b)	coefficient vector
e(V)	variance–covariance matrix of the estimators
e(Srep)	covariance matrix based on repeated measures (repeated() only)

Functions

e(sample)	marks estimation sample

References

Acock, A. C. 2006. *A Gentle Introduction to Stata*. College Station, TX: Stata Press.

Afifi, A. A., and S. P. Azen. 1979. *Statistical Analysis: A Computer-Oriented Approach*. 2nd ed. New York: Academic Press.

Altman, D. G. 1991. *Practical Statistics for Medical Research*. London: Chapman & Hall/CRC.

Box, G. E. P. 1954. Some theorems on quadratic forms applied in the study of analysis of variance problems, I. Effect of inequality of variance in the one-way classification. *Annals of Mathematical Statistics* 25: 290–302.

Box, J. F. 1978. *R. A. Fisher: The Life of a Scientist*. New York: Wiley.

Cobb, G. W. 1998. *Introduction to Design and Analysis of Experiments*. New York: Springer.

Edwards, A. L. 1985. *Multiple Regression and the Analysis of Variance and Covariance*. 2nd ed. New York: Freeman.

Fisher, R. A. 1925. *Statistical Methods for Research Workers*. Edinburgh: Oliver & Boyd.

———. 1935. *The Design of Experiments*. Edinburgh: Oliver & Boyd.

———. 1990. *Statistical Methods, Experimental Design, and Scientific Inference*. Oxford: Oxford University Press. Facsimile reprint of *Statistical Methods for Research Workers* (1925, 14/e 1970), *The Design of Experiments* (1935, 8/e 1966), and *Statistical Methods and Scientific Inference* (1956, 3/e 1973).

Geisser, S., and S. Greenhouse. 1958. An extension of Box's results on the use of the F distribution in multivariate analysis. *Annals of Mathematical Statistics* 29: 885–891.

Gleason, J. R. 1999. sg103: Within subjects (repeated measures) ANOVA, including between subjects factors. *Stata Technical Bulletin* 47: 40–45. Reprinted in *Stata Technical Bulletin Reprints*, vol. 8, pp. 236–243.

———. 2000. sg132: Analysis of variance from summary statistics. *Stata Technical Bulletin* 54: 42–46. Reprinted in *Stata Technical Bulletin Reprints*, vol. 9, pp. 328–332.

Higgins, J. E., and G. G. Koch. 1977. Variable selection and generalized chi-square analysis of categorical data applied to a large cross-sectional occupational health survey. *International Statistical Review* 45: 51–62.

Huynh, H. 1978. Some approximate tests for repeated measurement designs. *Psychometrika* 43: 161–175.

Huynh, H., and L. S. Feldt. 1976. Estimation of the Box correction for degrees of freedom from sample data in randomized block and split-plot designs. *Journal of Educational Statistics* 1: 69–82.

Kennedy, W. J., Jr., and J. E. Gentle. 1980. *Statistical Computing*. New York: Dekker.

Kuehl, R. O. 2000. *Design of Experiments: Statistical Principles of Research Design and Analysis*. 2nd ed. Belmont, CA: Duxbury.

Marchenko, Y. 2006. Estimating variance components in Stata. *Stata Journal* 6: 1–21.

Milliken, G. A., and D. E. Johnson. 1984. *Analysis of Messy Data, Volume 1: Designed Experiments*. New York: Nostrand Reinhold.

Rabe-Hesketh, S., and B. S. Everitt. 2004. *A Handbook of Statistical Analysis Using Stata*. 3rd ed. Boca Raton, FL: Chapman & Hall/CRC.

Scheffé, H. 1959. *The Analysis of Variance*. New York: Wiley.

Snedecor, G. W., and W. G. Cochran. 1989. *Statistical Methods*. 8th ed. Ames, IA: Iowa State University Press.

van Belle, G., L. D. Fisher, P. J. Heagerty, and T. Lumley. 2004. *Biostatistics: A Methodology for the Health Sciences*. 2nd ed. New York: Wiley.

Winer, B. J., D. R. Brown, and K. M. Michels. 1991. *Statistical Principles in Experimental Design*. 3rd ed. New York: McGraw–Hill.

Also See

[R] **anova postestimation** — Postestimation tools for anova

[R] **loneway** — Large one-way ANOVA, random effects, and reliability

[R] **oneway** — One-way analysis of variance

[R] **regress** — Linear regression

[MV] **manova** — Multivariate analysis of variance and covariance

Title

> **anova postestimation** — Postestimation tools for anova

Description

The following postestimation commands are of special interest after `anova`:

command	description
estat hettest	tests for heteroskedasticity
estat ovtest	Ramsey regression specification-error test for omitted variables
acprplot	augmented component-plus-residual plot
avplot	added-variable plot
avplots	all added-variable plots in one image
cprplot	component-plus-residual plot
lvr2plot	leverage-versus-squared-residual plot
rvfplot	residual-versus-fitted plot
rvpplot	residual-versus-predictor plot

For information about these commands, see [R] **regress postestimation**.

The following standard postestimation commands are also available:

command	description
adjust	adjusted predictions of $\mathbf{x}\beta$ or $\exp(\mathbf{x}\beta)$
estat	AIC, BIC, VCE, and estimation sample summary
estimates	cataloging estimation results
lincom	point estimates, standard errors, testing, and inference for linear combinations of coefficients
linktest	link test for model specification
predict	predictions, residuals, influence statistics, and other diagnostic measures
test	Wald tests for simple and composite linear hypotheses

See the corresponding entries in the *Stata Base Reference Manual* for details.

Special-interest postestimation commands

In addition to the common `estat` commands (see [R] **estat**), `estat hettest` and `estat ovtest` are also available. The syntax is the same as after `regress`, except that `estat hettest` does not allow the `rhs` option; see [R] **regress postestimation**.

Syntax for predict

`predict` after `anova` follows the same syntax as `predict` after `regress` and can provide predictions, residuals, standardized residuals, studentized residuals, the standard error of the residuals, the standard error of the prediction, the diagonal elements of the projection (hat) matrix, and Cook's D. See [R] **regress postestimation** for details.

Syntax for test after anova

<u>test</u>, test(*matname*) [<u>m</u>test[(*opt*)] matvlc(*matname*)]   syntax 1

<u>test</u> [*exp* = *exp*] [, <u>a</u>ccumulate <u>not</u>est matvlc(*matname*)]   syntax 2

<u>test</u> [*term* [*term* ...]] [/ *term* [*term* ...]] [, <u>s</u>ymbolic]   syntax 3

<u>test</u>, <u>showorder</u>   syntax 4

where *exp* must have references to coefficients enclosed in _coef[] (or synonym _b[]) and where *term* is as defined for anova.

syntax 1	test expression involving the coefficients of the underlying regression model; you provide information as a matrix
syntax 2	test expression involving the coefficients of the underlying regression model; you provide information as an equation
syntax 3	test effects and show symbolic forms
syntax 4	show underlying order of design matrix, which is useful when constructing *matname* argument of the test() option

Options for test after anova

test(*matname*) is required with the first syntax of test. The rows of *matname* specify linear combinations of the underlying design matrix of the ANOVA that are to be jointly tested. The columns correspond to the underlying design matrix (including the constant if it has not been suppressed). The column and row names of *matname* are ignored.

A listing of the constraints imposed by the test() option is presented before the table containing the tests. You should examine this table to verify that you have applied the linear combinations you desired. Typing test, showorder allows you to examine the ordering of the columns for the design matrix from the ANOVA.

mtest[(*opt*)] specifies that tests are performed for each condition separately. *opt* specifies the method for adjusting *p*-values for multiple testing. Valid values for *opt* are

<u>b</u>onferroni	Bonferroni's method
<u>h</u>olm	Holm's method
<u>s</u>idak	Šidák's method
<u>noadj</u>ust	no adjustment is to be made

Specifying mtest with no argument is equivalent to mtest(noadjust).

matvlc(*matname*), a programmer's option, saves the variance–covariance matrix of the linear combinations involved in the suite of tests. For the test $Lb = c$, what is returned in *matname* is LVL', where **V** is the estimated variance–covariance matrix of b.

accumulate allows a hypothesis to be tested jointly with the previously tested hypotheses.

notest suppresses the output. This option is useful when you are interested only in the joint test with additional hypotheses specified in a subsequent call of test, accumulate.

symbolic requests the symbolic form of the test rather than the test statistic. When this option is specified without any terms (test, symbolic), the symbolic form of the estimable functions is displayed.

showorder causes test to list the definition of each column in the design matrix. showorder is not allowed with any other option.

Remarks

Remarks are presented under the following headings:

> *Testing effects*
> *Obtaining symbolic forms*
> *Testing coefficients*

Testing effects

After fitting a model using anova, you can test for the significance of effects in the ANOVA table, as well as for effects that are not reported in the ANOVA table, by using the test command. You follow test by the list of effects that you wish to test. By default, test uses the residual mean squared error in the denominator of the F ratio. You can specify other error terms by using the slash notation, just as you would with anova.

▷ Example 1

Recall our byssinosis example (example 16) in [R] **anova**:

```
. anova prob workplace smokes race workplace*smokes workplace*race
> smokes*race workplace*smokes*race  [aweight=pop]
(sum of wgt is    5.4190e+03)
```

	Number of obs =	65	R-squared	=	0.8300
	Root MSE	= .025902	Adj R-squared =		0.7948

Source	Partial SS	df	MS	F	Prob > F
Model	.173646538	11	.015786049	23.53	0.0000
workplace	.097625175	2	.048812588	72.76	0.0000
smokes	.013030812	1	.013030812	19.42	0.0001
race	.001094723	1	.001094723	1.63	0.2070
workplace*smokes	.019690342	2	.009845171	14.67	0.0000
workplace*race	.001352516	2	.000676258	1.01	0.3718
smokes*race	.001662874	1	.001662874	2.48	0.1214
workplace*smokes*race	.000950841	2	.00047542	0.71	0.4969
Residual	.035557766	53	.000670901		
Total	.209204304	64	.003268817		

We can easily obtain a test on a particular term from the ANOVA table. Here are two examples:

```
. test smokes
```

Source	Partial SS	df	MS	F	Prob > F
smokes	.013030812	1	.013030812	19.42	0.0001
Residual	.035557766	53	.000670901		

```
. test smokes*race
```

Source	Partial SS	df	MS	F	Prob > F
smokes*race	.001662874	1	.001662874	2.48	0.1214
Residual	.035557766	53	.000670901		

Both these tests use residual error by default and agree with the ANOVA table produced earlier.

◁

❑ Technical Note

After `anova`, you can use the '/' syntax in `test` to perform tests with a variety of non-$\sigma^2 \mathbf{I}$ error structures. However, in most unbalanced models, the mean squares are not independent and do not have equal expectations under the null hypothesis. Also be warned that you assume responsibility for the validity of the test statistic.

❑

▷ Example 2

We return to the nested ANOVA example (example 9) in [R] **anova**, where five brands of machinery were compared in an assembly line. We can obtain appropriate tests for the nested terms using `test`, even if we had run the `anova` command without initially indicating the proper error terms.

```
. use http://www.stata-press.com/data/r10/machine
(machine data)
. anova output machine operator|machine
```

		Number of obs =	57	R-squared	= 0.8661
		Root MSE	= 1.47089	Adj R-squared =	0.8077

Source	Partial SS	df	MS	F	Prob > F
Model	545.822288	17	32.1071934	14.84	0.0000
machine	430.980792	4	107.745198	49.80	0.0000
operator|machine	101.353804	13	7.79644648	3.60	0.0009
Residual	84.3766582	39	2.16350406		
Total	630.198947	56	11.2535526		

In this ANOVA table, `machine` is tested with residual error. With this particular nested design, the appropriate error term for testing `machine` is `operator` nested within `machine`, which is easily obtained from `test`.

```
. test machine / operator|machine
```

Source	Partial SS	df	MS	F	Prob > F
machine	430.980792	4	107.745198	13.82	0.0001
operator|machine	101.353804	13	7.79644648		

This result from `test` matches what we obtained from our `anova` command.

◁

▷ Example 3

The other nested ANOVA example (example 10) in [R] **anova** was based on the skin rash data. The ANOVA table is presented here again. As before, we will use abbreviations of variable names in typing the commands.

```
. use http://www.stata-press.com/data/r10/rash
(skin rash data)
. anova response t / c|t / d|c|t / p|d|c|t /
```

	Number of obs =	384	R-squared	=	0.5040
	Root MSE =	11.6192	Adj R-squared =		0.3404

Source	Partial SS	df	MS	F	Prob > F
Model	39505.9896	95	415.852522	3.08	0.0000
treatment	4240.04167	1	4240.04167	9.79	0.0204
clinic|treatment	2599.48958	6	433.248264		
clinic|treatment	2599.48958	6	433.248264	0.93	0.4982
doctor|clinic|treatment	7429.58333	16	464.348958		
doctor|clinic|treatment	7429.58333	16	464.348958	1.32	0.2065
patient|doctor|clinic| treatment	25236.875	72	350.512153		
patient|doctor|clinic| treatment	25236.875	72	350.512153	2.60	0.0000
Residual	38882.00	288	135.006944		
Total	78387.9896	383	204.66838		

In practice, it is often beneficial to pool nonsignificant nested terms to increase the power of tests on remaining terms. One rule of thumb is to allow the pooling of a term whose p-value is larger than 0.25. In this skin rash example, the p-value for the test of clinic is 0.4982. This value indicates that the clinic effect is negligible and might be ignored. Currently, treatment is tested by clinic|treatment, which has only 6 degrees of freedom. If we pool the clinic and doctor terms and use this pooled estimate as the error term for treatment, we would have a term with 22 degrees of freedom.

Below are two tests: a test of treatment with the pooled clinic and doctor terms and a test of this pooled term by patient.

```
. test t / c|t d|c|t
```

Source	Partial SS	df	MS	F	Prob > F
treatment	4240.04167	1	4240.04167	9.30	0.0059
clinic|treatment doctor| clinic|treatment	10029.0729	22	455.866951		

```
. test c|t d|c|t / p|d|c|t
```

Source	Partial SS	df	MS	F	Prob > F
clinic|treatment doctor| clinic|treatment	10029.0729	22	455.866951	1.30	0.2014
patient|doctor|clinic| treatment	25236.875	72	350.512153		

In the first test, we included two terms after the forward slash (c|t and d|c|t). test after anova allows multiple terms both before and after the slash. The terms before the slash are combined and are then tested by the combined terms that follow the slash (or residual error if no slash is present).

The p-value for treatment using the pooled term is 0.0059. Originally, it was 0.0204. The increase in the power of the test is due to the increase in degrees of freedom for the pooled error term.

We can get identical results if we drop clinic from the anova model. (This dataset has unique numbers for each doctor so that there is no confusion of doctors when clinic is dropped.)

```
. anova response t / d|t / p|d|t /
```

	Number of obs =	384	R-squared	=	0.5040
	Root MSE =	11.6192	Adj R-squared =		0.3404

Source	Partial SS	df	MS	F	Prob > F
Model	39505.9896	95	415.852522	3.08	0.0000
treatment	4240.04167	1	4240.04167	9.30	0.0059
doctor|treatment	10029.0729	22	455.866951		
doctor|treatment	10029.0729	22	455.866951	1.30	0.2014
patient|doctor| treatment	25236.875	72	350.512153		
patient|doctor| treatment	25236.875	72	350.512153	2.60	0.0000
Residual	38882.00	288	135.006944		
Total	78387.9896	383	204.66838		

This output agrees with our earlier test results.

◁

In the following example, two terms from the anova are jointly tested (pooled).

▷ Example 4

In example 20 in [R] **anova**, we fitted the model anova drate region mage region*mage, continuous(mage). Now we test for the overall significance of region. The simple form of the test syntax is

test $\left[\,term\,\left[\,term\,\left[\ldots\right]\,\right]\,\right]$

```
. test region region*mage
```

Source	Partial SS	df	MS	F	Prob > F
region region*mage	1781.35344	6	296.89224	5.65	0.0002
Residual	2206.7263	42	52.5411023		

The overall F statistic associated with the region and region*mage terms is 5.65, and it is significant at the 0.02% level.

If test works, typing test followed by one term in our model should produce output that exactly matches that provided by the anova command. In the ANOVA output, the region term, by itself, had a sum of squares of 1166.15, which, based on 3 degrees of freedom, yielded an F statistic of 7.40 and a significance level of 0.0004.

```
. test region
```

Source	Partial SS	df	MS	F	Prob > F
region	1166.14735	3	388.715783	7.40	0.0004
Residual	2206.7263	42	52.5411023		

test yields the same result.

◁

Obtaining symbolic forms

test can also produce the symbolic form of the estimable functions and symbolic forms for particular tests.

▷ Example 5

After fitting an ANOVA model, we type test, symbolic to obtain the symbolic form of the estimable functions. For instance, returning to our blood pressure data, let's begin by reestimating systolic on drug, disease, and drug*disease:

```
. use http://www.stata-press.com/data/r10/systolic
(Systolic Blood Pressure Data)
. anova systolic drug disease drug*disease
```

		Number of obs =	58	R-squared	=	0.4560
		Root MSE	= 10.5096	Adj R-squared =		0.3259

Source	Partial SS	df	MS	F	Prob > F
Model	4259.33851	11	387.212591	3.51	0.0013
drug	2997.47186	3	999.157287	9.05	0.0001
disease	415.873046	2	207.936523	1.88	0.1637
drug*disease	707.266259	6	117.87771	1.07	0.3958
Residual	5080.81667	46	110.452536		
Total	9340.15517	57	163.862371		

To obtain the symbolic form of the estimable functions, type

```
. test, symbolic
_cons          r0
drug
        1      r1
        2      r2
        3      r3
        4      r0 - (r1+r2+r3)
disease
        1      r5
        2      r6
        3      r0 - (r5+r6)
drug*disease
    1   1      r8
    1   2      r9
    1   3      r1 - (r8+r9)
    2   1      r11
    2   2      r12
```

```
2   3        r2 - (r11+r12)
3   1        r14
3   2        r15
3   3        r3 - (r14+r15)
4   1        r5 - (r8+r11+r14)
4   2        r6 - (r9+r12+r15)
4   3        r0 - (r1+r2+r3+r5+r6-r8-r9-r11-r12-r14-r15)
```

◁

▷ Example 6

To obtain the symbolic form for a particular test, we type test *term* [*term* ...], symbolic. For instance, the symbolic form for the test of the main effect of drug is

```
. test drug, symbolic
_cons          0
drug
         1         r1
         2         r2
         3         r3
         4        -(r1+r2+r3)
disease
         1         0
         2         0
         3         0
drug*disease
      1  1         1/3 r1
      1  2         1/3 r1
      1  3         1/3 r1
      2  1         1/3 r2
      2  2         1/3 r2
      2  3         1/3 r2
      3  1         1/3 r3
      3  2         1/3 r3
      3  3         1/3 r3
      4  1        -1/3 (r1+r2+r3)
      4  2        -1/3 (r1+r2+r3)
      4  3        -1/3 (r1+r2+r3)
```

If we omit the symbolic option, we instead see the result of the test:

```
. test drug
```

Source	Partial SS	df	MS	F	Prob > F
drug	2997.47186	3	999.157287	9.05	0.0001
Residual	5080.81667	46	110.452536		

◁

Testing coefficients

The test command allows you to perform tests directly on the coefficients of the underlying regression model. When you specify the expressions, you must enclose references to coefficients in _coef[]; see [U] **13.5 Accessing coefficients and standard errors**. For instance, the coefficient on the first level of drug and the second level of disease is referred to as _coef[drug[1]*disease[2]]. If you were to omit the _coef[], Stata would think that you are trying to multiply two terms together rather than referring to the coefficient on the interaction of two terms.

▷ Example 7

Let's begin by testing whether the coefficient on the first `drug` is equal to the coefficient on the second in our blood pressure data. We have already fitted the model `anova systolic drug disease drug*disease`, and you can see the results of that estimation in example 4. Even though we have performed many tasks since we fitted the model, Stata still remembers, and we can perform tests at any time.

```
. test _coef[drug[1]]=_coef[drug[2]]
 ( 1)   drug[1] - drug[2] = 0
       F(  1,    46) =     0.12
             Prob > F =     0.7272
```

We find that the two coefficients are not significantly different, at least at any significance level smaller than 73%. Let's now add the constraint that the coefficient on the first drug interacted with the first disease is equal to the coefficient on the second drug, again interacted with the first disease. We do that by typing the new constraint and adding the `accumulate` option:

```
. test _coef[drug[1]*disease[1]]=_coef[drug[2]*disease[1]], accumulate
 ( 1)   drug[1] - drug[2] = 0
 ( 2)   drug[1]*disease[1] - drug[2]*disease[1] = 0
       F(  2,    46) =     0.08
             Prob > F =     0.9200
```

Let's continue. Our goal is to determine whether the first drug is significantly different from the second drug. So far, our test includes the equality of the two drug coefficients, along with the equality of the two drug coefficients when interacted with the first disease. We must add two more equations, one for each of the remaining two diseases.

```
. test _coef[drug[1]*disease[2]]=_coef[drug[2]*disease[2]], accumulate
 ( 1)   drug[1] - drug[2] = 0
 ( 2)   drug[1]*disease[1] - drug[2]*disease[1] = 0
 ( 3)   drug[1]*disease[2] - drug[2]*disease[2] = 0
       F(  3,    46) =     0.22
             Prob > F =     0.8806
. test _coef[drug[1]*disease[3]]=_coef[drug[2]*disease[3]], accumulate
 ( 1)   drug[1] - drug[2] = 0
 ( 2)   drug[1]*disease[1] - drug[2]*disease[1] = 0
 ( 3)   drug[1]*disease[2] - drug[2]*disease[2] = 0
 ( 4)   drug[1]*disease[3] - drug[2]*disease[3] = 0
       Constraint 4 dropped
       F(  3,    46) =     0.22
             Prob > F =     0.8806
```

The overall F statistic is 0.22, which is hardly significant. We cannot reject the hypothesis that the first drug has the same effect as the second drug.

You may notice that we also got the message "Constraint 4 dropped". For the technically inclined, the last constraint was unnecessary, given the normalization of the model. You need not worry about such problems because Stata handles them automatically.   ◁

The `test()` option of `test` provides a convenient alternative for testing coefficients. Instead of spelling out each coefficient involved in the test, a matrix representing the test provides the needed information. `test, showorder` shows the order of the terms in the anova corresponding to the order of the columns for the matrix argument of `test()`.

▷ Example 8

We repeat the last test of example 7 above with the `test()` option. First we view the definition and order of the columns underlying the ANOVA performed on the systolic data.

```
. test, showorder
Order of columns in the design matrix
      1: _cons
      2: (drug==1)
      3: (drug==2)
      4: (drug==3)
      5: (drug==4)
      6: (disease==1)
      7: (disease==2)
      8: (disease==3)
      9: (drug==1)*(disease==1)
     10: (drug==1)*(disease==2)
     11: (drug==1)*(disease==3)
     12: (drug==2)*(disease==1)
     13: (drug==2)*(disease==2)
     14: (drug==2)*(disease==3)
     15: (drug==3)*(disease==1)
     16: (drug==3)*(disease==2)
     17: (drug==3)*(disease==3)
     18: (drug==4)*(disease==1)
     19: (drug==4)*(disease==2)
     20: (drug==4)*(disease==3)
```

The first column corresponds to _cons, the constant in the model. The second through fifth columns correspond to the four levels of `drug`. The sixth through eighth columns correspond to the three levels of `disease`. Columns nine through 20 correspond to the interaction of `drug` and `disease`.

We construct the matrix `dr1vs2` with the same four constraints as the last test shown in example 7 and then use the option `test(dr1vs2)` to perform the test.

```
. mat dr1vs2 = (0,  1,-1,0,0,  0,0,0,  0,0,0, 0, 0, 0,0,0,0,0,0,0 \
>               0,  0, 0,0,0,  0,0,0,  1,0,0,-1, 0, 0,0,0,0,0,0,0 \
>               0,  0, 0,0,0,  0,0,0,  0,1,0, 0,-1, 0,0,0,0,0,0,0 \
>               0,  0, 0,0,0,  0,0,0,  0,0,1, 0, 0,-1,0,0,0,0,0,0)
. test , test(dr1vs2)
 ( 1)  drug[1] - drug[2] = 0
 ( 2)  drug[1]*disease[1] - drug[2]*disease[1] = 0
 ( 3)  drug[1]*disease[2] - drug[2]*disease[2] = 0
 ( 4)  drug[1]*disease[3] - drug[2]*disease[3] = 0
        Constraint 4 dropped
      F(  3,    46) =    0.22
            Prob > F =    0.8806
```

Here the effort involved with spelling out the coefficients is similar to that of constructing a matrix and using it in the `test()` option. When the test involving coefficients is more complicated, many people prefer using the `test()` option.

◁

(Continued on next page)

❏ Technical Note

You can use `test` to perform other, more complicated tests. In such cases, you will probably want to review the symbolic forms of particular tests, and you will certainly want to review the symbolic form of the estimable functions. We explained how to do that above.

Let's check that Stata gives the right answers by laboriously typing the gory details of the test for the main effect of drug. Stata already told us the symbolic form in the previous subsection. The obsessed among you have no doubt already worked through the algebra and established that Stata was correct. Here we spell out each coefficient instead of using the `test()` option. Using the `test()` option would have saved us some typing. Our chances of typing all the constraints correctly, however, are so small that we typed them into a do-file:

```
. do mainef
. test _coef[drug[1]]+
>        (_coef[drug[1]*disease[1]] + _coef[drug[1]*disease[2]] +
>         _coef[drug[1]*disease[3]] - _coef[drug[4]*disease[1]] -
>         _coef[drug[4]*disease[2]] - _coef[drug[4]*disease[3]])/3 = 0,
>        notest
```
  (*output omitted*)
```
. test _coef[drug[2]]+
>        (_coef[drug[2]*disease[1]] + _coef[drug[2]*disease[2]] +
>         _coef[drug[2]*disease[3]] - _coef[drug[4]*disease[1]] -
>         _coef[drug[4]*disease[2]] - _coef[drug[4]*disease[3]])/3 = 0,
>        accumulate notest
```
  (*output omitted*)
```
. test _coef[drug[3]]+
>        (_coef[drug[3]*disease[1]] + _coef[drug[3]*disease[2]] +
>         _coef[drug[3]*disease[3]] - _coef[drug[4]*disease[1]] -
>         _coef[drug[4]*disease[2]] - _coef[drug[4]*disease[3]])/3 = 0,
>        accumulate
 ( 1)   drug[1] + .3333333 drug[1]*disease[1] + .3333333 drug[1]*disease[2] + .3
> 333333 drug[1]*disease[3] - .3333333 drug[4]*disease[1] - .3333333 drug[4]*di
> sease[2] - .3333333 drug[4]*disease[3] = 0
 ( 2)   drug[2] + .3333333 drug[2]*disease[1] + .3333333 drug[2]*disease[2] + .3
> 333333 drug[2]*disease[3] - .3333333 drug[4]*disease[1] - .3333333 drug[4]*di
> sease[2] - .3333333 drug[4]*disease[3] = 0
 ( 3)   drug[3] + .3333333 drug[3]*disease[1] + .3333333 drug[3]*disease[2] + .3
> 333333 drug[3]*disease[3] - .3333333 drug[4]*disease[1] - .3333333 drug[4]*di
> sease[2] - .3333333 drug[4]*disease[3] = 0
        F( 3,   46) =    9.05
              Prob > F =    0.0001
end of do-file
```

We have our result. The F statistic has 3 degrees of freedom and is 9.05. This is the same result we obtained when we typed `test drug`. Typing `test drug` was much easier, however.

❏

Tests involving the underlying coefficients of an ANOVA model are often done to obtain various single-degree-of-freedom tests comparing different levels of a categorical variable. The `mtest` option presents each single-degree-of-freedom test in addition to the combined test. The `mtest` option is used in conjunction with the `test()` option.

▷ Example 9

Rencher (2000) illustrates single-degree-of-freedom contrasts for an ANOVA comparing the net weight of cans filled by five machines (labeled A–E). The data were originally obtained from Ostle and Mensing (1975). Rencher uses a cell-means ANOVA model approach for this problem. We could do the same by using the `noconstant` option of `anova`; see [R] **anova**. Instead, we obtain the same results by using the standard overparameterized ANOVA approach (i.e., we keep the constant in the model).

```
. use http://www.stata-press.com/data/r10/canfill
(Can Fill Data)
. list
```

	machine	weight
1.	A	11.95
2.	A	12.00
3.	A	12.25
4.	A	12.10
5.	B	12.18
6.	B	12.11
7.	C	12.16
8.	C	12.15
9.	C	12.08
10.	D	12.25
11.	D	12.30
12.	D	12.10
13.	E	12.10
14.	E	12.04
15.	E	12.02
16.	E	12.02

```
. anova weight machine
```

		Number of obs =	16	R-squared	=	0.4123
		Root MSE	= .087758	Adj R-squared =		0.1986

Source	Partial SS	df	MS	F	Prob > F
Model	.059426993	4	.014856748	1.93	0.1757
machine	.059426993	4	.014856748	1.93	0.1757
Residual	.084716701	11	.007701518		
Total	.144143694	15	.00960958		

```
. test, showorder
Order of columns in the design matrix
        1: _cons
        2: (machine==1)
        3: (machine==2)
        4: (machine==3)
        5: (machine==4)
        6: (machine==5)
```

The four single-degree-of-freedom tests of interest between the five machines are A and D versus B, C, and E; B and E versus C; A versus D; and B versus E. We place the orthogonal (unweighted) contrasts for these tests into the matrix C1 and use that with `test` to obtain the results.

```
. mat C1 = (0,3,-2,-2,3,-2 \ 0,0,1,-2,0,1 \ 0,1,0,0,-1,0 \ 0,0,1,0,0,-1)
. test, test(C1) mtest
( 1)   3 machine[1] - 2 machine[2] - 2 machine[3] + 3 machine[4] - 2 machine[5] = 0
( 2)   machine[2] - 2 machine[3] + machine[5] = 0
( 3)   machine[1] - machine[4] = 0
( 4)   machine[2] - machine[5] = 0
```

	F(df,11)	df	p	
(1)	0.75	1	0.4055	#
(2)	0.31	1	0.5916	#
(3)	4.47	1	0.0582	#
(4)	1.73	1	0.2150	#
all	1.93	4	0.1757	

 # unadjusted p-values

The `mtest` option causes `test` to produce the single-degree-of-freedom tests for each row of the matrix C1 provided in the `test()` option.

The significance values above are not adjusted for multiple comparisons. `mtest` takes an optional argument for specifying multiple comparison adjustments. We could have produced the Bonferroni adjusted significance values by using the option `mtest(bonferroni)`.

```
. test, test(C1) mtest(bonferroni)
( 1)   3 machine[1] - 2 machine[2] - 2 machine[3] + 3 machine[4] - 2 machine[5] = 0
( 2)   machine[2] - 2 machine[3] + machine[5] = 0
( 3)   machine[1] - machine[4] = 0
( 4)   machine[2] - machine[5] = 0
```

	F(df,11)	df	p	
(1)	0.75	1	1.0000	#
(2)	0.31	1	1.0000	#
(3)	4.47	1	0.2329	#
(4)	1.73	1	0.8601	#
all	1.93	4	0.1757	

 # Bonferroni adjusted p-values

◁

▷ Example 10

Here there are two factors, A and B, each with three levels. The levels are evenly spaced so that linear and quadratic contrasts are of interest.

```
. use http://www.stata-press.com/data/r10/twowaytrend
. anova Y A B A*B
```

	Number of obs = 36 R-squared = 0.9304				
	Root MSE = 2.6736 Adj R-squared = 0.9097				
Source	Partial SS	df	MS	F	Prob > F
Model	2578.55556	8	322.319444	45.09	0.0000
A	2026.72222	2	1013.36111	141.77	0.0000
B	383.722222	2	191.861111	26.84	0.0000
A*B	168.111111	4	42.0277778	5.88	0.0015
Residual	193	27	7.14814815		
Total	2771.55556	35	79.1873016		

```
. test, showorder
Order of columns in the design matrix
       1: _cons
       2: (A==1)
       3: (A==2)
       4: (A==3)
       5: (B==1)
       6: (B==2)
       7: (B==3)
       8: (A==1)*(B==1)
       9: (A==1)*(B==2)
      10: (A==1)*(B==3)
      11: (A==2)*(B==1)
      12: (A==2)*(B==2)
      13: (A==2)*(B==3)
      14: (A==3)*(B==1)
      15: (A==3)*(B==2)
      16: (A==3)*(B==3)
```

We create the matrices `Alin`, `Aquad`, `Blin`, and `Bquad` corresponding to the linear and quadratic single-degree-of-freedom contrasts for terms `A` and `B`. These are combined into matrices `A` and `B`. Then the four interaction contrasts are formed from the elementwise multiplication of appropriate pairs of the single-degree-of-freedom contrasts. We use `foreach` and the `vecdiag()` matrix function to automate the process. For this particular problem, the interaction contrasts are combined into two matrices `AxBlin` corresponding to the 2-degree-of-freedom test for factor `A` versus linear `B` and `AxBquad` corresponding to the 2-degree-of-freedom test for factor `A` versus quadratic `B`.

```
. mat Alin  =    0,  -3, 0,3,   0, 0,0,  -1,-1,-1, 0, 0, 0, 1, 1,1
. mat Aquad =    0,   3,-6,3,   0, 0,0,   1, 1, 1,-2,-2,-2, 1, 1,1
. mat Blin  =    0,   0, 0,0,  -3, 0,3,  -1, 0, 1,-1, 0, 1,-1, 0,1
. mat Bquad =    0,   0, 0,0,   3,-6,3,   1,-2, 1, 1,-2, 1, 1,-2,1
. mat A = Alin \ Aquad
. mat B = Blin \ Bquad
. foreach i in lin quad {
  2.        foreach j in lin quad {
  3.                mat A`i'B`j' = vecdiag(A`i''*B`j')
  4.        }
  5. }
. mat AxBlin  = AlinBlin  \ AquadBlin
. mat AxBquad = AlinBquad \ AquadBquad
```

We use the `test()` and `mtest` options of `test` to obtain the single-degree-of-freedom tests for linear and quadratic A and B.

```
. mat list A, nonames
A[2,16]
    0   -3    0    3    0    0    0   -1   -1   -1    0    0    0    1    1    1
    0    3   -6    3    0    0    0    1    1    1   -2   -2   -2    1    1    1
. test, test(A) mtest
 ( 1) - 3 A[1] + 3 A[3] - A[1]*B[1] - A[1]*B[2] - A[1]*B[3] + A[3]*B[1] + A[3]*
 > B[2] + A[3]*B[3] = 0
 ( 2)   3 A[1] - 6 A[2] + 3 A[3] + A[1]*B[1] + A[1]*B[2] + A[1]*B[3] - 2 A[2]*B[1]
 > - 2 A[2]*B[2] - 2 A[2]*B[3] + A[3]*B[1] + A[3]*B[2] + A[3]*B[3] = 0
```

	F(df,27)	df	p	
(1)	212.65	1	0.0000	#
(2)	70.88	1	0.0000	#
all	141.77	2	0.0000	

                    # unadjusted p-values

```
. mat list B, nonames
B[2,16]
    0    0    0    0   -3    0    3   -1    0    1   -1    0    1   -1    0    1
    0    0    0    0    3   -6    3    1   -2    1    1   -2    1    1   -2    1
. test, test(B) mtest
 ( 1) - 3 B[1] + 3 B[3] - A[1]*B[1] + A[1]*B[3] - A[2]*B[1] + A[2]*B[3] - A[3]*
 > B[1] + A[3]*B[3] = 0
 ( 2)   3 B[1] - 6 B[2] + 3 B[3] + A[1]*B[1] - 2 A[1]*B[2] + A[1]*B[3] + A[2]*B[1]
 > - 2 A[2]*B[2] + A[2]*B[3] + A[3]*B[1] - 2 A[3]*B[2] + A[3]*B[3] = 0
```

	F(df,27)	df	p	
(1)	26.17	1	0.0000	#
(2)	27.51	1	0.0000	#
all	26.84	2	0.0000	

                    # unadjusted p-values

All the above tests appear to be significant. In addition to presenting the single-degree-of-freedom tests, the combined tests for A and B are produced and agree with the original ANOVA results.

Now we explore the interaction between A and B.

```
. mat list AxBlin, nonames
AxBlin[2,16]
    0    0    0    0    0    0    0    1    0   -1    0    0    0   -1    0    1
    0    0    0    0    0    0    0   -1    0    1    2    0   -2   -1    0    1
```

```
. test, test(AxBlin) mtest
 ( 1)  A[1]*B[1] - A[1]*B[3] - A[3]*B[1] + A[3]*B[3] = 0
 ( 2) - A[1]*B[1] + A[1]*B[3] + 2 A[2]*B[1] - 2 A[2]*B[3] - A[3]*B[1] + A[3]*
> B[3] = 0
```

	F(df,27)	df	p
(1)	17.71	1	0.0003 #
(2)	0.07	1	0.7893 #
all	8.89	2	0.0011

unadjusted p-values

The 2-degree-of-freedom test of the interaction of A with the linear components of B is significant at the 0.0011 level. But, when we examine the two single-degree-of-freedom tests that compose this result, the significance is due to the linear A by linear B contrast (significance level of 0.0003). A significance value of 0.7893 for the quadratic A by linear B indicates that this factor is not significant for these data.

```
. mat list AxBquad, nonames

AxBquad[2,16]
    0   0   0   0   0   0   0  -1   2  -1   0   0   0   1  -2   1
    0   0   0   0   0   0   0   1  -2   1  -2   4  -2   1  -2   1
. test, test(AxBquad) mtest
 ( 1) - A[1]*B[1] + 2 A[1]*B[2] - A[1]*B[3] + A[3]*B[1] - 2 A[3]*B[2] + A[3]*B[
> 3] = 0
 ( 2)  A[1]*B[1] - 2 A[1]*B[2] + A[1]*B[3] - 2 A[2]*B[1] + 4 A[2]*B[2] - 2 A[2]
> *B[3] + A[3]*B[1] - 2 A[3]*B[2] + A[3]*B[3] = 0
```

	F(df,27)	df	p
(1)	2.80	1	0.1058 #
(2)	2.94	1	0.0979 #
all	2.87	2	0.0741

unadjusted p-values

The tests of A with the quadratic components of B does not fall below the 0.05 significance level.

◁

Methods and Formulas

All postestimation commands listed above are implemented as ado-files.

References

Ostle, B., and R. W. Mensing. 1975. *Statistics in Research*. 3rd ed. Ames, IA: Iowa State University Press.

Rencher, A. C. 2000. *Linear Models in Statistics*. New York: Wiley.

Also See

[R] **anova** — Analysis of variance and covariance

[R] **regress postestimation** — Postestimation tools for regress

[U] **20 Estimation and postestimation commands**

Title

areg — Linear regression with a large dummy-variable set

Syntax

areg *depvar* $[$*indepvars*$]$ $[$*if*$]$ $[$*in*$]$ $[$*weight*$]$, <u>ab</u>sorb(*varname*) $[$*options*$]$

options	description
Model	
* <u>ab</u>sorb(*varname*)	categorical variable to be absorbed
SE/Robust	
vce(*vcetype*)	*vcetype* may be ols, <u>r</u>obust, <u>cl</u>uster *clustvar*, <u>boot</u>strap, or jackknife
Reporting	
<u>l</u>evel(#)	set confidence level; default is level(95)

*absorb(*varname*) is required.

depvar and *indepvars* may contain time-series operators; see [U] **11.4.3 Time-series varlists**.

bootstrap, by, jackknife, rolling, statsby, and xi are allowed; see [U] **11.1.10 Prefix commands**.

Weights are not allowed with the bootstrap prefix.

aweights are not allowed with the jackknife prefix.

aweights, fweights, and pweights are allowed; see [U] **11.1.6 weight**.

See [U] **20 Estimation and postestimation commands** for more capabilities of estimation commands.

Description

areg fits a linear regression absorbing one categorical factor. areg is designed for datasets with many groups, but not a number of groups that increases with the sample size. See the command xtreg, fe in [XT] **xtreg** for an estimator that handles the case in which the number of groups increases with the sample size.

Options

 ┐ Model ┐

absorb(*varname*) specifies the categorical variable, which is to be included in the regression as if it were specified by dummy variables. absorb() is required.

 ┐ SE/Robust ┐

vce(*vcetype*) specifies the type of standard error reported, which includes types that are derived from asymptotic theory, that are robust to some kinds of misspecification, that allow for intragroup correlation, and that use bootstrap or jackknife methods; see [R] *vce_option*.

`vce(ols)`, the default, uses the standard variance estimator for ordinary least-squares regression.

Exercise caution when using the `vce(cluster clustvar)` option with `areg`. The effective number of degrees of freedom for the robust variance estimator is $n_g - 1$, where n_g is the number of clusters. Thus the number of levels of the `absorb()` variable should not exceed the number of clusters.

⌐Reporting⌐

`level(#)`; see [R] **estimation options**.

Remarks

Suppose that you have a regression model that includes among the explanatory variables a large number, k, of mutually exclusive and exhaustive dummies:

$$\mathbf{y} = \mathbf{X}\boldsymbol{\beta} + \mathbf{d}_1\gamma_1 + \mathbf{d}_2\gamma_2 + \cdots + \mathbf{d}_k\gamma_k + \boldsymbol{\epsilon}$$

For instance, the dummy variables $\mathbf{d}_i$ might indicate countries in the world or states of the United States. One solution would be to fit the model with `regress`, but this solution is possible only if k is small enough so that the total number of variables (the number of columns of $\mathbf{X}$ plus the number of $\mathbf{d}_i$s plus one for $\mathbf{y}$) is sufficiently small—meaning less than `matsize` (see [R] **matsize**). For problems with more variables than the largest possible value of `matsize` (40 for Small Stata, 800 for Stata/IC, and 11,000 for Stata/SE and Stata/MP), `regress` will not work. `areg` provides a way of obtaining estimates of $\boldsymbol{\beta}$—but not the γ_i's—in these cases. The effects of the dummy variables are said to be absorbed.

▷ Example 1

So that we can compare the results produced by `areg` with Stata's other regression commands, we will fit a model in which k is small. `areg`'s real use, however, is when k is large.

In our automobile data, we have a variable called `rep78` that is coded 1, 2, 3, 4, and 5, where 1 means poor and 5, excellent. Let us assume that we wish to fit a regression of `mpg` on `weight`, `gear_ratio`, and `rep78` (parameterized as a set of dummies). Since `rep78` assumes only five values, we can fit this model with `regress`. We first use `tabulate` to generate the dummies (see [R] **tabulate oneway** and [U] **25.2 Using indicator variables in estimation**).

```
. use http://www.stata-press.com/data/r10/auto
(1978 Automobile Data)
. tabulate rep78, gen(r)
```

Repair Record 1978	Freq.	Percent	Cum.
1	2	2.90	2.90
2	8	11.59	14.49
3	30	43.48	57.97
4	18	26.09	84.06
5	11	15.94	100.00
Total	69	100.00	

```
. regress mpg weight gear_ratio r1-r4
```

Source	SS	df	MS
Model	1575.97621	6	262.662702
Residual	764.226686	62	12.3262369
Total	2340.2029	68	34.4147485

Number of obs = 69
F(6, 62) = 21.31
Prob > F = 0.0000
R-squared = 0.6734
Adj R-squared = 0.6418
Root MSE = 3.5109

mpg	Coef.	Std. Err.	t	P>|t|	[95% Conf. Interval]	
weight	-.0051031	.0009206	-5.54	0.000	-.0069433	-.003263
gear_ratio	.901478	1.565552	0.58	0.567	-2.228015	4.030971
r1	-2.036937	2.740728	-0.74	0.460	-7.515574	3.4417
r2	-2.419822	1.764338	-1.37	0.175	-5.946682	1.107039
r3	-2.557432	1.370912	-1.87	0.067	-5.297846	.1829814
r4	-2.788389	1.395259	-2.00	0.050	-5.577472	.0006939
_cons	36.23782	7.01057	5.17	0.000	22.22389	50.25175

To fit the **areg** equivalent, we type

```
. areg mpg weight gear_ratio, absorb(rep78)
```

Linear regression, absorbing indicators

Number of obs = 69
F(2, 62) = 41.64
Prob > F = 0.0000
R-squared = 0.6734
Adj R-squared = 0.6418
Root MSE = 3.5109

mpg	Coef.	Std. Err.	t	P>|t|	[95% Conf. Interval]	
weight	-.0051031	.0009206	-5.54	0.000	-.0069433	-.003263
gear_ratio	.901478	1.565552	0.58	0.567	-2.228015	4.030971
_cons	34.05889	7.056383	4.83	0.000	19.95338	48.1644
rep78	F(4, 62) =		1.117	0.356	(5 categories)	

Both **regress** and **areg** display the same R^2 values, root mean squared error, and—for **weight** and **gear_ratio**—the same parameter estimates, standard errors, t statistics, significance levels, and confidence intervals. **areg**, however, does not report the coefficients for **rep78**, and—in fact, they are not even calculated. This computational trick makes the problem manageable when k is large. **areg** reports a test that the coefficients associated with **rep78** are jointly zero. Here this test has a significance level of 35.6%. This F test for **rep78** is the same that we would obtain after **regress** if we were to specify **test r1 r2 r3 r4** or, equivalently, **testparm r1-r4**; see [R] **test**.

The model F tests reported by **regress** and **areg** also differ. The **regress** command reports a test that all coefficients except that of the constant are equal to zero; thus, the dummies are included in this test. The **areg** output shows a test that all coefficients excluding the dummies and the constant are equal to zero. This is the same test that can be obtained after **regress** by typing **test weight gear_ratio**.

◁

❑ Technical Note

areg is designed for datasets with many groups, but not a number that grows with the sample size. Consider two different samples from the U.S. population. In the first sample, we have 10,000 individuals and we want to include an indicator for each of 50 states, whereas in the second sample we have 3 observations on each of 10,000 individuals and we want to include an indicator for each individual. areg was designed for datasets similar to the first sample in which we have a fixed number of groups, the 50 states. In the second sample, the number of groups, which is the number of individuals, grows as we include more individuals in the sample. For an estimator designed to handle the case in which the number of groups grows with the sample size, see the xtreg, fe command in [XT] **xtreg**.

Although the point estimates produced by areg and xtreg, fe are the same, the estimated VCEs differ when cluster() is specified because the commands make different assumptions about whether the number of groups increases with the sample size.

❑

❑ Technical Note

The intercept reported by areg deserves some explanation because, given k mutually exclusive and exhaustive dummies, it is arbitrary. areg identifies the model by choosing the intercept that makes the prediction calculated at the means of the independent variables equal to the mean of the dependent variable: $\overline{y} = \overline{x}\,\widehat{\beta}$.

```
. predict yhat
(option xb assumed; fitted values)
. summarize mpg yhat if rep78 < .
    Variable |     Obs       Mean    Std. Dev.      Min        Max
-------------+--------------------------------------------------------
         mpg |      69   21.28986   5.866408        12         41
        yhat |      69   21.28986   4.383224   11.58643   28.07367
```

We had to include if rep78 < . in our summarize command because we have missing values in our data. areg automatically dropped those missing values (as it should) in forming the estimates, but predict with the xb option will make predictions for cases with missing rep78 because it does not know that rep78 is really part of our model.

These predicted values do not include the absorbed effects (i.e., the $\mathbf{d}_i\gamma_i$). For predicted values that include these effects, use the xbd option of predict (see [R] **areg postestimation**) or see [XT] **xtreg**.

❑

▷ Example 2

areg, vce(robust) is a Huberized version of areg; see [P] **_robust**. Just as areg is equivalent to using regress with dummies, areg, vce(robust) is equivalent to using regress, vce(robust) with dummies. You can use areg, vce(robust) when you expect heteroskedastic or nonnormal errors. areg, vce(robust), like ordinary regression, assumes that the observations are independent, unless the vce(cluster clustvar) option is specified. If the vce(cluster clustvar) option is specified, this independence assumption is relaxed and only the clusters identified by equal values of clustvar are assumed to be independent.

Assume that we were to collect data by randomly sampling 10,000 doctors (from 100 hospitals) and then sampling 10 patients of each doctor, yielding a total dataset of 100,000 patients in a cluster sample. If in some regression we wished to include effects of the hospitals to which the doctors belonged, we would want to include a dummy variable for each hospital, adding 100 variables to our model. areg could fit this model by

```
. areg depvar patient_vars, absorb(hospital) vce(cluster doctor)
```

◁

Saved Results

areg saves the following in e():

Scalars

e(N)	number of observations	e(F_absorb)	F statistic for absorbed effect
e(tss)	total sum of squares		(when vce(robust) is not specified)
e(df_m)	model degrees of freedom	e(df_a)	degrees of freedom for absorbed effect
e(rss)	residual sum of squares	e(rmse)	root mean squared error
e(df_r)	residual degrees of freedom	e(ll)	log likelihood
e(r2)	R-squared	e(ll_0)	log likelihood, constant-only model
e(r2_a)	adjusted R-squared	e(N_clust)	number of clusters
e(F)	F statistic		

Macros

e(cmd)	areg	e(title)	title in estimation output
e(cmdline)	command as typed	e(clustvar)	name of cluster variable
e(depvar)	name of dependent variable	e(vce)	*vcetype* specified in vce()
e(absvar)	name of absorb variable	e(vcetype)	title used to label Std. Err.
e(wtype)	weight type	e(properties)	b V
e(wexp)	weight expression	e(predict)	program used to implement predict

Matrices

e(b)	coefficient vector	e(V)	variance–covariance matrix of the estimators

Functions

e(sample)	marks estimation sample

Methods and Formulas

areg is implemented as an ado-file.

areg begins by recalculating *depvar* and *indepvars* to have mean 0 within the groups specified by absorb(). The overall mean of each variable is then added back in. The adjusted *depvar* is then regressed on the adjusted *indepvars* with regress, yielding the coefficient estimates. The degrees of freedom of the variance–covariance matrix of the coefficients is then adjusted to account for the absorbed variables—this calculation yields the same results (up to numerical round-off error) as if the matrix had been calculated directly by the formulas given in [R] **regress**.

areg with vce(robust) works similarly, calling _robust after regress to produce the robust variance estimates; see [P] **_robust**. The model F test uses the robust variance estimates. There is, however, no simple computational means of obtaining a robust test of the absorbed dummies; thus, this test is not displayed when the vce(robust) option is specified.

The number of groups specified in absorb() are included in the degrees of freedom used in the finite-sample adjustment of the cluster–robust VCE estimator. This statement is only valid if the number of groups is small relative to the sample size. (Technically, the number of groups must remain fixed as the sample size grows.) For an estimator that allows the number of groups to grow with the sample size, see the xtreg, fe command in [XT] **xtreg**.

Reference

Blackwell, J. L., III. 2005. Estimation and testing of fixed-effect panel-data systems. *Stata Journal* 5: 202–207.

Also See

[R] **areg postestimation** — Postestimation tools for areg

[R] **regress** — Linear regression

[XT] **xtreg** — Fixed-, between-, and random-effects, and population-averaged linear models

[U] **20 Estimation and postestimation commands**

Title

areg postestimation — Postestimation tools for areg

Description

The following postestimation commands are available for `areg`:

command	description
estat	AIC, BIC, VCE, and estimation sample summary
estimates	cataloging estimation results
lincom	point estimates, standard errors, testing, and inference for linear combinations of coefficients
linktest	link test for model specification
lrtest	likelihood-ratio test
mfx	marginal effects or elasticities
nlcom	point estimates, standard errors, testing, and inference for nonlinear combinations of coefficients
predict	predictions, residuals, influence statistics, and other diagnostic measures
predictnl	point estimates, standard errors, testing, and inference for generalized predictions
test	Wald tests for simple and composite linear hypotheses
testnl	Wald tests of nonlinear hypotheses

See the corresponding entries in the *Stata Base Reference Manual* for details.

Syntax for predict

predict [*type*] *newvar* [*if*] [*in*] [, *statistic*]

where $y_j = \mathbf{x}_j\mathbf{b} + d_{\text{absorbvar}} + e_j$ and *statistic* is

statistic	description
Main	
xb	$\mathbf{x}_j\mathbf{b}$, fitted values; the default
stdp	standard error of the prediction
<u>d</u>residuals	$d_{\text{absorbvar}} + e_j = y_j - \mathbf{x}_j\mathbf{b}$
* xbd	$\mathbf{x}_j\mathbf{b} + d_{\text{absorbvar}}$
* d	$d_{\text{absorbvar}}$
* <u>residuals</u>	residual
* <u>sc</u>ore	score; equivalent to residuals

Unstarred statistics are available both in and out of sample; type `predict ... if e(sample) ...` if wanted only for the estimation sample. Starred statistics are calculated only for the estimation sample, even when `if e(sample)` is not specified.

Options for predict

Main

xb, the default, calculates the prediction of $\mathbf{x}_j\mathbf{b}$, the fitted values, by using the average effect of the absorbed variable. Also see xbd below.

stdp calculates the standard error of $\mathbf{x}_j\mathbf{b}$.

dresiduals calculates $y_j - \mathbf{x}_j\mathbf{b}$, which are the residuals plus the effect of the absorbed variable.

xbd calculates $\mathbf{x}_j\mathbf{b} + d_{\text{absorbvar}}$, which are the fitted values including the individual effects of the absorbed variable.

d calculates $d_{\text{absorbvar}}$, the individual coefficients for the absorbed variable.

residuals calculates the residuals, that is, $y_j - (\mathbf{x}_j\mathbf{b} + d_{\text{absorbvar}})$.

score is a synonym for residuals.

Methods and Formulas

All postestimation commands listed above are implemented as ado-files.

Also See

[R] **areg** — Linear regression with a large dummy-variable set

[U] **20 Estimation and postestimation commands**

Title

asclogit —	Alternative-specific conditional logit (McFadden's choice) model

Syntax

asclogit *depvar* [*indepvars*] [*if*] [*in*] [*weight*] , case(*varname*)

alternatives(*varname*) [*options*]

options	description
Model	
* case(*varname*)	use *varname* to identify cases
* alternatives(*varname*)	use *varname* to identify the alternatives available for each case
casevars(*varlist*)	case-specific variables
basealternative(*#*|*lbl*|*str*)	alternative used as base category
altwise	use alternative-wise deletion instead of casewise deletion
offset(*varname*)	include *varname* in model with coefficient constrained to 1
constraints(*constraints*)	apply specified linear constraints
collinear	keep collinear variables
SE/Robust	
vce(*vcetype*)	*vcetype* may be oim, robust, cluster *clustvar*, bootstrap, or jackknife
Reporting	
level(*#*)	set confidence level; default is level(95)
or	report odds ratios
noheader	do not display the header on the coefficient table
Max options	
maximize_options	control the maximization process; seldom used

* case(*varname*) and alternatives(*varname*) are required.

bootstrap, by, jackknife, statsby, and xi are allowed; see [U] **11.1.10 Prefix commands**.

Weights are not allowed with the bootstrap prefix.

fweights, iweights, and pweights are allowed (see [U] **11.1.6 weight**), but they are interpreted to apply to cases as a whole, not to individual observations. See *Use of weights* in [R] **clogit**.

See [U] **20 Estimation and postestimation commands** for more capabilities of estimation commands.

Description

asclogit fits McFadden's choice model, which is a specific case of the more general conditional logistic regression model (McFadden 1974). asclogit requires multiple observations for each case (individual or decision), where each observation represents an alternative that may be chosen. The cases are identified by the variable specified in the option case(), whereas the alternatives are identified by the variable specified in the option alternatives(). The outcome or chosen alternative is identified by a value of 1 in *depvar*, whereas zeros indicate the alternatives that were not chosen. There can be multiple alternatives chosen for each case.

asclogit allows two types of independent variables: alternative-specific variables and case-specific variables. Alternative-specific variables vary across both cases and alternatives and are specified in *indepvars*. Case-specific variables vary only across cases and are specified in the option casevars().

See [R] **clogit** for a more general application of conditional logistic regression. For example, clogit would be used when you have grouped data where each observation in a group may be a different individual, but all individuals in a group have a common characteristic. You may use clogit to obtain the same estimates as asclogit by specifying the case() variable as the group() variable in clogit and generating variables that interact the casevars() in asclogit with each alternative (in the form of an indicator variable), excluding the interaction variable associated with the base alternative. asclogit takes care of this data management burden for you. Also, for clogit each record (row in your data) is an observation, whereas in asclogit each case, consisting of several records (the alternatives) in your data, is an observation. This last point is important since asclogit will drop observations, by default, in a casewise fashion. That is, if there is at least one missing value in any of the variables for each record of a case, the entire case is dropped from estimation. To use alternative-wise deletion, specify the altwise option and only the records with missing values will be dropped from estimation.

Options

_____Model_____

case(*varname*) specifies the numeric variable that identifies each case. case() is required and must be integer valued.

alternatives(*varname*) specifies the variable that identifies the alternatives for each case. The number of alternatives can vary with each case; the maximum number of alternatives cannot exceed the limits of tabulate oneway; see [R] **tabulate oneway**. alternatives() is required and may be a numeric or a string variable.

casevars(*varlist*) specifies the case-specific numeric variables. These are variables that are constant for each case. If there are a maximum of J alternatives, there will be $J - 1$ sets of coefficients associated with the casevars().

basealternative(*# | lbl | str*) specifies the alternative that normalizes the latent-variable location (the level of utility). The base alternative may be specified as a number, label, or string depending on the storage type of the variable indicating alternatives. The default is the alternative with the highest frequency.

If vce(bootstrap) or vce(jackknife) is specified, you must specify the base alternative. This is to ensure that the same model is fitted with each call to asclogit.

altwise specifies that alternative-wise deletion be used when marking out observations due to missing values in your variables. The default is to use casewise deletion; that is, the entire group of observations making up a case is deleted if any missing values are encountered. This option does not apply to observations that are marked out by the if or in qualifier or the by prefix.

offset(*varname*), constraints(*numlist | matname*), collinear; see [R] **estimation options**.

_____SE/Robust_____

vce(*vcetype*) specifies the type of standard error reported, which includes types that are derived from asymptotic theory, that are robust to some kinds of misspecification, that allow for intragroup correlation, and that use bootstrap or jackknife methods; see [R] *vce_option*.

`level(#)`; see [R] **estimation options**.

`or` reports the estimated coefficients transformed to odds ratios, i.e., e^b rather than b. Standard errors and confidence intervals are similarly transformed. This option affects how results are displayed, not how they are estimated. `or` may be specified at estimation or when replaying previously estimated results.

`noheader` prevents the coefficient table header from being displayed.

Max options

maximize_options: <u>diff</u>icult, <u>tech</u>nique(*algorithm_spec*), <u>iter</u>ate(#), [<u>no</u>]<u>log</u>, <u>tra</u>ce, <u>grad</u>ient, showstep, <u>hess</u>ian, <u>shownr</u>tolerance, <u>tol</u>erance(#), <u>ltol</u>erance(#), <u>gtol</u>erance(#), <u>nrtol</u>erance(#), <u>nonrtol</u>erance, from(*init_specs*); see [R] **maximize**. These options are seldom used.

`technique(bhhh)` is not allowed.

The initial estimates must be specified as `from(`*matname* [`, copy`]`)`, where *matname* is the matrix containing the initial estimates and the `copy` option specifies that only the position of each element in *matname* is relevant. If `copy` is not specified, the column stripe of *matname* identifies the estimates.

Remarks

`asclogit` fits McFadden's choice model (McFadden 1974; for a brief introduction, see Greene 2003, sec. 21.7). In this model we have a set of unordered alternatives indexed by $1, 2, \ldots, J$. Let y_{ij}, $j = 1, \ldots, J$, be an indicator variable for the alternative actually chosen by the ith individual (case). That is, $y_{ij} = 1$ if individual i chose alternative j and $y_{ij} = 0$ otherwise. The independent variables come in two forms: alternative specific and case specific. Alternative-specific variables vary among the alternatives (as well as cases), and case-specific variables vary only among cases. Assume that we have p alternative-specific variables so that for case i we have a $J \times p$ matrix $\mathbf{X}_i$. Further, assume that we have q case-specific variables so that we have a $1 \times q$ vector $\mathbf{z}_i$ for case i. Our random utility model can the be expressed as

$$\mathbf{u}_i = \mathbf{X}_i \boldsymbol{\beta} + (\mathbf{z}_i \mathbf{A})' + \boldsymbol{\epsilon}_i$$

Here $\boldsymbol{\beta}$ is a $p \times 1$ vector of alternative-specific regression coefficients and $\mathbf{A} = (\boldsymbol{\alpha}_1, \ldots, \boldsymbol{\alpha}_J)$ is a $q \times J$ matrix of case-specific regression coefficients. The elements of the $J \times 1$ vector $\boldsymbol{\epsilon}_i$ are independent Type I (Gumbel type) extreme-value random variables with mean γ (the Euler–Mascheroni constant, approximately 0.577) and variance $\pi^2/6$. We must fix one of the $\boldsymbol{\alpha}_j$ to the constant vector to normalize the location. We set $\boldsymbol{\alpha}_k = 0$ where k is specified by the option `basealternative()`. The vector $\mathbf{u}_i$ quantifies the utility that the individual gains from the J alternatives. The alternative chosen by individual i is the one that maximizes utility.

▷ Example 1

We have data on 295 consumers and their choice of automobile. Each consumer chose among an American, Japanese, or European car; the variable `car` indicates the nationality of the car for each alternative. We want to explore the relationship between the choice of `car` to the consumer's sex (variable `sex`) and income (variable `income` in thousands of dollars). We also have information on the number of dealerships of each nationality in the consumer's city in the variable `dealer` that we want to include as a regressor. We assume that consumers' preferences are influenced by the number of dealerships in an area but that the number of dealerships is not influenced by consumer preferences (which we admit is a rather strong assumption). The variable dealer is an alternative-specific variable (X_i is a 3×1 vector in our previous notation), and sex and income are case-specific variables (z_i is a 1×2 vector). Each consumer's chosen car is indicated by the variable choice.

Let's list some of the data.

```
. use http://www.stata-press.com/data/r10/choice
. list id car choice dealer sex income in 1/12, sepby(id)
```

	id	car	choice	dealer	sex	income
1.	1	American	0	18	male	46.7
2.	1	Japan	0	8	male	46.7
3.	1	Europe	1	5	male	46.7
4.	2	American	1	17	male	26.1
5.	2	Japan	0	6	male	26.1
6.	2	Europe	0	2	male	26.1
7.	3	American	1	12	male	32.7
8.	3	Japan	0	6	male	32.7
9.	3	Europe	0	2	male	32.7
10.	4	American	0	18	female	49.2
11.	4	Japan	1	7	female	49.2
12.	4	Europe	0	4	female	49.2

We see, for example, that the first consumer, a male earning $46,700 per year, chose to purchase a European car even though there are more American and Japanese car dealers in his area. The fourth consumer, a female earning $49,200 per year, purchased a Japanese car.

We now fit our model.

```
. asclogit choice dealer, casevars(sex income) case(id) alternatives(car)
Iteration 0:   log likelihood = -273.55685
Iteration 1:   log likelihood = -252.75109
Iteration 2:   log likelihood = -250.78555
Iteration 3:   log likelihood =  -250.7794
Iteration 4:   log likelihood =  -250.7794
```

Alternative-specific conditional logit	Number of obs	=	885
Case variable: id	Number of cases	=	295
Alternative variable: car	Alts per case: min =		3
	avg =		3.0
	max =		3
	Wald chi2(5)	=	15.86
Log likelihood = -250.7794	Prob > chi2	=	0.0072

choice	Coef.	Std. Err.	z	P>|z|	[95% Conf. Interval]	
car						
dealer	.0680938	.0344465	1.98	0.048	.00058	.1356076
American	(base alternative)					
Japan						
sex	-.5346039	.3141564	-1.70	0.089	-1.150339	.0811314
income	.0325318	.012824	2.54	0.011	.0073973	.0576663
_cons	-1.352189	.6911829	-1.96	0.050	-2.706882	.0025049
Europe						
sex	.5704109	.4540247	1.26	0.209	-.3194612	1.460283
income	.032042	.0138676	2.31	0.021	.004862	.0592219
_cons	-2.355249	.8526681	-2.76	0.006	-4.026448	-.6840501

Displaying the results as odds ratios makes interpretation easier.

```
. asclogit, or noheader
```

choice	Odds Ratio	Std. Err.	z	P>|z|	[95% Conf. Interval]	
car						
dealer	1.070466	.0368737	1.98	0.048	1.00058	1.145232
American	(base alternative)					
Japan						
sex	.5859013	.1840647	-1.70	0.089	.3165294	1.084513
income	1.033067	.013248	2.54	0.011	1.007425	1.059361
Europe						
sex	1.768994	.8031669	1.26	0.209	.7265404	4.307178
income	1.032561	.0143191	2.31	0.021	1.004874	1.061011

These results indicate that men (sex = 1) are less likely to pick a Japanese car over an American car than women (odds ratio 0.59) but that men are more likely to choose a European car over an American car (odds ratio 1.77). Raising a person's income increases the likelihood that she purchases a Japanese or European car; interestingly, the effect of higher income is about the same for these two types of cars.

◁

❑ Technical Note

McFadden's choice model is related to multinomial logistic regression (see [R] **mlogit**). If all the independent variables are case specific, then the two models are identical. We verify this supposition by running the previous example without the alternative-specific variable, dealer.

```
. asclogit choice, casevars(sex income) case(id) alternatives(car) nolog
```

Alternative-specific conditional logit	Number of obs	=	885
Case variable: id	Number of cases	=	295
Alternative variable: car	Alts per case: min	=	3
	avg	=	3.0
	max	=	3
	Wald chi2(4)	=	12.53
Log likelihood = -252.72012	Prob > chi2	=	0.0138

choice	Coef.	Std. Err.	z	P>|z|	[95% Conf. Interval]	
American	(base alternative)					
Japan						
sex	-.4694799	.3114939	-1.51	0.132	-1.079997	.141037
income	.0276854	.0123666	2.24	0.025	.0034472	.0519236
_cons	-1.962652	.6216804	-3.16	0.002	-3.181123	-.7441807
Europe						
sex	.5388441	.4525279	1.19	0.234	-.3480942	1.425782
income	.0273669	.013787	1.98	0.047	.000345	.0543889
_cons	-3.180029	.7546837	-4.21	0.000	-4.659182	-1.700876

To run mlogit, we must rearrange the dataset. mlogit requires a dependent variable that indicates the choice—1, 2, or 3—for each individual. We will use car as our dependent variable for those observations that represent the choice actually chosen.

```
. keep if choice == 1
(590 observations deleted)

. mlogit car sex income
Iteration 0:   log likelihood =  -259.1712
Iteration 1:   log likelihood = -252.81165
Iteration 2:   log likelihood = -252.72014
Iteration 3:   log likelihood = -252.72012
```

Multinomial logistic regression	Number of obs	=	295
	LR chi2(4)	=	12.90
	Prob > chi2	=	0.0118
Log likelihood = -252.72012	Pseudo R2	=	0.0249

car	Coef.	Std. Err.	z	P>|z|	[95% Conf. Interval]	
Japan						
sex	-.4694799	.3114939	-1.51	0.132	-1.079997	.141037
income	.0276854	.0123666	2.24	0.025	.0034472	.0519236
_cons	-1.962652	.6216803	-3.16	0.002	-3.181123	-.7441807
Europe						
sex	.5388442	.4525278	1.19	0.234	-.3480941	1.425783
income	.0273669	.013787	1.98	0.047	.000345	.0543889
_cons	-3.180029	.7546837	-4.21	0.000	-4.659182	-1.700876

(car==American is the base outcome)

The results are the same except for the model statistic: asclogit uses a Wald test and mlogit uses a likelihood-ratio test. If you prefer the likelihood-ratio test, you can fit the constant-only model for asclogit followed by the full model and use [R] **lrtest**. The following example will carry this out.

```
. use http://www.stata-press.com/data/r10/choice, clear
. asclogit choice, case(id) alternatives(car)
. estimates store null
. asclogit choice, casevars(sex income) case(id) alternatives(car)
. lrtest null .
```

❑

❑ Technical Note

We force you to explicitly identify the case-specific variables in the casevars() option to ensure that the program behaves as you expect. For example, an if or in qualifier may drop observations in such a way that (what was expected to be) an alternative-specific variable turns into a case-specific variable. Here you would probably want asclogit to terminate instead of interacting the variable with the alternative indicators. This situation could also occur if asclogit drops cases, or observations if you use the altwise option, because of missing values.

❑

Saved Results

asclogit saves the following in e():

Scalars

e(N)	number of observations
e(N_case)	number of cases
e(k)	number of parameters
e(k_alt)	number of alternatives
e(k_indvars)	number of alternative-specific variables
e(k_casevars)	number of case-specific variables
e(k_eq)	number of equations in e(b)
e(k_eq_model)	number of equations in model Wald test
e(N_clust)	number of clusters
e(const)	constant indicator
e(i_base)	base alternative index
e(df_m)	model degrees of freedom
e(ll)	log likelihood
e(chi2)	χ^2
e(F)	F statistic
e(p)	significance
e(alt_min)	minimum number of alternatives
e(alt_max)	maximum number of alternatives
e(alt_avg)	average number of alternatives
e(rank)	rank of e(V)
e(ic)	number of iterations
e(rc)	return code
e(converged)	1 if converged, 0 otherwise

Macros
e(cmd)	asclogit
e(cmdline)	command as typed
e(depvar)	name of dependent variable
e(indvars)	alternative-specific independent variable
e(casevars)	case-specific variables
e(case)	variable defining cases
e(altvar)	variable defining alternatives
e(alteqs)	alternative equation names
e(alt#)	alternative labels
e(wtype)	weight type
e(wexp)	weight expression
e(title)	title in estimation output
e(clustvar)	name of cluster variable
e(offset)	offset
e(chi2type)	Wald, type of model χ^2 test
e(vce)	*vcetype* specified in vce()
e(vcetype)	title used to label Std. Err.
e(opt)	type of optimization
e(ml_method)	type of ml method
e(user)	name of likelihood-evaluator program
e(technique)	maximization technique
e(crittype)	optimization criterion
e(properties)	b V
e(datasignaturevars)	variables used in calculation of checksum
e(datasignature)	the checksum
e(estat_cmd)	program used to implement estat
e(predict)	program used to implement predict

Matrices
e(b)	coefficient vector
e(stats)	alternative statistics
e(altvals)	alternative values
e(altfreq)	alternative frequencies
e(ilog)	iteration log (up to 20 iterations)
e(gradient)	gradient vector
e(V)	variance–covariance matrix of the estimators

Functions
e(sample)	marks estimation sample

Methods and Formulas

asclogit is implemented as an ado-file.

In this model, we have a set of unordered alternatives indexed by $1, 2, \ldots, J$. Let $y_{ij}, j = 1, \ldots, J$, be an indicator variable for the alternative actually chosen by the ith individual (case). That is, $y_{ij} = 1$ if individual i chose alternative j and $y_{ij} = 0$ otherwise. The independent variables come in two forms: alternative specific and case specific. Alternative-specific variables vary among the alternatives (as well as cases), and case-specific variables vary only among cases. Assume that we have p alternative-specific variables so that for case i we have a $J \times p$ matrix $\mathbf{X}_i$. Further, assume that we have q case-specific variables so that we have a $1 \times q$ vector $\mathbf{z}_i$ for case i. The deterministic component of the random utility model can the be expressed as

$$\eta_i = \mathbf{X}_i\boldsymbol{\beta} + (\mathbf{z}_i\mathbf{A})'$$
$$= \mathbf{X}_i\boldsymbol{\beta} + (\mathbf{z}_i \otimes \mathbf{I}_J)\,\text{vec}(\mathbf{A}')$$
$$= (\mathbf{X}_i,\ \mathbf{z}_i \otimes \mathbf{I}_J)\begin{pmatrix}\boldsymbol{\beta} \\ \text{vec}(\mathbf{A}')\end{pmatrix}$$
$$= \mathbf{X}_i^*\boldsymbol{\beta}^*$$

As before $\boldsymbol{\beta}$ is a $p \times 1$ vector of alternative-specific regression coefficients and $\mathbf{A} = (\boldsymbol{\alpha}_1,\dots,\boldsymbol{\alpha}_J)$ is a $q \times J$ matrix of case-specific regression coefficients; remember that we must fix one of the $\boldsymbol{\alpha}_j$ to the constant vector to normalize the location. Here, $\mathbf{I}_J$ is the $J \times J$ identity matrix, vec() is the vector function that creates a vector from a matrix by placing each column of the matrix on top of the other (see [M-5] **vec( )**), and $\otimes$ is the Kronecker product (see [M-2] **op_kronecker**).

We have rewritten the linear equation so that it is a form that can be used by clogit, namely, $\mathbf{X}_i^*\boldsymbol{\beta}^*$, where

$$\mathbf{X}_i^* = (\mathbf{X}_i,\ \mathbf{z}_i \otimes \mathbf{I}_J)$$

$$\boldsymbol{\beta}^* = \begin{pmatrix}\boldsymbol{\beta} \\ \text{vec}(\mathbf{A}')\end{pmatrix}$$

With this in mind, see *Methods and Formulas* in [R] **clogit** for the computational details of the conditional logit model.

Daniel Little McFadden was born in 1937 in North Carolina. He studied physics, psychology, and economics at the University of Minnesota and has taught economics at Pittsburgh, Berkeley, and MIT. His contributions to logit models were triggered by a student's project on freeway routing decisions, and his work consistently links economic theory and applied problems. In 2000 he shared the Nobel Prize in Economics with James J. Heckman.

References

Greene, W. H. 2003. *Econometric Analysis*. 5th ed. Upper Saddle River, NJ: Prentice Hall.

McFadden, D. 1974. Conditional logit analysis of qualitative choice behavior. In *Frontiers in Econometrics*, ed. P. Zarembka, 105–142. New York: Academic Press.

Also See

[R] **asclogit postestimation** — Postestimation tools for asclogit

[R] **asmprobit** — Alternative-specific multinomial probit regression

[R] **asroprobit** — Alternative-specific rank-ordered probit regression

[R] **clogit** — Conditional (fixed-effects) logistic regression

[R] **logistic** — Logistic regression, reporting odds ratios

[R] **logit** — Logistic regression, reporting coefficients

[R] **nlogit** — Nested logit regression

[R] **ologit** — Ordered logistic regression

[U] **20 Estimation and postestimation commands**

Title

---
asclogit postestimation — Postestimation tools for asclogit

Description

The following postestimation commands are of special interest after `asclogit`:

commands	description
estat <u>alternatives</u>	alternative summary statistics
estat mfx	marginal effects

For information about these commands, see below.

The following standard postestimation commands are also available:

commands	description
estat	AIC, BIC, VCE, and estimation sample summary
estimates	cataloging estimation results
lincom	point estimates, standard errors, testing, and inference for linear combinations of coefficients
lrtest	likelihood-ratio test
nlcom	point estimates, standard errors, testing, and inference for nonlinear combinations of coefficients
predict	predicted probabilities, estimated linear predictor and its standard error
predictnl	point estimates, standard errors, testing, and inference for generalized predictions
test	Wald tests for simple and composite linear hypotheses
testnl	Wald tests of nonlinear hypotheses

See the corresponding entries in the *Stata Base Reference Manual* for details.

Special-interest postestimation commands

`estat alternatives` displays summary statistics about the alternatives in the estimation sample.

`estat mfx` computes probability marginal effects.

Syntax for predict

> predict [*type*] *newvar* [*if*] [*in*] [, *statistic options*]

> predict [*type*] { *stub** | *newvarlist* } [*if*] [*in*], <u>sc</u>ores

statistic	description
Main	
<u>pr</u>	probability that each alternative is chosen; the default
xb	linear prediction
stdp	standard error of the linear prediction

options	description
Main	
*k(*condspec*)	condition on *condspec* alternatives chosen by each case when computing probabilities
altwise	use alternative-wise deletion instead of casewise deletion when computing probabilities
<u>noo</u>ffset	ignore the offset() variable specified in asclogit

*k(*condspec*) may be used only with pr.

These statistics are available both in and out of sample; type predict ... if e(sample) ... if wanted only for the estimation sample.

Options for predict

⌐ Main ⌐

pr computes the probability of choosing each alternative conditioned on each case choosing k(*condspec*) alternatives. This is the default statistic with default k(1); one alternative per case is chosen.

xb computes the linear prediction.

stdp computes the standard error of the linear prediction.

k(# | observed) conditions the probability on choosing # alternatives per case, or use k(observed) to condition on the observed number of alternatives chosen. The default is k(1). This option may be used only with option pr.

altwise specifies that alternative-wise deletion be used when marking out observations due to missing values in your variables. The default is to use casewise deletion. Options xb and stdp always use alternative-wise deletion.

nooffset is relevant only if you specified offset(*varname*) for asclogit. It modifies the calculations made by predict so that they ignore the offset variable; the linear prediction is treated as $x\beta$ rather than as $x\beta$ + offset.

scores calculates the scores for each coefficient in e(b). This option requires a new variable list of length equal to the number of columns in e(b). Otherwise, use the *stub** option to have predict generate enumerated variables with prefix *stub*.

Syntax for estat alternatives

estat alternatives

Syntax for estat mfx

estat mfx $[if]$ $[in]$ [, options]

options	description
Main	
varlist(*varlist*)	display marginal effects for *varlist*
at(mean $[atlist]$ |median $[atlist]$)	calculate marginal effects at these values
k(#)	condition on the number of alternatives chosen to be #
Options	
level(#)	set confidence interval level; default is level(95)
nodiscrete	treat indicator variables as continuous
noesample	do not restrict calculation of means and medians to the estimation sample
nowght	ignore weights when calculating means and medians

Options for estat mfx

___ Main ___

varlist(*varlist*) specifies the variables for which to display marginal effects. The default is all variables.

at(mean $[atlist]$ \|median $[atlist]$) specifies the values at which the marginal effects are to be calculated. *atlist* is

$$\left[\,[\,alternative{:}variable\,=\,\#\,]\ [\,variable\,=\,\#\,]\ [\,alternative{:}offset\,=\,\#\,]\ [\,\dots\,]\,\right]$$

The default is to calculate the marginal effects at the means of the independent variables by using the estimation sample, at(mean). If offset() is used during estimation, the means of the offsets (by alternative) are computed by default.

After specifying the summary statistic, you can specify a series of specific values for variables. You can specify values for alternative-specific variables by alternative, or you can specify one value for all alternatives. You can specify only one value for case-specific variables. You specify values for the offset() variable (if present) the same way as for alternative-specific variables. For example, in the choice dataset (car choice), income is a case-specific variable, whereas dealer is an alternative-specific variable. The following would be a legal syntax for estat mfx:

. estat mfx, at(mean American:dealer=18 income=40)

When nodiscrete is not specified, at(mean $[atlist]$) or at(median $[atlist]$) has no effect on computing marginal effects for indicator variables, which are calculated as the discrete change in the simulated probability as the indicator variable changes from 0 to 1.

The mean and median computations respect any `if` or `in` qualifiers, so you can restrict the data over which the statistic is computed. You can even restrict the values to a specific case, e.g.,

. estat mfx if case==21

`k(#)` computes the probabilities conditioned on # alternatives chosen. The default is one alternative chosen.

Options

`level(#)` sets the confidence level; default is `level(95)`.

`nodiscrete` specifies that indicator variables be treated as continuous variables. An indicator variable is one that takes on the value 0 or 1 in the estimation sample. By default, the discrete change in the simulated probability is computed as the indicator variable changes from 0 to 1.

`noesample` specifies that the whole dataset be considered instead of only those marked in the `e(sample)` defined by the `asclogit` command.

`nowght` specifies that weights be ignored when calculating the medians.

Remarks

Remarks are presented under the following headings:

> *Predicted probabilities*
> *Obtaining estimation statistics*

Predicted probabilities

After fitting a McFadden's choice model with alternative-specific conditional logistic regression, you can use `predict` to obtain the estimated probability of alternative choices given case profiles.

▷ Example 1

For [R] **asclogit**, we fitted a model of consumer choice of automobile. The alternatives are nationality of the automobile manufacturer: American, Japanese, or European. There is one alternative-specific variable in the model, `dealer`, which contains the number of dealerships of each nationality in the consumer's city. The case-specific variables are `sex`, the consumer's sex, and `income`, the consumer's income in thousands of dollars.

```
. use http://www.stata-press.com/data/r10/choice
. asclogit choice dealer, casevars(sex income) case(id) alternatives(car)
  (output omitted)
. predict p
(option pr assumed; Pr(car))
. predict p2, k(2)
(option pr assumed; Pr(car))
. format p p2 %6.4f
```

(*Continued on next page*)

```
. list car choice dealer sex income p p2 in 1/9, sepby(id)
```

	car	choice	dealer	sex	income	p	p2
1.	American	0	18	male	46.7	0.6025	0.8589
2.	Japan	0	8	male	46.7	0.2112	0.5974
3.	Europe	1	5	male	46.7	0.1863	0.5437
4.	American	1	17	male	26.1	0.7651	0.9293
5.	Japan	0	6	male	26.1	0.1282	0.5778
6.	Europe	0	2	male	26.1	0.1067	0.4929
7.	American	1	12	male	32.7	0.6519	0.8831
8.	Japan	0	6	male	32.7	0.1902	0.5995
9.	Europe	0	2	male	32.7	0.1579	0.5174

◁

Obtaining estimation statistics

Here we will demonstrate the specialized estat subcommands after asclogit. Use estat
alternatives to obtain a table of alternative statistics. The table will contain the alternative values,
labels (if any), the number of cases in which each alternative is present, the frequency that the
alternative is selected, and the percent selected.

Use estat mfx to obtain marginal effects after asclogit.

▷ Example 2

We will continue with the automobile choice example, where we first list the alternative statistics
and then compute the marginal effects at the mean income in our sample, assuming that there are
five automobile dealers for each nationality. We will evaluate the probabilities for females since sex
is coded 0 for females, and we will be obtaining the discrete change from 0 to 1.

```
. estat alternatives
```

Alternatives summary for car

index	Alternative value	label	Cases present	Frequency selected	Percent selected
1	1	American	295	192	65.08
2	2	Japan	295	64	21.69
3	3	Europe	295	39	13.22

```
. estat mfx, at(dealer=0 sex=0) varlist(sex income)
```

Pr(choice = American|1 selected) = .41964329

variable	dp/dx	Std. Err.	z	P>|z|	[95% C.I.]		X
casevars							
sex*	.026238	.068311	0.38	0.701	-.107649	.160124	0
income	-.007891	.002674	-2.95	0.003	-.013132	-.00265	42.097

(*) dp/dx is for discrete change of indicator variable from 0 to 1

```
Pr(choice = Japan|1 selected) = .42696187
```

variable	dp/dx	Std. Err.	z	P>|z|	[95% C.I.]		X
casevars							
sex*	-.161164	.079238	-2.03	0.042	-.316468	-.005859	0
income	.005861	.002997	1.96	0.051	-.000014	.011735	42.097

(*) dp/dx is for discrete change of indicator variable from 0 to 1

```
Pr(choice = Europe|1 selected) = .15339484
```

variable	dp/dx	Std. Err.	z	P>|z|	[95% C.I.]		X
casevars							
sex*	.134926	.076556	1.76	0.078	-.015122	.284973	0
income	.00203	.001785	1.14	0.255	-.001469	.00553	42.097

(*) dp/dx is for discrete change of indicator variable from 0 to 1

The marginal effect of income indicates that there is a lower chance for a consumer to buy American automobiles with an increase in income. There is an indication that men have a higher preference for European automobiles than women but a lower preference for Japanese automobiles. We did not include the marginal effects for dealer since we view these as nuisance parameters, so we adjusted the probabilities by fixing dealer to a constant, 0.

◁

Saved Results

estat mfx saves the following in r():

Scalars

r(pr_alt)   scalars containing the computed probability of each alternative evaluated at the value that is labeled X in the table output. Here alt are the labels in the macro e(alteqs).

Matrices

r(alt)   matrices containing the computed marginal effects and associated statistics. There is one matrix for each alternative, where alt are the labels in the macro e(alteqs). Column 1 of each matrix contains the marginal effects; column 2, their standard errors; column 3, their z statistics; and columns 4 and 5, the confidence intervals. Column 6 contains the values of the independent variables used to compute the probabilities r(pr_alt).

Methods and Formulas

All postestimation commands listed above are implemented as ado-files.

(Continued on next page)

The deterministic component of the random utility model can the be expressed as

$$
\begin{aligned}
\boldsymbol{\eta} &= \mathbf{X}\boldsymbol{\beta} + (\mathbf{z}\mathbf{A})' \\
&= \mathbf{X}\boldsymbol{\beta} + (\mathbf{z} \otimes \mathbf{I}_J)\,\mathrm{vec}(\mathbf{A}') \\
&= (\mathbf{X},\ \mathbf{z} \otimes \mathbf{I}_J) \begin{pmatrix} \boldsymbol{\beta} \\ \mathrm{vec}(\mathbf{A}') \end{pmatrix} \\
&= \mathbf{X}^* \boldsymbol{\beta}^*
\end{aligned}
$$

where $\mathbf{X}$ is the $J \times p$ matrix containing the alternative-specific covariates, $\mathbf{z}$ is a $1 \times q$ vector of case-specific variables, $\boldsymbol{\beta}$ is a $p \times 1$ vector of alternative-specific regression coefficients, and $\mathbf{A} = (\boldsymbol{\alpha}_1, \ldots, \boldsymbol{\alpha}_J)$ is a $q \times J$ matrix of case-specific regression coefficients (with one of the $\boldsymbol{\alpha}_j$ fixed to the constant). Here $\mathbf{I}_J$ is the $J \times J$ identity matrix, vec() is the vector function that creates a vector from a matrix by placing each column of the matrix on top of the other (see [M-5] **vec( )**), and $\otimes$ is the Kronecker product (see [M-2] **op_kronecker**).

We have rewritten the linear equation so that it is a form that we all recognize, namely, $\boldsymbol{\eta} = \mathbf{X}^* \boldsymbol{\beta}^*$, where

$$
\mathbf{X}^* = (\mathbf{X},\ \mathbf{z} \otimes \mathbf{I}_J)
$$

$$
\boldsymbol{\beta}^* = \begin{pmatrix} \boldsymbol{\beta} \\ \mathrm{vec}(\mathbf{A}') \end{pmatrix}
$$

To compute the marginal effects, we use the derivative of the log likelihood $\partial \ell(\mathbf{y}|\boldsymbol{\eta}) / \partial \boldsymbol{\eta}$, where $\ell(\mathbf{y}|\boldsymbol{\eta}) = \log \mathrm{Pr}(\mathbf{y}|\boldsymbol{\eta})$ is the log of the probability of the choice indicator vector $\mathbf{y}$ given the linear predictor vector $\boldsymbol{\eta}$. Namely,

$$
\begin{aligned}
\frac{\partial \mathrm{Pr}(\mathbf{y}|\boldsymbol{\eta})}{\partial \mathrm{vec}(\mathbf{X}^*)'} &= \mathrm{Pr}(\mathbf{y}|\boldsymbol{\eta}) \frac{\partial \ell(\mathbf{y}|\boldsymbol{\eta})}{\partial \boldsymbol{\eta}'} \frac{\partial \boldsymbol{\eta}}{\partial \mathrm{vec}(\mathbf{X}^*)'} \\
&= \mathrm{Pr}(\mathbf{y}|\boldsymbol{\eta}) \frac{\partial \ell(\mathbf{y}|\boldsymbol{\eta})}{\partial \boldsymbol{\eta}'} \left(\boldsymbol{\beta}^{*\prime} \otimes \mathbf{I}_J \right)
\end{aligned}
$$

The standard errors of the marginal effects are computed using the delta method.

Also See

[R] **asclogit** — Alternative-specific conditional logit (McFadden's choice) model

[U] **20 Estimation and postestimation commands**

Title

asmprobit — Alternative-specific multinomial probit regression

Syntax

asmprobit *depvar* [*indepvars*] [*if*] [*in*] [*weight*] , case(*varname*)

alternatives(*varname*) [*options*]

options	description
Model	
* case(*varname*)	use *varname* to identify cases
* alternatives(*varname*)	use *varname* to identify the alternatives available for each case
casevars(*varlist*)	case-specific variables
constraints(*constraints*)	apply specified linear constraints
collinear	keep collinear variables
Model 2	
correlation(*correlation*)	correlation structure of the latent-variable errors; implies structural
stddev(*stddev*)	variance structure of the latent-variable errors; implies structural
structural	use the structural covariance parameterization; default is the differenced covariance parameterization
noconstant	suppress the alternative-specific constant terms
basealternative(*#* | *lbl* | *str*)	alternative used for normalizing location
scalealternative(*#* | *lbl* | *str*)	alternative used for normalizing scale
altwise	use alternative-wise deletion instead of casewise deletion
SE/Robust	
vce(*vcetype*)	*vcetype* may be oim, robust, cluster *clustvar*, opg, bootstrap, or jackknife
Reporting	
level(*#*)	set confidence level; default is level(95)
notransform	do not transform variance–covariance estimates to the standard deviation and correlation metric
Int options	
intmethod(*seqtype*)	type of quasi- or pseudouniform point set
intpoints(*#*)	number of points in each sequence
intburn(*#*)	starting index in the Hammersley or Halton sequence
intseed(*code* | *#*)	pseudouniform random-number seed
antithetics	use antithetic draws
initbhhh(*#*)	use the BHHH optimization algorithm for the first *#* iterations
Max options	
maximize_options	control the maximization process

correlation	description
<u>uns</u>tructured	one correlation parameter for each pair of alternatives; correlations with the basealternative() are zero
<u>exc</u>hangeable	one correlation parameter common to all pairs of alternatives; correlations with the basealternative() are zero
<u>ind</u>ependent	constrain all correlation parameters to zero
<u>pat</u>tern *matname*	user-specified matrix identifying the correlation pattern
<u>fix</u>ed *matname*	user-specified matrix identifying the fixed and free correlation parameters

stddev	description
<u>het</u>eroskedastic	estimate standard deviation for each alternative; standard deviations for basealternative() and scalealternative() set to one
<u>hom</u>oskedastic	all standard deviations are one
<u>pat</u>tern *matname*	user-specified matrix identifying the standard deviation pattern
<u>fix</u>ed *matname*	user-specified matrix identifying the fixed and free standard deviation parameters

seqtype	description
<u>ham</u>mersley	Hammersley point set
<u>hal</u>ton	Halton point set
<u>ran</u>dom	uniform pseudorandom point set

*case(*varname*) and alternatives(*varname*) are required.

bootstrap, by, jackknife, statsby, and xi are allowed; see [U] **11.1.10 Prefix commands**.

Weights are not allowed with the bootstrap prefix.

fweights, iweights, and pweights are allowed; see [U] **11.1.6 weight**.

See [U] **20 Estimation and postestimation commands** for more capabilities of estimation commands.

Description

asmprobit fits multinomial probit (MNP) models by using maximum simulated likelihood (MSL) implemented by the Geweke–Hajivassiliou–Keane (GHK) algorithm. By estimating the variance–covariance parameters of the latent-variable errors, the model allows you to relax the independence of irrelevant alternatives (IIA) property that is characteristic of the multinomial logistic model.

asmprobit requires multiple observations for each case (decision), where each observation represents an alternative that may be chosen. The cases are identified by the variable specified in the option case(), whereas the alternatives are identified by the variable specified in the option alternative(). The outcome (chosen alternative) is identified by a value of 1 in *depvar*, with 0 indicating the alternatives that were not chosen; only one alternative may be chosen for each case.

asmprobit allows two types of independent variables: alternative-specific variables and case-specific variables. Alternative-specific variables vary across both cases and alternatives and are specified in *indepvars*. Case-specific variables vary only across cases and are specified in the option casevars().

Options

> ⌐ Model ⌐

case(*varname*) specifies the variable that identifies each case. This variable identifies the individuals or entities making a choice. case() is required.

alternatives(*varname*) specifies the variable that identifies the alternatives available for each case. The number of alternatives can vary with each case; the maximum number of alternatives is 20. alternatives() is required.

casevars(*varlist*) specifies the case-specific variables that are constant for each case(). If there are a maximum of J alternatives, there will be $J - 1$ sets of coefficients associated with casevars().

constraints(*constraints*), collinear; see [R] **estimation options**.

> ⌐ Model 2 ⌐

correlation(*correlation*) specifies the correlation structure of the latent-variable errors. The correlation() option implies the structural covariance parameterization.

> correlation(unstructured) is the most general and has $J(J - 3)/2 + 1$ unique correlation parameters.

> correlation(exchangeable) provides for one correlation coefficient common to all latent variables, except the latent variable associated with the basealternative().

> correlation(independent) assumes that all correlations are zero.

> correlation(pattern *matname*) and correlation(fixed *matname*) give you more flexibility in defining the correlation structure. See *Variance structures* later in this entry for more information.

stddev(*stddev*) specifies the variance structure of the latent-variable errors. The stddev() option implies the structural covariance parameterization.

> stddev(heteroskedastic) is the most general and has $J - 2$ estimable parameters. The standard deviations of the latent-variable errors for the alternatives specified in basealternative() and scalealternative() are fixed to one.

> stddev(homoskedastic) constrains all the standard deviations to equal one.

> stddev(pattern *matname*) and stddev(fixed *matname*) give you added flexibility in defining the standard deviation parameters. See *Variance structures* later in this entry for more information.

structural requests the $J \times J$ structural covariance parameterization instead of the default $J-1 \times J-1$ differenced covariance parameterization (the covariance of the latent errors differenced with that of the base alternative). The differenced covariance parameterization will achieve the same maximum simulated likelihood regardless of the choice of basealternative() and scalealternative(). On the other hand, the structural covariance parameterization imposes more normalizations that may bound the model away from the maximum likelihood estimates and thus prevent convergence with some datasets or choices of basealternative() and scalealternative().

noconstant suppresses the $J - 1$ alternative-specific constant terms.

basealternative(*# | lbl | str*) specifies the alternative used to normalize the latent-variable location (also referred to as the level of utility). The base alternative may be specified as a number, label, or string. The standard deviation for the latent-variable error associated with the base alternative is fixed to one, and its correlations with all other latent-variable errors are set to zero. The default is the first alternative when sorted. If a fixed or pattern matrix is given in the stddev() and correlation() options, the basealternative() will be implied by the fixed standard deviations and correlations in the matrix specifications. `basealternative()` cannot be equal to `scalealternative()`.

`scalealternative(#|lbl|str)` specifies the alternative used to normalize the latent-variable scale (also referred to as the scale of utility). The scale alternative may be specified as a number, label, or string. The default is to use the second alternative when sorted. If a `fixed` or `pattern` matrix is given in the `stddev()` option, the `scalealternative()` will be implied by the fixed standard deviations in the matrix specification. `scalealternative()` cannot be equal to `basealternative()`.

If a `fixed` or `pattern` matrix is given for the `stddev()` option, the base alternative and scale alternative are implied by the standard deviations and correlations in the matrix specifications, and they need not be specified in the `basealternative()` and `scalealternative()` options.

`altwise` specifies that alternative-wise deletion be used when marking out observations due to missing values in your variables. The default is to use casewise deletion; that is, the entire group of observations making up a case is deleted if any missing values are encountered. This option does not apply to observations that are marked out by the `if` or `in` qualifier or the `by` prefix.

___
⌐ SE/Robust ⌐
___

`vce(vcetype)` specifies the type of standard error reported, which includes types that are derived from asymptotic theory, that are robust to some kinds of misspecification, that allow for intragroup correlation, and that use bootstrap or jackknife methods; see [R] *vce_option*.

If specifying `vce(bootstrap)` or `vce(jackknife)`, you must also specify `basealternative()` and `scalealternative()`.

___
⌐ Reporting ⌐
___

`level(#)`; see [R] **estimation options**.

`notransform` prevents retransforming the Cholesky-factored variance–covariance estimates to the correlation and standard deviation metric.

This option has no effect if `structural` is not specified because the default differenced variance–covariance estimates have no interesting interpretation as correlations and standard deviations. `notransform` also has no effect if the `correlation()` and `stddev()` options are specified with anything other than their default values. Here it is generally not possible to factor the variance–covariance matrix, so optimization is already performed using the standard deviation and correlation representations.

___
⌐ Int options ⌐
___

`intmethod(hammersley|halton|random)` specifies the method of generating the point sets used in the quasi–Monte Carlo integration of the multivariate normal density. `intmethod(hammersley)`, the default, uses the Hammersley sequence; `intmethod(halton)` uses the Halton sequence; and `intmethod(random)` uses a sequence of uniform random numbers.

`intpoints(#)` specifies the number of points to use in the quasi–Monte Carlo integration. If this option is not specified, the number of points is $50 \times J$ if `intmethod(hammersley)` or `intmethod(halton)` is used and $100 \times J$ if `intmethod(random)` is used. Larger values of `intpoints()` provide better approximations of the log likelihood, but at the cost of added computation time.

`intburn(#)` specifies where in the Hammersley or Halton sequence to start, which helps reduce the correlation between the sequences of each dimension. The default is 0. This option may not be specified with `intmethod(random)`.

intseed(*code* | *#*) specifies the seed to use for generating the uniform pseudorandom sequence. This option may be specified only with intmethod(random). *code* refers to a string that records the state of the random-number generator uniform(); see [D] **generate**. An integer value *#* may be used also. The default is to use the current seed value from Stata's uniform random-number generator, which can be obtained from c(seed).

antithetics specifies that antithetic draws be used. The antithetic draw for the $J - 1$ vector uniform-random variables, $\mathbf{x}$, is $1 - \mathbf{x}$.

initbhhh(*#*) specifies that the BHHH algorithm be used for the initial *#* optimization steps. This option is the only way to use the BHHH algorithm along with other optimization techniques. The algorithm switching feature of ml's technique() option cannot include bhhh.

⌐ Max options ⌐

maximize_options: difficult, technique(*algorithm_spec*), iterate(*#*), [no]log, trace, gradient, showstep, hessian, shownrtolerance, tolerance(*#*), ltolerance(*#*), gtolerance(*#*), nrtolerance(*#*), nonrtolerance, from(*init_specs*); see [R] **maximize**.

The following options may be particularly useful in obtaining convergence with asmprobit: difficult, technique(*algorithm_spec*), nrtolerance(*#*), nonrtolerance, and from(*init_specs*).

If technique() contains more than one algorithm specification, bhhh cannot be one of them. To use the BHHH algorithm with another algorithm, use the initbhhh() option and specify the other algorithm in technique().

Setting the optimization type to technique(bhhh) resets the default *vcetype* to vce(opg).

Remarks

Remarks are presented under the following headings:

> *Introduction*
> *Variance structures*

Introduction

The multinomial probit (MNP) model is used with discrete dependent variables that take on more than two outcomes that do not have a natural ordering. The stochastic error terms are assumed to have a multivariate normal distribution that is heteroskedastic and correlated. Say that you have a set of J unordered alternatives that are modeled by a regression of both case-specific and alternative-specific covariates. A "case" refers to the information on one decision maker. Underlying the model is the set of J latent variables (utilities),

$$\eta_{ij} = \mathbf{x}_{ij}\boldsymbol{\beta} + \mathbf{z}_i\boldsymbol{\alpha}_j + \xi_{ij} \tag{1}$$

where i denotes cases and j denotes alternatives. $\mathbf{x}_{ij}$ is a $1 \times p$ vector of alternative-specific variables, $\boldsymbol{\beta}$ is a $p \times 1$ vector of parameters, $\mathbf{z}_i$ is a $1 \times q$ vector of case-specific variables, $\boldsymbol{\alpha}_j$ is a $q \times 1$ vector of parameters for the jth alternative, and $\xi_i = (\xi_{i1}, \ldots, \xi_{iJ})$ is distributed multivariate normal with mean zero and covariance matrix $\boldsymbol{\Omega}$. The decision maker selects the alternative whose latent variable is highest.

Because the MNP model allows for a general covariance structure in ξ_{ij}, it does not impose the independence of irrelevant alternatives (IIA) property inherent in multinomial logistic and conditional logistic models. That is, the MNP model permits the odds of choosing one alternative over another to depend on the remaining alternatives. For example, consider the choice of travel mode between two cities: air, train, bus, or car, as a function of the travel mode cost, travel time (alternative-specific variables), and an individual's income (a case-specific variable). The odds of choosing air travel over a bus may not be independent of the train alternative since both bus and train travel are public ground transportation. That is, the probability of choosing air travel is $\Pr(\eta_{\text{air}} > \eta_{\text{bus}}, \eta_{\text{air}} > \eta_{\text{train}}, \eta_{\text{air}} > \eta_{\text{car}})$, and the two events $\eta_{\text{air}} > \eta_{\text{bus}}$ and $\eta_{\text{air}} > \eta_{\text{train}}$ may be correlated.

An alternative to MNP that will allow a nested correlation structure in ξ_{ij} is the nested logit model (see [R] **nlogit**).

The added flexibility of the MNP model does impose a significant computation burden because of the need to evaluate probabilities from the multivariate normal distribution. These probabilities are evaluated using simulation techniques because a closed-form solution does not exist. See *Methods and Formulas* for more information.

Not all the J sets of regression coefficients $\boldsymbol{\alpha}_j$ are identifiable, nor are all $J(J+1)/2$ elements of the variance–covariance matrix $\boldsymbol{\Omega}$. As described by Train (2003), the model requires normalization since both the location (level) and scale of the latent variable are irrelevant. Increasing the latent variables by a constant does not change which η_{ij} is the maximum for decision maker i, nor does multiplying them by a constant. To normalize location, we choose an alternative, indexed by k, say, and take the difference between the latent variable k and the $J - 1$ others,

$$
\begin{aligned}
v_{ijk} &= \eta_{ij} - \eta_{ik} \\
&= (\mathbf{x}_{ij} - \mathbf{x}_{ik})\boldsymbol{\beta} + \mathbf{z}_i(\boldsymbol{\alpha}_j - \boldsymbol{\alpha}_k) + \xi_{ij} - \xi_{ik} \\
&= \boldsymbol{\delta}_{ij'}\boldsymbol{\beta} + \mathbf{z}_i\boldsymbol{\gamma}_{j'} + \epsilon_{ij'} \\
&= \lambda_{ij'} + \epsilon_{ij'}
\end{aligned}
\tag{2}
$$

where $j' = j$ if $j < k$ and $j' = j - 1$ if $j > k$, so that $j' = 1, \ldots, J - 1$. One can now work with the $(J - 1) \times (J - 1)$ covariance matrix $\boldsymbol{\Sigma}_{(k)}$ for $\boldsymbol{\epsilon}'_i = (\epsilon_{i1}, \ldots, \epsilon_{i,J-1})$. The kth alternative here is the basealternative() in asmprobit. From (2), the probability that decision maker i chooses alternative k, for example, is

$$
\begin{aligned}
\Pr(i \text{ chooses } k) &= \Pr(v_{i1k} \leq 0, \ldots, v_{i,J-1,k} \leq 0) \\
&= \Pr(\epsilon_{i1} \leq -\lambda_{i1}, \ldots, \epsilon_{i,J-1} \leq -\lambda_{i,J-1})
\end{aligned}
$$

To normalize for scale, one of the diagonal elements of $\boldsymbol{\Sigma}_{(k)}$ must be fixed to a constant. In asmprobit this is the error variance for the alternative specified by scalealternative(). Thus there are a total of at most $J(J-1)/2 - 1$ identifiable variance–covariance parameters. See *Variance structures* below for more on this issue.

In fact, the model is slightly more general in that not all cases need to have faced all J alternatives. The model allows for situations in which some cases chose among all possible alternatives, whereas other cases were given a choice among a subset of them, and perhaps other cases were given a choice among a different subset. The number of observations for each case is equal to the number of alternatives faced.

The MNP model is often motivated using a random-utility consumer-choice framework. Equation (1) represents the utility that consumer i receives from good j. The consumer purchases the good for which the utility is highest. Because utility is ordinal, all that matters is the ranking of the utilities from the alternatives. Thus one must normalize for location and scale.

▷ Example 1

Application of MNP models is common in the analysis of transportation data. Greene (2003, sec. 21.7.8) uses travel-mode choice data between Sydney and Melbourne to demonstrate estimating parameters of various discrete-choice models. The data contain information on 210 individuals' choices of travel mode. The four alternatives are air, train, bus, and car, with indices 1, 2, 3, and 4, respectively. One alternative-specific variable is travelcost, a measure of generalized cost of travel that is equal to the sum of in-vehicle cost and a wagelike measure times the amount of time spent traveling. A second alternative-specific variable is the terminal time, termtime, which is zero for car transportation. Household income, income, is a case-specific variable.

```
. use http://www.stata-press.com/data/r10/travel
. list id mode choice travelcost termtime income in 1/12, sepby(id)
```

	id	mode	choice	travel~t	termtime	income
1.	1	air	0	70	69	35
2.	1	train	0	71	34	35
3.	1	bus	0	70	35	35
4.	1	car	1	30	0	35
5.	2	air	0	68	64	30
6.	2	train	0	84	44	30
7.	2	bus	0	85	53	30
8.	2	car	1	50	0	30
9.	3	air	0	129	69	40
10.	3	train	0	195	34	40
11.	3	bus	0	149	35	40
12.	3	car	1	101	0	40

The model of travel choice is

$$\eta_{ij} = \beta_1 \text{travelcost}_{ij} + \beta_2 \text{termtime}_{ij} + \alpha_{1j}\text{income}_i + \alpha_{0j} + \xi_{ij}$$

The alternatives can be grouped as air and ground travel. With this in mind, we set the air alternative to be the basealternative() and choose train as the scaling alternative. Since these are the first and second alternatives in variable mode, they are also the defaults.

(*Continued on next page*)

```
. asmprobit choice travelcost termtime, casevars(income) case(id) alternatives(mode)
```

(*output omitted*)

```
Alternative-specific multinomial probit        Number of obs      =          840
Case variable: id                              Number of cases    =          210

Alternative variable: mode                     Alts per case: min =            4
                                                              avg =          4.0
                                                              max =            4

Integration sequence:        Hammersley
Integration points:                 200        Wald chi2(5)       =        32.06
Log simulated-likelihood = -190.09419          Prob > chi2        =       0.0000
```

choice	Coef.	Std. Err.	z	P>|z|	[95% Conf. Interval]	
mode						
travelcost	-.0097707	.0027835	-3.51	0.000	-.0152261	-.0043152
termtime	-.0377034	.0094046	-4.01	0.000	-.0561361	-.0192708
air	(base alternative)					
train						
income	-.0291886	.0089232	-3.27	0.001	-.0466778	-.0116995
_cons	.5615485	.394619	1.42	0.155	-.2118906	1.334988
bus						
income	-.0127473	.0079269	-1.61	0.108	-.0282839	.0027892
_cons	-.0572738	.4791635	-0.12	0.905	-.9964169	.8818693
car						
income	-.0049067	.0077481	-0.63	0.527	-.0200927	.0102792
_cons	-1.833159	.81842	-2.24	0.025	-3.437233	-.2290856
/lnl2_2	-.5499745	.3903368	-1.41	0.159	-1.315021	.2150717
/lnl3_3	-.6008993	.3354232	-1.79	0.073	-1.258317	.056518
/l2_1	1.131589	.2125186	5.32	0.000	.7150604	1.548118
/l3_1	.9720683	.2352248	4.13	0.000	.5110362	1.4331
/l3_2	.5196988	.2860692	1.82	0.069	-.0409865	1.080384

```
(mode=air is the alternative normalizing location)
(mode=train is the alternative normalizing scale)

. estimates store full
```

By default, the differenced covariance parameterization is used, so the covariance matrix for this model is 3×3. There are two free variances to estimate and three correlations. To help ensure that the covariance matrix remains positive definite, asmprobit uses the square root transformation, where it optimizes on the Cholesky-factored variance–covariance. To ensure that the diagonal elements of the Cholesky estimates remain positive, we use the log transformation. The estimates labeled /lnl2_2 and /lnl3_3 in the coefficient table are the log-transformed diagonal elements of the Cholesky matrix. The estimates labeled /l2_1, /l3_1, and /l3_2 are the off-diagonal entries for elements $(2,1)$, $(3,1)$, and $(3,2)$ of the Cholesky matrix.

Although the transformed parameters of the differenced covariance parameterization are difficult to interpret, you can view them untransformed by using the estat command. Typing

. estat correlation

	train	bus	car
train	1.000		
bus	0.891	1.000	
car	0.790	0.895	1.000

Note: correlations are for alternatives differenced with air

gives the correlations, and typing

. estat covariance

	train	bus	car
train	2		
bus	1.600309	1.613382	
car	1.374712	1.39983	1.515656

Note: covariances are for alternatives differenced with air

gives the (co)variances.

◁

Variance structures

The matrix Ω has $J(J+1)/2$ distinct elements since it is symmetric. Selecting a base alternative, normalizing its error variance to one, and constraining the correlations between its error and the other errors reduces the number of estimable parameters by J. Moreover, selecting a scale alternative and normalizing its error variance to one reduces the number by one, as well. Hence, there are at most $m = J(J-1)/2 - 1$ estimable parameters in Ω.

In practice, estimating all m parameters can be difficult, so one must often place more restrictions on the parameters. The asmprobit command provides the option correlation() to specify restrictions on the $J(J-3)/2 + 1$ correlation parameters not already restricted as a result of choosing the base alternatives, and it provides stddev() to specify restrictions on the $J-2$ standard deviations not already restricted as a result of choosing the base and scale alternatives.

When the structural option is used, asmprobit fits the model by assuming that all m parameters can be estimated, which is equivalent to specifying correlation(unstructured) and stddev(heteroskedastic). The unstructured correlation structure means that all $J(J-3)/2 + 1$ of the remaining correlation parameters will be estimated, and the heteroskedastic specification means that all $J-2$ standard deviations will be estimated. With these default settings, the log likelihood is maximized with respect to the Cholesky decomposition of Ω, and then the parameters are transformed to the standard deviation and correlation form.

The correlation(exchangeable) option forces the $J(J-3)/2 + 1$ correlation parameters to be equal, and correlation(independent) forces all the correlations to be zero. Using the stddev(homoskedastic) option forces all J standard deviations to be one. These options may help in obtaining convergence for a model if the default options do not produce satisfactory results. In fact, when fitting a complex model, it may be advantageous to first fit a simple one and then proceed with removing the restrictions one at a time.

Advanced users may wish to specify alternative variance structures of their own choosing, and the next few paragraphs explain how to do so.

correlation(pattern matname) allows you to give the name of a $J \times J$ matrix that identifies a correlation structure. Sequential positive integers starting at 1 are used to identify each correlation parameter: if there are three correlation parameters, they are identified by 1, 2, and 3. The integers can be repeated to indicate that correlations with the same number should be constrained to be equal. A zero or a missing value (.) indicates that the correlation is to be set to zero. asmprobit considers only the elements of the matrix *below* the main diagonal.

Suppose that you have a model with four alternatives, numbered 1–4, and alternative 1 is the base. The unstructured and exchangeable correlation structures identified in the 4×4 lower triangular matrices are

$$
\begin{array}{cc}
\text{unstructured} & \text{exchangeable} \\
\begin{array}{c} \\ 1 \\ 2 \\ 3 \\ 4 \end{array}
\begin{array}{cccc} 1 & 2 & 3 & 4 \\ \end{array}
\left(\begin{array}{cccc}
\cdot & & & \\
0 & \cdot & & \\
0 & 1 & \cdot & \\
0 & 2 & 3 & \cdot
\end{array}\right)
&
\begin{array}{c} \\ 1 \\ 2 \\ 3 \\ 4 \end{array}
\begin{array}{cccc} 1 & 2 & 3 & 4 \\ \end{array}
\left(\begin{array}{cccc}
\cdot & & & \\
0 & \cdot & & \\
0 & 1 & \cdot & \\
0 & 1 & 1 & \cdot
\end{array}\right)
\end{array}
$$

asmprobit labels these correlation structures unstructured and exchangeable, even though the correlations corresponding to the base alternative are set to zero. More formally: these terms are appropriate when considering the $(J-1) \times (J-1)$ submatrix $\Sigma_{(k)}$ defined in the *Introduction* above.

You can also use the correlation(fixed *matname*) option to specify a matrix that specifies fixed and free parameters. Here the free parameters (those that are to be estimated) are identified by a missing value, and nonmissing values represent correlations that are to be taken as given. Below is a correlation structure that would set the correlations of alternative 1 to be 0.5:

$$
\begin{array}{c} \\ 1 \\ 2 \\ 3 \\ 4 \end{array}
\begin{array}{cccc} 1 & 2 & 3 & 4 \\ \end{array}
\left(\begin{array}{cccc}
\cdot & & & \\
0.5 & \cdot & & \\
0.5 & \cdot & \cdot & \\
0.5 & \cdot & \cdot & \cdot
\end{array}\right)
$$

The order of the elements of the pattern or fixed matrices must be the same as the numeric order of the alternative levels.

To specify the structure of the standard deviations—the diagonal elements of Ω—you can use the stddev(pattern *matname*) option, where *matname* is a $1 \times J$ matrix. Sequential positive integers starting at 1 are used to identify each standard deviation parameter. The integers can be repeated to indicate that standard deviations with the same number are to be constrained to be equal. A missing value indicates that the corresponding standard deviation is to be set to one. In the four-alternative example mentioned above, suppose that you wish to set the first and second standard deviations to one and that you wish to constrain the third and fourth standard deviations to be equal; the following pattern matrix will do that:

$$
\begin{array}{cccc} 1 & 2 & 3 & 4 \\ \end{array}
1 \left(\begin{array}{cccc} \cdot & \cdot & 1 & 1 \end{array}\right)
$$

Using the stddev(fixed *matname*) option allows you to identify the fixed and free standard deviations. Fixed standard deviations are entered as positive real numbers, and free parameters are identified with missing values. For example, to constrain the first and second standard deviations to equal one and to allow the third and fourth to be estimated, you would use this fixed matrix:

$$
\begin{array}{cccc} 1 & 2 & 3 & 4 \\ \end{array}
1 \left(\begin{array}{cccc} 1 & 1 & \cdot & \cdot \end{array}\right)
$$

When supplying either the pattern or the fixed matrices, you must ensure that the model is properly scaled. At least two standard deviations must be constant for the model to be scaled. A warning is issued if asmprobit detects that the model is not scaled.

The order of the elements of the `pattern` or `fixed` matrices must be the same as the numeric order of the alternative levels.

▷ Example 2

In example 1, we used the differenced covariance parameterization, the default. We now use the `structural` option to view the $J - 2$ standard deviation estimates and the $(J - 1)(J - 2)/2$ correlation estimates. Here we will fix the standard deviations for the `air` and `train` alternatives to 1 and the correlations between `air` and the rest of the alternatives to 0.

```
. asmprobit choice travelcost termtime, casevars(income) case(id) alternatives(mode)
> structural
```

(*output omitted*)

Alternative-specific multinomial probit	Number of obs	=	840
Case variable: id	Number of cases	=	210

Alternative variable: mode	Alts per case: min =	4
	avg =	4.0
	max =	4

Integration sequence:	Hammersley		
Integration points:	200	Wald chi2(5) =	32.05
Log simulated-likelihood = -190.09418		Prob > chi2 =	0.0000

choice	Coef.	Std. Err.	z	P>|z|	[95% Conf. Interval]	
mode						
travelcost	-.0097703	.0027834	-3.51	0.000	-.0152257	-.0043149
termtime	-.0377103	.0094092	-4.01	0.000	-.056152	-.0192687
air	(base alternative)					
train						
income	-.0291975	.0089246	-3.27	0.001	-.0466895	-.0117055
_cons	.5616448	.3946529	1.42	0.155	-.2118607	1.33515
bus						
income	-.01275	.0079266	-1.61	0.108	-.0282858	.0027858
_cons	-.0571664	.4791996	-0.12	0.905	-.9963803	.8820476
car						
income	-.0049085	.0077486	-0.63	0.526	-.0200955	.0102785
_cons	-1.833444	.8186343	-2.24	0.025	-3.437938	-.22895
/lnsigma3	-.2447428	.4953363	-0.49	0.621	-1.215584	.7260985
/lnsigma4	-.3309429	.6494493	-0.51	0.610	-1.60384	.9419543
/atanhr3_2	1.01193	.3890994	2.60	0.009	.249309	1.774551
/atanhr4_2	.5786576	.3940461	1.47	0.142	-.1936586	1.350974
/atanhr4_3	.8885204	.5600561	1.59	0.113	-.2091693	1.98621
sigma1	1	(base alternative)				
sigma2	1	(scale alternative)				
sigma3	.7829059	.3878017			.2965368	2.067
sigma4	.7182462	.4664645			.2011227	2.564989
rho3_2	.766559	.1604596			.244269	.9441061
rho4_2	.5216891	.2868027			-.1912734	.874283
rho4_3	.7106622	.277205			-.2061713	.9630403

(mode=air is the alternative normalizing location)
(mode=train is the alternative normalizing scale)

When comparing this output to that of example 1, we see that we have achieved the same log likelihood. That is, the structural parameterization using `air` as the base alternative and `train` as the scale alternative applied no restrictions on the model. This will not always be the case. We leave it up to you to try different base and scale alternatives, and you will see that not all the different combinations will achieve the same log likelihood. This is not true for the differenced covariance parameterization: it will always achieve the same log likelihood (and the maximum possible likelihood) regardless of the base and scale alternatives. This is why it is the default parameterization.

For an exercise, we can compute the differenced covariance displayed in example 1 by using the following ado-code.

```
. estat covariance
```

	air	train	bus	car
air	1			
train	0	1		
bus	0	.6001436	.6129416	
car	0	.3747012	.399619	.5158776

```
. return list

matrices:
            r(cov) :  4 x 4

. matrix cov = r(cov)

. matrix M = (1,-1,0,0 \ 1,0,-1,0 \ 1,0,0,-1)

. matrix cov1 = M*cov*M'

. matrix list cov1

symmetric cov1[3,3]
           r1          r2          r3
r1          2
r2  1.6001436   1.6129416
r3  1.3747012    1.399619   1.5158776
```

The slight difference in the regression coefficients between the example 1 and example 2 coefficient tables reflects the accuracy of the [M-5] **ghk( )** algorithm using 200 points from the Hammersley sequence.

We now fit the model using the exchangeable correlation matrix and compare the models with a likelihood-ratio test.

```
. asmprobit choice travelcost termtime, casevars(income) case(id) alternatives(mode)
> correlation(exchangeable)
```
  (*output omitted*)

Alternative-specific multinomial probit	Number of obs	=	840
Case variable: id	Number of cases	=	210

Alternative variable: mode	Alts per case: min =	4
	avg =	4.0
	max =	4

Integration sequence:	Hammersley		
Integration points:	200	Wald chi2(5) =	53.60
Log simulated-likelihood =	-190.4679	Prob > chi2 =	0.0000

choice	Coef.	Std. Err.	z	P>|z|	[95% Conf. Interval]	
mode						
travelcost	-.0084636	.0020452	-4.14	0.000	-.012472	-.0044551
termtime	-.0345394	.0072812	-4.74	0.000	-.0488103	-.0202684
air	(base alternative)					
train						
income	-.0290357	.0083226	-3.49	0.000	-.0453477	-.0127237
_cons	.5517445	.3719913	1.48	0.138	-.177345	1.280834
bus						
income	-.0132562	.0074133	-1.79	0.074	-.0277859	.0012735
_cons	-.0052517	.4337932	-0.01	0.990	-.8554708	.8449673
car						
income	-.0060878	.006638	-0.92	0.359	-.0190981	.0069224
_cons	-1.565918	.6633007	-2.36	0.018	-2.865964	-.265873
/lnsigmaP1	-.3557589	.1972809	-1.80	0.071	-.7424222	.0309045
/lnsigmaP2	-1.308596	.8872957	-1.47	0.140	-3.047663	.4304719
/atanhrP1	1.116589	.3765488	2.97	0.003	.3785667	1.854611
sigma1	1	(base alternative)				
sigma2	1	(scale alternative)				
sigma3	.7006416	.1382232			.4759596	1.031387
sigma4	.2701992	.2397466			.0474697	1.537983
rho3_2	.8063791	.131699			.3614621	.9521783
rho4_2	.8063791	.131699			.3614621	.9521783
rho4_3	.8063791	.131699			.3614621	.9521783

```
(mode=air is the alternative normalizing location)
(mode=train is the alternative normalizing scale)

. lrtest full .
```

Likelihood-ratio test	LR chi2(2) =	0.75
(Assumption: . nested in full)	Prob > chi2 =	0.6882

The likelihood-ratio test suggests that a common correlation is a plausible hypothesis, but this could be an artifact of the small sample size. The labeling of the standard deviation and correlation estimates has changed from /lnsigma and /atanhr, in the previous example, to /lnsigmaP and /atanhrP. The "P" identifies the parameter's index in the pattern matrices used by asmprobit. The pattern matrices are saved in e(stdpattern) and e(corpattern). ◁

❏ Technical Note

Another way to fit the model with the exchangeable correlation structure in example 2 is to use the `constraint` command to define the constraints on the rho parameters manually and then apply those.

```
. constraint 1 [atanhr3_2]_cons = [atanhr4_2]_cons
. constraint 2 [atanhr3_2]_cons = [atanhr4_3]_cons
. asmprobit choice travelcost termtime, casevars(income) case(id)
> alternatives(mode) constraints(1 2) structural
```

With this method, however, we must keep track of what parameterization of the rhos is used in estimation, and that depends on the options specified.

❏

▷ Example 3

In the last example, we used the `correlation(exchangeable)` option, reducing the number of correlation parameters from three to one. We can explore a two–correlation parameter model by specifying a `pattern` matrix in the `correlation()` option. Suppose that we wish to have the correlation between train and bus be equal to the correlation between bus and car and to have the standard deviations for the bus and car equations be equal. We will use air as the base category and train as the scale category.

```
. matrix define corpat = J(4, 4, .)
. matrix corpat[3,2] = 1
. matrix corpat[4,3] = 1
. matrix corpat[4,2] = 2
. matrix define stdpat = J(1, 4, .)
. matrix stdpat[1,3] = 1
. matrix stdpat[1,4] = 1
. asmprobit choice travelcost termtime, casevars(income) case(id) alternatives(mode)
> correlation(pattern corpat) stddev(pattern stdpat)
Iteration 0:    log simulated-likelihood = -201.33896
Iteration 1:    log simulated-likelihood = -201.00457  (backed up)
Iteration 2:    log simulated-likelihood = -200.80208  (backed up)
Iteration 3:    log simulated-likelihood = -200.79758  (backed up)
Iteration 4:    log simulated-likelihood = -200.55655  (backed up)
Iteration 5:    log simulated-likelihood =  -200.5421  (backed up)
Iteration 6:    log simulated-likelihood = -196.24925
  (output omitted )
Iteration 20:   log simulated-likelihood = -190.12874
Iteration 21:   log simulated-likelihood = -190.12871
Iteration 22:   log simulated-likelihood = -190.12871
```

Alternative-specific multinomial probit	Number of obs	=	840
Case variable: id	Number of cases	=	210

Alternative variable: mode

	Alts per case: min =	4
	avg =	4.0
	max =	4

Integration sequence:	Hammersley			
Integration points:	200	Wald chi2(5)	=	41.67
Log simulated-likelihood = -190.12871		Prob > chi2	=	0.0000

choice	Coef.	Std. Err.	z	P>|z|	[95% Conf. Interval]	
mode						
travelcost	-.0100335	.0026203	-3.83	0.000	-.0151692	-.0048979
termtime	-.0385731	.008608	-4.48	0.000	-.0554445	-.0217018
air	(base alternative)					
train						
income	-.029271	.0089739	-3.26	0.001	-.0468595	-.0116824
_cons	.56528	.4008037	1.41	0.158	-.2202809	1.350841
bus						
income	-.0124658	.0080043	-1.56	0.119	-.0281539	.0032223
_cons	-.0741685	.4763422	-0.16	0.876	-1.007782	.859445
car						
income	-.0046905	.0079934	-0.59	0.557	-.0203573	.0109763
_cons	-1.897931	.7912106	-2.40	0.016	-3.448675	-.3471867
/lnsigmaP1	-.197697	.2751269	-0.72	0.472	-.7369359	.3415418
/atanhrP1	.9704403	.3286981	2.95	0.003	.3262038	1.614677
/atanhrP2	.5830923	.3690419	1.58	0.114	-.1402165	1.306401
sigma1	1	(base alternative)				
sigma2	1	(scale alternative)				
sigma3	.8206185	.2257742			.4785781	1.407115
sigma4	.8206185	.2257742			.4785781	1.407115
rho3_2	.7488977	.1443485			.3151056	.9238482
rho4_2	.5249094	.2673598			-.1393048	.863362
rho4_3	.7488977	.1443485			.3151056	.9238482

(mode=air is the alternative normalizing location)
(mode=train is the alternative normalizing scale)

In the call to `asmprobit`, we did not need to specify the `basealternative()` and `scalealternative()` options because they are implied by the specifications of the pattern matrices.

◁

❑ Technical Note

If you experience convergence problems, try increasing `intpoints()`, specifying `antithetics`, specifying `technique(nr)` with `difficult`, or specifying a switching algorithm in the `technique()` option. As a last resort, you can use the `nrtolerance()` and `shownrtolerance` options. Changing the base and scale alternative in the model specification can also affect convergence if the `structural` option is used.

Because simulation methods are used to obtain multivariate normal probabilities, the estimates obtained have a limited degree of precision. Moreover, the solutions are particularly sensitive to the starting values used. Experimenting with different starting values may help in obtaining convergence, and doing so is a good way to verify previous results.

If you wish to use the BHHH algorithm along with another maximization algorithm, you must specify the `initbhhh(#)` option, where `#` is the number of BHHH iterations to use before switching to the algorithm specified in `technique()`. The BHHH algorithm uses an outer-product-of-gradients approximation for the Hessian, and `asmprobit` must perform the gradient calculations differently than for the other algorithms.

❑

❑ Technical Note

If there are no alternative-specific variables in your model, the variance–covariance matrix parameters are not identifiable. For such a model to converge, you would therefore need to use `correlation(independent)` and `stddev(homoskedastic)`. A better alternative is to use `mprobit`, which is geared specifically toward models with only case-specific variables. See [R] **mprobit**.

❑

Saved Results

`asmprobit` saves the following in `e()`:

Scalars

e(N)	number of observations
e(N_case)	number of cases
e(k)	number of parameters
e(k_alt)	number of alternatives
e(k_indvars)	number of alternative-specific variables
e(k_casevars)	number of case-specific variables
e(k_sigma)	number of variance estimates
e(k_rho)	number of correlation estimates
e(k_eq)	number of equations in e(b)
e(k_eq_model)	number of equations in model Wald test
e(N_clust)	number of clusters
e(const)	constant indicator
e(i_base)	base alternative index
e(i_scale)	scale alternative index
e(mc_points)	number of Monte Carlo replications
e(mc_burn)	starting sequence index
e(mc_antithetics)	antithetics indicator
e(df_m)	model degrees of freedom
e(ll)	log simulated-likelihood
e(chi2)	χ^2
e(p)	significance
e(fullcov)	unstructured covariance indicator
e(structcov)	1 if structured covariance; 0 otherwise
e(cholesky)	Cholesky-factored covariance indicator
e(alt_min)	minimum number of alternatives
e(alt_max)	maximum number of alternatives
e(alt_avg)	average number of alternatives
e(rank)	rank of e(V)
e(ic)	number of iterations
e(rc)	return code
e(converged)	1 if converged, 0 otherwise

Macros

e(cmd)	asmprobit
e(cmdline)	command as typed
e(depvar)	name of dependent variable
e(indvars)	alternative-specific independent variable
e(casevars)	case-specific variables
e(case)	variable defining cases
e(altvar)	variable defining alternatives
e(alteqs)	alternative equation names
e(alt#)	alternative labels
e(wtype)	weight type
e(wexp)	weight expression
e(title)	title in estimation output
e(clustvar)	name of cluster variable
e(correlation)	correlation structure
e(stddev)	variance structure
e(chi2type)	Wald, type of model χ^2 test
e(vce)	*vcetype* specified in vce()
e(vcetype)	title used to label Std. Err.
e(opt)	type of optimization
e(ml_method)	type of ml method
e(mc_method)	technique used to generate sequences
e(mc_seed)	random-number generator seed
e(user)	name of likelihood-evaluator program
e(technique)	maximization technique
e(crittype)	optimization criterion
e(properties)	b V
e(datasignaturevars)	variables used in calculation of checksum
e(datasignature)	the checksum
e(estat_cmd)	program used to implement estat
e(predict)	program used to implement predict

Matrices

e(b)	coefficient vector
e(stats)	alternative statistics
e(stdpattern)	variance pattern
e(stdfixed)	fixed and free standard deviations
e(altvals)	alternative values
e(altfreq)	alternative frequencies
e(ilog)	iteration log (up to 20 iterations)
e(gradient)	gradient vector
e(V)	variance–covariance matrix of the estimators
e(corpattern)	correlation structure
e(corfixed)	fixed and free correlations

Functions

e(sample)	marks estimation sample

Methods and Formulas

`asmprobit` is implemented as an ado-file.

The simulated maximum likelihood estimates for the MNP are obtained using `ml`; see [R] **ml**. The likelihood evaluator implements the GHK algorithm to approximate the multivariate distribution function (Geweke 1989; Hajivassiliou and McFadden 1998; Keane 1994). The technique is also described in detail by Genz (1992), but Genz describes a more general algorithm where both lower and upper bounds of integration are finite. We briefly describe the GHK simulator and refer you to Bolduc (1999) for the score computations.

As discussed earlier, the latent variables for a J-alternative model are $\eta_{ij} = \mathbf{x}_{ij}\boldsymbol{\beta} + \mathbf{z}_i\boldsymbol{\alpha}_j + \xi_{ij}$, for $j = 1, \ldots, J$, $i = 1, \ldots, n$, and $\boldsymbol{\xi}'_i = (\xi_{i,1}, \ldots, \xi_{i,J}) \sim \text{MVN}(\mathbf{0}, \boldsymbol{\Omega})$. The experimenter observes alternative k for the ith observation if $k = \arg\max(\eta_{ij}, j = 1, \ldots, J)$. Let

$$
\begin{aligned}
v_{ij'} &= \eta_{ij} - \eta_{ik} \\
&= (\mathbf{x}_{ij} - \mathbf{x}_{ik})\boldsymbol{\beta} + \mathbf{z}_i(\boldsymbol{\alpha}_j - \boldsymbol{\alpha}_k) + \xi_{ij} - \xi_{ik} \\
&= \boldsymbol{\delta}_{ij'}\boldsymbol{\beta} + \mathbf{z}_i\boldsymbol{\gamma}_{j'} + \epsilon_{ij'}
\end{aligned}
$$

where $j' = j$ if $j < k$ and $j' = j - 1$ if $j > k$, so that $j' = 1, \ldots, J - 1$. Further, $\epsilon_i = (\epsilon_{i1}, \ldots, \epsilon_{i,J-1}) \sim \text{MVN}(\mathbf{0}, \boldsymbol{\Sigma}_{(k)})$. $\boldsymbol{\Sigma}$ is indexed by k since it depends on the choice made. We denote the deterministic part of the model as $\lambda_{ij'} = \boldsymbol{\delta}_{ij'}\boldsymbol{\beta} + \mathbf{z}_j\boldsymbol{\gamma}_{j'}$, and the probability of this event is

$$
\begin{aligned}
\Pr(y_i = k) &= \Pr(v_{i1} \leq 0, \ldots, v_{i,J-1} \leq 0) \\
&= \Pr(\epsilon_{i1} \leq -\lambda_{i1}, \ldots, \epsilon_{i,J-1} \leq -\lambda_{i,J-1}) \\
&= (2\pi)^{-(J-1)/2} |\boldsymbol{\Sigma}_{(k)}|^{-1/2} \int_{-\infty}^{-\lambda_{i1}} \cdots \int_{-\infty}^{-\lambda_{i,J-1}} \exp\left(-\tfrac{1}{2}\mathbf{z}'\boldsymbol{\Sigma}_{(k)}^{-1}\mathbf{z}\right) d\mathbf{z}
\end{aligned}
\tag{3}
$$

Simulated likelihood

For clarity in the discussion that follows, we drop the index denoting case so that for an arbitrary observation $v' = (v_1, \ldots, v_{J-1})$, $\boldsymbol{\lambda}' = (\lambda_1, \ldots, \lambda_{J-1})$, and $\boldsymbol{\epsilon}' = (\epsilon_1, \ldots, \epsilon_{J-1})$.

The Cholesky-factored variance–covariance, $\boldsymbol{\Sigma} = \mathbf{LL}'$, is lower triangular,

$$
\mathbf{L} = \begin{pmatrix}
l_{11} & 0 & \cdots & 0 \\
l_{21} & l_{22} & \cdots & 0 \\
\vdots & \vdots & & \vdots \\
l_{J-1,1} & l_{J-1,2} & \cdots & l_{J-1,J-1}
\end{pmatrix}
$$

and the correlated latent-variable errors can be expressed as linear functions of uncorrelated normal variates, $\boldsymbol{\epsilon} = \mathbf{L}\boldsymbol{\zeta}$, where $\boldsymbol{\zeta}' = (\zeta_1, \ldots, \zeta_{J-1})$ and $\zeta_j \sim \text{iid } N(0,1)$. We now have $v = \boldsymbol{\lambda} + \mathbf{L}\boldsymbol{\zeta}$ and by defining

$$z_j = \begin{cases} -\dfrac{\lambda_1}{l_{11}} & \text{for } j = 1 \\[2ex] -\dfrac{\lambda_j + \sum_{i=1}^{j-1} l_{ji}\zeta_i}{l_{jj}} & \text{for } j = 2, \ldots, J-1 \end{cases} \qquad (4)$$

we can express the probability statement (3) as the product of conditional probabilities

$$\Pr(y_i = k) = \Pr\left(\zeta_1 \le z_1\right)\Pr\left(\zeta_2 \le z_2 \mid \zeta_1 \le z_1\right)\cdots$$
$$\Pr\left(\zeta_{J-1} \le z_{J-1} \mid \zeta_1 \le z_1, \ldots, \zeta_{J-2} \le z_{J-2}\right)$$

since

$$\Pr(v_1 \le 0) = \Pr(\lambda_1 + l_{11}\zeta_1 \le 0)$$
$$= \Pr\left(\zeta_1 \le -\frac{\lambda_1}{l_{11}}\right)$$
$$\Pr(v_2 \le 0) = \Pr(\lambda_2 + l_{21}\zeta_1 + l_{22}\zeta_2 \le 0)$$
$$= \Pr\left(\zeta_2 \le -\frac{\lambda_2 + l_{21}\zeta_1}{l_{22}} \mid \zeta_1 \le -\frac{\lambda_1}{l_{11}}\right)$$
$$\cdots$$

The Monte Carlo algorithm then must make draws from the truncated standard normal distribution. It does so by generating $J-1$ uniform variates, $\delta_j, j = 1, \ldots, J-1$ and computing

$$\widetilde{\zeta}_j = \begin{cases} \Phi^{-1}\left\{\delta_1 \Phi\left(-\dfrac{\lambda_1}{l_{11}}\right)\right\} & \text{for } j = 1 \\[3ex] \Phi^{-1}\left\{\delta_j \Phi\left(\dfrac{-\lambda_j - \sum_{i=1}^{j-1} l_{ji}\widetilde{\zeta}_i}{l_{jj}}\right)\right\} & \text{for } j = 2, \ldots, J-1 \end{cases}$$

Define $\widetilde{z}_j$ by replacing $\widetilde{\zeta}_i$ for ζ_i in (4) so that the simulated probability for the lth draw is

$$p_l = \prod_{j=1}^{J-1} \Phi(\widetilde{z}_j)$$

To increase accuracy, the bounds of integration, λ_j, are ordered so that the largest integration intervals are on the inside. The rows and columns of the variance–covariance matrix are pivoted accordingly (Genz 1992).

For a more detailed description of the GHK algorithm in Stata, see Gates (2006).

Repeated draws are made, say, N, and the simulated likelihood for the ith case, denoted $\widehat{L}_i$, is computed as

$$\widehat{L}_i = \frac{1}{N}\sum_{l=1}^{N} p_l$$

The overall simulated log-likelihood is $\sum_i \log \widehat{L}_i$.

If the true likelihood is L_i, the error bound on the approximation can be expressed as

$$|\widehat{L}_i - L_i| \leq V(L_i)D_N(\{\delta_i\})$$

where $V(L_i)$ is the total variation of L_i and D_N is the discrepancy, or nonuniformity, of the set of abscissas. For the uniform pseudorandom sequence, δ_i, the discrepancy is of order $O\{(\log \log N/N)^{1/2}\}$. The order of discrepancy can be improved by using quasirandom sequences.

Quasi–Monte Carlo integration is carried out by `asmprobit` by replacing the uniform deviates with either the Halton or the Hammersley sequences. These sequences spread the points more evenly than the uniform random sequence and have a smaller order of discrepancy, $O\left[\{(\log N)^{J-1}\}/N\right]$ and $O\left[\{(\log N)^{J-2}\}/N\right]$, respectively. The Halton sequence of dimension $J-1$ is generated from the first $J-1$ primes, p_k, so that on draw l we have $\mathbf{h}_l = \{r_{p_1}(l), r_{p_2}(l), \ldots, r_{p_{J-1}}(l)\}$, where

$$r_{p_k}(l) = \sum_{j=0}^{q} b_{jk}(l)p_k^{-j-1} \in (0, 1)$$

is the radical inverse function of l with base p_k so that $\sum_{j=0}^{q} b_{jk}(l)p_k^{j} = l$ where $p_k^q \leq l < p_k^{q+1}$ (Fang and Wang 1994).

This function is demonstrated with base $p_3 = 5$ and $l = 33$, which generates $r_5(33)$. Here $q = 2$, $b_{0,3}(33) = 3$, $b_{1,5}(33) = 1$, and $b_{2,5}(33) = 1$, so that $r_5(33) = 3/5 + 1/25 + 1/625$.

The Hammersley sequence uses an evenly spaced set of points with the first $J-2$ components of the Halton sequence

$$\mathbf{h}_l = \left\{\frac{2l-1}{2N}, r_{p_1}(l), r_{p_2}(l), \ldots, r_{p_{J-2}}(l)\right\}$$

for $l = 1, \ldots, N$.

For a more detailed description of the Halton and Hammersley sequences, see Drukker and Gates (2006).

Computations for the derivatives of the simulated likelihood are taken from Bolduc (1999). Bolduc gives the analytical first-order derivatives for the log of the simulated likelihood with respect to the regression coefficients and the parameters of the Cholesky-factored variance–covariance matrix. `asmprobit` uses these analytical first-order derivatives and numerical second-order derivatives.

References

Bolduc, D. 1999. A practical technique to estimate multinomial probit models in transportation. *Transportation Research Part B* 33: 63–79.

Bunch, D. S. 1991. Estimability of the multinomial probit model. *Transportation Research Part B* 25: 1–12.

Cappellari, L., and S. P. Jenkins. 2003. Multivariate probit regression using simulated maximum likelihood. *Stata Journal* 3: 278–294.

Drukker, D. M., and R. Gates. 2006. Generating Halton sequences using Mata. *Stata Journal* 6: 214–228.

Fang, K. T., and Y. Wang. 1994. *Number-Theoretic Methods in Statistics*. London: Chapman & Hall.

Gates, R. 2006. A Mata Geweke–Hajivassiliou–Keane multivariate normal simulator. *Stata Journal* 6: 190–213.

Genz, A. 1992. Numerical computation of multivariate normal probabilities. *Journal of Computational and Graphical Statistics* 1: 141–149.

Geweke, J. 1989. Bayesian inference in econometric models using Monte Carlo integration. *Econometrica* 57: 1317–1339.

Geweke, J., and M. Keane. 2001. Computationally intensive methods for integration in econometrics. In *Handbook of Econometrics*, ed. J. Heckman and E. Leamer, vol. 5, 3463–3568. Amsterdam: North–Holland.

Greene, W. H. 2003. *Econometric Analysis.* 5th ed. Upper Saddle River, NJ: Prentice Hall.

Haan, P., and A. Uhlendorff. 2006. Estimation of multinomial logit models with unobserved heterogeneity using maximum simulated likelihood. *Stata Journal* 6: 229–245.

Hajivassiliou, V., and D. McFadden. 1998. The method of simulated scores for the estimation of LDV models. *Econometrica* 66: 863–896.

Keane, M. 1994. A computationally practical simulation estimator for panel data. *Review of Economics and Statistics* 76: 648–672.

Train, K. E. 2003. *Discrete Choice Methods with Simulation.* Cambridge: Cambridge University Press.

Also See

[R] **asmprobit postestimation** — Postestimation tools for asmprobit

[R] **asclogit** — Alternative-specific conditional logit (McFadden's choice) model

[R] **asroprobit** — Alternative-specific rank-ordered probit regression

[R] **mprobit** — Multinomial probit regression

[R] **mlogit** — Multinomial (polytomous) logistic regression

[U] **20 Estimation and postestimation commands**

Title

> **asmprobit postestimation** — Postestimation tools for asmprobit

Description

The following postestimation commands are of special interest after `asmprobit`:

command	description
estat alternatives	alternative summary statistics
estat covariance	variance–covariance matrix of the alternatives
estat correlation	correlation matrix of the alternatives
estat mfx	marginal effects

For information about these commands, see below.

The following standard postestimation commands are also available:

command	description
estat	AIC, BIC, VCE, and estimation sample summary
estimates	cataloging estimation results
lincom	point estimates, standard errors, testing, and inference for linear combinations of coefficients
lrtest	likelihood-ratio test
nlcom	point estimates, standard errors, testing, and inference for nonlinear combinations of coefficients
predict	predicted probabilities, estimated linear predictor and its standard error
predictnl	point estimates, standard errors, testing, and inference for generalized predictions
test	Wald tests for simple and composite linear hypotheses
testnl	Wald tests of nonlinear hypotheses

See the corresponding entries in the *Stata Base Reference Manual* for details.

Special-interest postestimation commands

`estat alternatives` displays summary statistics about the alternatives in the estimation sample and provides a mapping between the index numbers that label the covariance parameters of the model and their associated values and labels for the alternative variable.

`estat covariance` computes the estimated variance–covariance matrix for the alternatives. The estimates are displayed, and the variance–covariance matrix is stored in `r(cov)`.

`estat correlation` computes the estimated correlation matrix for the alternatives. The estimates are displayed, and the correlation matrix is stored in `r(cor)`.

`estat mfx` computes the simulated probability marginal effects.

Syntax for predict

predict $\left[type \right]$ *newvar* $\left[if \right]$ $\left[in \right]$ $\left[, statistic \text{ altwise} \right]$

predict $\left[type \right]$ $\left\{ stub* | newvarlist \right\}$ $\left[if \right]$ $\left[in \right]$, <u>sc</u>ores

statistic	description
Main	
<u>pr</u>	probability alternative is chosen; the default
xb	linear prediction
stdp	standard error of the linear prediction

These statistics are available both in and out of sample; type predict ... if e(sample) ... if wanted only for the estimation sample.

Options for predict

    ⌐ Main ⌐

pr, the default, calculates the probability that alternative j is chosen in case i.

xb calculates the linear prediction $\mathbf{x}_{ij}\boldsymbol{\beta} + \mathbf{z}_i\boldsymbol{\alpha}_j$ for alternative j and case i.

stdp calculates the standard error of the linear predictor.

altwise specifies that alternative-wise deletion be used when marking out observations due to missing values in your variables. The default is to use casewise deletion. Options xb and stdp always use alternative-wise deletion.

scores calculates the scores for each coefficient in e(b). This option requires a new variable list of length equal to the number of columns in e(b). Otherwise, use the *stub** option to have predict generate enumerated variables with prefix *stub*.

Syntax for estat alternatives

estat <u>alternatives</u>

Syntax for estat covariance

estat <u>cov</u>ariance $\left[, \text{ } \underline{\text{for}}\text{mat}(string) \text{ } \underline{\text{bor}}\text{der}(string) \text{ left}(\#) \right]$

Options for estat covariance

format(*string*) sets the matrix display format. The default is format(%9.0g).

border(*string*) sets the matrix display border style. The default is border(all). See [P] **matlist**.

left(*#*) sets the matrix display left indent. The default is left(2). See [P] **matlist**.

Syntax for estat correlation

> estat correlation [, format(*string*) border(*string*) left(*#*)]

Options for estat correlation

format(*string*) sets the matrix display format. The default is format(%6.3f).

border(*string*) sets the matrix display border style. The default is border(all). See [P] **matlist**.

left(*#*) sets the matrix display left indent. The default is left(2). See [P] **matlist**.

Syntax for estat mfx

> estat mfx [*if*] [*in*] [, *options*]

options	description
Main	
varlist(*varlist*)	display marginal effects for *varlist*
at(mean [*atlist*]|median [*atlist*])	calculate marginal effects at these values
Options	
level(*#*)	set confidence interval level; default is level(95)
nodiscrete	treat indicator variables as continuous
noesample	do not restrict calculation of means and medians to the estimation sample
nowght	ignore weights when calculating means and medians

Options for estat mfx

      Main

varlist(*varlist*) specifies the variables for which to display marginal effects. The default is all variables.

at(mean [*atlist*]|median [*atlist*]) specifies the values at which the marginal effects are to be calculated. *atlist* is

> [[*alternative*:*variable* = *#*] [*variable* = *#*] [...]]

The default is to calculate the marginal effects at the means of the independent variables at the estimation sample, at(mean).

After specifying the summary statistic, you can specify a series of specific values for variables. You can specify values for alternative-specific variables by alternative, or you can specify one value for all alternatives. You can specify only one value for case-specific variables. For example, in the travel dataset, income is a case-specific variable, whereas termtime and travelcost are alternative-specific variables. The following would be a legal syntax for estat mfx:

> . estat mfx, at(mean air:termtime=50 travelcost=100 income=60)

When nodiscrete is not specified, at(mean [*atlist*]) or at(median [*atlist*]) has no effect on computing marginal effects for indicator variables, which are calculated as the discrete change in the simulated probability as the indicator variable changes from 0 to 1.

The mean and median computations respect any `if` and `in` qualifiers, so you can restrict the data over which the means or medians are computed. You can even restrict the values to a specific case; e.g.,

```
. estat mfx if case==21
```

`level(#)` specifies the confidence level, as a percentage, for confidence intervals. The default is `level(95)` or as set by `set level`; see [U] **20.7 Specifying the width of confidence intervals**.

`nodiscrete` specifies that indicator variables be treated as continuous variables. An indicator variable is one that takes on the value 0 or 1 in the estimation sample. By default, the discrete change in the simulated probability is computed as the indicator variable changes from 0 to 1.

`noesample` specifies that the whole dataset be considered instead of only those marked in the `e(sample)` defined by the `asmprobit` command.

`nowght` specifies that weights be ignored when calculating the means or medians.

Remarks

Remarks are presented under the following headings:

> *Predicted probabilities*
> *Obtaining estimation statistics*
> *Obtaining marginal effects*

Predicted probabilities

After fitting an alternative-specific multinomial probit model, you can use `predict` to obtain the simulated probabilities that an individual will choose each of the alternatives. When evaluating the multivariate normal probabilities via Monte Carlo simulation, `predict` uses the same method to generate the random sequence of numbers as the previous call to `asmprobit`. For example, if you specified `intmethod(Halton)` when fitting the model, `predict` also uses the Halton sequence.

▷ Example 1

In example 1 of [R] **asmprobit**, we fitted a model of individuals' travel-mode choices. We can obtain the simulated probabilities that an individual chooses each alternative by using `predict`:

```
. use http://www.stata-press.com/data/r10/travel
. asmprobit choice travelcost termtime, casevars(income) case(id) alternatives(mode)
  (output omitted)
. predict prob
(option pr assumed; Pr(mode))
```

(Continued on next page)

```
. list id mode prob choice in 1/12, sepby(id)
```

	id	mode	prob	choice
1.	1	air	.1494373	0
2.	1	train	.3292305	0
3.	1	bus	.1319851	0
4.	1	car	.3898138	1
5.	2	air	.2566063	0
6.	2	train	.2761403	0
7.	2	bus	.0116148	0
8.	2	car	.4556371	1
9.	3	air	.2098737	0
10.	3	train	.1082147	0
11.	3	bus	.1671145	0
12.	3	car	.5147865	1

◁

Obtaining estimation statistics

Once you have fitted a multinomial probit model, you can obtain the estimated variance or correlation matrices for the model alternatives by using the `estat` command.

▷ Example 2

To display the correlations of the errors in the latent-variable equations, we type

```
. estat correlation
```

	train	bus	car
train	1.000		
bus	0.891	1.000	
car	0.790	0.895	1.000

Note: correlations are for alternatives differenced with air

The covariance matrix can be displayed by typing

```
. estat covariance
```

	train	bus	car
train	2		
bus	1.600309	1.613382	
car	1.374712	1.39983	1.515656

Note: covariances are for alternatives differenced with air

◁

Obtaining marginal effects

The marginal effects are computed as the derivative of the simulated probability for an alternative with respect to an independent variable. A table of marginal effects is displayed for each alternative, with the table containing the marginal effect for each case-specific variable and for each alternative for each alternative-specific variable.

By default, the marginal effects are computed at the means of each continuous independent variable over the estimation sample. For indicator variables, the difference in the simulated probability evaluated at 0 and 1 is computed by default. Indicator variables will be treated as continuous variables if the nodiscrete option is used.

▷ Example 3

Continuing with our model from example 1, we obtain the marginal effects for alternatives air, train, bus, and car evaluated at the mean values of each independent variable. Recall that the variables travelcost and termtime are alternative specific, taking on different values for each alternative, so they have a separate marginal effect for each alternative.

```
. estat mfx
```

Pr(choice = air) = .29437379

variable	dp/dx	Std. Err.	z	P>|z|	[95% C.I.]		X
travelcost							
air	-.002689	.000677	-3.97	0.000	-.004015	-.001362	102.65
train	.000901	.000436	2.07	0.039	.000046	.001755	130.2
bus	.000376	.000271	1.39	0.166	-.000156	.000907	115.26
car	.001412	.00051	2.77	0.006	.000412	.002412	95.414
termtime							
air	-.010375	.00271	-3.83	0.000	-.015687	-.005063	61.01
train	.003475	.001638	2.12	0.034	.000265	.006686	35.69
bus	.00145	.001008	1.44	0.150	-.000524	.003425	41.657
car	.005449	.002164	2.52	0.012	.001207	.00969	0
casevars							
income	.00389	.001847	2.11	0.035	.00027	.00751	34.548

Pr(choice = train) = .29535865

variable	dp/dx	Std. Err.	z	P>|z|	[95% C.I.]		X
travelcost							
air	.000899	.000436	2.06	0.039	.000045	.001753	102.65
train	-.004081	.001466	-2.78	0.005	-.006954	-.001209	130.2
bus	.001278	.00063	2.03	0.042	.000043	.002513	115.26
car	.001904	.000887	2.15	0.032	.000166	.003642	95.414
termtime							
air	.003469	.001638	2.12	0.034	.000259	.00668	61.01
train	-.015749	.00247	-6.38	0.000	-.020589	-.010908	35.69
bus	.004932	.001593	3.10	0.002	.00181	.008053	41.657
car	.007348	.002229	3.30	0.001	.00298	.011716	0
casevars							
income	-.009568	.002223	-4.30	0.000	-.013925	-.005212	34.548

```
Pr(choice = bus) = .08879634
```

variable	dp/dx	Std. Err.	z	P>|z|	[95% C.I.]		X
travelcost							
air	.00038	.000274	1.39	0.165	-.000157	.000916	102.65
train	.001279	.00063	2.03	0.042	.000044	.002514	130.2
bus	-.003182	.001174	-2.71	0.007	-.005484	-.00088	115.26
car	.001524	.000675	2.26	0.024	.0002	.002848	95.414
termtime							
air	.001465	.001016	1.44	0.149	-.000527	.003457	61.01
train	.004935	.001591	3.10	0.002	.001817	.008053	35.69
bus	-.01228	.002803	-4.38	0.000	-.017774	-.006786	41.657
car	.00588	.002255	2.61	0.009	.001461	.010299	0
casevars							
income	.000434	.001461	0.30	0.766	-.002429	.003297	34.548

```
Pr(choice = car) = .32161847
```

variable	dp/dx	Std. Err.	z	P>|z|	[95% C.I.]		X
travelcost							
air	.00141	.00051	2.77	0.006	.000411	.002409	102.65
train	.001904	.000886	2.15	0.032	.000166	.003641	130.2
bus	.001523	.000676	2.25	0.024	.000199	.002848	115.26
car	-.004837	.001539	-3.14	0.002	-.007854	-.00182	95.414
termtime							
air	.005441	.002162	2.52	0.012	.001204	.009678	61.01
train	.007347	.002228	3.30	0.001	.00298	.011713	35.69
bus	.005879	.002256	2.61	0.009	.001456	.010301	41.657
car	-.018666	.003939	-4.74	0.000	-.026387	-.010945	0
casevars							
income	.005246	.002165	2.42	0.015	.001002	.00949	34.548

First we note that there is a separate marginal effects table for each alternative and that table begins by reporting the overall probability of choosing the alternative, e.g. .2944 for air travel. We see in the first table that a unit increase in terminal time for air travel from 61.01 minutes will result in a decrease in probability of choosing air travel (when the probability is evaluated at the mean of all variables) by approximately 0.01, with 95% confidence interval of about −0.016 to −0.005. Travel cost has a less negative effect of choosing air travel (at the average cost of 102.65). Alternatively, an increase in terminal time and travel cost for train, bus, or car from these mean values will increase the chance for air travel to be chosen. Also, with an increase in income from 34.5, it would appear that an individual would be more likely to choose air or automobile travel over bus or train. (While the marginal effect for bus travel is positive, it is not significant.)

◁

▷ Example 4

Plotting the simulated probability marginal effect evaluated over a range of values for an independent variable may be more revealing than a table of values. Below are the commands for generating the simulated probability marginal effect of air travel for increasing air travel terminal time. We fix all other independent variables at their medians.

```
. qui gen meff = .

. qui gen tt = .

. qui gen lb = .

. qui gen ub = .

. forvalues i=0/19 {
  2.          local termtime = 5+5*`i'
  3.          qui replace tt = `termtime' if _n == `i'+1
  4.          qui estat mfx, at(median air:termtime=`termtime') var(termtime)
  5.          mat air = r(air)
  6.          qui replace meff = air[1,1] if _n == `i'+1
  7.          qui replace lb = air[1,5] if _n == `i'+1
  8.          qui replace ub = air[1,6] if _n == `i'+1
  9.          qui replace prob = r(pr_air) if _n == `i'+1
 10. }

. label variable tt "terminal time"

. twoway (rarea lb ub tt, pstyle(ci)) (line meff tt, lpattern(solid)), name(meff)
> legend(off) title("  marginal effect of air travel" "terminal time and"
> "95% confidence interval", position(3))

. twoway line prob tt, name(prob) title("  probability of choosing" "air travel",
> position(3)) graphregion(margin(r+9)) ytitle("") xtitle("")

. graph combine prob meff, cols(1) graphregion(margin(l+5 r+5))
```

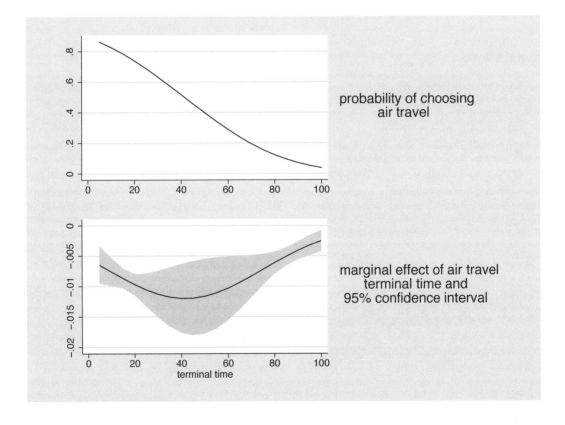

From the graphs we see that the simulated probability of choosing air travel decreases in an sigmoid fashion. The marginal effects display the rate of change in the simulated probability as a function of the air travel terminal time. The rate of change in the probability of choosing air travel decreases until the air travel terminal time reaches about 45, thereafter it increases.

◁

Saved Results

estat mfx saves the following in r():

Scalars

r(pr_*alt*)   scalars containing the computed probability of each alternative evaluated at the value that is labeled X in the table output. Here *alt* are the labels in the macro e(alteqs).

Matrices

r(*alt*)   matrices containing the computed marginal effects and associated statistics. There is one matrix for each alternative, where *alt* are the labels in the macro e(alteqs). Column 1 of each matrix contains the marginal effects; column 2, their standard errors; column 3, their z statistics; and columns 4 and 5, the confidence intervals. Column 6 contains the values of the independent variables used to compute the probabilities r(pr_*alt*).

Methods and Formulas

All postestimation commands listed above are implemented as ado-files.

Marginal effects

The marginal effects are computed as the derivative of the simulated probability with respect to each independent variable. A set of marginal effects are computed for each alternative; thus, for J alternatives, there will be J tables. Moreover, the alternative-specific variables will have J entries, one for each alternative in each table. The details of computing the effects are different for alternative-specific variables and case-specific variables, as well as for continuous and indicator variables.

We use the latent variable notation of asmprobit (see [R] **asmprobit**) for a J-alternative model and, for notational convenience, we will drop any subscripts involving observations. We then have the following linear functions $\eta_j = \mathbf{x}_j \boldsymbol{\beta} + \mathbf{z}\boldsymbol{\alpha}_j$, for $j = 1, \ldots, J$. Let k index the alternative of interest, and then

$$v_{j'} = \eta_j - \eta_k$$
$$= (\mathbf{x}_j - \mathbf{x}_k)\boldsymbol{\beta} + \mathbf{z}(\boldsymbol{\alpha}_j - \boldsymbol{\alpha}_k) + \epsilon_{j'}$$

where $j' = j$ if $j < k$ and $j' = j - 1$ if $j > k$, so that $j' = 1, \ldots, J - 1$ and $\epsilon_{j'} \sim \text{MVN}(\mathbf{0}, \boldsymbol{\Sigma})$. Denote $p_k = \Pr(v_1 \leq 0, \ldots, v_{J-1} \leq 0)$ as the simulated probability of choosing alternative k given profile $\mathbf{x}_k$ and $\mathbf{z}$. The marginal effects are then $\partial p_k / \partial \mathbf{x}_k$, $\partial p_k / \partial \mathbf{x}_j$, and $\partial p_k / \partial \mathbf{z}$, where $k = 1, \ldots, J$, $j \neq k$. asmprobit analytically computes the first-order derivatives of the simulated probability with respect to the v's, and the marginal effects for $\mathbf{x}$'s and $\mathbf{z}$ are obtained via the chain rule. The standard errors for the marginal effects are computed using the delta method.

Also See

[R] **asmprobit** — Alternative-specific multinomial probit regression

[U] **20 Estimation and postestimation commands**

Title

> **asroprobit** — Alternative-specific rank-ordered probit regression

Syntax

asroprobit *depvar* [*indepvars*] [*if*] [*in*] [*weight*] , case(*varname*)

    <u>alt</u>ernatives(*varname*) [*options*]

options	description
Model	
* case(*varname*)	use *varname* to identify cases
* <u>alt</u>ernatives(*varname*)	use *varname* to identify the alternatives available for each case
<u>casev</u>ars(*varlist*)	case-specific variables
<u>constr</u>aints(*constraints*)	apply specified linear constraints
<u>col</u>linear	keep collinear variables
Model 2	
<u>corr</u>elation(*correlation*)	correlation structure of the latent-variable errors; implies structural
<u>std</u>dev(*stddev*)	variance structure of the latent-variable errors; implies structural
<u>str</u>uctural	use the structural covariance parameterization; default is the differenced covariance parameterization
<u>noc</u>onstant	suppress the alternative-specific constant terms
<u>base</u>alternative(*#* | *lbl* | *str*)	alternative used for normalizing location
<u>scale</u>alternative(*#* | *lbl* | *str*)	alternative used for normalizing scale
altwise	use alternative-wise deletion instead of casewise deletion
<u>rev</u>erse	interpret the lowest rank in *depvar* as the best; the default is the highest rank is the best
SE/Robust	
vce(*vcetype*)	*vcetype* may be oim, <u>r</u>obust, <u>cl</u>uster *clustvar*, opg, <u>boot</u>strap, or <u>jack</u>knife
Reporting	
<u>l</u>evel(*#*)	set confidence level; default is level(95)
<u>notran</u>sform	do not transform variance–covariance estimates to the standard deviation and correlation metric

(*Continued on next page*)

Int options

intmethod(*seqtype*)	type of quasi- or pseudouniform sequence
intpoints(*#*)	number of points in each sequence
intburn(*#*)	starting index in the Hammersley or Halton sequence
intseed(*code* | *#*)	pseudouniform random-number seed
antithetics	use antithetic draws
initbhhh(*#*)	use the BHHH optimization algorithm for the first *#* iterations

Max options

maximize_options	control the maximization process

correlation	description
unstructured	one correlation parameter for each pair of alternatives; correlations with the basealternative() are zero
exchangeable	one correlation parameter common to all pairs of alternatives; correlations with the basealternative() are zero
independent	constrain all correlation parameters to zero
pattern *matname*	user-specified matrix identifying the correlation pattern
fixed *matname*	user-specified matrix identifying the fixed and free correlation parameters

stddev	description
heteroskedastic	estimate standard deviation for each alternative; standard deviations for basealternative() and scalealternative() set to one
homoskedastic	all standard deviations are one
pattern *matname*	user-specified matrix identifying the standard deviation pattern
fixed *matname*	user-specified matrix identifying the fixed and free standard deviations

seqtype	description
hammersley	Hammersley point set
halton	Halton point set
random	uniform pseudorandom point set

*case(*varname*) and alternatives(*varname*) are required.

bootstrap, by, jackknife, statsby, and xi are allowed; see [U] **11.1.10 Prefix commands**.

Weights are not allowed with the bootstrap prefix.

fweights, iweights, and pweights are allowed; see [U] **11.1.6 weight**.

See [U] **20 Estimation and postestimation commands** for more capabilities of estimation commands.

Description

asroprobit fits rank-ordered probit (ROP) models by using maximum simulated likelihood (MSL). The model allows you to relax the independence of irrelevant alternatives (IIA) property that is characteristic of the rank-ordered logistic model by estimating the variance–covariance parameters of the latent-variable errors. Each unique identifier in the variable case() has multiple alternatives identified in the variable alternatives(), and *depvar* contains the ranked alternatives made by each case. Only the order in the ranks, not the magnitude of their differences, is assumed to be relevant. By default, the largest rank indicates the more desirable alternative. Use the reverse option if the lowest rank should be interpreted as the more desirable alternative. Tied ranks are allowed, but they increase the computation time since all permutations of the tied ranks are used in computing the likelihood for each case. asroprobit allows two types of independent variables: alternative-specific variables, in which the values of each variable vary with each alternative, and case-specific variables, which vary with each case.

The estimation technique of asroprobit is nearly identical to that of asmprobit, and the two routines share many of the same options; see [R] **asmprobit**.

Options

―――――⌐ Model ⌐―――

case(*varname*) specifies the variable that identifies each case. This variable identifies the individuals or entities making a choice. case() is required.

alternatives(*varname*) specifies the variable that identifies the alternatives available for each case. The number of alternatives can vary with each case; the maximum number of alternatives is 20. alternatives() is required.

casevars(*varlist*) specifies the case-specific variables that are constant for each case(). If there are a maximum of J alternatives, there will be $J - 1$ sets of coefficients associated with casevars().

constraints(*constraints*), collinear; see [R] **estimation options**.

―――――⌐ Model 2 ⌐――

correlation(*correlation*) specifies the correlation structure of the latent-variable errors. The correlation() option implies the structural covariance parameterization.

correlation(unstructured) is the most general and has $J(J - 3)/2 + 1$ unique correlation parameters.

correlation(exchangeable) provides for one correlation coefficient common to all latent variables, except the latent variable associated with the basealternative().

correlation(independent) assumes that all correlations are zero.

correlation(pattern *matname*) and correlation(fixed *matname*) give you more flexibility in defining the correlation structure. See *Variance structures* in [R] **asmprobit** for more information.

stddev(*stddev*) specifies the variance structure of the latent-variable errors. The stddev() option implies the structural covariance parameterization.

stddev(heteroskedastic) is the most general and has $J - 2$ estimable parameters. The standard deviations of the latent-variable errors for the alternatives specified in basealternative() and scalealternative() are fixed to one.

stddev(homoskedastic) constrains all the standard deviations to equal one.

stddev(pattern *matname*) and stddev(fixed *matname*) give you added flexibility in defining the standard deviation parameters. See *Variance structures* in [R] **asmprobit** for more information.

structural requests the $J \times J$ structural covariance parameterization instead of the default $J-1 \times J-1$ differenced covariance parameterization (the covariance of the latent errors differenced with that of the base alternative). The differenced covariance parameterization will achieve the same maximum simulated likelihood regardless of the choice of basealternative() and scalealternative(). On the other hand, the structural covariance parameterization imposes more normalizations that may bound the model away from its maximum likelihood and thus prevent convergence with some datasets or choices of basealternative() and scalealternative().

noconstant suppresses the $J - 1$ alternative-specific constant terms.

basealternative(#|*lbl*|*str*) specifies the alternative used to normalize the latent-variable location (also referred to as the level of utility). The base alternative may be specified as a number, label, or string. The standard deviation for the latent-variable error associated with the base alternative is fixed to one, and its correlations with all other latent-variable errors are set to zero. The default is the first alternative when sorted. If a fixed or pattern matrix is given in the stddev() and correlation() options, the basealternative() will be implied by the fixed standard deviations and correlations in the matrix specifications. basealternative() cannot be equal to scalealternative().

scalealternative(#|*lbl*|*str*) specifies the alternative used to normalize the latent-variable scale (also referred to as the scale of utility). The scale alternative may be specified as a number, label, or string. The default is to use the second alternative when sorted. If a fixed or pattern matrix is given in the stddev() option, the scalealternative() will be implied in the fixed standard deviations in the matrix specification. scalealternative() cannot be equal to basealternative().

If a fixed or pattern matrix is given for the stddev() option, the base alternative and scale alternative are implied by the standard deviations and correlations in the matrix specifications, and they need not be specified in the basealternative() and scalealternative() options.

altwise specifies that alternative-wise deletion be used when marking out observations due to missing values in your variables. The default is to use casewise deletion; that is, the entire group of observations making up a case is deleted if any missing values are encountered. This option does not apply to observations that are marked out by the if or in qualifier or the by prefix.

reverse directs asroprobit to interpret the rank in *depvar* that is smallest in value as the preferred alternative. By default, the rank that is largest in value is the favored alternative.

SE/Robust

vce(*vcetype*) specifies the type of standard error reported, which includes types that are derived from asymptotic theory, that are robust to some kinds of misspecification, that allow for intragroup correlation, and that use bootstrap or jackknife methods; see [R] *vce_option*.

If specifying vce(bootstrap) or vce(jackknife), you must also specify basealternative() and scalealternative().

Reporting

level(#); see [R] **estimation options**.

notransform prevents retransforming the Cholesky-factored variance–covariance estimates to the correlation and standard deviation metric.

This option has no effect if `structural` is not specified because the default differenced variance–covariance estimates have no interesting interpretation as correlations and standard deviations. `notransform` also has no effect if the `correlation()` and `stddev()` options are specified with anything other than their default values. Here it is generally not possible to factor the variance–covariance matrix, so optimization is already performed using the standard deviation and correlation representations.

⌐ Int options ⌐

`intmethod(hammersley | halton | random)` specifies the method of generating the point sets used in the quasi–Monte Carlo integration of the multivariate normal density. `intmethod(hammersley)`, the default, uses the Hammersley sequence; `intmethod(halton)` uses the Halton sequence; and `intmethod(random)` uses a sequence of uniform random numbers.

`intpoints(#)` specifies the number of points to use in the quasi–Monte Carlo integration. If this option is not specified, the number of points is $50 \times J$ if `intmethod(hammersley)` or `intmethod(halton)` is used and $100 \times J$ if `intmethod(random)` is used. Larger values of `intpoints()` provide better approximations of the log likelihood, but at the cost of added computation time.

`intburn(#)` specifies where in the Halton or Hammersley sequence to start, which helps reduce the correlation between the sequences of each dimension. The default is 0. This option may not be specified with `intmethod(random)`.

`intseed(code | #)` specifies the seed to use for generating the uniform pseudorandom sequence. This option may be specified only with `intmethod(random)`. *code* refers to a string that records the state of the random-number generator `uniform()`; see [D] **generate**. An integer value # may be used also. The default is to use the current seed value from Stata's uniform random-number generator, which can be obtained from `c(seed)`.

`antithetics` specifies that antithetic draws be used. The antithetic draw for the $J - 1$ vector uniform-random variables, `x`, is $1 - \mathbf{x}$.

`initbhhh(#)` specifies that the BHHH algorithm be used for the initial # optimization steps. This option is the only way to use the BHHH algorithm along with other optimization techniques. The algorithm switching feature of `ml`'s `technique()` option cannot include `bhhh`.

⌐ Max options ⌐

maximize_options: <u>difficult</u>, <u>tech</u>nique(*algorithm_spec*), <u>iter</u>ate(*#*), [<u>no</u>]<u>log</u>, <u>trace</u>, gradient, showstep, <u>hessian</u>, <u>shownr</u>tolerance, <u>tol</u>erance(*#*), <u>ltol</u>erance(*#*), <u>gtol</u>erance(*#*), <u>nrtol</u>erance(*#*), <u>nonrtol</u>erance, from(*init_specs*); see [R] **maximize**.

The following options may be particularly useful in obtaining convergence with `asroprobit`: `difficult`, `technique(`*algorithm_spec*`)`, `nrtolerance(#)`, `nonrtolerance`, and `from(`*init_specs*`)`.

If `technique()` contains more than one algorithm specification, `bhhh` cannot be one of them. To use the BHHH algorithm with another algorithm, use the `initbhhh()` option and specify the other algorithm in `technique()`.

Setting the optimization type to `technique(bhhh)` resets the default *vcetype* to `vce(opg)`.

Remarks

The mathematical description and numerical computations of the rank-ordered probit model are similar to that of the multinomial probit model. The only difference is that the dependent variable of the rank-ordered probit model is ordinal, showing preferences among alternatives, as opposed to the binary dependent variable of the multinomial probit model, indicating a chosen alternative. We will describe how the likelihood of a ranking is computed using the latent-variable framework here, but for details of the latent-variable parameterization of these models and the method of maximum simulated likelihood, see [R] **asmprobit**.

Consider the latent-variable parameterization of a J alternative rank-ordered probit model. Using the notation from `asmprobit` we have variables η_{ij}, $j = 1, \ldots, J$, such that

$$\eta_{ij} = \mathbf{x}_{ij}\boldsymbol{\beta} + \mathbf{z}_i\boldsymbol{\alpha}_j + \xi_{ij}$$

Here the $\mathbf{x}_{ij}$ are the alternative-specific independent variables, the $\mathbf{z}_i$ are the case-specific variables, and the ξ_{ij} are multivariate normal with mean zero and covariance $\boldsymbol{\Omega}$. Without loss of generality, assume that individual i ranks the alternatives in order of the alternative indices $j = 1, 2, \ldots, J$, so the alternative J is the preferred alternative and alternative 1 is the least preferred alternative. The probability of this ranking given $\boldsymbol{\beta}$ and $\boldsymbol{\alpha}_j$ is the probability that $\eta_{i,J-1} - \eta_{i,J} \leq 0$ and $\eta_{i,J-2} - \eta_{i,J-1} \leq 0$, $\ldots$, and $\eta_{i,1} - \eta_{i,2} \leq 0$.

▷ Example 1

Long and Freese (2006) provide an example of a rank-ordered logit model with alternative-specific variables. We use this dataset to demonstrate `asroprobit`. The data come from the Wisconsin Longitudinal Study. This is a study of 1957 Wisconsin high school graduates that were asked to rate their relative preference of four job characteristics: esteem, a job other people regard highly; variety, a job that is not repetitive and allows you to do a variety of things; autonomy, a job where your supervisor does not check on you frequently; and security, a job with a low risk of being laid off. The case-specific covariates are gender, `female`, an indicator variable for females, and `score`, a score on a general mental ability test measured in standard deviations. The alternative-specific variables are `high` and `low`, which indicate whether the respondent's current job is high or low in esteem, variety, autonomy, or security. This approach provides three states for a respondent's current job status for each alternative, $(1, 0)$, $(0, 1)$, and $(0, 0)$, using the notation (`high`, `low`). The score $(1, 1)$ is omitted since the respondent's current job cannot be considered both high and low in one of the job characteristics. The $(0, 0)$ score would indicate that the respondent's current job does not rank high or low (is neutral) in a job characteristic. The alternatives are ranked such that 1 is the preferred alternative and 4 is the least preferred.

```
. use http://www.stata-press.com/data/r10/wlsrank
(1992 Wisconsin Longitudinal Study data on job values)

. list id jobchar rank female score high low in 1/12, sepby(id)
```

	id	jobchar	rank	female	score	high	low
1.	1	security	1	1	.0492111	0	0
2.	1	autonomy	4	1	.0492111	0	0
3.	1	variety	1	1	.0492111	0	0
4.	1	esteem	3	1	.0492111	0	0
5.	5	security	2	1	2.115012	1	0
6.	5	variety	2	1	2.115012	1	0
7.	5	esteem	2	1	2.115012	1	0
8.	5	autonomy	1	1	2.115012	0	0
9.	7	autonomy	1	0	1.701852	1	0
10.	7	variety	1	0	1.701852	0	1
11.	7	esteem	4	0	1.701852	0	0
12.	7	security	1	0	1.701852	0	0

The three cases listed have tied ranks. `asroprobit` will allow ties, but at the cost of increased computation time. To evaluate the likelihood of the first observation, `asroprobit` must compute

$$\Pr(\texttt{esteem} = 3,\ \texttt{variety} = 1,\ \texttt{autonomy} = 4,\ \texttt{security} = 2) +$$
$$\Pr(\texttt{esteem} = 3,\ \texttt{variety} = 2,\ \texttt{autonomy} = 4,\ \texttt{security} = 1)$$

and both of these probabilities are estimated using simulation. In fact, the full dataset contains 7,237 tied ranks and `asroprobit` takes a great deal of time to estimate the parameters. For exposition, we estimate the rank-ordered probit model by using the cases without ties. These cases are marked in the variable `noties`.

The model of job preference is

$$\eta_{ij} = \beta_1 \texttt{high}_{ij} + \beta_2 \texttt{low}_{ij} + \alpha_{1j} \texttt{female}_i + \alpha_{2j} \texttt{score}_i + \alpha_{0j} + \xi_{ij}$$

for $j = 1, 2, 3, 4$. The base alternative will be `esteem`, so $\alpha_{01} = \alpha_{11} = \alpha_{21} = 0$.

```
. asroprobit rank high low if noties, casevars(female score) case(id)
> alternatives(jobchar) reverse
note: variable high has 107 cases that are not alternative-specific: there is
        no within-case variability
note: variable low has 193 cases that are not alternative-specific: there is
        no within-case variability
Iteration 0:    log simulated-likelihood = -1103.2768
Iteration 1:    log simulated-likelihood = -1089.3361  (backed up)
Iteration 2:    log simulated-likelihood = -1085.8939  (backed up)
Iteration 3:    log simulated-likelihood = -1083.1802  (backed up)
Iteration 4:    log simulated-likelihood =  -1082.741  (backed up)
Iteration 5:    log simulated-likelihood = -1082.4945  (backed up)
Iteration 6:    log simulated-likelihood = -1082.4849
Iteration 7:    log simulated-likelihood = -1082.4664
Iteration 8:    log simulated-likelihood =  -1082.154
Iteration 9:    log simulated-likelihood = -1082.0463  (backed up)
Iteration 10:   log simulated-likelihood = -1081.9785
Iteration 11:   log simulated-likelihood = -1081.3655
Iteration 12:   log simulated-likelihood = -1080.8175
Iteration 13:   log simulated-likelihood = -1080.7346
Iteration 14:   log simulated-likelihood = -1080.6862
Iteration 15:   log simulated-likelihood = -1080.6528
Iteration 16:   log simulated-likelihood = -1080.4176
Iteration 17:   log simulated-likelihood = -1080.3399
Iteration 18:   log simulated-likelihood = -1080.3304
Iteration 19:   log simulated-likelihood = -1080.3075
Iteration 20:   log simulated-likelihood = -1080.2206
Iteration 21:   log simulated-likelihood = -1080.2206

Alternative-specific rank-ordered probit     Number of obs       =      1660
Case variable: id                            Number of cases     =       415

Alternative variable: jobchar                Alts per case: min =         4
                                                            avg =       4.0
                                                            max =         4

Integration sequence:      Hammersley
Integration points:             200          Wald chi2(8)        =     34.01
Log simulated-likelihood = -1080.2206        Prob > chi2         =    0.0000
```

rank	Coef.	Std. Err.	z	P>|z|	[95% Conf. Interval]	
jobchar						
high	.3741029	.0925685	4.04	0.000	.192672	.5555337
low	-.0697443	.1093317	-0.64	0.524	-.2840305	.1445419
esteem	(base alternative)					
variety						
female	.1351487	.1843088	0.73	0.463	-.2260899	.4963873
score	.1405482	.0977567	1.44	0.151	-.0510515	.3321479
_cons	1.735016	.1451343	11.95	0.000	1.450558	2.019474
autonomy						
female	.2561828	.1679565	1.53	0.127	-.0730059	.5853715
score	.1898853	.0875668	2.17	0.030	.0182575	.361513
_cons	.7009797	.1227336	5.71	0.000	.4604262	.9415333
security						
female	.232622	.2057547	1.13	0.258	-.1706497	.6358938
score	-.1780076	.1102115	-1.62	0.106	-.3940181	.038003
_cons	1.343766	.1600059	8.40	0.000	1.030161	1.657372
/lnl2_2	.1805151	.0757296	2.38	0.017	.0320878	.3289424
/lnl3_3	.4843091	.0793343	6.10	0.000	.3288168	.6398014
/l2_1	.6062037	.1169368	5.18	0.000	.3770117	.8353957
/l3_1	.4509217	.1431183	3.15	0.002	.1704151	.7314283
/l3_2	.2289447	.1226081	1.87	0.062	-.0113627	.4692521

(jobchar=esteem is the alternative normalizing location)
(jobchar=variety is the alternative normalizing scale)

We specified the `reverse` option since a rank of 1 is the highest preference. The variance–covariance estimates are for the Cholesky-factored variance–covariance for the latent-variable errors differenced with that of alternative `esteem`. We can view the estimated correlations by entering

```
. estat correlation
```

	vari~y	auto~y	secu~y
variety	1.000		
autonomy	0.452	1.000	
security	0.265	0.240	1.000

Note: correlations are for alternatives differenced with esteem

and typing

```
. estat covariance
```

	variety	autonomy	security
variety	2		
autonomy	.8573015	1.80229	
security	.6376996	.5475882	2.890048

Note: covariances are for alternatives differenced with esteem gives the (co)variances. [R] **mprobit** explains that if the latent-variable errors are independent then the correlations in the differenced parameterization should be ~0.5 and the variances should be ~2.0, which seems to be the case here.

The coefficient estimates for the probit models can be difficult to interpret because of the normalization for location and scale. The regression estimates for the case-specific variables will be relative to the base alternative and the regression estimates for both the case-specific and alternative-specific variables are affected by the scale normalization. The more pronounced the heteroskedasticity and correlations the more pronounced the resulting estimate differences when choosing alternatives to normalize for location and scale. However, when using the differenced covariance structure you will obtain the same model likelihood regardless of which alternatives you choose as the base and scale alternatives. For model interpretation you can examine the estimated probabilities and marginal effects by using postestimation routines `predict` and `estat mfx`. See [R] **asroprobit postestimation**.

◁

Saved Results

`asroprobit` saves the following in `e()`:

Scalars

e(N)	number of observations
e(N_case)	number of cases
e(N_ties)	number of ties
e(k)	number of parameters
e(k_alt)	number of alternatives
e(k_indvars)	number of alternative-specific variables
e(k_casevars)	number of case-specific variables
e(k_sigma)	number of variance estimates
e(k_rho)	number of correlation estimates
e(k_eq)	number of equations in e(b)
e(k_eq_model)	number of equations in model Wald test
e(N_clust)	number of clusters
e(const)	constant indicator
e(i_base)	base alternative index
e(i_scale)	scale alternative index
e(mc_points)	number of Monte Carlo replications
e(mc_burn)	starting sequence index
e(mc_antithetics)	antithetics indicator
e(df_m)	model degrees of freedom
e(ll)	log simulated-likelihood
e(reverse)	1 if minimum rank is best, 0 if maximum rank is best
e(chi2)	χ^2
e(p)	significance
e(fullcov)	unstructured covariance indicator
e(structcov)	1 if structured covariance; 0 otherwise
e(cholesky)	Cholesky-factored covariance indicator
e(alt_min)	minimum number of alternatives
e(alt_max)	maximum number of alternatives
e(alt_avg)	average number of alternatives
e(rank)	rank of e(V)
e(ic)	number of iterations
e(rc)	return code
e(converged)	1 if converged, 0 otherwise

Macros
    e(cmd)   asroprobit
    e(cmdline)   command as typed
    e(depvar)   name of dependent variable
    e(indvars)   alternative-specific independent variable
    e(casevars)   case-specific variables
    e(case)   variable defining cases
    e(altvar)   variable defining alternatives
    e(alteqs)   alternative equation names
    e(alt#)   alternative labels
    e(wtype)   weight type
    e(wexp)   weight expression
    e(title)   title in estimation output
    e(clustvar)   name of cluster variable
    e(correlation)   correlation structure
    e(stddev)   variance structure
    e(chi2type)   Wald, type of model χ^2 test
    e(vce)   *vcetype* specified in vce()
    e(vcetype)   title used to label Std. Err.
    e(opt)   type of optimization
    e(ml_method)   type of ml method
    e(mc_method)   technique used to generate sequences
    e(mc_seed)   random-number generator seed
    e(user)   name of likelihood-evaluator program
    e(technique)   maximization technique
    e(crittype)   optimization criterion
    e(properties)   b V
    e(datasignaturevars) variables used in calculation of checksum
    e(datasignature)   the checksum
    e(estat_cmd)   program used to implement estat
    e(predict)   program used to implement predict

Matrices
    e(b)   coefficient vector
    e(stats)   alternative statistics
    e(stdpattern)   variance pattern
    e(stdfixed)   fixed and free standard deviations
    e(altvals)   alternative values
    e(altfreq)   alternative frequencies
    e(ilog)   iteration log (up to 20 iterations)
    e(gradient)   gradient vector
    e(V)   variance–covariance matrix of the estimators
    e(corpattern)   correlation structure
    e(corfixed)   fixed and free correlations

Functions
    e(sample)   marks estimation sample

Methods and Formulas

asroprobit is implemented as an ado-file.

From a computational perspective, asroprobit is similar to asmprobit and the two programs share many numerical tools. Therefore, we will use the notation from *Methods and Formulas* in [R] **asmprobit** to discuss the rank-ordered probit probability model.

The latent variables for a J-alternative model are $\eta_{ij} = \mathbf{x}_{ij}\boldsymbol{\beta} + \mathbf{z}_i\boldsymbol{\alpha}_j + \xi_{ij}$, for $j = 1, \ldots, J$, $i = 1, \ldots, n$, and $\boldsymbol{\xi}'_i = (\xi_{i,1}, \ldots, \xi_{i,J}) \sim \text{MVN}(\mathbf{0}, \boldsymbol{\Omega})$. Without loss of generality, assume for the ith observation that an individual ranks the alternatives in the order of their numeric indices, $\mathbf{y}_i = (J, J - 1, \ldots, 1)$, so the first alternative is the most preferred and the last alternative is the least preferred. We can then difference the latent variables such that

$$
\begin{aligned}
v_{ik} &= \eta_{i,k+1} - \eta_{i,k} \\
&= (\mathbf{x}_{i,k+1} - \mathbf{x}_{i,k})\boldsymbol{\beta} + \mathbf{z}_i(\boldsymbol{\alpha}_{k+1} - \boldsymbol{\alpha}_k) + \xi_{i,k+1} - \xi_{ik} \\
&= \boldsymbol{\delta}_{ik}\boldsymbol{\beta} + \mathbf{z}_i\boldsymbol{\gamma}_k + \epsilon_{ik}
\end{aligned}
$$

for $k = 1, \ldots, J - 1$ and where $\epsilon_i = (\epsilon_{i1}, \ldots, \epsilon_{i,J-1}) \sim \text{MVN}(\mathbf{0}, \boldsymbol{\Sigma}_{(i)})$. $\boldsymbol{\Sigma}$ is indexed by i since it is specific to the ranking of individual i. We denote the deterministic part of the model as $\lambda_{ik} = \boldsymbol{\delta}_{ik}\boldsymbol{\beta} + \mathbf{z}_j\boldsymbol{\gamma}_k$, and the probability of this event is

$$
\begin{aligned}
\Pr(\mathbf{y}_i) &= \Pr(v_{i1} \leq 0, \ldots, v_{i,J-1} \leq 0) \\
&= \Pr(\epsilon_{i1} \leq -\lambda_{i1}, \ldots, \epsilon_{i,J-1} \leq -\lambda_{i,J-1}) \\
&= (2\pi)^{-(J-1)/2} |\boldsymbol{\Sigma}_{(i)}|^{-1/2} \int_{-\infty}^{-\lambda_{i1}} \cdots \int_{-\infty}^{-\lambda_{i,J-1}} \exp\left(-\tfrac{1}{2}\mathbf{z}'\boldsymbol{\Sigma}_{(i)}^{-1}\mathbf{z}\right) d\mathbf{z}
\end{aligned}
$$

The integral has the same form as (3) of *Methods and Formulas* in [R] **asmprobit**. See [R] **asmprobit** for details on evaluating this integral numerically using simulation.

asroprobit handles tied ranks by enumeration. For k tied ranks it will generate $k!$ rankings, where ! is the factorial operator $k! = k(k-1)(k-2)\cdots(2)(1)$. For two sets of tied ranks of size k_1 and k_2, asroprobit will generate $k_1!k_2!$ rankings. The total probability is the sum of the probability of each ranking. For example, if there are two tied ranks such that $\mathbf{y}_i = (J, J, J - 2, \ldots, 1)$, then asroprobit will evaluate $\Pr(\mathbf{y}_i) = \Pr(\mathbf{y}_i^{(1)}) + \Pr(\mathbf{y}_i^{(2)})$, where $\mathbf{y}_i^{(1)} = (J, J - 1, J - 2, \ldots, 1)$ and $\mathbf{y}_i^{(2)} = (J - 1, J, J - 2, \ldots, 1)$.

Reference

Long, J. S., and J. Freese. 2006. *Regression Models for Categorical Dependent Variables Using Stata*. 2nd ed. College Station, TX: Stata Press.

Also See

[R] **asroprobit postestimation** — Postestimation tools for asroprobit

[R] **asmprobit** — Alternative-specific multinomial probit regression

[R] **mprobit** — Multinomial probit regression

[R] **mlogit** — Multinomial (polytomous) logistic regression

[R] **oprobit** — Ordered probit regression

[U] **20 Estimation and postestimation commands**

Title

> **asroprobit postestimation** — Postestimation tools for asroprobit

Description

The following postestimation commands are of special interest after `asroprobit`:

command	description
estat alternatives	alternative summary statistics
estat covariance	variance–covariance matrix of the alternatives
estat correlation	correlation matrix of the alternatives
estat mfx	marginal effects

For information about these commands, see below.

The following standard postestimation commands are also available:

command	description
estat	AIC, BIC, VCE, and estimation sample summary
estimates	cataloging estimation results
lincom	point estimates, standard errors, testing, and inference for linear combinations of coefficients
lrtest	likelihood-ratio test
nlcom	point estimates, standard errors, testing, and inference for nonlinear combinations of coefficients
predict	predicted probabilities, estimated linear predictor and its standard error
predictnl	point estimates, standard errors, testing, and inference for generalized predictions
test	Wald tests for simple and composite linear hypotheses
testnl	Wald tests of nonlinear hypotheses

See the corresponding entries in the *Stata Base Reference Manual* for details.

Special-interest postestimation commands

`estat alternatives` displays summary statistics about the alternatives in the estimation sample. The command also provides a mapping between the index numbers that label the covariance parameters of the model and their associated values and labels for the alternative variable.

`estat covariance` computes the estimated variance–covariance matrix for the alternatives. The estimates are displayed, and the variance–covariance matrix is stored in `r(cov)`.

`estat correlation` computes the estimated correlation matrix for the alternatives. The estimates are displayed, and the correlation matrix is stored in `r(cor)`.

`estat mfx` computes marginal effects of a simulated probability of a set of ranked alternatives. The probability is stored in `r(pr)`, the matrix of rankings is stored in `r(ranks)`, and the matrix of marginal-effect statistics is stored in `r(mfx)`.

Syntax for predict

predict [*type*] *newvar* [*if*] [*in*] [, *statistic* altwise]

predict [*type*] { *stub** | *newvarlist* } [*if*] [*in*], scores

statistic	description
Main	
pr	probability of each ranking, by case; the default
pr1	probability that each alternative is preferred
xb	linear prediction
stdp	standard error of the linear prediction

These statistics are available both in and out of sample; type predict ... if e(sample) ... if wanted only for the estimation sample.

Options for predict

⌐ Main ⌐

pr, the default, calculates the probability of each ranking. For each case, one probability is computed for the ranks in e(depvar).

pr1 calculates the probability that each alternative is preferred.

xb calculates the linear prediction $\mathbf{x}_{ij}\boldsymbol{\beta} + \mathbf{z}_i\boldsymbol{\alpha}_j$ for alternative j and case i.

stdp calculates the standard error of the linear predictor.

altwise specifies that alternative-wise deletion be used when marking out observations due to missing values in your variables. The default is to use casewise deletion. Options xb and stdp always use alternative-wise deletion.

scores calculates the scores for each coefficient in e(b). This option requires a new variable list of length equal to the number of columns in e(b). Otherwise, use the *stub** option to have predict generate enumerated variables with prefix *stub*.

Syntax for estat alternatives

estat alternatives

Syntax for estat covariance

estat covariance [, format(*string*) border(*string*) left(*#*)]

Options for estat covariance

format(*string*) sets the matrix display format. The default is format(%9.0g).

border(*string*) sets the matrix display border style. The default is border(all). See [P] **matlist**.

left(*#*) sets the matrix display left indent. The default is left(2). See [P] **matlist**.

Syntax for estat correlation

> estat correlation [, format(*string*) border(*string*) left(*#*)]

Options for estat correlation

format(*string*) sets the matrix display format. The default is format(%6.3f).

border(*string*) sets the matrix display border style. The default is border(all). See [P] **matlist**.

left(*#*) sets the matrix display left indent. The default is left(2). See [P] **matlist**.

Syntax for estat mfx

> estat mfx [*if*] [*in*] [, *options*]

options	description
Main	
varlist(*varlist*)	display marginal effects for *varlist*
at(median [*atlist*])	calculate marginal effects at these values
rank(*atlist*)	calculate marginal effects for the simulated probability of these ranked alternatives
Options	
level(*#*)	set confidence interval level; default is level(95)
nodiscrete	treat indicator variables as continuous
noesample	do not restrict calculation of the medians to the estimation sample
nowght	ignore weights when calculating medians

Options for estat mfx

⌐ Main ⌐

varlist(*varlist*) specifies the variables for which to display marginal effects. The default is all variables.

at(median [[*alternative*:*variable* = *#*] [*variable* = *#*] [...]]) specifies the values at which the marginal effects are to be calculated. The default is to compute the marginal effects at the medians of the independent variables by using the estimation sample, at(median). You can also specify specific values for variables. Values for alternative-specific variables can be specified by alternative, or you can specify one value for all alternatives. You can specify only one value for case-specific variables. For example, in the wlsrank dataset, female and score are case-specific variables, whereas high and low are alternative-specific variables. The following would be a legal syntax for estat mfx:

. estat mfx, at(median high=0 esteem:high=1 low=0 security:low=1 female=1)

When nodiscrete is not specified, at(median [*atlist*]) has no effect on computing marginal effects for indicator variables, which are calculated as the discrete change in the simulated probability as the indicator variable changes from 0 to 1.

The median computations respect any if or in qualifiers, so you can restrict the data over which the medians are computed. You can even restrict the values to a specific case; e.g.,

. estat mfx if case==13

rank(*alternative* = # *alternative* = # $\left[\ldots \right]$) specifies the ranks for the alternatives. The default is to rank the calculated latent variables. Alternatives excluded from rank() are omitted from the analysis. You must therefore specify at least two alternatives in rank(). You may have tied ranks in the rank specification. Only the order in the ranks is relevant.

⌐ Options ∟

level(#) sets the confidence level; default is level(95).

nodiscrete specifies that indicator variables be treated as continuous variables. An indicator variable is one that takes on the value 0 or 1 in the estimation sample. By default, the discrete change in the simulated probability is computed as the indicator variable changes from 0 to 1.

noesample specifies that the whole dataset be considered instead of only those marked in the e(sample) defined by the asroprobit command.

nowght specifies that weights be ignored when calculating the medians.

Remarks

Remarks are presented under the following headings:

> *Predicted probabilities*
> *Obtaining estimation statistics*

Predicted probabilities

After fitting an alternative-specific rank-ordered probit model, you can use predict to obtain the probabilities of alternative rankings or the probabilities of each alternative being preferred. When evaluating the multivariate normal probabilities via (quasi) Monte Carlo, predict uses the same method to generate the (quasi) random sequence of numbers as the previous call to asroprobit. For example, if you specified intmethod(halton) when fitting the model, predict also uses Halton sequences.

▷ Example 1

In the example of [R] **asroprobit**, we fitted a model of job characteristic preferences. This is a study of 1957 Wisconsin high school graduates that were asked to rate their relative preference of four job characteristics: esteem, a job other people regard highly; variety, a job that is not repetitive and allows you to do a variety of things; autonomy, a job where your supervisor does not check on you frequently; and security, a job with a low risk of being laid off. The case-specific covariates are gender, female, an indicator variable for females, and score, a score on a general mental ability test measured in standard deviations. The alternative-specific variables are high and low, which indicate whether the respondent's current job is high or low in esteem, variety, autonomy, or security. This approach provides three states for a respondent's current job status for each alternative, $(1, 0)$, $(0, 1)$, and $(0, 0)$, using the notation (high, low). The score $(1, 1)$ is omitted since the respondent's current job cannot be considered both high and low in one of the job characteristics. The $(0, 0)$ score would indicate that the respondent's current job does not rank high or low (is neutral) in a job characteristic. The alternatives are ranked such that 1 is the preferred alternative and 4 is the least preferred.

We can obtain the probabilities of the observed alternative rankings, option pr, and the probability of each alternative being preferred, option pr1, by using predict:

```
. use http://www.stata-press.com/data/r10/wlsrank
(1992 Wisconsin Longitudinal Study data on job values)

. asroprobit rank high low if noties, casevars(female score) case(id)
> alternatives(jobchar) reverse
(output omitted)

. keep if e(sample)
(11244 observations deleted)

. predict prob, pr

. predict prob1, pr1

. list id jobchar prob prob1 rank female score high low in 1/12, sepby(id)
```

	id	jobchar	prob	prob1	rank	female	score	high	low
1.	13	security	.0421807	.2784269	3	0	.3246512	0	1
2.	13	autonomy	.0421807	.1029036	1	0	.3246512	0	0
3.	13	variety	.0421807	.6026725	2	0	.3246512	1	0
4.	13	esteem	.0421807	.0160111	4	0	.3246512	0	1
5.	19	autonomy	.0942025	.1232488	4	1	.0492111	0	0
6.	19	esteem	.0942025	.0140261	3	1	.0492111	0	0
7.	19	security	.0942025	.4601368	1	1	.0492111	1	0
8.	19	variety	.0942025	.4025715	2	1	.0492111	0	0
9.	22	esteem	.1414177	.0255264	4	1	1.426412	1	0
10.	22	variety	.1414177	.4549441	1	1	1.426412	0	0
11.	22	security	.1414177	.2629494	3	1	1.426412	0	0
12.	22	autonomy	.1414177	.2566032	2	1	1.426412	1	0

The variable prob is constant for each case since it contains the probability of the ranking in variable rank. On the other hand, the variable prob1 contains the estimated probability of each alternative being preferred. For each case the sum of the values in prob1 will be approximately 1.0. They do not add up to exactly 1.0 because of approximations due to the GHK algorithm.

◁

Obtaining estimation statistics

For examples of the specialized estat subcommands covariance and correlation, see [R] **asmprobit postestimation**. The entry also has a good example of computing marginal effects after asmprobit that is applicable to asroprobit. Below we will elaborate further on marginal effects after asroprobit where we manipulate the rank() option.

▷ Example 2

We will continue with the preferred job characteristics example where we first compute the marginal effects for case id = 13.

```
. estat mfx if id==13, rank(security=3 autonomy=1 variety=2 esteem=4)

Pr(esteem=4 variety=2 autonomy=1 security=3) = .04218068
```

variable	dp/dx	Std. Err.	z	P>|z|	[95% C.I.]	X
high*						
esteem	-.008713	.001964	-4.44	0.000	-.012562 -.004864	0
variety	-.009102	.003127	-2.91	0.004	-.015231 -.002973	1
autonomy	.025535	.007029	3.63	0.000	.011758 .039313	0
security	-.003745	.001394	-2.69	0.007	-.006477 -.001013	0
low*						
esteem	.001614	.002646	0.61	0.542	-.003572 .0068	1
variety	.001809	.003012	0.60	0.548	-.004094 .007712	0
autonomy	-.003849	.006104	-0.63	0.528	-.015813 .008115	0
security	.000582	.000985	0.59	0.554	-.001348 .002513	1
casevars						
female*	.009767	.009064	1.08	0.281	-.007998 .027533	0
score	.008587	.004488	1.91	0.056	-.00021 .017384	.32465

(*) dp/dx is for discrete change of indicator variable from 0 to 1

Next we compute the marginal effects for the probability that autonomy is preferred given the profile of case id = 13.

```
. estat mfx if id==13, rank(security=2 autonomy=1 variety=2 esteem=2)

Pr(esteem=3 variety=4 autonomy=1 security=2) +
Pr(esteem=4 variety=3 autonomy=1 security=2) +
Pr(esteem=2 variety=4 autonomy=1 security=3) +
Pr(esteem=4 variety=2 autonomy=1 security=3) +
Pr(esteem=2 variety=3 autonomy=1 security=4) +
Pr(esteem=3 variety=2 autonomy=1 security=4) = .10276103
```

variable	dp/dx	Std. Err.	z	P>|z|	[95% C.I.]	X
high*						
esteem	-.003524	.001258	-2.80	0.005	-.005989 -.001059	0
variety	-.036203	.00894	-4.05	0.000	-.053724 -.018681	1
autonomy	.057279	.013801	4.15	0.000	.030231 .084328	0
security	-.0128	.002665	-4.80	0.000	-.018024 -.007576	0
low*						
esteem	.000518	.000833	0.62	0.534	-.001116 .002151	1
variety	.006409	.010588	0.61	0.545	-.014343 .027161	0
autonomy	-.008818	.013766	-0.64	0.522	-.035799 .018163	0
security	.002314	.003697	0.63	0.531	-.004932 .009561	1
casevars						
female*	.013839	.021607	0.64	0.522	-.028509 .056188	0
score	.017917	.011062	1.62	0.105	-.003764 .039598	.32465

(*) dp/dx is for discrete change of indicator variable from 0 to 1

The probability computed by estat mfx matches the probability computed by predict, pr1 only within three digits. This outcome is because of how the computation is carried out and the numeric inaccuracy of the GHK simulator using a Hammersley point set of length 200. The computation carried out by estat mfx literally computes all six probabilities listed in the header of the MFX table and sums them. The computation by predict, pr1 is the same as predict after asmprobit (multinomial probit): it computes the probability that autonomy is chosen, thus requiring only one call to the GHK simulator. Hence, there is a difference in the reported values even though the two probability statements are equivalent.

◁

Saved Results

estat mfx saves the following in r():

Scalars
    r(pr)   scalar containing the computed probability of the ranked alternatives.

Matrices
    r(ranks) column vector containing the alternative ranks. The rownames identify the alternatives.

    r(mfx)   matrix containing the computed marginal effects and associated statistics. Column 1 of the matrix contains the marginal effects; column 2, their standard errors; column 3, their z statistics; and columns 4 and 5, the confidence intervals. Column 6 contains the values of the independent variables used to compute the probabilities r(pr).

Methods and Formulas

All postestimation commands listed above are implemented as ado-files.

Also See

[R] **asroprobit** — Alternative-specific rank-ordered probit regression

[R] **asmprobit** — Alternative-specific multinomial probit regression

[U] **20 Estimation and postestimation commands**

Title

> **BIC note** — Calculating and interpreting BIC

Description

This entry discusses a statistical issue that arises when using the Bayesian information criterion (BIC) to compare models.

Stata calculates BIC, assuming $N = $ `e(N)` —we will explain—but sometimes it would be better if a different N were used. Commands that calculate BIC have an option `n()`, allowing you to specify the N to be used.

In summary,

1. If you are comparing results estimated by the same estimation command, using the default BIC calculation is probably fine. There is an issue, but most researchers would ignore it.

2. If you are comparing results estimated by different estimation commands, you need to be on your guard.

   a. If the different estimation commands share the same definitions of observations, independence, and the like, you are back in case 1.

   b. If they differ in these regards, you need to think about the value of N that should be used. For example, `logit` and `xtlogit` differ in that the former assumes independent observations and the latter, independent panels.

   c. If estimation commands differ in the events being used over which the likelihood function is calculated, the information criteria may not be comparable at all. We say information *criteria* because this would apply equally to the Akaike information criterion (AIC) as well as BIC. For instance, `streg` and `stcox` produce such incomparable results. The events used by `streg` are the actual survival times, whereas the events used by `stcox` are failures within risk pools, conditional on the times at which failures occurred.

Remarks

Remarks are presented under the following headings:

Background
The problem of determining N
The problem of conformable likelihoods
The first problem does not arise with AIC; the second problem does
Calculating BIC correctly

Background

The AIC and the BIC are two popular measures for comparing maximum likelihood models. AIC and BIC are defined as

$$\text{AIC} = -2 \times \ln(\text{likelihood}) + 2 \times k$$

$$\text{BIC} = -2 \times \ln(\text{likelihood}) + \ln(N) \times k$$

where

$$k = \text{model degrees of freedom}$$

$$N = \text{number of observations}$$

We are going to discuss AIC along with BIC because AIC has some of the problems that BIC has, but not all.

AIC and BIC can be viewed as measures that combine fit and complexity. Fit is measured negatively by $-2 \times \ln(\text{likelihood})$; the larger the value, the worse the fit. Complexity is measured positively, either by $2 \times k$ (AIC) or $\ln(N) \times k$ (BIC).

Given two models fitted on the same data, the model with the smaller value of the information criterion is considered to be better.

There is a substantial literature on these measures; see Akaike (1974); Raftery (1996); Sakamoto, Ishiguro, and Kitagawa (1986); and Schwarz (1978).

When Stata calculates the above measures, it uses the rank of e(V) for k and it uses e(N) for N. e(V) and e(N) are Stata notation for results stored by the estimation command. e(V) is the variance–covariance matrix of the estimated parameters, and e(N) is the number of observations in the dataset used in calculating the result.

The problem of determining N

The difference between AIC and BIC is that AIC uses the constant 2 to weight k, whereas BIC uses $\ln(N)$.

Determining that value of N should be used is problematic. Despite appearances, the definition "N is the number of observations" is not easy to make operational. N does not appear in the likelihood function itself, N is not the output of a standard statistical formula, and what is an observation is often subjective.

▷ Example 1

Often, what is meant by N is obvious. Consider a simple logit model. What is meant by N is the number of observations that are statistically independent and that corresponds to M, the number of observations in the dataset used in the calculation. We will write $N = M$.

But now assume that the same dataset has a grouping variable and the data are thought to be clustered within group. To keep the problem simple, let's pretend that there are G groups and m observations within group, so that $M = G \times m$. Because you are worried about intragroup correlation, you fit your model with xtlogit, grouping on the grouping variable. Now you wish to calculate BIC. What is the N that should be used? $N = M$ or $N = G$?

That is a deep question. If the observations really are independent, then you should use $N = M$. If the observations within group are not just correlated but are duplicates of one another, and they had to be so, then you should use $M = G$. Between those two extremes, you should probably use a number between N and G, but determining what that number should be from measured correlations is difficult. Using $N = M$ is conservative in that, if anything, it overweights complexity. Conservativeness, however, is subjective, too: using $N = G$ could be considered more conservative in that fewer constraints are being placed on the data.

When the estimated correlation is high, our reaction would be that using $N = G$ is probably more reasonable. Our first reaction, however, would be that using BIC to compare models is probably a misuse of the measure.

Stata uses $N = M$. An informal survey of web-based literature suggests that $N = M$ is the popular choice.

There is another reason, not so good, to choose $N = M$. It makes across-model comparisons more likely to be valid when performed without thinking about the issue. Say that you wish to compare the logit and xtlogit results. Thus, you need to calculate

$$\mathrm{BIC}_p = -2 \times \ln(\mathrm{likelihood}_p) + \ln(N_p) \times k$$

$$\mathrm{BIC}_x = -2 \times \ln(\mathrm{likelihood}_x) + \ln(N_x) \times k$$

Whatever N you use, you must use the same N in both formulas. Stata's choice of $N = M$ at least meets that test.

◁

▷ Example 2

In the above example, using $N = M$ is reasonable. Now let's look at when using $N = M$ is wrong even if popular.

Consider a model fitted by stcox. Using $N = M$ is certainly wrong if for no other reason than M is not even a well-defined number. The same data can be represented by different datasets with different numbers of observations. For example, in one dataset, there might be 1 observation per subject. In another, the same subjects could have two records each, the first recording the first half of the time at risk and the second recording the remaining part. All statistics calculated by Stata on either dataset would be the same, but M would be different.

Deciding on the right definition, however, is difficult. Viewed one way, N in the Cox regression case should be the number of risk pools, R, because the Cox regression calculation is made on the basis of the independent risk pools. Viewed another way, N should be the number of subjects, N_{subj}, because, even though the likelihood function is based on risk pools, the parameters estimated are at the subject level.

You can decide which argument you prefer.

For parametric survival models, in single-record data, $N = M$ is unambiguously correct. For multirecord data, there is an argument for $N = M$ and for $N = N_{\mathrm{subj}}$.

◁

The problem of conformable likelihoods

The problem of conformable likelihoods does not concern N. Researchers sometimes use information criteria such as BIC and AIC to make comparisons across models. For that to be valid, the likelihoods must be conformable; i.e., the likelihoods must all measure the same thing.

It is common to think of the likelihood function as the $\Pr(\mathrm{data} \mid \mathrm{parameters})$, but in fact, the likelihood is

$$\Pr(\mathrm{particular\ events\ in\ the\ data} \mid \mathrm{parameters})$$

You must ensure that the events are the same.

For instance, they are not the same in the semiparametric Cox regression and the various parametric survival models. In Cox regression, the events are, at each failure time, that the subjects observed to fail in fact failed, given that failures occurred at those times. In the parametric models, the events are that each subject failed exactly when the subject was observed to fail.

The formula for AIC and BIC is

$$\text{measure} = -2 \times \ln(\text{likelihood}) + \text{complexity}$$

When you are comparing models, if the likelihoods are measuring different events, even if the models obtain estimates of the same parameters, differences in the information measures are irrelevant.

The first problem does not arise with AIC; the second problem does

Regardless of model, the problem of defining N never arises with AIC because N is not used in the AIC calculation. AIC uses a constant 2 to weight complexity as measured by k, rather than $\ln(N)$.

For both AIC and BIC, however, the likelihood functions must be conformable; i.e., they must be measuring the same event.

Calculating BIC correctly

When using BIC to compare results, and especially when using BIC to compare results from different models, you should think carefully about how N should be defined. Then specify that number by using the n() option:

```
. estimates stats full sub, n(74)
```

Model	Obs	ll(null)	ll(model)	df	AIC	BIC
full	102	-45.03321	-20.59083	4	49.18167	58.39793
sub	102	-45.03321	-27.17516	3	60.35031	67.26251

Note: N = 74 used in calculating BIC

Both estimates stats and estat ic allow the n() option; see [R] **estimates stats** and [R] **estat**.

Methods and Formulas

AIC and BIC are defined as

$$\text{AIC} = -2 \times \ln(\text{likelihood}) + 2 \times k$$

$$\text{BIC} = -2 \times \ln(\text{likelihood}) + \ln(N) \times k$$

where k is the model degrees of freedom calculated as the rank of variance–covariance matrix of the parameters e(V) and N is the number of observations used in estimation or, more precisely, the number of independent terms in the likelihood. Operationally, N is defined as e(N) unless option n() is specified.

References

Akaike, H. 1974. A new look at the statistical model identification. *IEEE Transactions on Automatic Control* 19: 716–723.

Raftery, A. 1996. Bayesian model selection in social research. In *Sociological Methodology*, ed. P. V. Marsden, 111–163. Oxford: Blackwell.

Sakamoto, Y., M. Ishiguro, and G. Kitagawa. 1986. *Akaike Information Criterion Statistics.* Dordrecht, The Netherlands: Reidel.

Schwarz, G. 1978. Estimating the dimension of a model. *Annals of Statistics* 6: 461–464.

Also See

[R] **estat** — Postestimation statistics

[R] **estimates stats** — Model statistics

Title

---
binreg — Generalized linear models: Extensions to the binomial family
---

Syntax

binreg *depvar* [*indepvars*] [*if*] [*in*] [*weight*] [, *options*]

options	description
Model	
<u>nocon</u>stant	suppress constant term
or	use logit link and report odds ratios
rr	use log link and report risk ratios
hr	use log-complement link and report health ratios
rd	use identity link and report risk differences
n(*#* | *varname*)	use *#* or *varname* for number of trials
<u>exposure</u>(*varname*)	include ln(*varname*) in model with coefficient constrained to 1
<u>off</u>set(*varname*)	include *varname* in model with coefficient constrained to 1
mu(*varname*)	use *varname* as the initial estimate for the mean of *depvar*
<u>init</u>(*varname*)	synonym for mu(*varname*)
SE/Robust	
vce(*vcetype*)	*vcetype* may be eim, <u>r</u>obust, <u>c</u>luster *clustvar*, oim, opg, <u>boot</u>strap, jackknife, hac *kernel*, jackknife1, or <u>unbiased</u>
t(*varname*)	variable name corresponding to time
<u>vf</u>actor(*#*)	multiply variance matrix by scalar *#*
disp(*#*)	quasi-likelihood multiplier
<u>scale</u>(x2 | dev | *#*)	set the scale parameter; default is scale(1)
Reporting	
<u>level</u>(*#*)	set confidence level; default is level(95)
<u>coef</u>ficients	report nonexponentiated coefficients
Max options	
irls	use iterated, reweighted least-squares optimization; the default
ml	use maximum likelihood optimization
maximize_options	control the maximization process; seldom used
fisher(*#*)	Fisher scoring steps
search	search for good starting values

depvar and *indepvars* may contain time-series operators; see [U] **11.4.3 Time-series varlists**.
bootstrap, by, jackknife, rolling, statsby, and xi are allowed; see [U] **11.1.10 Prefix commands**.
Weights are not allowed with the bootstrap prefix.
aweights are not allowed with the jackknife prefix.
fweights, aweights, iweights, and pweights are allowed; see [U] **11.1.6 weight**.
See [U] **20 Estimation and postestimation commands** for more capabilities of estimation commands.

174

Description

binreg fits generalized linear models for the binomial family. It estimates odds ratios, risk ratios, health ratios, and risk differences. The available links are

Option	Implied link	Parameter
or	logit	odds ratios $= \exp(\beta)$
rr	log	risk ratios $= \exp(\beta)$
hr	log complement	health ratios $= \exp(\beta)$
rd	identity	risk differences $= \beta$

Estimates of odds, risk, and health ratios are obtained by exponentiating the appropriate coefficients. The option or produces the same results as Stata's logistic command, and or coefficients yields the same results as the logit command. When no link is specified, or is assumed.

Options

_____ Model _____

noconstant; see [R] **estimation options**.

or requests the logit link and results in odds ratios if coefficients is not specified.

rr requests the log link and results in risk ratios if coefficients is not specified.

hr requests the log-complement link and results in health ratios if coefficients is not specified.

rd requests the identity link and results in risk differences.

n(# | *varname*) specifies either a constant integer to use as the denominator for the binomial family or a variable that holds the denominator for each observation.

exposure(*varname*), offset(*varname*); see [R] **estimation options**.

mu(*varname*) specifies *varname* containing an initial estimate for the mean of *depvar*. This option can be useful if you encounter convergence difficulties. init(*varname*) is a synonym.

_____ SE/Robust _____

vce(*vcetype*) specifies the type of standard error reported, which includes types that are robust to some kinds of misspecification, that allow for intragroup correlation, that are derived from asymptotic theory, and that use bootstrap or jackknife methods; see [R] *vce_option*.

vce(eim), the default, uses the expected information matrix for the variance estimator.

binreg also allows the following:

vce(hac *kernel* [#]) specifies that a heteroskedasticity- and autocorrelation-consistent (HAC) variance estimate be used. HAC refers to the general form for combining weighted matrices to form the variance estimate. There are three kernels built into binreg. *kernel* is a user-written program or one of

$$\underline{\text{nw}}\text{est} \mid \underline{\text{gall}}\text{ant} \mid \underline{\text{and}}\text{erson}$$

If # is not specified, $N - 2$ is assumed.

vce(jackknife1) specifies that the one-step jackknife estimate of variance be used.

vce(unbiased) specifies that the unbiased sandwich estimate of variance be used.

t(*varname*) specifies the variable name corresponding to time; see [TS] **tsset**. binreg does not always need to know t(), though it does if vce(hac ...) is specified. In that case, you can either specify the time variable with t(), or you can tsset your data before calling binreg. When the time variable is required, binreg assumes that the observations are spaced equally over time.

vfactor(#) specifies a scalar by which to multiply the resulting variance matrix. This option allows users to match output with other packages, which may apply degrees of freedom or other small-sample corrections to estimates of variance.

disp(#) multiplies the variance of *depvar* by # and divides the deviance by #. The resulting distributions are members of the quasilikelihood family.

scale(x2|dev|#) overrides the default scale parameter. This option is allowed only with Hessian (information matrix) variance estimates.

By default, scale(1) is assumed for the discrete distributions (binomial, Poisson, and negative binomial), and scale(x2) is assumed for the continuous distributions (Gaussian, gamma, and inverse Gaussian).

scale(x2) specifies that the scale parameter be set to the Pearson chi-squared (or generalized chi-squared) statistic divided by the residual degrees of freedom, which was recommended by McCullagh and Nelder (1989) as a good general choice for continuous distributions.

scale(dev) sets the scale parameter to the deviance divided by the residual degrees of freedom. This option provides an alternative to scale(x2) for continuous distributions and overdispersed or underdispersed discrete distributions.

scale(#) sets the scale parameter to #.

___ Reporting ___

level(#), noconstant; see [R] **estimation options**.

coefficients displays the nonexponentiated coefficients and corresponding standard errors and confidence intervals. This option has no effect when the rd option is specified, as it always presents the nonexponentiated coefficients.

___ Max options ___

irls requests iterated, reweighted least-squares (IRLS) optimization of the deviance instead of Newton–Raphson optimization of the log likelihood. This option is the default.

ml requests that optimization be carried out by using Stata's ml command.

maximize_options: technique(*algorithm_spec*), [no]log, trace, gradient, showstep, hessian, shownrtolerance, difficult, iterate(#), tolerance(#), ltolerance(#), gtolerance(#), nrtolerance(#), nonrtolerance, from(*init_specs*); see [R] **maximize**. These options are seldom used.

Setting the optimization method to ml, with technique() set to something other than BHHH, changes the *vcetype* to vce(oim). Specifying technique(bhhh) changes *vcetype* to vce(opg).

fisher(#) specifies the number of Newton–Raphson steps that should use the Fisher scoring Hessian or expected information matrix (EIM) before switching to the observed information matrix (OIM). This option is available only if ml is specified and is useful only for Newton–Raphson optimization.

search specifies that the command search for good starting values. This option is available only if ml is specified and is useful only for Newton–Raphson optimization.

Remarks

Wacholder (1986) suggests methods for estimating risks ratios and risk differences from prospective binomial data. These estimates are obtained by selecting the proper link functions in the generalized linear-model framework. (See *Methods and Formulas* for details; also see [R] **glm**.)

▷ Example 1

Wacholder (1986) presents an example, using data from Wright et al. (1983), of an investigation of the relationship between alcohol consumption and the risk of a low-birthweight baby. Covariates examined included whether the mother smoked (yes or no), mother's social class (three levels), and drinking frequency (light, moderate, or heavy). The data for the 18 possible categories determined by the covariates are illustrated below.

Let's first describe the data and list a few observations.

```
. use http://www.stata-press.com/data/r10/binreg
. list
```

	cat	d	n	alc	smo	soc
1.	1	11	84	3	1	1
2.	2	5	79	2	1	1
3.	3	11	169	1	1	1
4.	4	6	28	3	2	1
5.	5	3	13	2	2	1
6.	6	1	26	1	2	1
7.	7	4	22	3	1	2
8.	8	3	25	2	1	2
9.	9	12	162	1	1	2
10.	10	4	17	3	2	2
11.	11	2	7	2	2	2
12.	12	6	38	1	2	2
13.	13	0	14	3	1	3
14.	14	1	18	2	1	3
15.	15	12	91	1	1	3
16.	16	7	19	3	2	3
17.	17	2	18	2	2	3
18.	18	8	70	1	2	3

Each observation corresponds to one of the 18 covariate structures. The number of low-birthweight babies from n in each category is given by the variable d.

We begin by estimating risk ratios:

```
. xi: binreg d I.soc I.alc I.smo, n(n) rr
I.soc             _Isoc_1-3         (naturally coded; _Isoc_1 omitted)
I.alc             _Ialc_1-3         (naturally coded; _Ialc_1 omitted)
I.smo             _Ismo_1-2         (naturally coded; _Ismo_1 omitted)
Iteration 1:   deviance =    14.2879
Iteration 2:   deviance =     13.607
Iteration 3:   deviance =   13.60503
Iteration 4:   deviance =   13.60503
```

```
Generalized linear models                          No. of obs       =          18
Optimization       : MQL Fisher scoring            Residual df      =          12
                     (IRLS EIM)                     Scale parameter =           1
Deviance           =    13.6050268                 (1/df) Deviance =    1.133752
Pearson            =    11.51517095                (1/df) Pearson  =   .9595976

Variance function: V(u) = u*(1-u/n)                [Binomial]
Link function    : g(u) = ln(u/n)                  [Log]
                                                   BIC              =   -21.07943
```

		EIM					
d	Risk Ratio	Std. Err.	z	P>|z|	[95% Conf. Interval]	
_Isoc_2	1.340001	.3127382	1.25	0.210	.848098	2.11721
_Isoc_3	1.349487	.3291488	1.23	0.219	.8366715	2.176619
_Ialc_2	1.191157	.3265354	0.64	0.523	.6960276	2.038503
_Ialc_3	1.974078	.4261751	3.15	0.002	1.293011	3.013884
_Ismo_2	1.648444	.332875	2.48	0.013	1.109657	2.448836

By default, Stata reports the risk ratios (the exponentiated regression coefficients) estimated by the model. We can see that the risk ratio comparing heavy drinkers with light drinkers, after adjusting for smoking and social class, is $\exp(0.6801017) = 1.9740785$. That is, mothers who drink heavily during their pregnancy have approximately twice the risk of delivering low-birthweight babies as mothers who are light drinkers.

The nonexponentiated coefficients can be obtained with the `coefficients` option:

```
. xi: binreg d I.soc I.alc I.smo, n(n) rr coefficients
I.soc            _Isoc_1-3          (naturally coded; _Isoc_1 omitted)
I.alc            _Ialc_1-3          (naturally coded; _Ialc_1 omitted)
I.smo            _Ismo_1-2          (naturally coded; _Ismo_1 omitted)

Iteration 1:   deviance =    14.2879
Iteration 2:   deviance =     13.607
Iteration 3:   deviance =   13.60503
Iteration 4:   deviance =   13.60503

Generalized linear models                          No. of obs       =          18
Optimization       : MQL Fisher scoring            Residual df      =          12
                     (IRLS EIM)                     Scale parameter =           1
Deviance           =    13.6050268                 (1/df) Deviance =    1.133752
Pearson            =    11.51517095                (1/df) Pearson  =   .9595976

Variance function: V(u) = u*(1-u/n)                [Binomial]
Link function    : g(u) = ln(u/n)                  [Log]
                                                   BIC              =   -21.07943
```

		EIM					
d	Coef.	Std. Err.	z	P>|z|	[95% Conf. Interval]	
_Isoc_2	.2926702	.2333866	1.25	0.210	-.1647591	.7500994
_Isoc_3	.2997244	.2439066	1.23	0.219	-.1783238	.7777726
_Ialc_2	.1749248	.274133	0.64	0.523	-.362366	.7122156
_Ialc_3	.6801017	.2158856	3.15	0.002	.2569737	1.10323
_Ismo_2	.4998317	.2019329	2.48	0.013	.1040505	.8956129
_cons	-2.764079	.2031606	-13.61	0.000	-3.162266	-2.365891

Risk differences are obtained with the `rd` option:

```
. xi: binreg d I.soc I.alc I.smo, n(n) rd
I.soc              _Isoc_1-3          (naturally coded; _Isoc_1 omitted)
I.alc              _Ialc_1-3          (naturally coded; _Ialc_1 omitted)
I.smo              _Ismo_1-2          (naturally coded; _Ismo_1 omitted)

Iteration 1:    deviance =   18.67277
Iteration 2:    deviance =   14.94364
Iteration 3:    deviance =   14.9185
Iteration 4:    deviance =   14.91762
Iteration 5:    deviance =   14.91758
Iteration 6:    deviance =   14.91758
Iteration 7:    deviance =   14.91758

Generalized linear models                  No. of obs       =         18
Optimization     : MQL Fisher scoring      Residual df      =         12
                   (IRLS EIM)              Scale parameter  =          1
Deviance         =  14.91758277           (1/df) Deviance  =   1.243132
Pearson          =  12.60353235           (1/df) Pearson   =   1.050294

Variance function: V(u) = u*(1-u/n)        [Binomial]
Link function    : g(u) = u/n             [Identity]

                                           BIC              =  -19.76688
```

d	Risk Diff.	EIM Std. Err.	z	P>|z|	[95% Conf. Interval]	
_Isoc_2	.0263817	.0232124	1.14	0.256	-.0191137	.0718771
_Isoc_3	.0365553	.0268668	1.36	0.174	-.0161026	.0892132
_Ialc_2	.0122539	.0257713	0.48	0.634	-.0382569	.0627647
_Ialc_3	.0801291	.0302878	2.65	0.008	.020766	.1394921
_Ismo_2	.0542415	.0270838	2.00	0.045	.0011582	.1073248
_cons	.059028	.0160693	3.67	0.000	.0275327	.0905232

The risk difference between heavy drinkers and light drinkers is simply the value of the coefficient for _Ialc_3 = 0.0801291. Because the risk differences are obtained directly from the coefficients estimated by using the identity link, the `coefficients` option has no effect here.

Health ratios are obtained with the `hr` option. The health ratios (exponentiated coefficients for the log-complement link) are reported directly.

(*Continued on next page*)

```
. xi: binreg d I.soc I.alc I.smo, n(n) hr
I.soc               _Isoc_1-3        (naturally coded; _Isoc_1 omitted)
I.alc               _Ialc_1-3        (naturally coded; _Ialc_1 omitted)
I.smo               _Ismo_1-2        (naturally coded; _Ismo_1 omitted)

Iteration 1:   deviance =  21.15233
Iteration 2:   deviance =  15.16467
Iteration 3:   deviance =  15.13205
Iteration 4:   deviance =  15.13114
Iteration 5:   deviance =  15.13111
Iteration 6:   deviance =  15.13111
Iteration 7:   deviance =  15.13111
```

```
Generalized linear models                   No. of obs          =          18
Optimization      : MQL Fisher scoring      Residual df         =          12
                    (IRLS EIM)              Scale parameter =              1
Deviance          =   15.13110545           (1/df) Deviance =        1.260925
Pearson           =   12.84203917           (1/df) Pearson  =         1.07017

Variance function: V(u) = u*(1-u/n)         [Binomial]
Link function     : g(u) = ln(1-u/n)        [Log complement]

                                            BIC                 =   -19.55336
```

d	HR	EIM Std. Err.	z	P>|z|	[95% Conf. Interval]	
_Isoc_2	.9720541	.024858	-1.11	0.268	.9245342	1.022017
_Isoc_3	.9597182	.0290412	-1.36	0.174	.9044535	1.01836
_Ialc_2	.9871517	.0278852	-0.46	0.647	.9339831	1.043347
_Ialc_3	.9134243	.0325726	-2.54	0.011	.8517631	.9795493
_Ismo_2	.9409983	.0296125	-1.93	0.053	.8847125	1.000865

(HR) Health ratios

To see the nonexponentiated coefficients, we can specify the `coefficients` option.   ◁

Saved Results

binreg saves the following in e():

Scalars

e(N)	number of observations	e(N_clust)	number of clusters
e(k)	number of parameters	e(deviance)	deviance
e(k_eq)	number of equations in e(b)	e(deviance_s)	scaled deviance
e(k_eq_model)	number of equations in model Wald test	e(deviance_p)	Pearson deviance
		e(deviance_ps)	scaled Pearson deviance
e(df)	residual degrees of freedom	e(dispers)	dispersion
e(phi)	model scale parameter	e(dispers_s)	scaled dispersion
e(disp)	dispersion parameter	e(dispers_p)	Pearson dispersion
e(df_m)	model degrees of freedom	e(dispers_ps)	scaled Pearson dispersion
e(aic)	model AIC, if ml	e(vf)	factor set by vfactor(),
e(bic)	model BIC		1 if not set
e(ll)	log likelihood, if ml	e(rc)	return code

Macros

e(cmd)	binreg	e(offset)	offset
e(cmdline)	command as typed	e(cons)	noconstant or not set
e(depvar)	name of dependent variable	e(hac_kernel)	HAC kernel
e(eform)	eform() option implied by or, rr, hr, or rd	e(hac_lag)	HAC lag
		e(vce)	*vcetype* specified in vce()
e(varfunc)	name of variance function used	e(vcetype)	title used to label Std. Err.
e(varfunct)	Binomial	e(opt)	type of optimization
e(varfuncf)	variance function	e(opt1)	optimization title, line 1
e(link)	link function used by glm	e(opt2)	optimization title, line 2
e(linkt)	link title	e(ml_method)	type of ml method
e(linkf)	link form	e(user)	name of likelihood-evaluator program
e(m)	number of binomial trials	e(technique)	maximization technique
e(wtype)	weight type	e(crittype)	optimization criterion
e(wexp)	weight expression	e(properties)	b V
e(title_fl)	family–link title	e(predict)	program used to implement predict
e(clustvar)	name of cluster variable		

Matrices

e(b)	coefficient vector	e(V)	variance–covariance matrix of the estimators
e(ilog)	iteration log (up to 20 iterations)		
e(gradient)	gradient vector		

Functions

e(sample)	marks estimation sample

Methods and Formulas

binreg is implemented as an ado-file.

Let π_i be the probability of success for the ith observation, $i = 1, \dots, N$, and let $X\beta$ be the linear predictor. The link function relates the covariates of each observation to its respective probability through the linear predictor.

In logistic regression, the logit link is used:

$$\ln\left(\frac{\pi}{1-\pi}\right) = X\beta$$

The regression coefficient β_k represents the change in the logarithm of the odds associated with a one-unit change in the value of the X_k covariate; thus, $\exp(\beta_k)$ is the ratio of the odds associated with a change of one unit in X_k.

For risk differences, the identity link $\pi = X\beta$ is used. The regression coefficient β_k represents the risk difference associated with a change of one unit in X_k. When using the identity link, you can obtain fitted probabilities outside the interval $(0, 1)$. As suggested by Wacholder, at each iteration, fitted probabilities are checked for range conditions (and put back in range if necessary). For example, if the identity link results in a fitted probability that is smaller than $1e-4$, the probability is replaced with $1e-4$ before the link function is calculated.

A similar adjustment is made for the logarithmic link, which is used for estimating the risk ratio, $\ln(\pi) = X\beta$, where $\exp(\beta_k)$ is the risk ratio associated with a change of one unit in X_k, and for the log-complement link used to estimate the probability of no disease or health, where $\exp(\beta_k)$ represents the "health ratio" associated with a change of one unit in X_k.

References

Hardin, J., and M. Cleves. 1999. sbe29: Generalized linear models: Extensions to the binomial family. *Stata Technical Bulletin* 50: 21–25. Reprinted in *Stata Technical Bulletin Reprints*, vol. 9, pp. 140–146.

Kleinbaum, D. G., and M. Klein. 2002. *Logistic Regression: A Self-Learning Text*. 2nd ed. New York: Springer.

McCullagh, P., and J. A. Nelder. 1989. *Generalized Linear Models*. 2nd ed. London: Chapman & Hall/CRC.

Wacholder, S. 1986. Binomial regression in GLIM: Estimating risk ratios and risk differences. *American Journal of Epidemiology* 123: 174–184.

Wright, J. T., E. J. Waterson, I. G. Barrison, P. J. Toplis, I. G. Lewis, M. G. Gordon, K. D. MacRae, N. F. Morris, and I. M. Murray-Lyon. 1983. Alcohol consumption, pregnancy and low birthweight. *Lancet* 1: 663–665.

Also See

[R] **binreg postestimation** — Postestimation tools for binreg

[R] **glm** — Generalized linear models

[U] **20 Estimation and postestimation commands**

Title

binreg postestimation — Postestimation tools for binreg

Description

The following postestimation commands are available for binreg:

command	description
adjust[1]	adjusted predictions of $\mathbf{x}\beta$, probabilities, or $\exp(\mathbf{x}\beta)$
estat	AIC, BIC, VCE, and estimation sample summary
estimates	cataloging estimation results
lincom	point estimates, standard errors, testing, and inference for linear combinations of coefficients
linktest	link test for model specification
mfx	marginal effects or elasticities
nlcom	point estimates, standard errors, testing, and inference for nonlinear combinations of coefficients
predict	predictions, residuals, influence statistics, and other diagnostic measures
predictnl	point estimates, standard errors, testing, and inference for generalized predictions
test	Wald tests for simple and composite linear hypotheses
testnl	Wald tests of nonlinear hypotheses

[1] adjust is not appropriate with time-series operators.

See the corresponding entries in the *Stata Base Reference Manual* for details.

(*Continued on next page*)

Syntax for predict

predict [*type*] *newvar* [*if*] [*in*] [, *statistic options*]

statistic	description
Main	
<u>mu</u>	$g^{-1}(\mathbf{x}\widehat{\beta})$; the default
xb	linear prediction $\eta = \mathbf{x}\widehat{\beta}$
<u>eta</u>	synonym for xb
stdp	standard error of the linear prediction
<u>a</u>nscombe	Anscombe (1953) residuals
<u>c</u>ooksd	Cook's distance
<u>d</u>eviance	deviance residuals
hat	diagonals of the "hat" matrix as an analog to simple linear regression
likelihood	weighted average of the standardized deviance and standard Pearson residuals
<u>p</u>earson	Pearson residuals
<u>r</u>esponse	differences between the observed and fitted outcomes
<u>sc</u>ore	first derivative of the log likelihood with respect to $\mathbf{x}_j\beta$
<u>w</u>orking	working residuals

options	description
Options	
<u>noo</u>ffset	modify calculations to ignore the offset variable
<u>stand</u>ardized	multiply residual by the factor $(1-h)^{1/2}$
<u>stud</u>entized	multiply residual by one over the square root of the estimated scale parameter
<u>mod</u>ified	modify denominator of residual to be a reasonable estimate of the variance of *depvar*
<u>adj</u>usted	adjust deviance residual to make the convergence to the limiting normal distribution faster

These statistics are available both in and out of sample; type predict ... if e(sample) ... if wanted only for the estimation sample.

Options for predict

⌐ **Main** ⌐

mu, the default, specifies that predict calculate $g^{-1}(\mathbf{x}\widehat{\beta})$, the inverse link of the linear prediction.

xb calculates the linear prediction $\eta = \mathbf{x}\widehat{\beta}$.

eta is a synonym for xb.

stdp calculates the standard error of the linear prediction.

anscombe calculates the Anscombe (1953) residuals to produce residuals that closely follow a normal distribution.

cooksd calculates Cook's distance, which measures the aggregate change in the estimated coefficients when each observation is left out of the estimation.

deviance calculates the deviance residuals, which are recommended by McCullagh and Nelder (1989) and others as having the best properties for examining goodness of fit of a GLM. They are approximately normally distributed if the model is correct and may be plotted against the fitted values or against a covariate to inspect the model's fit. Also see the pearson option below.

hat calculates the diagonals of the "hat" matrix as an analog to simple linear regression.

likelihood calculates a weighted average of the standardized deviance and standardized Pearson (described below) residuals.

pearson calculates the Pearson residuals, which often have markedly skewed distributions for nonnormal family distributions. Also see the deviance option above.

response calculates the differences between the observed and fitted outcomes.

score calculates the equation-level score, $\partial \ln L / \partial(\mathbf{x}_j \boldsymbol{\beta})$.

working calculates the working residuals, which are response residuals weighted according to the derivative of the link function.

Options

nooffset is relevant only if you specified offset(*varname*) for binreg. It modifies the calculations made by predict so that they ignore the offset variable; the linear prediction is treated as $\mathbf{x}_j \mathbf{b}$ rather than as $\mathbf{x}_j \mathbf{b} + \text{offset}_j$.

standardized requests that the residual be multiplied by the factor $(1 - h)^{-1/2}$, where h is the diagonal of the hat matrix. This step is done to take into account the correlation between *depvar* and its predicted value.

studentized requests that the residual be multiplied by one over the square root of the estimated scale parameter.

modified requests that the denominator of the residual be modified to be a reasonable estimate of the variance of *depvar*. The base residual is multiplied by the factor $(k/w)^{-1/2}$, where k is either one or the user-specified dispersion parameter and w is the specified weight (or one if left unspecified).

adjusted adjusts the deviance residual to make the convergence to the limiting normal distribution faster. The adjustment deals with adding to the deviance residual a higher-order term depending on the variance function family. This option is allowed only when deviance is specified.

Methods and Formulas

All postestimation commands listed above are implemented as an ado-file.

References

Anscombe, F. J. 1953. Contribution of discussion paper by H. Hotelling 'New light on the correlation coefficient and its transform'. *Journal of the Royal Statistical Society, Series B* 15: 229–230.

McCullagh, P., and J. A. Nelder. 1989. *Generalized Linear Models*. 2nd ed. London: Chapman & Hall/CRC.

Also See

[R] **binreg** — Generalized linear models: Extensions to the binomial family

[U] **20 Estimation and postestimation commands**

Title

> **biprobit** — Bivariate probit regression

Syntax

Bivariate probit model

> biprobit *depvar*$_1$ *depvar*$_2$ [*varlist*] [*if*] [*in*] [*weight*] [, *options*]

Seemingly unrelated bivariate probit model

> biprobit *equation*$_1$ *equation*$_2$ [*if*] [*in*] [*weight*] [, *su_options*]

where *equation*$_1$ and *equation*$_2$ are specified as

> ([*eqname*:] *depvar* [=] [*varlist*] [, <u>nocon</u>stant <u>off</u>set(*varname*)])

options	description
Model	
<u>nocons</u>tant	suppress constant term
<u>partial</u>	fit partial observability model
offset1(*varname*)	offset variable for first equation
offset2(*varname*)	offset variable for second equation
<u>constr</u>aints(*constraints*)	apply specified linear constraints
<u>coll</u>inear	keep collinear variables
SE/Robust	
vce(*vcetype*)	*vcetype* may be oim, <u>r</u>obust, <u>cl</u>uster *clustvar*, opg, <u>boot</u>strap, or <u>jackk</u>nife
Reporting	
<u>level</u>(*#*)	set confidence level; default is level(95)
noskip	perform likelihood-ratio test
Max options	
maximize_options	control the maximization process; seldom used

su_options	description
Model	
<u>partial</u>	fit partial observability model
<u>constr</u>aints(*constraints*)	apply specified linear constraints
<u>col</u>linear	keep collinear variables
SE/Robust	
vce(*vcetype*)	*vcetype* may be oim, <u>r</u>obust, <u>cl</u>uster *clustvar*, opg, <u>boot</u>strap, or <u>jack</u>knife
Reporting	
<u>l</u>evel(#)	set confidence level; default is level(95)
noskip	perform likelihood-ratio test
Max options	
maximize_options	control the maximization process; seldom used

depvar$_1$, *depvar*$_2$, *varlist*, and *depvar* may contain time-series operators; see [U] **11.4.3 Time-series varlists**.
bootstrap, by, jackknife, rolling, statsby, svy, and xi are allowed; see [U] **11.1.10 Prefix commands**.
Weights are not allowed with the bootstrap prefix.
vce(), noskip, and weights are not allowed with the svy prefix.
pweights, fweights, and iweights are allowed; see [U] **11.1.6 weight**.
See [U] **20 Estimation and postestimation commands** for more capabilities of estimation commands.

Description

biprobit fits maximum-likelihood two-equation probit models—either a bivariate probit or a seemingly unrelated probit (limited to two equations).

Options

> ___ Model ___

noconstant; see [R] **estimation options**.

partial specifies that the partial observability model be fitted. This particular model commonly has poor convergence properties, so we recommend that you use the difficult option if you want to fit the Poirier partial observability model; see [R] **ml**.

This model computes the product of the two dependent variables so that you do not have to replace each with the product.

offset1(*varname*), offset2(*varname*), constraints(*constraints*), collinear; see [R] **estimation options**.

> ___ SE/Robust ___

vce(*vcetype*) specifies the type of standard error reported, which includes types that are derived from asymptotic theory, that are robust to some kinds of misspecification, that allow for intragroup correlation, and that use bootstrap or jackknife methods; see [R] ***vce_option***.

`level(#)`; see [R] **estimation options**.

`noskip` specifies that a full maximum-likelihood model with only a constant for the regression equation be fitted. This model is not displayed but is used as the base model to compute a likelihood-ratio test for the model test statistic displayed in the estimation header. By default, the overall model test statistic is an asymptotically equivalent Wald test of all the parameters in the regression equation being zero (except the constant). For many models, this option can substantially increase estimation time.

maximize_options: <u>diff</u>icult, <u>tech</u>nique(*algorithm_spec*), <u>iter</u>ate(*#*), [<u>no</u>]<u>log</u>, <u>tr</u>ace, gradient, showstep, <u>hess</u>ian, <u>shownr</u>tolerance, <u>tol</u>erance(*#*), <u>ltol</u>erance(*#*), <u>gtol</u>erance(*#*), <u>nrtol</u>erance(*#*), <u>nonrtol</u>erance, from(*init_specs*); see [R] **maximize**. These options are seldom used.

Setting the optimization type to `technique(bhhh)` resets the default *vcetype* to `vce(opg)`.

Remarks

For a good introduction to the bivariate probit models, see Greene (2003, 710–714) and Pindyck and Rubinfeld (1998). Poirier (1980) explains the partial observability model. Van de Ven and Van Pragg (1981) explain the probit model with sample selection; see [R] **heckprob** for details.

▷ Example 1

We use the data from Pindyck and Rubinfeld (1998, 332). In this dataset, the variables are whether children attend private school (`private`), number of years the family has been at the present residence (`years`), log of property tax (`logptax`), log of income (`loginc`), and whether the head of the household voted for an increase in property taxes (`vote`).

We wish to model the bivariate outcomes of whether children attend private school and whether the head of the household voted for an increase in property tax based on the other covariates.

```
. use http://www.stata-press.com/data/r10/school

. biprobit private vote years logptax loginc

Fitting comparison equation 1:

Iteration 0:   log likelihood = -31.967097
Iteration 1:   log likelihood = -31.454068
Iteration 2:   log likelihood = -31.448959
Iteration 3:   log likelihood = -31.448958

Fitting comparison equation 2:

Iteration 0:   log likelihood = -63.036914
Iteration 1:   log likelihood = -58.581911
Iteration 2:   log likelihood = -58.497419
Iteration 3:   log likelihood = -58.497288

Comparison:    log likelihood = -89.946246

Fitting full model:

Iteration 0:   log likelihood = -89.946246
Iteration 1:   log likelihood = -89.258897
Iteration 2:   log likelihood = -89.254028
Iteration 3:   log likelihood = -89.254028
```

Bivariate probit regression

```
Number of obs   =        95
Wald chi2(6)    =      9.59
Prob > chi2     =    0.1431
```

Log likelihood = -89.254028

	Coef.	Std. Err.	z	P>|z|	[95% Conf. Interval]	
private						
years	-.0118884	.0256778	-0.46	0.643	-.0622159	.0384391
logptax	-.1066962	.6669782	-0.16	0.873	-1.413949	1.200557
loginc	.3762037	.5306484	0.71	0.478	-.663848	1.416255
_cons	-4.184694	4.837817	-0.86	0.387	-13.66664	5.297253
vote						
years	-.0168561	.0147834	-1.14	0.254	-.0458309	.0121188
logptax	-1.288707	.5752266	-2.24	0.025	-2.416131	-.1612839
loginc	.998286	.4403565	2.27	0.023	.1352031	1.861369
_cons	-.5360573	4.068509	-0.13	0.895	-8.510188	7.438073
/athrho	-.2764525	.2412099	-1.15	0.252	-.7492153	.1963102
rho	-.2696186	.2236753			-.6346806	.1938267

Likelihood-ratio test of rho=0:   chi2(1) =   1.38444   Prob > chi2 = 0.2393

The output shows several iteration logs. The first iteration log corresponds to running the univariate probit model for the first equation, and the second log corresponds to running the univariate probit for the second model. If $\rho = 0$, the sum of the log likelihoods from these two models will equal the log likelihood of the bivariate probit model; this sum is printed in the iteration log as the comparison log likelihood.

The final iteration log is for fitting the full bivariate probit model. A likelihood-ratio test of the log likelihood for this model and the comparison log likelihood is presented at the end of the output. If we had specified the vce(robust) option, this test would be presented as a Wald test instead of as a likelihood-ratio test.

We could have fitted the same model by using the seemingly unrelated syntax as

```
. biprobit (private=years logptax loginc) (vote=years logptax loginc)
```

◁

Saved Results

biprobit saves the following in e():

Scalars

e(N)	number of observations	e(N_clust)	number of clusters
e(k)	number of parameters	e(chi2)	χ^2
e(k_eq)	number of equations	e(chi2_c)	χ^2 for comparison test
e(k_dv)	number of dependent variables	e(p)	significance
e(k_aux)	number of auxiliary parameters	e(rho)	ρ
e(k_eq_model)	number of equations in model Wald test	e(rank)	rank of e(V)
		e(rank0)	rank of e(V) for constant-only model
e(df_m)	model degrees of freedom		
e(ll)	log likelihood	e(ic)	number of iterations
e(ll_0)	log likelihood, constant-only model (noskip only)	e(rc)	return code
		e(converged)	1 if converged, 0 otherwise
e(ll_c)	log likelihood, comparison model		

Macros

e(cmd)	biprobit	e(vce)	*vcetype* specified in vce()
e(cmdline)	command as typed	e(vcetype)	title used to label Std. Err.
e(depvar)	names of dependent variables	e(diparm#)	display transformed parameter #
e(wtype)	weight type	e(opt)	type of optimization
e(wexp)	weight expression	e(ml_method)	type of ml method
e(title)	title in estimation output	e(user)	name of likelihood-evaluator program
e(clustvar)	name of cluster variable		
e(offset1)	offset for first equation	e(technique)	maximization technique
e(offset2)	offset for second equation	e(crittype)	optimization criterion
e(chi2type)	Wald or LR; type of model χ^2 test	e(properties)	b V
e(chi2_ct)	Wald or LR; type of model χ^2 test corresponding to e(chi2_c)	e(predict)	program used to implement predict

Matrices

e(b)	coefficient vector	e(V)	variance–covariance matrix of the estimators
e(ilog)	iteration log (up to 20 iterations)		
e(gradient)	gradient vector		

Functions

e(sample)	marks estimation sample

Methods and Formulas

`biprobit` is implemented as an ado-file.

The log likelihood, $\ln L$, is given by

$$
\xi_j^\beta = x_j \beta + \text{offset}_j^\beta
$$

$$
\xi_j^\gamma = z_j \gamma + \text{offset}_j^\gamma
$$

$$
q_{1j} = \begin{cases} 1 & \text{if } y_{1j} \neq 0 \\ -1 & \text{otherwise} \end{cases}
$$

$$
q_{2j} = \begin{cases} 1 & \text{if } y_{2j} \neq 0 \\ -1 & \text{otherwise} \end{cases}
$$

$$
\rho_j^* = q_{1j} q_{2j} \rho
$$

$$
\ln L = \sum_{j=1}^n w_j \ln \Phi_2 \left(q_{1j} \xi_j^\beta, q_{2j} \xi_j^\gamma, \rho_j^* \right)
$$

where $\Phi_2()$ is the cumulative bivariate normal distribution function (with mean $\begin{bmatrix} 0 & 0 \end{bmatrix}'$) and w_j is an optional weight for observation j. This derivation assumes that

$$
y_{1j}^* = x_j \beta + \epsilon_{1j} + \text{offset}_j^\beta
$$

$$
y_{2j}^* = z_j \gamma + \epsilon_{2j} + \text{offset}_j^\gamma
$$

$$
E(\epsilon_1) = E(\epsilon_2) = 0
$$

$$
\text{Var}(\epsilon_1) = \text{Var}(\epsilon_2) = 1
$$

$$
\text{Cov}(\epsilon_1, \epsilon_2) = \rho
$$

where y_{1j}^* and y_{2j}^* are the unobserved latent variables; instead, we observe only $y_{ij} = 1$ if $y_{ij}^* > 0$ and $y_{ij} = 0$ otherwise (for $i = 1, 2$).

In the maximum likelihood estimation, ρ is not directly estimated, but atanh ρ is

$$
\text{atanh } \rho = \frac{1}{2} \ln \left(\frac{1 + \rho}{1 - \rho} \right)
$$

From the form of the likelihood, if $\rho = 0$, then the log likelihood for the bivariate probit models is equal to the sum of the log likelihoods of the two univariate probit models. A likelihood-ratio test may therefore be performed by comparing the likelihood of the full bivariate model with the sum of the log likelihoods for the univariate probit models.

References

Greene, W. H. 2003. *Econometric Analysis*. 5th ed. Upper Saddle River, NJ: Prentice Hall.

Hardin, J. W. 1996. sg61: Bivariate probit models. *Stata Technical Bulletin* 33: 15–20. Reprinted in *Stata Technical Bulletin Reprints*, vol. 6, pp. 152–158.

Heckman, J. 1979. Sample selection bias as a specification error. *Econometrica* 47: 153–161.

Pindyck, R. S., and D. L. Rubinfeld. 1998. *Econometric Models and Economic Forecasts*. 4th ed. New York: McGraw–Hill.

Poirier, D. 1980. Partial observability in bivariate probit models. *Journal of Econometrics* 12: 209–217.

Van de Ven, W. P. M. M., and B. M. S. Van Pragg. 1981. The demand for deductibles in private health insurance: A probit model with sample selection. *Journal of Econometrics* 17: 229–252.

Also See

[R] **biprobit postestimation** — Postestimation tools for biprobit

[R] **mprobit** — Multinomial probit regression

[R] **probit** — Probit regression

[SVY] **svy estimation** — Estimation commands for survey data

[U] **20 Estimation and postestimation commands**

Title

---
biprobit postestimation — Postestimation tools for biprobit

Description

The following postestimation commands are available for `biprobit`:

command	description
adjust[1]	adjusted predictions of $\mathbf{x}\beta$ or probabilities
estat	AIC, BIC, VCE, and estimation sample summary
estat (svy)	postestimation statistics for survey data
estimates	cataloging estimation results
lincom	point estimates, standard errors, testing, and inference for linear combinations of coefficients
lrtest[2]	likelihood-ratio test
mfx	marginal effects or elasticities
nlcom	point estimates, standard errors, testing, and inference for nonlinear combinations of coefficients
predict	predictions, residuals, influence statistics, and other diagnostic measures
predictnl	point estimates, standard errors, testing, and inference for generalized predictions
suest	seemingly unrelated estimation
test	Wald tests for simple and composite linear hypotheses
testnl	Wald tests of nonlinear hypotheses

[1] `adjust` is not appropriate with time-series operators.

[2] `lrtest` is not appropriate with `svy` estimation results.

See the corresponding entries in the *Stata Base Reference Manual* for details, but see [SVY] **estat** for details about `estat` (svy).

(Continued on next page)

Syntax for predict

predict $[type]$ *newvar* $[if]$ $[in]$ $[,$ *statistic* <u>nooff</u>set$]$

predict $[type]$ $\{$ *stub** $|$ *newvar*$_{\mathrm{eq1}}$ *newvar*$_{\mathrm{eq2}}$ *newvar*$_{\mathrm{athrho}}$ $\}$ $[if]$ $[in]$, <u>sc</u>ores

statistic	description
Main	
<u>p</u>11	$\Phi_2(\mathbf{x}_j\mathbf{b}, \mathbf{z}_j\mathbf{g}, \rho)$, predicted probability $\Pr(y_{1j} = 1, y_{2j} = 1)$; the default
p10	$\Phi_2(\mathbf{x}_j\mathbf{b}, -\mathbf{z}_j\mathbf{g}, -\rho)$, predicted probability $\Pr(y_{1j} = 1, y_{2j} = 0)$
p01	$\Phi_2(-\mathbf{x}_j\mathbf{b}, \mathbf{z}_j\mathbf{g}, -\rho)$, predicted probability $\Pr(y_{1j} = 0, y_{2j} = 1)$
p00	$\Phi_2(-\mathbf{x}_j\mathbf{b}, -\mathbf{z}_j\mathbf{g}, \rho)$, predicted probability $\Pr(y_{1j} = 0, y_{2j} = 0)$
pmarg1	$\Phi(\mathbf{x}_j\mathbf{b})$, marginal success probability for equation 1
pmarg2	$\Phi(\mathbf{z}_j\mathbf{g})$, marginal success probability for equation 2
pcond1	$\Phi_2(\mathbf{x}_j\mathbf{b}, \mathbf{z}_j\mathbf{g}, \rho)/\Phi(\mathbf{z}_j\mathbf{g})$, conditional probability of success for equation 1
pcond2	$\Phi_2(\mathbf{x}_j\mathbf{b}, \mathbf{z}_j\mathbf{g}, \rho)/\Phi(\mathbf{x}_j\mathbf{b})$, conditional probability of success for equation 2
xb1	$\mathbf{x}_j\mathbf{b}$, linear prediction for equation 1
xb2	$\mathbf{z}_j\mathbf{g}$, linear prediction for equation 2
stdp1	standard error of the linear prediction for equation 1
stdp2	standard error of the linear prediction for equation 2

where $\Phi()$ is the standard normal-distribution function and $\Phi_2()$ is the bivariate standard normal-distribution function.

These statistics are available both in and out of sample; type predict ... if e(sample) ... if wanted only for the estimation sample.

Options for predict

    Main

p11, the default, calculates the bivariate predicted probability $\Pr(y_{1j} = 1, y_{2j} = 1)$.

p10 calculates the bivariate predicted probability $\Pr(y_{1j} = 1, y_{2j} = 0)$.

p01 calculates the bivariate predicted probability $\Pr(y_{1j} = 0, y_{2j} = 1)$.

p00 calculates the bivariate predicted probability $\Pr(y_{1j} = 0, y_{2j} = 0)$.

pmarg1 calculates the univariate (marginal) predicted probability of success $\Pr(y_{1j} = 1)$.

pmarg2 calculates the univariate (marginal) predicted probability of success $\Pr(y_{2j} = 1)$.

pcond1 calculates the conditional (on success in equation 2) predicted probability of success $\Pr(y_{1j} = 1, y_{2j} = 1)/\Pr(y_{2j} = 1)$.

pcond2 calculates the conditional (on success in equation 1) predicted probability of success $\Pr(y_{1j} = 1, y_{2j} = 1)/\Pr(y_{1j} = 1)$.

xb1 calculates the probit linear prediction $\mathbf{x}_j\mathbf{b}$.

xb2 calculates the probit linear prediction $\mathbf{z}_j\mathbf{g}$.

stdp1 calculates the standard error of the linear prediction for equation 1.

stdp2 calculates the standard error of the linear prediction for equation 2.

nooffset is relevant only if you specified offset1(*varname*) or offset2(*varname*) for biprobit. It modifies the calculations made by predict so that they ignore the offset variables; the linear predictions are treated as $\mathbf{x}_j\mathbf{b}$ rather than as $\mathbf{x}_j\mathbf{b} + \text{offset}_{1j}$ and $\mathbf{z}_j\boldsymbol{\gamma}$ rather than as $\mathbf{z}_j\boldsymbol{\gamma} + \text{offset}_{2j}$.

scores calculates equation-level score variables.

The first new variable will contain $\partial \ln L / \partial(\mathbf{x}_j\boldsymbol{\beta})$.

The second new variable will contain $\partial \ln L / \partial(\mathbf{z}_j\boldsymbol{\gamma})$.

The third new variable will contain $\partial \ln L / \partial(\text{atanh}\,\rho)$.

Methods and Formulas

All postestimation commands listed above are implemented as ado-files.

Also See

[R] **biprobit** — Bivariate probit regression

[U] **20 Estimation and postestimation commands**

Title

bitest — Binomial probability test

Syntax

Binomial probability test

> bitest *varname* == $\#_p$ $\left[\textit{if} \right]$ $\left[\textit{in} \right]$ $\left[\textit{weight} \right]$ $\left[\textit{, detail} \right]$

Immediate form of binomial probability test

> bitesti $\#_N$ $\#_{\mathrm{succ}}$ $\#_p$ $\left[\textit{, detail} \right]$

by is allowed with bitest; see [D] **by**.

fweights are allowed with bitest; see [U] **11.1.6 weight**.

Description

bitest performs exact hypothesis tests for binomial random variables. The null hypothesis is that the probability of a success on a trial is $\#_p$. The total number of trials is the number of nonmissing values of *varname* (in bitest) or $\#_N$ (in bitesti). The number of observed successes is the number of 1s in *varname* (in bitest) or $\#_{\mathrm{succ}}$ (in bitesti). *varname* must contain only 0s, 1s, and missing.

bitesti is the immediate form of bitest; see [U] **19 Immediate commands** for a general introduction to immediate commands.

Option

_____| Advanced |_____

detail shows the probability of the observed number k_{obs} of successes, the probability of the number k_{opp} of successes on the opposite tail of the distribution that is used to compute the two-sided p-value, and the probability of the point next to k_{opp}. This information can be safely ignored. See the technical note below for details.

Remarks

Remarks are presented under the following headings:

> *bitest*
> *bitesti*

bitest

▷ Example 1

We test 15 university students for high levels of one measure of visual quickness which, from other evidence, we believe is present in 30% of the nonuniversity population. Included in our data is quick, taking on the values 1 ("success") or 0 ("failure") depending on the outcome of the test.

```
. use http://www.stata-press.com/data/r10/quick
. bitest quick == 0.3
    Variable |        N   Observed k   Expected k   Assumed p   Observed p
-------------+-------------------------------------------------------------
       quick |       15            7          4.5     0.30000      0.46667

Pr(k >= 7)              = 0.131143  (one-sided test)
Pr(k <= 7)              = 0.949987  (one-sided test)
Pr(k <= 1 or k >= 7)    = 0.166410  (two-sided test)
```

The first part of the output reveals that, assuming a true probability of success of 0.3, the expected number of successes is 4.5 and that we observed seven. Said differently, the assumed frequency under the null hypothesis H_0 is 0.3, and the observed frequency is 0.47.

The first line under the table is a one-sided test; it is the probability of observing seven or more successes conditional on $p = 0.3$. It is a test of $H_0: p = 0.3$ versus the alternative hypothesis $H_A: p > 0.3$. Said in English, the alternative hypothesis is that more than 30% of university students score at high levels on this test of visual quickness. The p-value for this hypothesis test is 0.13.

The second line under the table is a one-sided test of H_0 versus the opposite alternative hypothesis $H_A: p < 0.3$.

The third line is the two-sided test. It is a test of H_0 versus the alternative hypothesis $H_A: p \neq 0.3$.

◁

❏ Technical Note

The p-value of a hypothesis test is the probability (calculated assuming H_0 is true) of observing any outcome as extreme or more extreme than the observed outcome, extreme meaning in the direction of the alternative hypothesis. In example 1, the outcomes $k = 8, 9, \ldots, 15$ are clearly "more extreme" than the observed outcome $k_{obs} = 7$ when considering the alternative hypothesis $H_A: p \neq 0.3$. However, outcomes with only a few successes are also in the direction of this alternative hypothesis. For two-sided hypotheses, outcomes with k successes are considered "as extreme or more extreme" than the observed outcome k_{obs} if $\Pr(k) \leq \Pr(k_{obs})$. Here $\Pr(k = 0)$ and $\Pr(k = 1)$ are both less than $\Pr(k = 7)$, so they are included in the two-sided p-value.

The detail option allows you to see the probability (assuming that H_0 is true) of the observed successes ($k = 7$) and the probability of the boundary point ($k = 1$) of the opposite tail used for the two-sided p-value.

```
. bitest quick == 0.3, detail
    Variable |        N   Observed k   Expected k   Assumed p   Observed p
-------------+-------------------------------------------------------------
       quick |       15            7          4.5     0.30000      0.46667

Pr(k >= 7)              = 0.131143  (one-sided test)
Pr(k <= 7)              = 0.949987  (one-sided test)
Pr(k <= 1 or k >= 7)    = 0.166410  (two-sided test)

Pr(k == 7)             = 0.081130  (observed)
Pr(k == 2)             = 0.091560
Pr(k == 1)             = 0.030520  (opposite extreme)
```

Also shown is the probability of the point next to the boundary point. This probability, namely $\Pr(k = 2) = 0.092$, is certainly close to the probability of the observed outcome $\Pr(k = 7) = 0.081$, so some people might argue that $k = 2$ should be included in the two-sided p-value. Statisticians (at least some we know) would reply that the p-value is a precisely defined concept and that this is an arbitrary "fuzzification" of its definition. When you compute exact p-values according to the precise definition of a p-value, your Type I error is never more than what you say it is—so no one can criticize you for being anticonservative. Including the point $k = 2$ is being overly conservative since it makes the p-value larger yet. But it is your choice; being overly conservative, at least in statistics, is always safe. Know that `bitest` and `bitesti` always keep to the precise definition of a p-value, so if you wish to include this extra point, you must do so by hand or by using the `r()` saved results; see *Saved Results* below.

❑

bitesti

▷ Example 2

The binomial test is a function of two statistics and one parameter: N, the number of observations; k_{obs}, the number of observed successes; and p, the assumed probability of a success on a trial. For instance, in a city of $N = 2,500,000$, we observe $k_{obs} = 36$ cases of a particular disease when the population rate for the disease is $p = 0.00001$.

```
. bitesti 2500000 36 .00001
         N    Observed k    Expected k    Assumed p    Observed p

   2500000           36            25       0.00001       0.00001
   Pr(k >= 36)                = 0.022458  (one-sided test)
   Pr(k <= 36)                = 0.985448  (one-sided test)
   Pr(k <= 14 or k >= 36) = 0.034859  (two-sided test)
```

◁

▷ Example 3

Boice and Monson (1977) present data on breast cancer cases and person-years of observations for women with tuberculosis who were repeatedly exposed to multiple x-ray fluoroscopies and for women with tuberculosis who were not. The data are

	Exposed	Not Exposed	Total
Breast cancer	41	15	56
Person-years	28,010	19,017	47,027

We can thus test whether x-ray fluoroscopic examinations are associated with breast cancer; the assumed rate of exposure is $p = 28010/47027$.

```
. bitesti 56 41 28010/47027
         N    Observed k    Expected k    Assumed p    Observed p

        56           41      33.35446       0.59562       0.73214
   Pr(k >= 41)                = 0.023830  (one-sided test)
   Pr(k <= 41)                = 0.988373  (one-sided test)
   Pr(k <= 25 or k >= 41) = 0.040852  (two-sided test)
```

◁

Saved Results

bitest and bitesti save the following in r():

Scalars

r(N)	number N of trials	r(k_opp)	opposite extreme k
r(P_p)	assumed probability p of success	r(P_k)	probability of observed k (detail only)
r(k)	observed number k of successes	r(P_oppk)	probability of opposite extreme k (detail only)
r(p_l)	lower one-sided p-value	r(k_nopp)	k next to opposite extreme (detail only)
r(p_u)	upper one-sided p-value	r(P_noppk)	probability of k next to opposite extreme
r(p)	two-sided p-value		(detail only)

Methods and Formulas

bitest and bitesti are implemented as ado-files.

Let N, k_{obs}, and p be, respectively, the number of observations, the observed number of successes, and the assumed probability of success on a trial. The expected number of successes is Np, and the observed probability of success on a trial is k_{obs}/N.

bitest and bitesti compute exact p-values based on the binomial distribution. The upper one-sided p-value is

$$\Pr(k \geq k_{\mathrm{obs}}) = \sum_{m=k_{\mathrm{obs}}}^{N} \binom{N}{m} p^m (1-p)^{N-m}$$

The lower one-sided p-value is

$$\Pr(k \leq k_{\mathrm{obs}}) = \sum_{m=0}^{k_{\mathrm{obs}}} \binom{N}{m} p^m (1-p)^{N-m}$$

If $k_{\mathrm{obs}} \geq Np$, the two-sided p-value is

$$\Pr(k \leq k_{\mathrm{opp}} \text{ or } k \geq k_{\mathrm{obs}})$$

where k_{opp} is the largest number $\leq Np$ such that $\Pr(k = k_{\mathrm{opp}}) \leq \Pr(k = k_{\mathrm{obs}})$. If $k_{\mathrm{obs}} < Np$, the two-sided p-value is

$$\Pr(k \leq k_{\mathrm{obs}} \text{ or } k \geq k_{\mathrm{opp}})$$

where k_{opp} is the smallest number $\geq Np$ such that $\Pr(k = k_{\mathrm{opp}}) \leq \Pr(k = k_{\mathrm{obs}})$.

References

Boice, J. D., and R. R. Monson. 1977. Breast cancer in women after repeated fluoroscopic examinations of the chest. *Journal of the National Cancer Institute* 59: 823–832.

Hoel, P. G. 1984. *Introduction to Mathematical Statistics.* 5th ed. New York: Wiley.

Also See

[R] **ci** — Confidence intervals for means, proportions, and counts

[R] **prtest** — One- and two-sample tests of proportions

Title

bootstrap — Bootstrap sampling and estimation

Syntax

bootstrap *exp_list* [, *options eform_option*] : *command*

options	description
Main	
<u>r</u>eps(*#*)	perform *#* bootstrap replications; default is reps(50)
Options	
<u>str</u>ata(*varlist*)	variables identifying strata
<u>si</u>ze(*#*)	draw samples of size *#*; default is _N
<u>cl</u>uster(*varlist*)	variables identifying resampling clusters
<u>id</u>cluster(*newvar*)	create new cluster ID variable
<u>sa</u>ving(*filename*, ...)	save results to *filename*; save statistics in double precision; save results to *filename* every *#* replications
bca	compute acceleration for BC_a confidence intervals
mse	use MSE formula for variance estimation
Reporting	
<u>l</u>evel(*#*)	set confidence level; default is level(95)
<u>nota</u>ble	suppress table of results
<u>noh</u>eader	suppress table header
<u>nol</u>egend	suppress table legend
<u>v</u>erbose	display the full table legend
nodots	suppress the replication dots
<u>noi</u>sily	display any output from *command*
<u>tr</u>ace	trace the *command*
<u>title</u>(*text*)	use *text* as title for bootstrap results
Advanced	
nodrop	do not drop observations
nowarn	do not warn when e(sample) is not set
force	do not check for *weights* or svy commands; seldom used
reject(*exp*)	identify invalid results
seed(*#*)	set random-number seed to *#*
† group(*varname*)	ID variable for groups within cluster()

† group() does not appear in the dialog box.

weights are not allowed in *command*.

See [U] **20 Estimation and postestimation commands** for more capabilities of estimation commands.

exp_list contains	(*name*: *elist*)
	elist
	eexp
elist contains	*newvar* = (*exp*)
	(*exp*)
eexp is	*specname*
	[*eqno*]*specname*
specname is	_b
	_b[]
	_se
	_se[]
eqno is	# #
	name

exp is a standard Stata expression; see [U] **13 Functions and expressions**.

Distinguish between [ ], which are to be typed, and [ ], which indicate optional arguments.

Description

bootstrap performs bootstrap estimation. Typing

> . bootstrap *exp_list*, reps(#): *command*

executes *command* multiple times, bootstrapping the statistics in *exp_list* by resampling observations (with replacement) from the data in memory # times. This method is commonly referred to as the nonparametric bootstrap.

command defines the statistical command to be executed. Most Stata commands and user-written programs can be used with bootstrap, as long as they follow standard Stata syntax; see [U] **11 Language syntax**. If the bca option is supplied, *command* must also work with jackknife; see [R] **jackknife**. The by prefix may not be part of *command*.

exp_list specifies the statistics to be collected from the execution of *command*. If *command* changes the contents in e(b), *exp_list* is optional and defaults to _b.

Since bootstrapping is a random process, if you want to be able to reproduce results, set the random-number seed by specifying the seed(#) option or typing

> . set seed #

where # is a seed of your choosing, before running bootstrap; see [D] **generate**.

Many estimation commands allow the vce(bootstrap) option. For those commands, we recommend using vce(bootstrap) over bootstrap since the estimation command already handles clustering and other model-specific details for you. The bootstrap prefix command is intended for use with nonestimation commands, such as summarize, user-written commands, or functions of coefficients.

bs and bstrap are synonyms for bootstrap.

Options

Main

reps(#) specifies the number of bootstrap replications to be performed. The default is 50. A total of 50–200 replications are generally adequate for estimates of standard error and thus are adequate for normal-approximation confidence intervals; see Mooney and Duval (1993, 11). Estimates of confidence intervals using the percentile or bias-corrected methods typically require 1,000 or more replications.

Options

strata(*varlist*) specifies the variables that identify strata. If this option is specified, bootstrap samples are taken independently within each stratum.

size(#) specifies the size of the samples to be drawn. The default is _N, meaning to draw samples of the same size as the data. If specified, # must be less than or equal to the number of observations within strata().

If cluster() is specified, the default size is the number of clusters in the original dataset. For unbalanced clusters, resulting sample sizes will differ from replication to replication. For cluster sampling, # must be less than or equal to the number of clusters within strata().

cluster(*varlist*) specifies the variables that identify resampling clusters. If this option is specified, the sample drawn during each replication is a bootstrap sample of clusters.

idcluster(*newvar*) creates a new variable containing a unique identifier for each resampled cluster. This option requires that cluster() also be specified.

saving(*filename*[, *suboptions*]) creates a Stata data file (.dta file) consisting of, for each statistic in *exp_list*, a variable containing the bootstrap replicates.

double specifies that the results for each replication be stored as doubles, meaning 8-byte reals. By default, they are stored as floats, meaning 4-byte reals. This option may be used without the saving() option to compute the variance estimates using double precision.

every(#) specifies that results be written to disk every #th replication. every() should be specified only in conjunction with saving() when *command* takes a long time for each replication. This option will allow recovery of partial results should some other software crash your computer. See [P] **postfile**.

replace specifies that *filename* be overwritten, if it exists. This option does not appear in the dialog box.

bca specifies that bootstrap estimate the acceleration of each statistic in *exp_list*. This estimate is used to construct BC_a confidence intervals. Type estat bootstrap, bca to display the BC_a confidence interval generated by the bootstrap command.

mse specifies that bootstrap compute the variance by using deviations of the replicates from the observed value of the statistics based on the entire dataset. By default, bootstrap computes the variance by using deviations from the average of the replicates.

Reporting

level(#); see [R] **estimation options**.

notable suppresses the display of the table of results.

noheader suppresses the display of the table header. This option implies nolegend. This option may also be specified when replaying estimation results.

nolegend suppresses the display of the table legend. This option may also be specified when replaying estimation results.

verbose specifies that the full table legend be displayed. By default, coefficients and standard errors are not displayed. This option may also be specified when replaying estimation results.

nodots suppresses display of the replication dots. By default, one dot character is displayed for each successful replication. A red 'x' is displayed if *command* returns an error or if one of the values in *exp_list* is missing.

noisily specifies that any output from *command* be displayed. This option implies the nodots option.

trace causes a trace of the execution of *command* to be displayed. This option implies the noisily option.

title(*text*) specifies a title to be displayed above the table of bootstrap results. The default title is the title saved in e(title) by an estimation command, or if e(title) is not filled in, Bootstrap results is used. title() may also be specified when replaying estimation results.

⌐ Advanced ⌐

nodrop prevents observations outside e(sample) and the *if* and *in* conditions from being dropped before the data are resampled.

nowarn suppresses the display of a warning message when *command* does not set e(sample).

force suppresses the restriction that *command* not specify weights or be a svy command. This is a rarely used option. Use it only if you know what you are doing.

reject(*exp*) identifies an expression that indicates when results should be rejected. When *exp* is true, the resulting values are reset to missing values.

seed(*#*) sets the random-number seed. Specifying this option is equivalent to typing the following command prior to calling bootstrap:

. set seed #

The following options are available with bootstrap but are not shown in the dialog box:

eform_option causes the coefficient table to be displayed in exponentiated form: For each coefficient, $\exp(b)$ rather than b is displayed. Standard errors and confidence intervals are also transformed. Display of the intercept, if any, is suppressed.

command determines which of the following are allowed (eform(*string*) and eform are always allowed).

eform_option	description
eform(*string*)	use *string* for the column title
eform	exponentiated coefficient, *string* is "exp(b)"
hr	hazard ratio, *string* is "Haz. Ratio"
irr	incidence-rate ratio, *string* is "IRR"
or	odds ratio, *string* is "Odds ratio"
rrr	relative-risk ratio, *string* is "RRR"

group(*varname*) recreates *varname* containing a unique identifier for each group across the resampled clusters. This option requires that idcluster() also be specified.

This option is useful for maintaining unique group identifiers when sampling clusters with replacement. Suppose that cluster 1 contains 3 groups. If the `idcluster(newclid)` option is specified and cluster 1 is sampled multiple times, `newclid` uniquely identifies each copy of cluster 1. If `group(newgroupid)` is also specified, `newgroupid` uniquely identifies each copy of each group.

Remarks

Remarks are presented under the following headings:

> *Introduction*
> *Regression coefficients*
> *Expressions*
> *Combining bootstrap datasets*
> *A note about macros*
> *Achieved significance level*
> *Bootstrapping a ratio*
> *Warning messages and e(sample)*
> *Bootstrapping statistics from data with a complex structure*

Introduction

With few assumptions, bootstrapping provides a way of estimating standard errors and other measures of statistical precision (Efron 1979, Efron and Stein 1981, Efron 1982, Efron and Tibshirani 1986, and Efron and Tibshirani 1993; also see Davison and Hinkley 1997, Guan 2003, Mooney and Duval 1993, Poi 2004, and Stine 1990). It provides a way to obtain such measures when no formula is otherwise available or when available formulas make inappropriate assumptions.

To illustrate bootstrapping, suppose that you have a dataset containing N observations and an estimator that, when applied to the data, produces certain statistics. You draw, with replacement, N observations from the N-observation dataset. In this random drawing, some of the original observations will appear once, some more than once, and some not at all. Using the resampled dataset, you apply the estimator and collect the statistics. This process is repeated many times, each time a new random sample is drawn and the statistics are recalculated.

This process builds a dataset of replicated statistics. From these data, you can calculate the standard error by using the standard formula for the sample standard deviation

$$\widehat{se} = \left\{ \frac{1}{k-1} \sum (\widehat{\theta}_i - \overline{\theta})^2 \right\}^{1/2}$$

where $\widehat{\theta}_i$ is the statistic calculated using the ith bootstrap sample and k is the number of replications. This formula gives an estimate of the standard error of the statistic, according to Hall and Wilson (1991). Although the average, $\overline{\theta}$, of the bootstrapped estimates is used in calculating the standard deviation, it is not used as the estimated value of the statistic itself. Instead, the original observed value $\widehat{\theta}$ of the statistic is used, meaning the value of the statistic computed using the original N observations.

You might think that $\overline{\theta}$ is a better estimate of the parameter than $\widehat{\theta}$, but it is not. If the statistic is biased, bootstrapping exaggerates the bias. In fact, the bias can be estimated as $\overline{\theta} - \widehat{\theta}$ (Efron 1982, 33). Knowing this, you might be tempted to subtract this estimate of bias from $\widehat{\theta}$ to produce an unbiased statistic. The bootstrap bias estimate has an indeterminate amount of random error, so this unbiased estimator may have greater mean squared error than the biased estimator (Mooney and Duval 1993, Hinkley 1978). Thus $\widehat{\theta}$ is the best point estimate of the statistic.

The logic behind the bootstrap is that all measures of precision come from a statistic's sampling distribution. When the statistic is estimated on a sample of size N from some population, the sampling distribution tells you the relative frequencies of the values of the statistic. The sampling distribution, in turn, is determined by the distribution of the population and the formula used to estimate the statistic.

Sometimes the sampling distribution can be derived analytically. For instance, if the underlying population is distributed normally and you calculate means, the sampling distribution for the mean is also normal but has a smaller variance than that of the population. In other cases, deriving the sampling distribution is difficult, as when means are calculated from nonnormal populations. Sometimes, as in the case of means, it is not too difficult to derive the sampling distribution as the sample size goes to infinity ($N \rightarrow \infty$). However, such asymptotic distributions may not perform well when applied to finite samples.

If you knew the population distribution, you could obtain the sampling distribution by simulation: you could draw random samples of size N, calculate the statistic, and make a tally. Bootstrapping does precisely this, but it uses the observed distribution of the sample in place of the true population distribution. Thus the bootstrap procedure hinges on the assumption that the observed distribution is a good estimate of the underlying population distribution. In return, the bootstrap produces an estimate, called the bootstrap distribution, of the sampling distribution. From this, you can estimate the standard error of the statistic, produce confidence intervals, etc.

The accuracy with which the bootstrap distribution estimates the sampling distribution depends on the number of observations in the original sample and the number of replications in the bootstrap. A crudely estimated sampling distribution is adequate if you are only going to extract, say, a standard error. A better estimate is needed if you want to use the 2.5th and 97.5th percentiles of the distribution to produce a 95% confidence interval. To extract many features simultaneously about the distribution, an even better estimate is needed. Generally, replications on the order of 1,000 produce very good estimates, but only 50–200 replications are needed for estimates of standard errors. See Poi (2004) for a method to choose the number of bootstrap replications.

Regression coefficients

▷ Example 1

Let's say that we wish to compute bootstrap estimates for the standard errors of the coefficients from the following regression:

```
. use http://www.stata-press.com/data/r10/auto
(1978 Automobile Data)

. regress mpg weight gear foreign
```

Source	SS	df	MS		
Model	1629.67805	3	543.226016	Number of obs =	74
Residual	813.781411	70	11.6254487	F(3, 70) =	46.73
				Prob > F =	0.0000
				R-squared =	0.6670
				Adj R-squared =	0.6527
Total	2443.45946	73	33.4720474	Root MSE =	3.4096

mpg	Coef.	Std. Err.	t	P>|t|	[95% Conf. Interval]	
weight	-.006139	.0007949	-7.72	0.000	-.0077245	-.0045536
gear_ratio	1.457113	1.541286	0.95	0.348	-1.616884	4.53111
foreign	-2.221682	1.234961	-1.80	0.076	-4.684735	.2413715
_cons	36.10135	6.285984	5.74	0.000	23.56435	48.63835

To run the bootstrap, we simply prefix the above regression command with the `bootstrap` command (specifying its options before the colon separator). We must set the random-number seed before calling `bootstrap`.

```
. bootstrap, reps(100) seed(1): regress mpg weight gear foreign
(running regress on estimation sample)

Bootstrap replications (100)
```

```
..................................................   50
..................................................  100

Linear regression                   Number of obs    =         74
                                    Replications     =        100
                                    Wald chi2(3)     =     111.96
                                    Prob > chi2      =     0.0000
                                    R-squared        =     0.6670
                                    Adj R-squared    =     0.6527
                                    Root MSE         =     3.4096
```

mpg	Observed Coef.	Bootstrap Std. Err.	z	P>|z|	Normal-based [95% Conf. Interval]	
weight	-.006139	.0006498	-9.45	0.000	-.0074127	-.0048654
gear_ratio	1.457113	1.297786	1.12	0.262	-1.086501	4.000727
foreign	-2.221682	1.162728	-1.91	0.056	-4.500587	.0572236
_cons	36.10135	4.71779	7.65	0.000	26.85465	45.34805

The displayed confidence interval is based on the assumption that the sampling (and hence bootstrap) distribution is approximately normal (see *Methods and Formulas* below). Since this confidence interval is based on the standard error, it is a reasonable estimate if normality is approximately true, even for a few replications. Other types of confidence intervals are available after `bootstrap`; see [R] **bootstrap postestimation**.

We could instead supply names to our expressions when we run `bootstrap`. For example,

```
. bootstrap diff=(_b[weight]-_b[gear]): regress mpg weight gear foreign
```

would bootstrap a statistic, named `diff`, equal to the difference between the coefficients on `weight` and `gear_ratio`.

◁

Expressions

▷ Example 2

When we use `bootstrap`, the list of statistics can contain complex expressions, as long as each expression is enclosed in parentheses. For example, to bootstrap the range of a variable x, we could type

```
. bootstrap range=(r(max)-r(min)), reps(1000): summarize x
```

Of course, we could also bootstrap the minimum and maximum and later compute the range.

```
. bootstrap max=r(max) min=r(min), reps(1000) saving(mybs): summarize x
. use mybs, clear
(bootstrap: summarize)
. generate range = max - min
. bstat range, stat(19.5637501)
```

The difference between the maximum and minimum of x in the sample is 19.5637501.

The stat() option to bstat specifies the observed value of the statistic (range) to be summarized. This option is useful when, as shown above, the statistic of ultimate interest is not specified directly to bootstrap but instead is calculated by other means.

Here the observed values of r(max) and r(min) are saved as characteristics of the dataset created by bootstrap and are thus available for retrieval by bstat; see [R] **bstat**. The observed range, however, is unknown to bstat, so it must be specified.

◁

Combining bootstrap datasets

You can combine two datasets from separate runs of bootstrap by using append (see [D] **append**) and then get the bootstrap statistics for the combined datasets by running bstat. The runs must have been performed independently (having different starting random-number seeds), and the original dataset, command, and bootstrap statistics must have been all the same.

A note about macros

In the previous example, we executed the command

```
. bootstrap max=r(max) min=r(min), reps(1000) saving(mybs): summarize x
```

We did not enclose r(max) and r(min) in single quotes, as we would in most other contexts, because it would not produce what was intended:

```
. bootstrap 'r(max)' 'r(min)', reps(1000) saving(mybs): summarize x
```

To understand why, note that 'r(max)', like any reference to a local macro, will evaluate to a literal string containing the contents of r(max) *before* bootstrap is even executed. Typing the command above would appear to Stata as if we had typed

```
. bootstrap 14.5441234 33.4393293, reps(1000) saving(mybs): summarize x
```

Even worse, the current contents of r(min) and r(max) could be empty, producing an even more confusing result. To avoid this outcome, refer to statistics by name (e.g., r(max)) and not by value (e.g., 'r(max)').

(Continued on next page)

Achieved significance level

▷ Example 3

Suppose that we wish to estimate the *achieved significance level* (ASL) of a test statistic by using the bootstrap. ASL is another name for *p*-value. An example is

$$\text{ASL} = \Pr\left(\widehat{\theta}^* \geq \widehat{\theta}\,|H_0\right)$$

for an upper-tailed, alternative hypothesis, where H_0 denotes the null hypothesis, $\widehat{\theta}$ is the observed value of the test statistic, and $\widehat{\theta}^*$ is the random variable corresponding to the test statistic, assuming that H_0 is true.

Here we will compare the mean miles per gallon (mpg) between foreign and domestic cars by using the two-sample *t* test with unequal variances. The following results indicate the *p*-value to be 0.0034 for the two-sided test using Satterthwaite's approximation. Thus assuming that mean mpg is the same for foreign and domestic cars, we would expect to observe a *t* statistic more extreme (in absolute value) than 3.1797 in about 0.3% of all possible samples of the type that we observed. Thus we have evidence to reject the null hypothesis that the means are equal.

```
. use http://www.stata-press.com/data/r10/auto
(1978 Automobile Data)

. ttest mpg, by(foreign) unequal

Two-sample t test with unequal variances
```

Group	Obs	Mean	Std. Err.	Std. Dev.	[95% Conf. Interval]	
Domestic	52	19.82692	.657777	4.743297	18.50638	21.14747
Foreign	22	24.77273	1.40951	6.611187	21.84149	27.70396
combined	74	21.2973	.6725511	5.785503	19.9569	22.63769
diff		-4.945804	1.555438		-8.120053	-1.771556

```
        diff = mean(Domestic) - mean(Foreign)                    t =  -3.1797
Ho: diff = 0                         Satterthwaite's degrees of freedom =   30.5463

     Ha: diff < 0                     Ha: diff != 0                   Ha: diff > 0
 Pr(T < t) = 0.0017            Pr(|T| > |t|) = 0.0034            Pr(T > t) = 0.9983
```

We also place the value of the test statistic in a scalar for later use.

```
. scalar tobs = r(t)
```

Efron and Tibshirani (1993, 224) describe an alternative to Satterthwaite's approximation that estimates the ASL by bootstrapping the statistic from the test of equal means. Their idea is to recenter the two samples to the combined sample mean so that the data now conform to the null hypothesis but that the variances within the samples remain unchanged.

```
. summarize mpg, meanonly
. scalar omean = r(mean)
. summarize mpg if foreign==0, meanonly
. replace mpg = mpg - r(mean) + scalar(omean) if foreign==0
mpg was int now float
(52 real changes made)
. summarize mpg if foreign==1, meanonly
. replace mpg = mpg - r(mean) + scalar(omean) if foreign==1
(22 real changes made)
. sort foreign
. by foreign: summarize mpg
```

```
-> foreign = Domestic
    Variable │       Obs        Mean    Std. Dev.        Min         Max
─────────────┼──────────────────────────────────────────────────────────
         mpg │        52     21.2973     4.743297    13.47037    35.47038

-> foreign = Foreign
    Variable │       Obs        Mean    Std. Dev.        Min         Max
─────────────┼──────────────────────────────────────────────────────────
         mpg │        22     21.2973     6.611187    10.52457    37.52457
```

Each sample (foreign and domestic) is a stratum, so the bootstrapped samples must have the same number of foreign and domestic cars as the original dataset. This requirement is facilitated by the strata() option to bootstrap. By typing the following, we bootstrap the test statistic using the modified dataset and save the values in bsauto2.dta:

```
. keep mpg foreign
. set seed 1
. bootstrap t=r(t), rep(1000) strata(foreign) saving(bsauto2) nodots: ttest mpg,
> by(foreign) unequal
```

```
Warning:  Since ttest is not an estimation command or does not set e(sample),
          bootstrap has no way to determine which observations are used in
          calculating the statistics and so assumes that all observations are
          used.  This means no observations will be excluded from the
          resampling due to missing values or other reasons.

          If the assumption is not true, press Break, save the data, and drop
          the observations that are to be excluded.  Be sure that the dataset
          in memory contains only the relevant data.
```

```
Bootstrap results

Number of strata   =            2              Number of obs    =          74
                                               Replications     =        1000

       command:  ttest mpg, by(foreign) unequal
             t:  r(t)
```

	Observed Coef.	Bootstrap Std. Err.	z	P>│z│	Normal-based [95% Conf. Interval]	
t	1.75e-07	1.036437	0.00	1.000	-2.031379	2.031379

We can use the data in bsauto2.dta to estimate ASL via the fraction of bootstrapped test statistics that are more extreme than 3.1797.

```
. use http://www.stata-press.com/data/r10/bsauto2, clear
(bootstrap: ttest)

. generate indicator = abs(t)>=abs(scalar(tobs))

. summarize indicator, meanonly

. display "ASLboot = " r(mean)
ASLboot = .005
```

The result is $\text{ASL}_{\text{boot}} = 0.005$. Assuming that the mean mpg is the same between foreign and domestic cars, we would expect to observe a t statistic more extreme (in absolute value) than 3.1797 in about 0.5% of all possible samples of the type we observed. This finding is still strong evidence to reject the hypothesis that the means are equal.

◁

Bootstrapping a ratio

▷ Example 4

Suppose that we wish to produce a bootstrap estimate of the ratio of two means. Since summarize saves results for only one variable, we must call summarize twice to compute the means. Actually, we could use collapse to compute the means in one call, but calling summarize twice is much faster. Thus we will have to write a small program that will return the results we want.

We write the program below and save it to a file called ratio.ado (see [U] **17 Ado-files**). Our program takes two variable names as input and saves them in the local macros y (first variable) and x (second variable). It then computes one statistic: the mean of 'y' divided by the mean of 'x'. This value is returned as a scalar in r(ratio). ratio also returns the ratio of the number of observations used to the mean for each variable.

```
program myratio, rclass
        version 10
        args y x
        confirm var 'y'
        confirm var 'x'
        tempname ymean yn
        summarize 'y', meanonly
        scalar 'ymean' = r(mean)
        return scalar n_'y' = r(N)
        summarize 'x', meanonly
        return scalar n_'x' = r(N)
        return scalar ratio = 'ymean'/r(mean)
end
```

Remember to test any newly written commands before using them with bootstrap.

```
. use http://www.stata-press.com/data/r10/auto, clear
(1978 Automobile Data)

. summarize price
```

Variable	Obs	Mean	Std. Dev.	Min	Max
price	74	6165.257	2949.496	3291	15906

```
. scalar mean1=r(mean)
```

```
. summarize weight
    Variable |        Obs        Mean    Std. Dev.       Min         Max
      weight |         74    3019.459    777.1936        1760        4840
. scalar mean2=r(mean)
. di scalar(mean1)/scalar(mean2)
2.0418412
. myratio price weight
. return list
scalars:
              r(ratio) =  2.041841210168278
           r(n_weight) =  74
            r(n_price) =  74
```

The results of running `bootstrap` on our program are

```
. use http://www.stata-press.com/data/r10/auto
(1978 Automobile Data)
. set seed 1
. bootstrap ratio=r(ratio), reps(1000) nowarn nodots: myratio price weight
Bootstrap results                               Number of obs     =         74
                                                Replications      =       1000

      command:  myratio price weight
        ratio:  r(ratio)
```

	Observed Coef.	Bootstrap Std. Err.	z	P>|z|	Normal-based [95% Conf. Interval]	
ratio	2.041841	.0942932	21.65	0.000	1.85703	2.226652

As mentioned previously, we should specify the `saving()` option if we wish to save the bootstrap dataset.

◁

Warning messages and e(sample)

`bootstrap` is not meant to be used with weighted calculations. `bootstrap` determines the presence of weights by parsing the prefixed command with standard syntax. However, commands like `stcox` and `streg` require that weights be specified in `stset`, and some user commands may allow weights to be specified by using an option instead of the standard syntax. Both cases pose a problem for `bootstrap` because it cannot determine the presence of weights under these circumstances. In these cases, we can only assume that you know what you are doing.

`bootstrap` does not know which variables of the dataset in memory matter to the calculation at hand. You can speed their execution by dropping unnecessary variables because, otherwise, they are included in each bootstrap sample.

You should thus drop observations with missing values. Leaving in missing values causes no problem in one sense because all Stata commands deal with missing values gracefully. It does, however, cause a statistical problem. Bootstrap sampling is defined as drawing, with replacement, samples of size N from a set of N observations. `bootstrap` determines N by counting the number of observations in memory, not counting the number of nonmissing values on the relevant variables. The result is that too many observations are resampled; the resulting bootstrap samples, since they are drawn from a population with missing values, are of unequal sizes.

If the number of missing values relative to the sample size is small, this will make little difference. If you have many missing values, however, you should first drop the observations that contain them.

▷ Example 5

To illustrate, we use the previous example but replace some of the values of `price` with missing values. The number of values of `price` used to compute the mean for each bootstrap is not constant. This is the purpose of the `Warning` message.

```
. use http://www.stata-press.com/data/r10/auto
(1978 Automobile Data)
. replace price = . if inlist(_n,1,3,5,7)
(4 real changes made, 4 to missing)
. set seed 1
. bootstrap ratio=r(ratio) np=r(n_price) nw=r(n_weight), reps(100) nodots:
> myratio price weight
```

```
Warning:  Since myratio is not an estimation command or does not set
          e(sample), bootstrap has no way to determine which observations are
          used in calculating the statistics and so assumes that all
          observations are used.  This means no observations will be excluded
          from the resampling due to missing values or other reasons.

          If the assumption is not true, press Break, save the data, and drop
          the observations that are to be excluded.  Be sure that the dataset
          in memory contains only the relevant data.
```

```
Bootstrap results                          Number of obs     =         74
                                           Replications      =        100

        command:  myratio price weight
          ratio:  r(ratio)
             np:  r(n_price)
             nw:  r(n_weight)
```

	Observed Coef.	Bootstrap Std. Err.	z	P>|z|	Normal-based [95% Conf. Interval]	
ratio	2.063051	.0893669	23.09	0.000	1.887896	2.238207
np	70	1.872178	37.39	0.000	66.3306	73.6694
nw	74	.	.	.	.	.

◁

Bootstrapping statistics from data with a complex structure

Here we describe how to bootstrap statistics from data with a complex structure, e.g., longitudinal or panel data, or matched data. `bootstrap`, however, is not designed to work with complex survey data. It is important to include all necessary information about the structure of the data in the `bootstrap` syntax to obtain correct bootstrap estimates for standard errors and confidence intervals.

`bootstrap` offers several options identifying the specifics of the data. These options are `strata()`, `cluster()`, `idcluster()`, and `group()`. The usage of `strata()` was described in example 3 above. Below we demonstrate several examples that require specifying the other three options.

▷ Example 6

Suppose that the `auto` data in example 1 above are clustered by `rep78`. We want to obtain bootstrap estimates for the standard errors of the difference between the coefficients on `weight` and `gear_ratio`, taking into account clustering.

We supply the `cluster(rep78)` option to `bootstrap` to request resampling from clusters rather than from observations in the dataset.

```
. use http://www.stata-press.com/data/r10/auto, clear
(1978 Automobile Data)

. keep if rep78<.
(5 observations deleted)

. bootstrap diff=(_b[weight]-_b[gear]), seed(1) cluster(rep78): regress mpg weight
> gear foreign
(running regress on estimation sample)

Bootstrap replications (50)
———+——— 1 ———+——— 2 ———+——— 3 ———+——— 4 ———+——— 5
..................................................   50

Linear regression                        Number of obs    =        69
                                         Replications     =        50

    command:  regress mpg weight gear foreign
       diff:  _b[weight]-_b[gear]

                        (Replications based on 5 clusters in rep78)
```

	Observed Coef.	Bootstrap Std. Err.	z	P>|z|	Normal-based [95% Conf. Interval]
diff	-1.910396	1.876778	-1.02	0.309	-5.588812 1.768021

We drop missing values in `rep78` before issuing the command since `bootstrap` does not allow missing values in `cluster()`. See the section above about using `bootstrap` when variables contain missing values.

We can also obtain these same results by using the following syntax:

```
. bootstrap diff=(_b[weight]-_b[gear]), seed(1): regress mpg weight gear foreign,
> vce(cluster rep78)
```

When only clustered information is provided to the command, `bootstrap` can pick up the `vce(cluster clustvar)` option from the main command and use it to resample from clusters.

◁

▷ Example 7

Suppose now that we have matched data and want to use `bootstrap` to obtain estimates of the standard errors of the exponentiated difference between two coefficients (or, equivalently, the ratio of two odds ratios) estimated by `clogit`. Consider the example of matched case–control data on birthweight of infants described in example 2 in [R] **clogit**.

The infants are paired by being matched on mother's age. All groups, defined by the `pairid` variable, have 1 : 2 matching. `clogit` requires that the matching information, `pairid`, be supplied to the `group()` (or, equivalently, `strata()`) option to be used in computing the parameter estimates. Since the data are matched, we need to resample from groups rather than from the whole dataset. However, simply supplying the grouping variable `pairid` in `cluster()` is not enough with `bootstrap`, as it is with clustered data.

```
. use http://www.stata-press.com/data/r10/lowbirth, clear
(Applied Logistic Regression, Hosmer & Lemeshow)
. bootstrap ratio=exp(_b[smoke]-_b[ptd]), seed(1) cluster(pairid): clogit low lwt smoke
> ptd ht ui race2 race3, group(pairid)
(running clogit on estimation sample)

Bootstrap replications (50)
---+--- 1 ---+--- 2 ---+--- 3 ---+--- 4 ---+--- 5
.................................................. 50

Bootstrap results                      Number of obs     =      112
                                       Replications      =       50

       command:  clogit low lwt smoke ptd ht ui race2 race3, group(pairid)
         ratio:  exp(_b[smoke]-_b[ptd])

                          (Replications based on 56 clusters in pairid)
```

	Observed Coef.	Bootstrap Std. Err.	z	P>|z|	Normal-based [95% Conf. Interval]
ratio	.6654095	17.71791	0.04	0.970	-34.06106 35.39188

For the syntax above, imagine that the first pair was sampled twice during a replication. Then the bootstrap sample has four subjects with `pairid` equal to one, which clearly violates the original 1 : 2 matching design. As a result, the estimates of the coefficients obtained from this bootstrap sample will be incorrect.

Therefore, in addition to resampling from groups, we need to ensure that resampled groups are uniquely identified in each of the bootstrap samples. The `idcluster(newcluster)` option is designed for this. It requests that at each replication `bootstrap` create the new variable, *newcluster*, containing unique identifiers for all resampled groups. Thus, to make sure that the correct matching is preserved during each replication, we need to specify the grouping variable in `cluster()`, supply a variable name to `idcluster()`, and use this variable as the grouping variable with `clogit` as we demonstrate below.

```
. bootstrap ratio=exp(_b[smoke]-_b[ptd]), seed(1) cluster(pairid) idcluster(newpairid):
> clogit low lwt smoke ptd ht ui race2 race3, group(newpairid)
(running clogit on estimation sample)

Bootstrap replications (50)
---+--- 1 ---+--- 2 ---+--- 3 ---+--- 4 ---+--- 5
.................................................. 50

Bootstrap results                      Number of obs     =      112
                                       Replications      =       50

       command:  clogit low lwt smoke ptd ht ui race2 race3, group(newpairid)
         ratio:  exp(_b[smoke]-_b[ptd])

                          (Replications based on 56 clusters in pairid)
```

	Observed Coef.	Bootstrap Std. Err.	z	P>|z|	Normal-based [95% Conf. Interval]
ratio	.6654095	7.919441	0.08	0.933	-14.85641 16.18723

Note the difference between the estimates of the bootstrap standard error for the two specifications of the `bootstrap` syntax.

◁

❑ Technical Note

Similarly, when you have panel (longitudinal) data, all resampled panels must be unique in each of the bootstrap samples to obtain correct bootstrap estimates of statistics. Therefore, both cluster(*panelvar*) and idcluster(*newpanelvar*) must be specified with bootstrap, and i(*newpanelvar*) must be used with the main command.

❑

▷ Example 8

Continuing with our birthweight data, suppose that we have more information about doctors supervising women's pregnancies. We believe that the data on the pairs of infants from the same doctor may be correlated and want to adjust standard errors for possible correlation among the pairs. clogit offers the vce(cluster *clustvar*) option to do this.

Let's add a cluster variable to our dataset. One thing to keep in mind is that to use vce(cluster *clustvar*), groups in group() must be nested within clusters.

```
. use http://www.stata-press.com/data/r10/lowbirth, clear
(Applied Logistic Regression, Hosmer & Lemeshow)
. set seed 12345
. by pairid, sort: egen byte doctor = total(int(2*uniform()+1)*(_n == 1))
. clogit low lwt smoke ptd ht ui race2 race3, group(pairid) vce(cluster doctor)
Iteration 0:   log pseudolikelihood = -26.768693
Iteration 1:   log pseudolikelihood = -25.810476
Iteration 2:   log pseudolikelihood = -25.794296
Iteration 3:   log pseudolikelihood = -25.794271
Iteration 4:   log pseudolikelihood = -25.794271
```

```
Conditional (fixed-effects) logistic regression    Number of obs   =        112
                                                   Wald chi2(1)    =          .
                                                   Prob > chi2     =          .
Log pseudolikelihood = -25.794271                  Pseudo R2       =     0.3355
```

(Std. Err. adjusted for 2 clusters in doctor)

low	Coef.	Robust Std. Err.	z	P>|z|	[95% Conf. Interval]	
lwt	-.0183757	.0217802	-0.84	0.399	-.0610641	.0243128
smoke	1.400656	.0085545	163.73	0.000	1.38389	1.417423
ptd	1.808009	.938173	1.93	0.054	-.0307765	3.646794
ht	2.361152	1.587013	1.49	0.137	-.7493362	5.47164
ui	1.401929	.8568119	1.64	0.102	-.2773913	3.08125
race2	.5713643	.0672593	8.49	0.000	.4395385	.7031902
race3	-.0253148	.9149785	-0.03	0.978	-1.81864	1.76801

To obtain correct bootstrap standard errors of the exponentiated difference between the two coefficients in this example, we need to make sure that both resampled clusters and groups within resampled clusters are unique in each of the bootstrap samples. To achieve this, bootstrap needs the information about clusters in cluster(), the variable name of the new identifier for clusters in idcluster(), and the information about groups in group(). We demonstrate the corresponding syntax of bootstrap below.

```
. bootstrap ratio=exp(_b[smoke]-_b[ptd]), seed(1) cluster(doctor) idcluster(uidoctor)
> group(pairid): clogit low lwt smoke ptd ht ui race2 race3, group(pairid)
(running clogit on estimation sample)

Bootstrap replications (50)
────┼─── 1 ──┼── 2 ──┼── 3 ───┼── 4 ──┼── 5
...................................................  50

Bootstrap results                        Number of obs     =        112
                                         Replications      =         50

      command:  clogit low lwt smoke ptd ht ui race2 race3, group(pairid)
        ratio:  exp(_b[smoke]-_b[ptd])

                             (Replications based on 2 clusters in doctor)
```

	Observed Coef.	Bootstrap Std. Err.	z	P>|z|	Normal-based [95% Conf. Interval]	
ratio	.6654095	.3156251	2.11	0.035	.0467956	1.284023

In the above syntax, although we specify group(pairid) with clogit, it is not the group identifiers of the original pairid variable that are used to compute parameter estimates from bootstrap samples. The way bootstrap works is that, at each replication, the clusters defined by doctor are resampled and the new variable, uidoctor, uniquely identifying resampled clusters is created. After that, another new variable uniquely identifying the (uidoctor, group) combination is created and renamed to have the same name as the grouping variable, pairid. This newly defined grouping variable is then used by clogit to obtain the parameter estimates from this bootstrap sample of clusters. After all replications are performed, the original values of the grouping variable are restored.

◁

❑ Technical Note

The same logic must be used when running bootstrap with commands designed for panel (longitudinal) data that allow specifying the option cluster(*clustervar*). To ensure that the combination of (*clustervar*, *panelvar*) values are unique in each of the bootstrap samples, cluster(*clustervar*), idcluster(*newclustervar*), and group(*panelvar*) must be specified with bootstrap, and i(*panelvar*) must be used with the main command.

❑

Bradley Efron was born in 1938 in Minnesota and studied mathematics and statistics at Caltech and Stanford; he has lived in northern California since 1960. He has worked on empirical Bayes, survival analysis, exponential families, bootstrap and jackknife methods, and confidence intervals in conjunction with applied work in biostatistics, astronomy, and physics.

Saved Results

bootstrap saves the following in e():

Scalars

e(N_reps)	number of complete replications	e(N)	sample size
e(N_misreps)	number of incomplete replications	e(N_strata)	number of strata
e(level)	confidence level for bootstrap CIs	e(N_clust)	number of clusters

Macros

e(cmdname)	command name from *command*	e(prefix)	bootstrap
e(cmd)	same as e(cmdname) or bootstrap	e(strata)	strata variables
e(command)	*command*	e(cluster)	cluster variables
e(cmdline)	command as typed	e(seed)	initial random-number seed
e(exp#)	expression for the #th statistic	e(size)	from the size(#) option
e(mse)	mse if specified	e(title)	title in estimation output
e(vce)	bootstrap	e(properties)	b V
e(vcetype)	title used to label Std. Err.		

Matrices

e(b)	observed statistics
e(V)	bootstrap variance–covariance matrix
e(b_bs)	bootstrap estimates
e(reps)	number of nonmissing results
e(bias)	estimated biases
e(se)	estimated standard errors
e(z0)	median biases
e(accel)	estimated accelerations
e(ci_normal)	normal-approximation CIs
e(ci_percentile)	percentile CIs
e(ci_bc)	bias-corrected CIs
e(ci_bca)	bias-corrected and accelerated CIs

When *exp_list* is _b, bootstrap will also carry forward most of the results already in e() from *command*.

Methods and Formulas

bootstrap is implemented as an ado-file.

Let $\widehat{\theta}$ be the observed value of the statistic, that is, the value of the statistic calculated with the original dataset. Let $i = 1, 2, \ldots, k$ denote the bootstrap samples, and let $\widehat{\theta}_i$ be the value of the statistic from the ith bootstrap sample.

When the mse option is specified, the standard error is estimated as

$$\widehat{se}_{\mathrm{MSE}} = \left\{ \frac{1}{k} \sum_{i=1}^{k} (\widehat{\theta}_i - \widehat{\theta})^2 \right\}^{1/2}$$

Otherwise, the standard error is estimated as

$$\widehat{se} = \left\{ \frac{1}{k-1} \sum_{i=1}^{k} (\widehat{\theta}_i - \overline{\theta})^2 \right\}^{1/2}$$

where

$$\bar{\theta} = \frac{1}{k} \sum_{i=1}^{k} \widehat{\theta}_i$$

The variance–covariance matrix is similarly computed. The bias is estimated as

$$\widehat{\text{bias}} = \bar{\theta} - \widehat{\theta}$$

Confidence intervals with nominal coverage rates $1 - \alpha$ are calculated according to the following formulas. The normal-approximation method yields the confidence intervals

$$\left[\widehat{\theta} - z_{1-\alpha/2}\, \widehat{\text{se}},\ \widehat{\theta} + z_{1-\alpha/2}\, \widehat{\text{se}} \right]$$

where $z_{1-\alpha/2}$ is the $(1 - \alpha/2)$th quantile of the standard normal distribution. If the `mse` option is specified, `bootstrap` will report the normal confidence interval using $\widehat{\text{se}}_{\text{MSE}}$ instead of $\widehat{\text{se}}$. `estat bootstrap` only uses $\widehat{\text{se}}$ in the normal confidence interval.

The percentile method yields the confidence intervals

$$\left[\theta^*_{\alpha/2},\ \theta^*_{1-\alpha/2} \right]$$

where θ^*_p is the pth quantile (the $100p$th percentile) of the bootstrap distribution $(\widehat{\theta}_1, \ldots, \widehat{\theta}_k)$.

Let

$$z_0 = \Phi^{-1}\{\#(\widehat{\theta}_i \le \widehat{\theta})/k\}$$

where $\#(\widehat{\theta}_i \le \widehat{\theta})$ is the number of elements of the bootstrap distribution that are less than or equal to the observed statistic and Φ is the standard cumulative normal. z_0 is known as the median bias of $\widehat{\theta}$. Let

$$a = \frac{\sum_{i=1}^{n}(\bar{\theta}_{(\cdot)} - \widehat{\theta}_{(i)})^3}{6\{\sum_{i=1}^{n}(\bar{\theta}_{(\cdot)} - \widehat{\theta}_{(i)})^2\}^{3/2}}$$

where $\widehat{\theta}_{(i)}$ are the leave-one-out (jackknife) estimates of $\widehat{\theta}$ and $\bar{\theta}_{(\cdot)}$ is their mean. This expression is known as the jackknife estimate of acceleration for $\widehat{\theta}$. Let

$$p_1 = \Phi\left\{ z_0 + \frac{z_0 - z_{1-\alpha/2}}{1 - a(z_0 - z_{1-\alpha/2})} \right\}$$

$$p_2 = \Phi\left\{ z_0 + \frac{z_0 + z_{1-\alpha/2}}{1 - a(z_0 + z_{1-\alpha/2})} \right\}$$

where $z_{1-\alpha/2}$ is the $(1-\alpha/2)$th quantile of the normal distribution. The bias-corrected and accelerated (BC_a) method yields confidence intervals

$$\left[\theta^*_{p_1},\ \theta^*_{p_2} \right]$$

where θ^*_p is the pth quantile of the bootstrap distribution as defined previously. The bias-corrected (but not accelerated) method is a special case of BC_a with $a = 0$.

References

Davison, A. C., and D. V. Hinkley. 1997. *Bootstrap Methods and Their Application.* Cambridge: Cambridge University Press.

Efron, B. 1979. Bootstrap methods: Another look at the jackknife. *Annals of Statistics* 7: 1–26.

——. 1982. *The Jackknife, the Bootstrap and Other Resampling Plans.* Philadelphia: Society for Industrial and Applied Mathematics.

Efron, B., and C. Stein. 1981. The jackknife estimate of variance. *Annals of Statistics* 9: 586–596.

Efron, B., and R. J. Tibshirani. 1986. Bootstrap measures for standard errors, confidence intervals, and other measures of statistical accuracy. *Statistical Science* 1: 54–77.

——. 1993. *An Introduction to the Bootstrap.* New York: Chapman & Hall/CRC.

Gleason, J. R. 1997. ip18: A command for randomly resampling a dataset. *Stata Technical Bulletin* 37: 17–22. Reprinted in *Stata Technical Bulletin Reprints*, vol. 7, pp. 77–83.

——. 1999. ip18.1: Update to resample. *Stata Technical Bulletin* 52: 9–10. Reprinted in *Stata Technical Bulletin Reprints*, vol. 9, p. 119.

Gould, W. W. 1994. ssi6.2: Faster and easier bootstrap estimation. *Stata Technical Bulletin* 21: 24–33. Reprinted in *Stata Technical Bulletin Reprints*, vol. 4, pp. 211–223.

Guan, W. 2003. From the help desk: Bootstrapped standard errors. *Stata Journal* 3: 71–80.

Hall, P., and S. R. Wilson. 1991. Two guidelines for bootstrap hypothesis testing. *Biometrics* 47: 757–762.

Hamilton, L. C. 1991. ssi2: Bootstrap programming. *Stata Technical Bulletin* 4: 18–27. Reprinted in *Stata Technical Bulletin Reprints*, vol. 1, pp. 208–220.

——. 1992. *Regression with Graphics: A Second Course in Applied Statistics.* Belmont, CA: Duxbury.

——. 2006. *Statistics with Stata (Updated for Version 9).* Belmont, CA: Duxbury.

Hinkley, D. V. 1978. Improving the jackknife with special reference to correlation estimation. *Biometrika* 65: 13–22.

Holmes, S., C. Morris, and R. Tibshirani. 2003. Bradley Efron: A conversation with good friends. *Statistical Science* 18: 268–281.

Mooney, C. Z., and R. D. Duval. 1993. *Bootstrapping: A Nonparametric Approach to Statistical Inference.* Newbury Park, CA: Sage.

Poi, B. P. 2004. From the help desk: Some bootstrapping techniques. *Stata Journal* 4: 312–328.

Stine, R. 1990. An introduction to bootstrap methods: Examples and ideas. In *Modern Methods of Data Analysis*, ed. J. Fox and J. S. Long, 353–373. Newbury Park, CA: Sage.

Also See

[R] **bootstrap postestimation** — Postestimation tools for bootstrap

[R] **jackknife** — Jackknife estimation

[R] **permute** — Monte Carlo permutation tests

[R] **simulate** — Monte Carlo simulations

[U] **13.5 Accessing coefficients and standard errors**

[U] **13.6 Accessing results from Stata commands**

[U] **20 Estimation and postestimation commands**

Title

> **bootstrap postestimation** — Postestimation tools for bootstrap

Description

The following postestimation command is of special interest after `bootstrap`:

command	description
estat bootstrap	percentile-based and bias-corrected CI tables

For information about `estat bootstrap`, see below.

The following standard postestimation commands are also available:

command	description
*adjust	adjusted predictions of $\mathbf{x}\beta$, probabilities, or $\exp(\mathbf{x}\beta)$
estat	AIC, BIC, VCE, and estimation sample summary
estimates	cataloging estimation results
*hausman	Hausman's specification test
*lincom	point estimates, standard errors, testing, and inference for linear combinations of coefficients
*mfx	marginal effects or elasticities
*nlcom	point estimates, standard errors, testing, and inference for nonlinear combinations of coefficients
*predict	predictions, residuals, influence statistics, and other diagnostic measures
*predictnl	point estimates, standard errors, testing, and inference for generalized predictions
*test	Wald tests for simple and composite linear hypotheses
*testnl	Wald tests of nonlinear hypotheses

*This postestimation command is allowed if it may be used after *command*.

See the corresponding entries in the *Stata Base Reference Manual* for details.

Special-interest postestimation command

`estat bootstrap` displays a table of confidence intervals for each statistic from a bootstrap analysis.

Syntax for predict

The syntax of `predict` (and even if `predict` is allowed) following `bootstrap` depends upon the *command* used with `bootstrap`. If `predict` is not allowed, neither are `adjust`, `mfx`, or `predictnl`.

Syntax for estat bootstrap

estat <u>boot</u>strap [, *options*]

options	description
bc	bias-corrected CIs; the default
bca	bias-corrected and accelerated (BC$_a$) CIs
<u>normal</u>	normal based CIs
<u>p</u>ercentile	percentile CIs
all	all available CIs
<u>noh</u>eader	suppress table header
<u>nol</u>egend	suppress table legend
<u>v</u>erbose	display the full table legend

bc, bca, normal, and percentile may be used together.

Options for estat bootstrap

bc is the default and displays bias-corrected confidence intervals.

bca displays bias-corrected and accelerated confidence intervals. This option assumes that you also specified the bca option on the bootstrap prefix command.

normal displays normal approximation confidence intervals.

percentile displays percentile confidence intervals.

all displays all available confidence intervals.

noheader suppresses display of the table header. This option implies nolegend.

nolegend suppresses display of the table legend, which identifies the rows of the table with the expressions they represent.

verbose requests that the full table legend be displayed.

Remarks

▷ Example 1

The estat bootstrap postestimation command produces a table containing the observed value of the statistic, an estimate of its bias, the bootstrap standard error, and up to four different confidence intervals.

If we were interested merely in getting bootstrap standard errors for the model coefficients, we could use the bootstrap prefix with our estimation command. If we were interested in performing a thorough bootstrap analysis of the model coefficients, we could use the estat bootstrap postestimation command after fitting the model with the bootstrap prefix.

Using example 1 from [R] **bootstrap**, we need many more replications for the confidence interval types other than the normal based, so let's rerun the estimation command. We will reset the random-number seed—in case we wish to reproduce the results—increase the number of replications, and save the bootstrap distribution as a dataset called bsauto.dta.

```
. use http://www.stata-press.com/data/r10/auto
(1978 Automobile Data)

. set seed 1

. bootstrap _b, reps(1000) saving(bsauto) bca: regress mpg weight gear foreign
  (output omitted )

. estat bootstrap, all
```

Linear regression   Number of obs   =   74
                                               Replications   =   1000

mpg	Observed Coef.	Bias	Bootstrap Std. Err.	[95% Conf. Interval]		
weight	-.00613903	.0000567	.000628	-.0073699	-.0049082	(N)
				-.0073044	-.0048548	(P)
				-.0074355	-.004928	(BC)
				-.0075282	-.0050258	(BCa)
gear_ratio	1.4571134	.1051696	1.4554785	-1.395572	4.309799	(N)
				-1.262111	4.585372	(P)
				-1.523927	4.174376	(BC)
				-1.492223	4.231356	(BCa)
foreign	-2.2216815	-.0196361	1.2023286	-4.578202	.1348393	(N)
				-4.442199	.2677989	(P)
				-4.155504	.6170642	(BC)
				-4.216531	.5743973	(BCa)
_cons	36.101353	-.502281	5.4089441	25.50002	46.70269	(N)
				24.48569	46.07086	(P)
				25.59799	46.63227	(BC)
				25.85658	47.02108	(BCa)

(N)   normal confidence interval
(P)   percentile confidence interval
(BC)   bias-corrected confidence interval
(BCa) bias-corrected and accelerated confidence interval

The estimated standard errors here differ from our previous estimates using only 100 replications by, respectively, 8%, 3%, 11%, and 6%; see example 1 of [R] **bootstrap**. So much for our advice that 50–200 replications are good enough to estimate standard errors. Well, the more replications the better—that advice you should believe.

Which of the methods to compute confidence intervals should we use? If the statistic is unbiased, the percentile (P) and bias-corrected (BC) methods should give similar results. The bias-corrected confidence interval will be the same as the percentile confidence interval when the observed value of the statistic is equal to the median of the bootstrap distribution. Thus for unbiased statistics, the two methods should give similar results as the number of replications becomes large. For biased statistics, the bias-corrected method should yield confidence intervals with better coverage probability (closer to the nominal value of 95% or whatever was specified) than the percentile method. For statistics with variances that vary as a function of the parameter of interest, the bias-corrected and accelerated method (BC_a) will typically have better coverage probability than the others.

When the bootstrap distribution is approximately normal, all these methods should give similar confidence intervals as the number of replications becomes large. If we examine the normality of these bootstrap distributions using, say, the pnorm command (see [R] **diagnostic plots**), we see that they closely follow a normal distribution. Thus in this case, the normal approximation would also be a valid choice. The chief advantage of the normal-approximation method is that it (supposedly) requires fewer replications than the other methods. Of course, it should be used only when the bootstrap distribution exhibits normality.

We can load `bsauto.dta` containing the bootstrap distributions for these coefficients:

```
. use bsauto
(bootstrap: regress)
. describe *
```

variable name	storage type	display format	value label	variable label
_b_weight	float	%9.0g		_b[weight]
_b_gear_ratio	float	%9.0g		_b[gear_ratio]
_b_foreign	float	%9.0g		_b[foreign]
_b_cons	float	%9.0g		_b[_cons]

We can now run other commands, such as `pnorm`, on the bootstrap distributions. As with all standard estimation commands, we can use the `bootstrap` command to replay its output table. The default variable names assigned to the statistics in *exp_list* are _bs_1, _bs_2, ..., and each variable is labeled with the associated expression. The naming convention for the extended expressions _b and _se is to prepend _b_ and _se_, respectively, onto the name of each element of the coefficient vector. Here the first coefficient is _b[weight], so `bootstrap` named it _b_weight.

◁

Methods and Formulas

All postestimation commands listed above are implemented as ado-files.

Also See

[R] **bootstrap** — Bootstrap sampling and estimation

[U] **20 Estimation and postestimation commands**

Title

> **boxcox** — Box–Cox regression models

Syntax

boxcox *depvar* [*indepvars*] [*if*] [*in*] [*weight*] [, *options*]

options	description
Model	
noconstant	suppress constant term
model(lhsonly)	left-hand-side Box–Cox model; the default
model(rhsonly)	right-hand-side Box–Cox model
model(lambda)	both sides Box–Cox model with same parameter
model(theta)	both sides Box–Cox model with different parameters
notrans(*varlist*)	nontransformed independent variables
Reporting	
level(*#*)	set confidence level; default is level(95)
lrtest	perform likelihood-ratio test
Max options	
nolog	suppress full-model iteration log
nologlr	suppress restricted-model lrtest iteration log
maximize_options	control the maximization process; seldom used

depvar and *indepvars* may contain time-series operators; see [U] **11.4.3 Time-series varlists**.
bootstrap, by, jackknife, rolling, statsby, and xi are allowed; see [U] **11.1.10 Prefix commands**.
Weights are not allowed with the bootstrap prefix.
fweights and iweights are allowed; see [U] **11.1.6 weight**.
See [U] **20 Estimation and postestimation commands** for more capabilities of estimation commands.

Description

boxcox finds the maximum likelihood estimates of the parameters of the Box–Cox transform, the coefficients on the independent variables, and the standard deviation of the normally distributed errors for a model in which *depvar* is regressed on *indepvars*. You can fit the following models:

Option	Estimates
lhsonly	$y_j^{(\theta)} = \beta_1 x_{1j} + \beta_2 x_{2j} + \cdots + \beta_k x_{kj} + \epsilon_j$
rhsonly	$y_j\ = \beta_1 x_{1j}^{(\lambda)} + \beta_2 x_{2j}^{(\lambda)} + \cdots + \beta_k x_{kj}^{(\lambda)} + \epsilon_j$
rhsonly notrans()	$y_j\ = \beta_1 x_{1j}^{(\lambda)} + \beta_2 x_{2j}^{(\lambda)} + \cdots + \beta_k x_{kj}^{(\lambda)} + \gamma_1 z_{1j} + \cdots + \gamma_l z_{lj} + \epsilon_j$
lambda	$y_j^{(\lambda)} = \beta_1 x_{1j}^{(\lambda)} + \beta_2 x_{2j}^{(\lambda)} + \cdots + \beta_k x_{kj}^{(\lambda)} + \epsilon_j$
lambda notrans()	$y_j^{(\lambda)} = \beta_1 x_{1j}^{(\lambda)} + \beta_2 x_{2j}^{(\lambda)} + \cdots + \beta_k x_{kj}^{(\lambda)} + \gamma_1 z_{1j} + \cdots + \gamma_l z_{lj} + \epsilon_j$
theta	$y_j^{(\theta)} = \beta_1 x_{1j}^{(\lambda)} + \beta_2 x_{2j}^{(\lambda)} + \cdots + \beta_k x_{kj}^{(\lambda)} + \epsilon_j$
theta notrans()	$y_j^{(\theta)} = \beta_1 x_{1j}^{(\lambda)} + \beta_2 x_{2j}^{(\lambda)} + \cdots + \beta_k x_{kj}^{(\lambda)} + \gamma_1 z_{1j} + \cdots + \gamma_l z_{lj} + \epsilon_j$

Any variable to be transformed must be strictly positive.

Options

⌐ Model ⌐

noconstant; see [R] **estimation options**.

model(lhsonly | rhsonly | lambda | theta) specifies which of the four models to fit.

model(lhsonly) applies the Box–Cox transform to *depvar* only. model(lhsonly) is the default.

model(rhsonly) applies the transform to the *indepvars* only.

model(lambda) applies the transform to both *depvar* and *indepvars*, and they are transformed by the same parameter.

model(theta) applies the transform to both *depvar* and *indepvars*, but this time, each side is transformed by a separate parameter.

notrans(*varlist*) specifies that the variables in *varlist* be included as nontransformed independent variables.

⌐ Reporting ⌐

level(*#*); see [R] **estimation options**.

lrtest specifies that a likelihood-ratio test of significance be performed and reported for each independent variable.

⌐ Max options ⌐

nolog suppresses the iteration log when fitting the full model.

nologlr suppresses the iteration log when fitting the restricted models required by the lrtest option.

maximize_options: iterate(*#*) and from(*init_specs*); see [R] **maximize**.

Model	Initial value specification
lhsonly	from(θ_0, copy)
rhsonly	from(λ_0, copy)
lambda	from(λ_0, copy)
theta	from($\lambda_0\ \theta_0$, copy)

Remarks

Remarks are presented under the following headings:

Introduction
Theta model
Lambda model
Left-hand-side-only model
Right-hand-side-only model

Introduction

The Box–Cox transform

$$y^{(\lambda)} = \frac{y^\lambda - 1}{\lambda}$$

has been widely used in applied data analysis. Box and Cox (1964) developed the transformation and argued that the transformation could make the residuals more closely normal and less heteroskedastic. Cook and Weisberg (1982) discuss the transform in this light. Since the transform embeds several popular functional forms, it has received some attention as a method for testing functional forms, in particular,

$$y^{(\lambda)} = \begin{cases} y - 1 & \text{if } \lambda = 1 \\ \ln(y) & \text{if } \lambda = 0 \\ 1 - 1/y & \text{if } \lambda = -1 \end{cases}$$

Davidson and MacKinnon (1993) discuss this use of the transform. Atkinson (1985) also gives a good general treatment.

Theta model

boxcox obtains the maximum likelihood estimates of the parameters for four different models. The most general of the models, the `theta` model, is

$$y_j^{(\theta)} = \beta_0 + \beta_1 x_{1j}^{(\lambda)} + \beta_2 x_{2j}^{(\lambda)} + \cdots + \beta_k x_{kj}^{(\lambda)} + \gamma_1 z_{1j} + \gamma_2 z_{2j} + \cdots + \gamma_l z_{lj} + \epsilon_j$$

where $\epsilon \sim N(0, \sigma^2)$. Here the dependent variable, y, is subject to a Box–Cox transform with parameter θ. Each of the *indepvars*, $x_1, x_2, \ldots, x_k$, is transformed by a Box–Cox transform with parameter λ. The $z_1, z_2, \ldots, z_l$ specified in the `notrans()` option are independent variables that are not transformed.

Box and Cox (1964) argued that this transformation would leave behind residuals that more closely follow a normal distribution than those produced by a simple linear regression model. Bear in mind that the normality of ϵ is assumed and that boxcox obtains maximum likelihood estimates of the $k + l + 4$ parameters under this assumption. boxcox does not choose λ and θ so that the residuals are approximately normally distributed. If you are interested in this type of transformation to normality, see the official Stata commands lnskew0 and bcskew0 in [R] **lnskew0**. However, those commands work on a more restrictive model in which none of the independent variables is transformed.

▷ Example 1

Consider an example using the auto data.

```
. use http://www.stata-press.com/data/r10/auto
(1978 Automobile Data)
. boxcox mpg weight price, notrans(foreign) model(theta) lrtest
Fitting comparison model
Iteration 0:   log likelihood = -234.39434
Iteration 1:   log likelihood = -228.26891
Iteration 2:   log likelihood = -228.26777
Iteration 3:   log likelihood = -228.26777

Fitting full model
Iteration 0:   log likelihood = -194.13727
   (output omitted)

Fitting comparison models for LR tests
Iteration 0:   log likelihood = -179.58214
Iteration 1:   log likelihood = -177.59036
Iteration 2:   log likelihood = -177.58739
Iteration 3:   log likelihood = -177.58739

Iteration 0:   log likelihood = -203.92855
Iteration 1:   log likelihood = -201.30202
Iteration 2:   log likelihood = -201.18359
Iteration 3:   log likelihood = -201.18233
Iteration 4:   log likelihood = -201.18233

Iteration 0:   log likelihood = -178.83799
Iteration 1:   log likelihood = -175.98405
Iteration 2:   log likelihood = -175.97931
Iteration 3:   log likelihood = -175.97931
```

```
                                            Number of obs   =         74
                                            LR chi2(4)      =     105.19
Log likelihood = -175.67343                 Prob > chi2     =      0.000
```

mpg	Coef.	Std. Err.	z	P>|z|	[95% Conf. Interval]
/lambda	.7601686	.6289991	1.21	0.227	-.472647 1.992984
/theta	-.7189313	.3244439	-2.22	0.027	-1.35483 -.0830331

Estimates of scale-variant parameters

	Coef.	chi2(df)	P>chi2(df)	df of chi2
Notrans				
foreign	-.0114338	3.828	0.050	1
_cons	1.3774			
Trans				
weight	-.000239	51.018	0.000	1
price	-6.18e-06	0.612	0.434	1
/sigma	.0143509			

Test H0:	Restricted log likelihood	chi2	Prob > chi2
theta=lambda = -1	-181.64479	11.94	0.001
theta=lambda = 0	-178.2406	5.13	0.023
theta=lambda = 1	-194.13727	36.93	0.000

The output is composed of the iteration logs and three distinct tables. The first table contains a standard header for a maximum likelihood estimator and a standard output table for the Box–Cox transform parameters. The second table contains the estimates of the scale-variant parameters. The third table contains the output from likelihood-ratio tests on three standard functional form specifications.

If we were to interpret this output, the right-hand-side transformation would not significantly add to the regression, whereas the left-hand-side transformation would make the 5% but not the 1% cutoff. `price` is certainly not significant, and `foreign` lies right on the 5% cutoff. `weight` is clearly significant. The output also shows that the linear and multiplicative inverse specifications are both strongly rejected. A natural log specification can be rejected at the 5% level but not at the 1% level.

◁

❑ Technical Note

Spitzer (1984) showed that the Wald tests of the joint significance of the coefficients of the right-hand-side variables, either transformed or untransformed, are not invariant to changes in the scale of the transformed dependent variable. Davidson and MacKinnon (1993) also discuss this point. This problem demonstrates that Wald statistics can be manipulated in nonlinear models. Lafontaine and White (1986) analyze this problem numerically, and Phillips and Park (1988) analyze it by using Edgeworth expansions. See Drukker (2000b) for a more detailed discussion of this issue. Since the parameter estimates and their Wald tests are not scale invariant, no Wald tests or confidence intervals are reported for these parameters. However, when the `lrtest` option is specified, likelihood-ratio tests are performed and reported. Schlesselman (1971) showed that, if a constant is included in the model, the parameter estimates of the Box–Cox transforms are scale invariant. For this reason, we strongly recommend that you not use the `noconstant` option.

The `lrtest` option does not perform a likelihood-ratio test on the constant, so no value for this statistic is reported. Unless the data are properly scaled, the restricted model does not often converge. For this reason, no likelihood-ratio test on the constant is performed by the `lrtest` option. However, if you have a special interest in performing this test, you can do so by fitting the constrained model separately. If problems with convergence are encountered, rescaling the data by their means may help.

❑

Lambda model

A less general model than the one above is called the `lambda` model. It specifies that the same parameter be used in both the left-hand side and right-hand side transformations. Specifically,

$$y_j^{(\lambda)} = \beta_0 + \beta_1 x_{1j}^{(\lambda)} + \beta_2 x_{2j}^{(\lambda)} + \cdots + \beta_k x_{kj}^{(\lambda)} + \gamma_1 z_{1j} + \gamma_2 z_{2j} + \cdots + \gamma_l z_{lj} + \epsilon_j$$

where $\epsilon \sim N(0, \sigma^2)$. Here the *depvar* variable, y, and each of the *indepvars*, $x_1, x_2, \ldots, x_k$, are transformed by a Box–Cox transform with the common parameter λ. Again the $z_1, z_2, \ldots, z_l$ are independent variables that are not transformed.

Left-hand-side-only model

Even more restrictive than a common transformation parameter is transforming the dependent variable only. Since the dependent variable is on the left-hand side of the equation, this model is known as the lhsonly model. Here you are estimating the parameters of the model

$$y_j^{(\theta)} = \beta_0 + \beta_1 x_{1j} + \beta_2 x_{2j} + \cdots + \beta_k x_{kj} + \epsilon_j$$

where $\epsilon \sim N(0, \sigma^2)$. Here only the *depvar*, y, is transformed by a Box–Cox transform with the parameter θ.

▷ Example 2

Here we again hypothesize mpg to be a function of weight, price, and foreign in a Box–Cox model in which only mpg is subject to the transform.

```
. boxcox mpg weight price foreign, model(lhs) lrtest nolog nologlr
Fitting comparison model
Fitting full model
Fitting comparison models for LR tests
```

		Number of obs	=	74
		LR chi2(3)	=	105.04
Log likelihood = -175.74705		Prob > chi2	=	0.000

mpg	Coef.	Std. Err.	z	P>|z|	[95% Conf. Interval]
/theta	-.7826999	.281954	-2.78	0.006	-1.33532 -.2300802

Estimates of scale-variant parameters

	Coef.	chi2(df)	P>chi2(df)	df of chi2
Notrans				
weight	-.0000294	58.056	0.000	1
price	-4.66e-07	0.469	0.493	1
foreign	-.0097564	4.644	0.031	1
_cons	1.249845			
/sigma	.0118454			

Test H0:	Restricted log likelihood	LR statistic chi2	P-value Prob > chi2
theta = -1	-176.04312	0.59	0.442
theta = 0	-179.54104	7.59	0.006
theta = 1	-194.13727	36.78	0.000

This model rejects both linear and log specifications of mpg but fails to reject the hypothesis that $1/\text{mpg}$ is linear in the independent variables. These findings are in line with what an engineer would have expected. In engineering terms, gallons per mile represents actual energy consumption, and energy consumption should be approximately linear in weight.

◁

Right-hand-side-only model

The fourth model leaves the *depvar* alone and transforms a subset of the *indepvars* using the parameter λ. This is the `rhsonly` model. In this model, the *depvar*, y, is given by

$$y_j = \beta_0 + \beta_1 x_{1j}^{(\lambda)} + \beta_2 x_{2j}^{(\lambda)} + \cdots + \beta_k x_{kj}^{(\lambda)} + \gamma_1 z_{1j} + \gamma_2 z_{2j} + \cdots + \gamma_l z_{lj} + \epsilon_j$$

where $\epsilon \sim N(0, \sigma^2)$. Here each of the *indepvars*, $x_1, x_2, \ldots, x_k$, are transformed by a Box–Cox transform with the parameter λ. Again the $z_1, z_2, \ldots, z_l$ are independent variables that are not transformed.

▷ Example 3

Here is an example with the `rhsonly` model. `price` and `foreign` are not included in the list of covariates. (You are invited to use the auto data and check that they fare no better here than above.)

```
. boxcox mpg weight, model(rhs) lrtest nolog nologlr

Fitting full model

Fitting comparison models for LR tests

Comparison model for LR test on weight is a linear regression.
Lambda is not identified in the restricted model.
```

		Number of obs	=	74
		LR chi2(2)	=	82.90
Log likelihood = -192.94368		Prob > chi2	=	0.000

mpg	Coef.	Std. Err.	z	P>|z|	[95% Conf. Interval]
/lambda	-.4460916	.6551107	-0.68	0.496	-1.730085 .8379018

Estimates of scale-variant parameters

	Coef.	chi2(df)	P>chi2(df)	df of chi2
Notrans				
_cons	1359.092			
Trans				
weight	-614.3876	82.901	0.000	1
/sigma	3.281854			

Test H0:	Restricted log likelihood	LR statistic chi2	P-value Prob > chi2
lambda = -1	-193.2893	0.69	0.406
lambda = 0	-193.17892	0.47	0.493
lambda = 1	-195.38869	4.89	0.027

The interpretation of the output is similar to that in all the cases above, with one caveat. As requested, a likelihood-ratio test was performed on the lone independent variable. However, when it is dropped to form the constrained model, the comparison model is not a right-hand-side-only Box–Cox model but rather a simple linear regression on a constant model. When weight is dropped, there are no longer any transformed variables. Hence, λ is not identified, and it must also be dropped. This process leaves a linear regression on a constant as the "comparison model". It also implies that the test statistic has 2 degrees of freedom instead of 1. At the top of the output, a more concise warning informs you of this point.

A similar identification issue can also arise in the `lambda` and `theta` models when only one independent variable is specified. In these cases, warnings also appear on the output.

◁

Saved Results

boxcox saves the following in e():

Scalars

e(N)	number of observations	e(ll_tm1)	log likelihood of model $\lambda=\theta=-1$
e(ll)	log likelihood	e(chi2_tm1)	LR of $\lambda=\theta=-1$ vs. full model
e(chi2)	LR statistic of full vs. comparison	e(p_tm1)	p-value of $\lambda=\theta=-1$ vs. full model
e(df_m)	full model degrees of freedom	e(ll_t0)	log likelihood of model $\lambda=\theta=0$
e(ll0)	log likelihood of the restricted model	e(chi2_t0)	LR of $\lambda=\theta=0$ vs. full model
e(df_r)	restricted model degrees of freedom	e(p_t0)	p-value of $\lambda=\theta=0$ vs. full model
e(ll_t1)	log likelihood of model $\lambda=\theta=1$	e(ic)	number of iterations
e(chi2_t1)	LR of $\lambda=\theta=1$ vs. full model	e(rc)	return code
e(p_t1)	p-value of $\lambda=\theta=1$ vs. full model		

Macros

e(cmd)	boxcox	e(ntrans)	yes if nontransformed *indepvars*
e(cmdline)	command as typed	e(chi2type)	LR; type of model χ^2 test
e(depvar)	name of dependent variable	e(lrtest)	lrtest if requested
e(model)	lhsonly, rhsonly, lambda, or theta	e(properties)	b V
e(wtype)	weight type	e(predict)	program used to implement
e(wexp)	weight expression		predict

Matrices

e(b)	coefficient vector	e(df)	degrees of freedom of LR tests on
e(V)	variance–covariance matrix of		*indepvars*
	the estimators (see note below)	e(chi2m)	LR statistics for tests on *indepvars*
e(pm)	p-values for LR tests on *indepvars*		

Functions

e(sample)	marks estimation sample

e(V) contains all zeros, except for the elements that correspond to the parameters of the Box–Cox transform.

Methods and Formulas

boxcox is implemented as an ado-file.

In the internal computations,

$$y^{(\lambda)} = \begin{cases} \frac{y^{\lambda}-1}{\lambda} & \text{if } |\lambda| > 10^{-10} \\ \\ \ln(y) & \text{otherwise} \end{cases}$$

The unconcentrated log likelihood for the `theta` model is

$$\ln L = \left(\frac{-N}{2}\right)\left\{\ln(2\pi) + \ln(\sigma^2)\right\} + (\theta - 1)\sum_{i=1}^{N}\ln(y_i) - \left(\frac{1}{2\sigma^2}\right)\text{SSR}$$

where

$$\text{SSR} = \sum_{i=1}^{N}(y_i^{(\theta)} - \beta_0 + \beta_1 x_{i1}^{(\lambda)} + \beta_2 x_{i2}^{(\lambda)} + \cdots + \beta_k x_{ik}^{(\lambda)} + \gamma_1 z_{i1} + \gamma_2 z_{i2} + \cdots + \gamma_l z_{il})^2$$

Writing the SSR in matrix form,

$$\text{SSR} = (\mathbf{Y}^{(\theta)} - \mathbf{X}^{(\lambda)}\mathbf{b}' - \mathbf{Z}\mathbf{g}')'(\mathbf{Y}^{(\theta)} - \mathbf{X}^{(\lambda)}\mathbf{b}' - \mathbf{Z}\mathbf{g}')$$

where $\mathbf{Y}^{(\theta)}$ is an $N \times 1$ vector of elementwise transformed data, $\mathbf{X}^{(\lambda)}$ is an $N \times k$ matrix of elementwise transformed data, $\mathbf{Z}$ is an $N \times l$ matrix of untransformed data, $\mathbf{b}$ is a $1 \times k$ vector of coefficients, and $\mathbf{g}$ is a $1 \times l$ vector of coefficients. Letting

$$\mathbf{W}_\lambda = \left(\mathbf{X}^{(\lambda)}\ \mathbf{Z}\right)$$

be the horizontal concatenation of $\mathbf{X}^{(\lambda)}$ and $\mathbf{Z}$ and

$$\mathbf{d}' = \begin{pmatrix} \mathbf{b}' \\ \mathbf{g}' \end{pmatrix}$$

be the vertical concatenation of the coefficients yields

$$\text{SSR} = (\mathbf{Y}^{(\theta)} - \mathbf{W}_\lambda\mathbf{d}')'(\mathbf{Y}^{(\theta)} - \mathbf{W}_\lambda\mathbf{d}')$$

For given values of λ and θ, the solutions for $\mathbf{d}'$ and σ^2 are

$$\widehat{\mathbf{d}}' = (\mathbf{W}_\lambda'\mathbf{W}_\lambda)^{-1}\mathbf{W}_\lambda'\mathbf{Y}^{(\theta)}$$

and

$$\widehat{\sigma}^2 = \frac{1}{N}\left(\mathbf{Y}^{(\theta)} - \mathbf{W}_\lambda\widehat{\mathbf{d}}'\right)'\left(\mathbf{Y}^{(\theta)} - \mathbf{W}_\lambda\widehat{\mathbf{d}}'\right)$$

Substituting these solutions into the log-likelihood function yields the concentrated log-likelihood function

$$\ln L_c = \left(-\frac{N}{2}\right)\left\{\ln(2\pi) + 1 + \ln(\widehat{\sigma}^2)\right\} + (\theta - 1)\sum_{i=1}^{N}\ln(y_i)$$

Similar calculations yield the concentrated log-likelihood function for the `lambda` model,

$$\ln L_c = \left(-\frac{N}{2}\right) \left\{ \ln(2\pi) + 1 + \ln(\widehat{\sigma}^2) \right\} + (\lambda - 1) \sum_{i=1}^{N} \ln(y_i)$$

the `lhsonly` model,

$$\ln L_c = \left(-\frac{N}{2}\right) \left\{ \ln(2\pi) + 1 + \ln(\widehat{\sigma}^2) \right\} + (\theta - 1) \sum_{i=1}^{N} \ln(y_i)$$

and the `rhsonly` model,

$$\ln L_c = \left(-\frac{N}{2}\right) \left\{ \ln(2\pi) + 1 + \ln(\widehat{\sigma}^2) \right\}$$

where $\widehat{\sigma}^2$ is specific to each model and is defined analogously to that in the `theta` model.

References

Atkinson, A. C. 1985. *Plots, Transformations, and Regression: An Introduction to Graphical Methods of Diagnostic Regression Analysis.* Oxford: Oxford University Press.

Box, G. E. P., and D. R. Cox. 1964. An analysis of transformations. *Journal of the Royal Statistical Society, Series B* 26: 211–243.

Carroll, R. J., and D. Ruppert. 1988. *Transformation and Weighting in Regression.* New York: Chapman & Hall.

Cook, R. D., and S. Weisberg. 1982. *Residuals and Influence in Regression.* New York: Chapman & Hall.

Davidson, R., and J. G. MacKinnon. 1993. *Estimation and Inference in Econometrics.* Oxford: Oxford University Press.

Drukker, D. M. 2000a. sg130: Box–Cox regression models. *Stata Technical Bulletin* 54: 27–36. Reprinted in *Stata Technical Bulletin Reprints,* vol. 9, pp. 307–319.

——. 2000b. sg131: On the manipulability of Wald statistics in Box–Cox regression models. *Stata Technical Bulletin* 54: 36–42. Reprinted in *Stata Technical Bulletin Reprints,* vol. 9, pp. 319–327.

Lafontaine, F., and K. J. White. 1986. Obtaining any Wald statistic you want. *Economics Letters* 21: 35–40.

Phillips, P. C. B., and J. Y. Park. 1988. On the formulation of Wald tests of nonlinear restrictions. *Econometrica* 56: 1065–1083.

Schlesselman, J. J. 1971. Power families: A note on the Box and Cox transformation. *Journal of the Royal Statistical Society, Series B* 33: 307–311.

Spitzer, J. J. 1984. Variance estimates in models with the Box–Cox transformation: Implications for estimation and hypothesis testing. *Review of Economics and Statistics* 66: 645–652.

Also See

[R] **boxcox postestimation** — Postestimation tools for boxcox

[R] **regress** — Linear regression

[R] **lnskew0** — Find zero-skewness log or Box–Cox transform

[U] **20 Estimation and postestimation commands**

Title

> **boxcox postestimation** — Postestimation tools for boxcox

Description

The following postestimation commands are available for `boxcox`:

command	description
estat	AIC, BIC, VCE, and estimation sample summary
estimates	cataloging estimation results
*lincom	point estimates, standard errors, testing, and inference for linear combinations of coefficients
*nlcom	point estimates, standard errors, testing, and inference for nonlinear combinations of coefficients
predict	predictions, residuals, influence statistics, and other diagnostic measures
*test	Wald tests for simple and composite linear hypotheses
*testnl	Wald tests of nonlinear hypotheses

*Inference is only valid for hypotheses concerning λ and θ.

See the corresponding entries in the *Stata Base Reference Manual* for details.

Syntax for predict

> predict [*type*] *newvar* [*if*] [*in*] [, *statistic*]

statistic	description
Main	
xbt	transformed linear prediction; the default
yhat	predicted value of y
residuals	residuals

These statistics are available both in and out of sample; type predict ... if e(sample) ... if wanted only for the estimation sample.

Options for predict

Main

xbt, the default, calculates the "linear" prediction. For all the models except model(lhsonly), all the *indepvars* except those specified in the notrans() option are transformed.

yhat calculates the predicted value of y.

residuals calculates the residuals after the predicted value of y has been subtracted from the actual value.

Remarks

boxcox estimates variances only for the λ and θ parameters (see the Technical Note in [R] **boxcox**), so the extent to which postestimation commands can be used following boxcox is limited. Formulas used in lincom, nlcom, test, and testnl are dependent on the estimated variances. Therefore, the use of these commands is limited and generally applicable only to inferences on the λ and θ coefficients.

Methods and Formulas

All postestimation commands listed above are implemented as ado-files.

Also See

[R] **boxcox** — Box–Cox regression models

[R] **lnskew0** — Find zero-skewness log or Box–Cox transform

[U] **20 Estimation and postestimation commands**

Title

> **brier** — Brier score decomposition

Syntax

> brier *outcomevar forecastvar* $[if]$ $[in]$ $[$, group(*#*) $]$

by is allowed; see [D] **by**.

Description

brier computes the Yates, Sanders, and Murphy decompositions of the Brier Mean Probability Score. *outcomevar* contains 0/1 values reflecting the actual outcome of the experiment, and *forecastvar* contains the corresponding probabilities as predicted by, say, logit, probit, or a human forecaster.

Option

> Main

group(*#*) specifies the number of groups that will be used to compute the decomposition. group(10) is the default.

Remarks

You have a binary (0/1) response and a formula that predicts the corresponding probabilities of having observed a positive outcome (1). If the probabilities were obtained from logistic regression, there are many methods that assess goodness of fit (see, for instance, estat gof in [R] **logistic**). However, the probabilities might be computed from a published formula or from a model fitted on another sample, both completely unrelated to the data at hand, or perhaps the forecasts are not from a formula at all. In any case, you now have a *test dataset* consisting of the forecast probabilities and observed outcomes. Your test dataset might, for instance, record predictions made by a meteorologist on the probability of rain along with a variable recording whether it actually rained.

The Brier score is an aggregate measure of disagreement between the observed outcome and a prediction—the average squared error difference. The Brier score decomposition is a partition of the Brier score into components that suggest reasons for discrepancy. These reasons fall roughly into three groups: (1) lack of overall calibration between the average predicted probability and the actual probability of the event in your data, (2) misfit of the data in groups defined within your sample, and (3) inability to match actual 0 and 1 responses.

Problem (1) refers to simply overstating or understating the probabilities.

Problem (2) refers to what is standardly called a goodness-of-fit test: the data are grouped, and the predictions for the group are compared with the outcomes.

Problem (3) refers to an individual-level measure of fit. Imagine that the grouped outcomes are predicted on average correctly but that within the group, the outcomes are poorly predicted.

Using logit or probit analysis to fit your data will guarantee that there is no lack of fit due to (1), and a good model fitter will be able to avoid problem (2). Problem (3) is inherent in any prediction exercise.

236

▷ Example 1

We have data on the outcomes of 20 basketball games (`win`) and the probability of victory predicted by a local pundit (`for`).

```
. use http://www.stata-press.com/data/r10/bball

. summarize win for
```

Variable	Obs	Mean	Std. Dev.	Min	Max
win	20	.65	.4893605	0	1
for	20	.4785	.2147526	.15	.9

```
. brier win for, group(5)
```

Mean probability of outcome	0.6500	
of forecast	0.4785	
Correlation	0.5907	
ROC area	0.8791	p = 0.0030
Brier score	0.1828	
Spiegelhalter's z-statistic	-0.6339	p = 0.7369
Sanders-modified Brier score	0.1861	
Sanders resolution	0.1400	
Outcome index variance	0.2275	
Murphy resolution	0.0875	
Reliability-in-the-small	0.0461	
Forecast variance	0.0438	
Excess forecast variance	0.0285	
Minimum forecast variance	0.0153	
Reliability-in-the-large	0.0294	
2*Forecast-Outcome-Covar	0.1179	

The mean probabilities of forecast and outcome are simply the mean of the predicted probabilities and the actual outcomes (wins/losses). The correlation is the product-moment correlation between them.

The Brier score measures the total difference between the event (winning) and the forecast probability of that event as an average squared difference. As a benchmark, a perfect forecaster would have a Brier score of 0, a perfect misforecaster (predicts probability of win is 1 when loses and 0 when wins) would have a Brier score of 1, and a fence-sitter (forecasts every game as 50/50) would have a Brier score of .25. Our pundit is doing reasonably well.

Spiegelhalter's Z statistic is a standard normal test statistic for testing whether an individual Brier score is extreme. The ROC area is the area under the receiver operating curve, and the associated test is a test of whether it is greater than 0.5. The more accurate the forecast probabilities, the larger the ROC area.

The Sanders-modified Brier score measures the difference between a grouped forecast measure and the event, where the data are grouped by sorting the sample on the forecast and dividing it into approximately equally sized groups. The difference between the modified and the unmodified score is typically minimal. For this and the other statistics that require grouping—the Sanders and Murphy resolutions and reliability-in-the-small—to be well-defined, group boundaries are chosen so as not to allocate observations with the same forecast probability to different groups. This task is done by grouping on the forecast using `xtile, n(#)`, with # being the number of groups; see [D] **pctile**.

The Sanders resolution measures error that arises from statistical considerations in evaluating the forecast for a group. A group with all positive or all negative outcomes would have a Sanders resolution of 0; it would most certainly be feasible to predict exactly what happened to each member of the group. If the group had 40% positive responses, on the other hand, a forecast that assigned $p = .4$ to each member of the group would be a good one, and yet, there would be "errors" in the squared difference sense. The "error" would be $(1 - .4)^2$ or $(0 - .4)^2$ for each member. The Sanders resolution is the average across groups of such "expected" errors. The .1400 value in our data from an overall Brier score of .1828 or .1861 suggests that a substantial portion of the "error" in our data is inherent.

Outcome index variance is just the variance of the outcome variable. This is the expected value of the Brier score if all the forecast probabilities were merely the average observed outcome. Remember that a fence-sitter has an expected Brier score of .25; a smarter fence sitter (who would guess $p = .65$ for these data) would have a Brier score of .2275.

The Murphy resolution measures the variation in the average outcomes across groups. If all groups have the same frequency of positive outcomes, little information in any forecast is possible, and the Murphy resolution is 0. If groups differ markedly, the Murphy resolution is as large as .25. The .0875 means that there is some variation but not a lot, and .0875 is probably higher than in most real cases. If you had groups in your data that varied between 40% and 60% positive outcomes, the Murphy resolution would be .01; between 30% and 70%, it would be .04.

Reliability-in-the-small measures the error that comes from the average forecast within group not measuring the average outcome within group—a classical goodness-of-fit measure, with 0 meaning a perfect fit and 1 meaning a complete lack of fit. The calculated value of 0.0461 shows some amount of lack of fit. Remember, the number is squared, and we are saying that probabilities could be just more than $\sqrt{.0461} = .215$ or 21.5% off.

Forecast variance measures the amount of discrimination being attempted—that is, the variation in the forecasted probabilities. A small number indicates a fence-sitter making constant predictions. If the forecasts were from a logistic regression model, forecast variance would tend to increase with the amount of information available. Our pundit shows considerable forecast variance of .0438 (standard deviation $\sqrt{.0438} = .2093$), which is in line with the reliability-in-the-small, suggesting that the forecaster is attempting as much variation as is available in these data.

Excess forecast variance is the amount of actual forecast variance over a theoretical minimum. The theoretical minimum—called the minimum forecast variance—corresponds to forecasts of p_0 for observations ultimately observed to be negative responses and p_1 for observations ultimately observed to be positive outcomes. Moreover, p_0 and p_1 are set to the average forecasts made for the ultimate negative and positive outcomes. These predictions would be just as good as the predictions the forecaster did make, and any variation in the actual forecast probabilities above this is useless. If this number is large, above 1%–2%, then the forecaster may be attempting more than is possible. The .0285 in our data suggests this possibility.

Reliability-in-the-large measures the discrepancy between the mean forecast and the observed fraction of positive outcomes. This discrepancy will be 0 for forecasts made by most statistical models—at least when measured on the same sample used for estimation—since they, by design, reproduce sample means. For our human pundit, the .0294 says that there is a $\sqrt{.0294}$, or 17-percentage-point, difference. (This difference can also be found by calculating the difference in the averages of the observed outcomes and forecast probabilities: $.65 - .4785 = .17$.) That difference, however, is not significant, as we would see if we typed `ttest win=for`; see [R] **ttest**. If these data were larger and the bias persisted, this difference would be a critical shortcoming of the forecast.

Twice the forecast-outcome covariance is a measure of how accurately the forecast corresponds to the outcome. The concept is similar to that of R-squared in linear regression.

◁

Saved Results

brier saves the following in r():

Scalars

r(p_roc)	significance of ROC area	r(murphy)	Murphy resolution
r(roc_area)	ROC area	r(relinsm)	reliability in the small
r(z)	Spiegelhalter's z statistic	r(Var_f)	forecast variance
r(p)	significance of z statistic	r(Var_fex)	excess forecast variance
r(brier)	Brier score	r(Var_fmin)	minimum forecast variance
r(brier_s)	Sanders-modified Brier score	r(relinla)	reliability in the large
r(sanders)	Sanders resolution	r(cov_2f)	$2 \times$ forecast-outcome-covariance
r(oiv)	outcome index variance		

Methods and Formulas

brier is implemented as an ado-file.

See Wilks (1995, 259–263) or Schmidt and Griffith (1998) for a discussion of the Brier score.

Let d_j, $j = 1, \ldots, N$ be the observed outcomes with $d_j = 0$ or $d_j = 1$, and let f_j be the corresponding forecasted probabilities that d_j is 1, $0 \leq f_j \leq 1$. Assume that the data are ordered so that $f_{j+1} \geq f_j$ (brier sorts the data to obtain this order). Divide the data into K nearly equally sized groups, group 1 containing observations 1 through $j_2 - 1$, group 2 observations j_2 through $j_3 - 1$, and so on.

Define

$$\overline{f}_0 = \text{average } f_j \text{ among } d_j = 0$$

$$\overline{f}_1 = \text{average } f_j \text{ among } d_j = 1$$

$$\overline{f} = \text{average } f_j$$

$$\overline{d} = \text{average } d_j$$

$$\widetilde{f}_k = \text{average } f_j \text{ in group } k$$

$$\widetilde{d}_k = \text{average } d_j \text{ in group } k$$

$$\widetilde{n}_k = \text{number of observations in group } k$$

The Brier score is $\sum_j (d_j - f_j)^2 / N$.

The Sanders-modified Brier score is $\sum_j (d_j - \widetilde{f}_{k(j)})^2 / N$.

Let p_j denote the true but unknown probability that $d_j = 1$. Under the null hypothesis that $p_j = f_j$ for all j, Spiegelhalter (1986) determined that the expectation and variance of the Brier score is given by the following:

(Continued on next page)

$$E(\text{Brier}) = \frac{1}{N} \sum_{j=1}^{N} f_j (1 - f_j)$$

$$\text{Var}(\text{Brier}) = \frac{1}{N^2} \sum_{j=1}^{N} f_j (1 - f_j)(1 - 2f_j)^2$$

Denoting the observed value of the Brier score by $O(\text{Brier})$, Spiegelhalter's Z statistic is given by

$$Z = \frac{O(\text{Brier}) - E(\text{Brier})}{\sqrt{\text{Var}(\text{Brier})}}$$

The corresponding p-value is given by the upper-tail probability of Z under the standard normal distribution.

The area under the ROC curve is estimated by applying the trapezoidal rule to the empirical ROC curve. This area is Wilcoxon's test statistic, so the corresponding p-value is just that of a one-sided Wilcoxon test of the null hypothesis that the distribution of predictions is constant across the two outcomes.

The Sanders resolution is $\sum_k \widetilde{n}_k \{\widetilde{d}_k (1 - \widetilde{d}_k)\}/N$.

The outcome index variance is $\overline{d}(1 - \overline{d})$.

The Murphy resolution is $\sum_k \widetilde{n}_k (\widetilde{d}_k - \overline{d})^2/N$.

Reliability-in-the-small is $\sum_k \widetilde{n}_k (\widetilde{d}_k - \widetilde{f}_k)^2/N$.

The forecast variance is $\sum_j (f_j - \overline{f})^2/N$.

The minimum forecast variance is $\left\{ \sum_{j \in F}(f_j - \overline{f}_0)^2 + \sum_{j \in S}(f_j - \overline{f}_1)^2 \right\}/N$, where F is the set of observations for which $d_j = 0$ and S is the complement.

The excess forecast variance is the difference between the forecast variance and the minimum forecast variance.

Reliability-in-the-large is $(\overline{f} - \overline{d})^2$.

Twice the outcome covariance is $2(\overline{f}_1 - \overline{f}_0)\overline{d}(1 - \overline{d})$.

Glenn Wilson Brier (1913–1998) was an American meteorological statistician who, after taking degrees in physics and statistics, was for many years head of meteorological statistics at the U.S. Weather Bureau, Washington, DC. In the latter part of his career, he was associated with Colorado State University. Brier worked especially on verification and evaluation of predictions and forecasts, statistical decision making, the statistical theory of turbulence, the analysis of weather modification experiments, and the application of permutation techniques.

Acknowledgment

We thank Richard Goldstein for his contributions to this improved version of `brier`.

References

Brier, G. W. 1950. Verification of forecasts expressed in terms of probability. *Monthly Weather Review* 78: 1–3.

Goldstein, R. 1996. sg55: Extensions to the brier command. *Stata Technical Bulletin* 32: 21–22. Reprinted in *Stata Technical Bulletin Reprints*, vol. 6, pp. 133–134.

Hadorn, D., E. B. Keeler, W. H. Rogers, and R. Brook. 1993. *Assessing the Performance of Mortality Prediction Models*. N-3599-HCFA. Santa Monica, CA: The Rand Corporation.

Holloway, L., and P. Mielke. 1998. Glenn Wilson Brier 1913–1998. *Bulletin of the American Meteorological Society* 79: 1438–1439.

Jolliffe, I. T., and D. B. Stephenson, ed. 2003. *Forecast Verification: A Practitioner's Guide in Atmospheric Science*. Chichester, UK: Wiley.

Murphy, A. H. 1973. A new vector partition of the probability score. *Journal of Applied Meteorology* 12: 595–600.

——. 1997. Forecast verification. In *Economic Value of Weather and Climate Forecasts*, ed. R. W. Katz and A. H. Murphy, 19–74. Cambridge: Cambridge University Press.

Redelmeier, D. A., D. A. Bloch, and D. H. Hickam. 1991. Assessing predictive accuracy: How to compare Brier scores. *Journal of Clinical Epidemiology* 44: 1141–1146.

Rogers, W. H. 1992. sbe9: Brier score decomposition. *Stata Technical Bulletin* 10: 20–22. Reprinted in *Stata Technical Bulletin Reprints*, vol. 2, pp. 92–94.

Sanders, F. 1963. On subjective probability forecasting. *Journal of Applied Meteorology* 2: 191–201.

Schmidt, C. H., and J. L. Griffith. 1998. Multivariate classification rules: Calibration and discrimination. In *Encyclopedia of Biostatistics*, ed. P. Armitage and T. Colton, 2844–2850. New York: Wiley.

Spiegelhalter, D. J. 1986. Probabilistic prediction in patient management and clinical trials. *Statistics in Medicine* 5: 421–433.

Von Storch, H., and F. W. Zwiers. 1999. *Statistical Analysis in Climate Research*. Cambridge: Cambridge University Press.

Wilks, D. S. 1995. *Statistical Methods in the Atmospheric Sciences*. San Diego: Academic Press.

Yates, J. F. 1982. External correspondence: Decompositions of the mean probability score. *Organizational Behavior and Human Performance* 30: 132–156.

Also See

[R] **logistic** — Logistic regression, reporting odds ratios

[R] **logit** — Logistic regression, reporting coefficients

[R] **probit** — Probit regression

Title

> **bsample** — Sampling with replacement

Syntax

bsample $\left[\,exp\,\right]\,\left[\,if\,\right]\,\left[\,in\,\right]\,\left[\,,\,options\,\right]$

where *exp* is a standard Stata expression; see [U] **13 Functions and expressions**.

options	description
strata(*varlist*)	variables identifying strata
cluster(*varlist*)	variables identifying resampling clusters
idcluster(*newvar*)	create new cluster ID variable
weight(*varname*)	replace *varname* with frequency weights

Description

bsample draws random samples with replacement from the data in memory.

exp specifies the size of the sample, which must be less than or equal to the number of sampling units in the data. The observed number of units is the default when *exp* is not specified.

For a simple random sample (SRS) of the observations, *exp* must be less than or equal to _N (the number of observations in the data; see [U] **13.4 System variables (_variables)**).

For stratified SRS, *exp* must be less than or equal to _N within the strata identified by the strata() option.

For clustered sampling, *exp* must be less than or equal to N_c (the number of clusters identified by the cluster() option).

For stratified sampling of clusters, *exp* must be less than or equal to N_c within the strata identified by the strata() option.

Observations that do not meet the optional *if* and *in* criteria are dropped (not sampled).

Options

strata(*varlist*) specifies the variables identifying strata. If strata() is specified, bootstrap samples are selected within each stratum.

cluster(*varlist*) specifies the variables identifying resampling clusters. If cluster() is specified, the sample drawn during each replication is a bootstrap sample of clusters.

idcluster(*newvar*) creates a new variable containing a unique identifier for each resampled cluster.

weight(*varname*) specifies a variable in which the sampling frequencies will be placed. *varname* must be an existing variable, which will be replaced. After bsample, *varname* can be used as an fweight in any Stata command that accepts fweights, which can speed up resampling for commands like regress and summarize. This option cannot be combined with idcluster().

By default, bsample replaces the data in memory with the sampled observations; however, specifying the weight() option causes only the specified *varname* to be changed.

242

Remarks

Below is a series of examples illustrating how `bsample` is used with various sampling schemes.

▷ Example 1: Simple random sampling

We have data on the characteristics of hospital patients and wish to draw a simple random sample of 200 patients. We type

```
. use http://www.stata-press.com/data/r10/bsample1
. bsample 200
. count
200
```

◁

▷ Example 2: Stratified samples with equal sizes

Among the variables in our dataset is `female`, an indicator for the female patients. To get a stratified simple random sample of 200 female patients and 200 male patients, we type

```
. use http://www.stata-press.com/data/r10/bsample1, clear
. bsample 200, strata(female)
. tab female
```

female	Freq.	Percent	Cum.
male	200	50.00	50.00
female	200	50.00	100.00
Total	400	100.00	

◁

▷ Example 3: Stratified samples with unequal sizes

To sample 300 females and 200 males, we must generate a variable that is 300 for females and 200 for males and then use this variable in *exp* when we call `bsample`.

```
. use http://www.stata-press.com/data/r10/bsample1, clear
. gen nsamp = cond(female,300,200)
. bsample nsamp, strata(female)
. tab female
```

female	Freq.	Percent	Cum.
male	200	40.00	40.00
female	300	60.00	100.00
Total	500	100.00	

◁

▷ Example 4: Samples satisfying a condition

For a simple random sample of 200 female patients, we type

```
. use http://www.stata-press.com/data/r10/bsample1, clear
. bsample 200 if female
. tab female
```

female	Freq.	Percent	Cum.
female	200	100.00	100.00
Total	200	100.00	

◁

▷ Example 5: Generating frequency weights

To identify the sampled observations using frequency weights instead of dropping unsampled observations, we use the `weight()` option (we will need to supply it an existing variable name) and type

```
. use http://www.stata-press.com/data/r10/bsample1, clear
. set seed 1234
. gen fw = .
(5810 missing values generated)
. bsample 200 if female, weight(fw)
. tabulate fw female
```

fw	female		Total
	male	female	
0	2,392	3,221	5,613
1	0	194	194
2	0	3	3
Total	2,392	3,418	5,810

Note that $(194 \times 1) + (3 \times 2) = 200$.

◁

▷ Example 6: Oversampling observations

`bsample` requires the expression in *exp* to evaluate to a number that is less than or equal to the number of observations. To sample twice as many male and female patients as there are already in memory, we must expand the data before using `bsample`. For example,

```
. use http://www.stata-press.com/data/r10/bsample1, clear
. set seed 1234
. expand 2
(5810 observations created)
. bsample, strata(female)
```

```
. tab female
```

female	Freq.	Percent	Cum.
male	4,784	41.17	41.17
female	6,836	58.83	100.00
Total	11,620	100.00	

◁

▷ Example 7: Stratified oversampling with unequal sizes

To sample twice as many female patients as male patients, we must expand the records for the female patients because there are less than twice as many of them as there are male patients, but first put the number of observed male patients in a local macro. After expanding the female records, we generate a variable that contains the number of observations to sample within the two groups.

```
. use http://www.stata-press.com/data/r10/bsample1, clear
. set seed 1234
. count if !female
  2392
. local nmale = r(N)
. expand 2 if female
(3418 observations created)
. gen nsamp = cond(female,2*`nmale',`nmale')
. bsample nsamp, strata(female)
. tab female
```

female	Freq.	Percent	Cum.
male	2,392	33.33	33.33
female	4,784	66.67	100.00
Total	7,176	100.00	

◁

▷ Example 8: Oversampling of clusters

For clustered data, sampling more clusters than are present in the original dataset requires more than just expanding the data. To illustrate, suppose we wanted a bootstrap sample of eight clusters from a dataset consisting of five clusters of observations.

```
. use http://www.stata-press.com/data/r10/bsample2, clear
. tabstat x, stat(n mean) by(group)
Summary for variables: x
      by categories of: group
```

group	N	mean
A	15	-.3073028
B	10	-.00984
C	11	.0810985
D	11	-.1989179
E	29	-.095203
Total	76	-.1153269

bsample will complain if we simply expand the dataset.

```
. use http://www.stata-press.com/data/r10/bsample2, clear
. expand 3
(152 observations created)
. bsample 8, cluster(group)
resample size must not be greater than number of clusters
r(498);
```

Expanding the data will only partly solve the problem. We also need a new variable that uniquely identifies the copied clusters. We use the expandcl command to accomplish both these tasks; see [D] **expandcl**.

```
. use http://www.stata-press.com/data/r10/bsample2, clear
. set seed 1234
. expandcl 2, generate(expgroup) cluster(group)
(76 observations created)
. tabstat x, stat(n mean) by(expgroup)
Summary for variables: x
     by categories of: expgroup
```

expgroup	N	mean
1	15	-.3073028
2	15	-.3073028
3	10	-.00984
4	10	-.00984
5	11	.0810985
6	11	.0810985
7	11	-.1989179
8	11	-.1989179
9	29	-.095203
10	29	-.095203
Total	152	-.1153269

```
. gen fw = .
(152 missing values generated)
. bsample 8, cluster(expgroup) weight(fw)
. tabulate fw group
```

fw	A	B	group C	D	E	Total
0	15	10	0	0	29	54
1	15	10	22	22	0	69
2	0	0	0	0	29	29
Total	30	20	22	22	58	152

The results from tabulate on the generated frequency weight variable versus the original cluster ID (group) show us that the bootstrap sample contains one copy from cluster A, one copy of cluster B, two copies of cluster C, two copies of cluster D, and two copies of cluster E $(1 + 1 + 2 + 2 + 2 = 8)$.

◁

▷ Example 9: Stratified oversampling of clusters

Suppose that we have a dataset containing two strata with five clusters in each stratum, but the cluster identifiers are not unique between the strata. To get a stratified bootstrap sample with eight clusters in each stratum, we first use `expandcl` to expand the data and get a new cluster ID variable. We use `cluster(strid group)` in the call to `expandcl`; this action will uniquely identify the $2 * 5 = 10$ clusters across the strata.

```
. use http://www.stata-press.com/data/r10/bsample2, clear
. set seed 1234
. tab group strid
```

	strid		
group	1	2	Total
A	7	8	15
B	5	5	10
C	5	6	11
D	5	6	11
E	14	15	29
Total	36	40	76

```
. expandcl 2, generate(expgroup) cluster(strid group)
(76 observations created)
```

Now we can use `bsample` with the expanded data, stratum ID variable, and new cluster ID variable.

```
. gen fw = .
(152 missing values generated)
. bsample 8, cluster(expgroup) str(strid) weight(fw)
. by strid, sort: tabulate fw group
```

-> strid = 1

			group			
fw	A	B	C	D	E	Total
0	0	5	0	5	14	24
1	14	5	10	5	0	34
2	0	0	0	0	14	14
Total	14	10	10	10	28	72

-> strid = 2

			group			
fw	A	B	C	D	E	Total
0	8	10	0	6	0	24
1	8	0	6	6	15	35
2	0	0	6	0	15	21
Total	16	10	12	12	30	80

The results from `by strid: tabulate` on the generated frequency weight variable versus the original cluster ID (`group`) show us how many times each cluster was sampled for each stratum. For stratum 1, the bootstrap sample contains two copies of cluster A, one copy of cluster B, two copies of cluster C, one copy of cluster D, and two copies of cluster E ($2 + 1 + 2 + 1 + 2 = 8$). For stratum 2, the bootstrap sample contains one copy of cluster A, zero copies of cluster B, three copies of cluster C, one copy of cluster D, and three copies of cluster E ($1 + 0 + 3 + 1 + 3 = 8$). ◁

Methods and Formulas

bsample is implemented as an ado-file.

Also See

[R] **bootstrap** — Bootstrap sampling and estimation

[R] **bstat** — Report bootstrap results

[R] **simulate** — Monte Carlo simulations

[D] **sample** — Draw random sample

Title

---

bstat — Report bootstrap results

Syntax

Bootstrap statistics from variables

    bstat $[\,varlist\,]$ $[\,if\,]$ $[\,in\,]$ $[\,,\ options\,]$

Bootstrap statistics from file

    bstat $[\,namelist\,]$ $[\,$using *filename*$\,]$ $[\,if\,]$ $[\,in\,]$ $[\,,\ options\,]$

options	description
Main	
<u>stat</u>(*vector*)	observed values for each statistic
accel(*vector*)	acceleration values for each statistic
mse	use MSE formula for variance estimation
Reporting	
<u>l</u>evel(*#*)	set confidence level; default is level(95)
n(*#*)	# of observations from which bootstrap samples were taken
notable	suppress table of results
<u>noh</u>eader	suppress table header
<u>nol</u>egend	suppress table legend
<u>v</u>erbose	display the full table legend
<u>ti</u>tle(*text*)	use *text* as title for bootstrap results

See [U] **20 Estimation and postestimation commands** for more capabilities of estimation commands.

Description

bstat is a programmer's command that computes and displays estimation results from bootstrap statistics.

For each variable in *varlist* (the default is all variables), bstat computes a covariance matrix, estimates bias, and constructs several different confidence intervals (CIs). The following CIs are constructed by bstat:

    1. Normal CIs (using the normal approximation)

    2. Percentile CIs

    3. Bias-corrected (BC) CIs

    4. Bias-corrected and accelerated (BC_a) CIs (optional)

estat bootstrap displays a table of one or more of the above confidence intervals; see [R] **bootstrap postestimation**.

If there are bootstrap estimation results in e(), bstat replays them. If given the using modifier, bstat uses the data in *filename* to compute the bootstrap statistics while preserving the data currently in memory. Otherwise, bstat uses the data in memory to compute the bootstrap statistics.

249

The following options may be used to replay estimation results from `bstat`:

level(*#*) notable noheader nolegend verbose title(*text*)

For all other options and the qualifiers `using`, `if`, and `in`, `bstat` requires a bootstrap dataset.

Options

> Main

`stat`(*vector*) specifies the observed value of each statistic (i.e., the value of the statistic using the original dataset).

`accel`(*vector*) specifies the acceleration of each statistic, which is used to construct BC_a CIs.

`mse` specifies that `bstat` compute the variance by using deviations of the replicates from the observed value of the statistics. By default, `bstat` computes the variance by using deviations from the average of the replicates.

> Reporting

`level`(*#*); see [R] **estimation options**.

`n`(*#*) specifies the number of observations from which bootstrap samples were taken. This value is used in no calculations but improves the table header when this information is not saved in the bootstrap dataset.

`notable` suppresses the display of the output table.

`noheader` suppresses the display of the table header. This option implies `nolegend`.

`nolegend` suppresses the display of the table legend.

`verbose` specifies that the full table legend be displayed. By default, coefficients and standard errors are not displayed.

`title`(*text*) specifies a title to be displayed above the table of bootstrap results; the default title is `Bootstrap results`.

Remarks

Remarks are presented under the following headings:

> *Bootstrap datasets*
> *Creating a bootstrap dataset*

Bootstrap datasets

Although `bstat` allows you to specify the observed value and acceleration of each bootstrap statistic via the `stat()` and `accel()` options, programmers may be interested in what `bstat` uses when these options are not supplied.

When working from a bootstrap dataset, `bstat` first checks the data characteristics (see [P] **char**) that it understands:

`_dta[bs_version]` identifies the version of the bootstrap dataset. This characteristic may be empty (not defined), 2, or 3; otherwise, `bstat` will quit and display an error message. This version tells `bstat` which other characteristics to look for in the bootstrap dataset.

bstat uses the following characteristics from version 3 bootstrap datasets:
        _dta[N]
        _dta[N_strata]
        _dta[N_cluster]
        _dta[command]
        varname[observed]
        varname[acceleration]
        varname[expression]

bstat uses the following characteristics from version 2 bootstrap datasets:
        _dta[N]
        _dta[N_strata]
        _dta[N_cluster]
        varname[observed]
        varname[acceleration]

An empty bootstrap dataset version implies that the dataset was created by the bstrap command in a version of Stata earlier than Stata 8. Here bstat expects *varname*[bstrap] to contain the observed value of the statistic identified by *varname* (*varname*[observed] in version 2). All other characteristics are ignored.

_dta[N] is the number of observations in the observed dataset. This characteristic may be overruled by specifying the n() option.

_dta[N_strata] is the number of strata in the observed dataset.

_dta[N_cluster] is the number of clusters in the observed dataset.

_dta[command] is the command used to compute the observed values of the statistics.

varname[observed] is the observed value of the statistic identified by *varname*. To specify a different value, use the stat() option.

varname[acceleration] is the estimate of acceleration for the statistic identified by *varname*. To specify a different value, use the accel() option.

varname[expression] is the expression or label that describes the statistic identified by *varname*.

Creating a bootstrap dataset

Suppose that we are interested in obtaining bootstrap statistics by resampling the residuals from a regression (which is not possible with the bootstrap command). After loading some data, we run a regression, save some results relevant to the bstat command, and save the residuals in a new variable named res.

```
. use http://www.stata-press.com/data/r10/auto
(1978 Automobile Data)
```

(*Continued on next page*)

```
. regress mpg weight length
```

Source	SS	df	MS		
Model	1616.08062	2	808.040312		
Residual	827.378835	71	11.653223		
Total	2443.45946	73	33.4720474		

Number of obs = 74
F(2, 71) = 69.34
Prob > F = 0.0000
R-squared = 0.6614
Adj R-squared = 0.6519
Root MSE = 3.4137

mpg	Coef.	Std. Err.	t	P>|t|	[95% Conf. Interval]
weight	-.0038515	.001586	-2.43	0.018	-.0070138 -.0006891
length	-.0795935	.0553577	-1.44	0.155	-.1899736 .0307867
_cons	47.88487	6.08787	7.87	0.000	35.746 60.02374

```
. matrix b = e(b)
. local n = e(N)
. predict res, residuals
```

We can resample the residual values in `res` by generating a random observation ID (`rid`), generate a new response variable (`y`), and run the original regression with the new response variables.

```
. set seed 54321
. gen rid = int(_N*uniform())+1
. matrix score double y = b
. replace y = y + res[rid]
(74 real changes made)
. regress y weight length
```

Source	SS	df	MS		
Model	1773.23548	2	886.617741		
Residual	608.747732	71	8.57391172		
Total	2381.98321	73	32.629907		

Number of obs = 74
F(2, 71) = 103.41
Prob > F = 0.0000
R-squared = 0.7444
Adj R-squared = 0.7372
Root MSE = 2.9281

y	Coef.	Std. Err.	t	P>|t|	[95% Conf. Interval]
weight	-.0059938	.0013604	-4.41	0.000	-.0087064 -.0032813
length	-.0127875	.0474837	-0.27	0.788	-.1074673 .0818924
_cons	42.23195	5.22194	8.09	0.000	31.8197 52.6442

Instead of programming this resampling inside a loop, it is much more convenient to write a short program and use the `simulate` command; see [R] **simulate**. In the following, mysim_r requires the user to specify a coefficient vector and a residual variable. mysim_r then retrieves the list of predictor variables (removing _cons from the list), generates a new temporary response variable using the resampled residuals, and regresses the new response variable on the predictors.

```
program mysim_r
        version 10
        syntax name(name=bvector), res(varname)
        tempvar y rid
        local xvars : colnames `bvector'
        local cons _cons
        local xvars : list xvars - cons
        matrix score double `y' = `bvector'
        gen long `rid' = int(_N*uniform()) + 1
        replace `y' = `y' + `res'[`rid']
        regress `y' `xvars'
end
```

We can now give `mysim_r` a test run, but we first set the random-number seed (to reproduce results).

```
. set seed 54321
. mysim_r b, res(res)
(74 real changes made)
```

Source	SS	df	MS			
Model	1773.23548	2	886.617741			
Residual	608.747732	71	8.57391172			
Total	2381.98321	73	32.629907			

```
                                      Number of obs =      74
                                      F(  2,    71) =  103.41
                                      Prob > F      =  0.0000
                                      R-squared     =  0.7444
                                      Adj R-squared =  0.7372
                                      Root MSE      =  2.9281
```

__000000	Coef.	Std. Err.	t	P>|t|	[95% Conf.	Interval]
weight	-.0059938	.0013604	-4.41	0.000	-.0087064	-.0032813
length	-.0127875	.0474837	-0.27	0.788	-.1074673	.0818924
_cons	42.23195	5.22194	8.09	0.000	31.8197	52.6442

Now that we have a program that will compute the results we want, we can use `simulate` to generate a bootstrap dataset and `bstat` to display the results.

```
. set seed 54321
. simulate, reps(200) nodots: mysim_r b, res(res)
      command:  mysim_r b, res(res)
. bstat, stat(b) n('n')
Bootstrap results
```

```
                                   Number of obs     =      74
                                   Replications      =     200
```

	Observed Coef.	Bootstrap Std. Err.	z	P>|z|	Normal-based [95% Conf. Interval]	
_b_weight	-.0038515	.0015715	-2.45	0.014	-.0069316	-.0007713
_b_length	-.0795935	.0552415	-1.44	0.150	-.1878649	.0286779
_b_cons	47.88487	6.150069	7.79	0.000	35.83096	59.93879

Finally, we see that `simulate` created some of the data characteristics recognized by `bstat`. All we need to do is to correctly specify the version of the bootstrap dataset, and `bstat` will automatically use the relevant data characteristics.

```
. char list
_dta[seed]:                  X681014b5c43f462544a474abacbdd93d12a1
_dta[command]:               mysim_r b, res(res)
_b_weight[is_eexp]:          1
_b_weight[colname]:          weight
_b_weight[coleq]:            _
_b_weight[expression]:       _b[weight]
_b_length[is_eexp]:          1
_b_length[colname]:          length
_b_length[coleq]:            _
_b_length[expression]:       _b[length]
_b_cons[is_eexp]:            1
_b_cons[colname]:            _cons
_b_cons[coleq]:              _
_b_cons[expression]:         _b[_cons]
```

```
. char _dta[bs_version] 3
. bstat, stat(b) n('n')
```

Bootstrap results   Number of obs   =   74
                                                       Replications   =   200

        command: mysim_r b, res(res)

	Observed Coef.	Bootstrap Std. Err.	z	P>|z|	Normal-based [95% Conf. Interval]	
weight	-.0038515	.0015715	-2.45	0.014	-.0069316	-.0007713
length	-.0795935	.0552415	-1.44	0.150	-.1878649	.0286779
_cons	47.88487	6.150069	7.79	0.000	35.83096	59.93879

See Poi (2004) for another example of residual resampling.

Saved Results

bstat saves the following in e():

Scalars
e(N_reps)	number of complete replications	e(N)	sample size
e(N_misreps)	number of incomplete replications	e(N_strata)	number of strata
e(level)	confidence level for bootstrap CIs	e(N_clust)	number of clusters

Macros
e(cmd)	bstat	e(title)	title in estimation output
e(cmdline)	command as typed	e(vce)	bootstrap
e(command)	from _dta[command]	e(vcetype)	title used to label Std. Err.
e(prefix)	bootstrap	e(mse)	mse if specified
e(exp#)	expression for the #th statistic	e(properties)	b V

Matrices
e(b)	observed statistics
e(V)	bootstrap variance–covariance matrix
e(b_bs)	bootstrap estimates
e(reps)	number of nonmissing results
e(bias)	estimated biases
e(se)	estimated standard errors
e(z0)	median biases
e(accel)	estimated accelerations
e(ci_normal)	normal-approximation CIs
e(ci_percentile)	percentile CIs
e(ci_bc)	bias-corrected CIs
e(ci_bca)	bias-corrected and accelerated CIs

Methods and Formulas

bstat is implemented as an ado-file.

Reference

Poi, B. P. 2004. From the help desk: Some bootstrapping techniques. *Stata Journal* 4: 312–328.

Also See

[R] **bootstrap** — Bootstrap sampling and estimation

[R] **bsample** — Sampling with replacement

Title

centile — Report centile and confidence interval

Syntax

centile [*varlist*] [*if*] [*in*] [, *options*]

options	description
Main	
<u>c</u>entile(*numlist*)	report specified centiles; default is centile(50)
Options	
<u>cc</u>i	binomial exact; conservative confidence interval
<u>n</u>ormal	normal, based on observed centiles
<u>m</u>eansd	normal, based on mean and standard deviation
<u>l</u>evel(*#*)	set confidence level; default is level(95)

by is allowed; see [D] **by**.

Description

centile estimates specified centiles and calculates confidence intervals. If no *varlist* is specified, centile calculates centiles for all the variables in the dataset. If centile() is not specified, medians (centile(50)) are reported.

Options

> _Main_

centile(*numlist*) specifies the centiles to be reported. The default is to display the 50th centile. Specifying centile(5) requests that the fifth centile be reported. Specifying centile(5 50 95) requests that the 5th, 50th, and 95th centiles be reported. Specifying centile(10(10)90) requests that the 10th, 20th, ..., 90th centiles be reported; see [U] **11.1.8 numlist**.

> _Options_

cci (conservative confidence interval) forces the confidence limits to fall exactly on sample values. Confidence intervals displayed with the cci option are slightly wider than those with the default (nocci) option.

normal causes the confidence interval to be calculated by using a formula for the standard error of a normal-distribution quantile given by Kendall and Stuart (1969, 237). The normal option is useful when you want empirical centiles—that is, centiles based on sample order statistics rather than on the mean and standard deviation—and are willing to assume normality.

meansd causes the centile and confidence interval to be calculated based on the sample mean and standard deviation, and it assumes normality.

level(*#*) specifies the confidence level, as a percentage, for confidence intervals. The default is level(95) or as set by set level; see [R] **level**.

Remarks

The qth centile of a continuous random variable, X, is defined as the value of C_q, which fulfills the condition $\Pr(X \leq C_q) = q/100$. The value of q must be in the range $0 < q < 100$, though q is not necessarily an integer. By default, `centile` estimates C_q for the variables in *varlist* and for the values of q given in `centile(numlist)`. It makes no assumptions about the distribution of X, and, if necessary, uses linear interpolation between neighboring sample values. Extreme centiles (for example, the 99th centile in samples smaller than 100) are fixed at the minimum or maximum sample value. An "exact" confidence interval for C_q is also given, using the binomial-based method described below in *Methods and Formulas* and in Conover (1999, 143–148). Again linear interpolation is used to improve the accuracy of the estimated confidence limits, but extremes are fixed at the minimum or maximum sample value.

You can prevent `centile` from interpolating when calculating binomial-based confidence intervals by specifying `cci`. The resulting intervals are generally wider than with the default; that is, the coverage (confidence level) tends to be greater than the nominal value (given as usual by `level(#)`, by default 95%).

If the data are believed to be normally distributed (a common case), there are two alternative methods for estimating centiles. If `normal` is specified, C_q is calculated, as just described, but its confidence interval is based on a formula for the standard error (s.e.) of a normal-distribution quantile given by Kendall and Stuart (1969, 237). If `meansd` is alternatively specified, C_q is estimated as $\overline{x} + z_q \times s$, where $\overline{x}$ and s are the sample mean and standard deviation, and z_q is the qth centile of the standard normal distribution (e.g., $z_{95} = 1.645$). The confidence interval is derived from the s.e. of the estimate of C_q.

▷ Example 1

Using the `auto.dta`, we estimate the 5th, 50th, and 95th centiles of the variable `price`:

```
. use http://www.stata-press.com/data/r10/auto
(1978 Automobile Data)

. format price %8.2fc

. centile price, centile(5 50 95)
```

Variable	Obs	Percentile	Centile	— Binom. Interp. — [95% Conf. Interval]	
price	74	5	3,727.75	3,291.23	3,914.16
		50	5,006.50	4,593.57	5,717.90
		95	13,498.00	11,061.53	15,865.30

`summarize` produces somewhat different results from `centile`; see *Methods and Formulas*.

```
. summarize price, detail
```

	Price			
	Percentiles	Smallest		
1%	3291	3291		
5%	3748	3299		
10%	3895	3667	Obs	74
25%	4195	3748	Sum of Wgt.	74
50%	5006.5		Mean	6165.257
		Largest	Std. Dev.	2949.496
75%	6342	13466		
90%	11385	13594	Variance	8699526
95%	13466	14500	Skewness	1.653434
99%	15906	15906	Kurtosis	4.819188

The confidence limits produced by using the cci option are slightly wider than those produced without this option:

```
. centile price, c(5 50 95) cci
```

| | | | | — Binomial Exact — |
Variable	Obs	Percentile	Centile	[95% Conf. Interval]
price	74	5	3,727.75	3,291.00 3,955.00
		50	5,006.50	4,589.00 5,719.00
		95	13,498.00	10,372.00 15,906.00

If we are willing to assume that price is normally distributed, we could include either the normal or the meansd option:

```
. centile price, c(5 50 95) normal
```

| | | | | — Normal, based on observed centiles — |
Variable	Obs	Percentile	Centile	[95% Conf. Interval]
price	74	5	3,727.75	3,211.19 4,244.31
		50	5,006.50	4,096.68 5,916.32
		95	13,498.00	5,426.81 21,569.19

```
. centile price, c(5 50 95) meansd
```

| | | | | — Normal, based on mean and std. dev.— |
Variable	Obs	Percentile	Centile	[95% Conf. Interval]
price	74	5	1,313.77	278.93 2,348.61
		50	6,165.26	5,493.24 6,837.27
		95	11,016.75	9,981.90 12,051.59

With the normal option, the centile estimates are, by definition, the same as before. The confidence intervals for the 5th and 50th centiles are similar to the previous ones, but the interval for the 95th centile is different. The results using the meansd option also differ from both previous sets of estimates.

We can use sktest (see [R] **sktest**) to check the correctness of the normality assumption:

```
. sktest price
```

Skewness/Kurtosis tests for Normality

| | | | | ——— joint ——— |
Variable	Pr(Skewness)	Pr(Kurtosis)	adj chi2(2)	Prob>chi2
price	0.000	0.013	21.77	0.0000

sktest reveals that price is definitely not normally distributed, so the normal assumption is not reasonable, and the normal and meansd options are not appropriate for these data. We should rely on the results from the default choice, which does not assume normality. If the data are normally distributed, however, the precision of the estimated centiles and their confidence intervals will be ordered (best) meansd > normal > [default] (worst). The normal option is useful when we really do want empirical centiles (that is, centiles based on sample order statistics rather than on the mean and standard deviation) but are willing to assume normality.

◁

Saved Results

centile saves the following in r():

Scalars

r(N)	number of observations	r(lb_#)	#-requested centile lower confidence bound
r(n_cent)	number of centiles requested	r(ub_#)	#-requested centile upper confidence bound
r(c_#)	value of # centile		

Macros

r(centiles)   centiles requested

Methods and Formulas

centile is implemented as an ado-file.

Default case

The calculation is based on the method of Mood and Graybill (1963, 408). Let $x_1 \leq x_2 \leq \cdots \leq x_n$ be a sample of size n arranged in ascending order. Denote the estimated qth centile of the xs as c_q. We require that $0 < q < 100$. Let $R = (n+1)q/100$ have integer part r and fractional part f; that is, $r = \text{int}(R)$ and $f = R - r$. (If R is itself an integer, then $r = R$ and $f = 0$.) Note that $0 \leq r \leq n$. For convenience, define $x_0 = x_1$ and $x_{n+1} = x_n$. C_q is estimated by

$$c_q = x_r + f \times (x_{r+1} - x_r)$$

that is, c_q is a weighted average of x_r and x_{r+1}. Loosely speaking, a (conservative) $p\%$ confidence interval for C_q involves finding the observations ranked t and u, which correspond, respectively, to the $\alpha = (100 - p)/200$ and $1 - \alpha$ quantiles of a binomial distribution with parameters n and $q/100$, i.e., $\text{B}(n, q/100)$. More precisely, define the ith value $(i = 0, \ldots, n)$ of the cumulative binomial distribution function as $F_i = \text{Pr}(S \leq i)$, where S has distribution $\text{B}(n, q/100)$. For convenience, let $F_{-1} = 0$ and $F_{n+1} = 1$. t is found such that $F_t \leq \alpha$ and $F_{t+1} > \alpha$, and u is found such that $1 - F_u \leq \alpha$ and $1 - F_{u-1} > \alpha$.

With the cci option in force, the (conservative) confidence interval is (x_{t+1}, x_{u+1}), and its actual coverage probability is $F_u - F_t$.

The default case uses linear interpolation on the F_i as follows. Let

$$g = (\alpha - F_t)/(F_{t+1} - F_t)$$
$$h = \{\alpha - (1 - F_u)\}/\{(1 - F_{u-1}) - (1 - F_u)\}$$
$$= (\alpha - 1 + F_u)/(F_u - F_{u-1})$$

The interpolated lower and upper confidence limits (c_{qL}, c_{qU}) for C_q are

$$c_{qL} = x_{t+1} + g \times (x_{t+2} - x_{t+1})$$
$$c_{qU} = x_{u+1} - h \times (x_{u+1} - x_u)$$

Suppose that we want a 95% confidence interval for the median of a sample of size 13. $n = 13$, $q = 50$, $p = 95$, $\alpha = .025$, $R = 14 \times 50/100 = 7$, and $f = 0$. Therefore, the median is the 7th observation. Some example data, x_i, and the values of F_i are as follows:

i	F_i	$1 - F_i$	x_i	i	F_i	$1 - F_i$	x_i
0	0.0001	0.9999	–	7	0.7095	0.2905	33
1	0.0017	0.9983	5	8	0.8666	0.1334	37
2	0.0112	0.9888	7	9	0.9539	0.0461	45
3	0.0461	0.9539	10	10	0.9888	0.0112	59
4	0.1334	0.8666	15	11	0.9983	0.0017	77
5	0.2905	0.7095	23	12	0.9999	0.0001	104
6	0.5000	0.5000	28	13	1.0000	0.0000	211

The median is $x_7 = 33$. Also, $F_2 \le .025$ and $F_3 > .025$, so $t = 2$; $1 - F_{10} \le .025$ and $1 - F_9 > .025$, so $u = 10$. The conservative confidence interval is therefore

$$(c_{50L}, c_{50U}) = (x_{t+1}, x_{u+1}) = (x_3, x_{11}) = (10, 77)$$

with actual coverage $F_{10} - F_2 = .9888 - .0112 = .9776$ (97.8% confidence). For the interpolation calculation, we have

$$g = (.025 - .0112)/(.0461 - .0112) = .395$$
$$h = (.025 - 1 + .9888)/(.9888 - .9539) = .395$$

So,

$$c_{50L} = x_3 + .395 \times (x_4 - x_3) = 10 + .395 \times 5 = 11.98$$
$$c_{50U} = x_{11} - .395 \times (x_{11} - x_{10}) = 77 - .395 \times 18 = 69.89$$

Normal case

The value of c_q is as above. Its s.e. is given by the formula

$$s_q = \sqrt{q(100 - q)} \Big/ \left\{ 100 n Z(c_q; \bar{x}, s) \right\}$$

where $\bar{x}$ and s are the mean and s.d. of the x_i, and

$$Z(Y; \mu, \sigma) = \left(1/\sqrt{2\pi\sigma^2}\right) e^{-(Y-\mu)^2/2\sigma^2}$$

is the density function of a normally distributed variable Y with mean μ and s.d. σ. The confidence interval for C_q is $(c_q - z_{100(1-\alpha)} s_q, c_q + z_{100(1-\alpha)} s_q)$.

meansd case

The value of c_q is $\bar{x} + z_q \times s$. Its s.e. is given by the formula

$$s_q^\star = s\sqrt{1/n + z_q^2/(2n - 2)}$$

The confidence interval for C_q is $(c_q - z_{100(1-\alpha)} \times s_q^\star, c_q + z_{100(1-\alpha)} \times s_q^\star)$.

Acknowledgment

centile was written by Patrick Royston of the MRC Clinical Trials Unit, London.

References

Conover, W. J. 1999. *Practical Nonparametric Statistics*. 3rd ed. New York: Wiley.

Kendall, M. G., and A. Stuart. 1969. *The Advanced Theory of Statistics: Distribution Theory, Vol. I*. 3rd ed. London: Griffin.

Mood, A. M., and F. A. Graybill. 1963. *Introduction to the Theory of Statistics*. 2nd ed. New York: McGraw–Hill.

Newson, R. 2000. snp16: Robust confidence intervals for median and other percentile differences between two groups. *Stata Technical Bulletin* 58: 30–35. Reprinted in *Stata Technical Bulletin Reprints*, vol. 10, pp. 324–331.

Royston, J. P. 1992. sg7: Centile estimation command. *Stata Technical Bulletin* 8: 12–15. Reprinted in *Stata Technical Bulletin Reprints*, vol. 2, pp. 122–125.

Also See

[R] **ci** — Confidence intervals for means, proportions, and counts

[R] **summarize** — Summary statistics

[D] **pctile** — Create variable containing percentiles

Title

> **ci** — Confidence intervals for means, proportions, and counts

Syntax

Syntax for ci

> ci [*varlist*] [*if*] [*in*] [*weight*] [, *options*]

Immediate command for variable distributed as normal

> cii $\#_{\text{obs}}$ $\#_{\text{mean}}$ $\#_{\text{sd}}$ [, *ciin_option*]

Immediate command for variable distributed as binomial

> cii $\#_{\text{obs}}$ $\#_{\text{succ}}$ [, *ciib_options*]

Immediate command for variable distributed as Poisson

> cii $\#_{\text{exposure}}$ $\#_{\text{events}}$, <u>p</u>oisson [*ciip_options*]

options	description
Main	
<u>binomial</u>	binomial 0/1 variables; compute exact confidence intervals
<u>p</u>oisson	Poisson variables; compute exact confidence intervals
<u>e</u>xposure(*varname*)	exposure variable; implies poisson
<u>exact</u>	calculate exact confidence intervals; the default
<u>wald</u>	calculate Wald confidence intervals
<u>wilson</u>	calculate Wilson confidence intervals
<u>agresti</u>	calculate Agresti–Coull confidence intervals
<u>jeffreys</u>	calculate Jeffreys confidence intervals
<u>total</u>	add output for all groups combined (for use with by only)
<u>separator</u>(#)	draw separator line after every # variables; default is separator(5)
<u>level</u>(#)	set confidence level; default is level(95)

by is allowed with ci; see [D] **by**.

aweights and fweights are allowed, but aweights may not be specified with options binomial or
poisson; see [U] **11.1.6 weight**.

ciin_option	description
<u>level</u>(#)	set confidence level; default is level(95)

ciib_options	description
level(#)	set confidence level; default is level(95)
exact	calculate exact confidence intervals; the default
wald	calculate Wald confidence intervals
wilson	calculate Wilson confidence intervals
agresti	calculate Agresti–Coull confidence intervals
jeffreys	calculate Jeffreys confidence intervals

ciip_options	description
*poisson	numbers are Poisson-distributed counts
level(#)	set confidence level; default is level(95)

*poisson is required.

Description

ci computes standard errors and confidence intervals for each of the variables in *varlist*.

cii is the immediate form of ci; see [U] **19 Immediate commands** for a general discussion of immediate commands.

In the binomial and Poisson variants of cii, the second number specified ($\#_{\text{succ}}$ or $\#_{\text{events}}$) must be an integer or between 0 and 1. If the number is between 0 and 1, Stata interprets it as the fraction of successes or events and converts it to an integer number representing the number of successes or events. The computation then proceeds as if two integers had been specified.

Options

___| Main |_____

binomial tells ci that the variables are 0/1 variables and that binomial confidence intervals will be calculated. (cii produces binomial confidence intervals when only two numbers are specified.)

poisson specifies that the variables (or numbers for cii) are Poisson-distributed counts; exact Poisson confidence intervals will be calculated.

exposure(*varname*) is used only with poisson. You do not need to specify poisson if you specify exposure(); poisson is assumed. *varname* contains the total exposure (typically a time or an area) during which the number of events recorded in *varlist* were observed.

exact, wald, wilson, agresti, and jeffreys specify that variables are 0/1 and specify how binomial confidence intervals are to be calculated.

exact is the default and specifies exact (also known in the literature as Clopper–Pearson [1934]) binomial confidence intervals.

wald specifies calculation of Wald confidence intervals.

wilson specifies calculation of Wilson confidence intervals.

`agresti` specifies calculation of Agresti–Coull confidence intervals.

`jeffreys` specifies calculation of Jeffreys confidence intervals.

See Brown, Cai, and DasGupta (2001) for a discussion and comparison of the different binomial confidence intervals.

`total` is for use with the `by` prefix. It requests that, in addition to output for each by-group, output be added for all groups combined.

`separator(#)` specifies how often separation lines should be inserted into the output. The default is `separator(5)`, meaning that a line is drawn after every five variables. `separator(10)` would draw the line after every 10 variables. `separator(0)` suppresses the separation line.

`level(#)` specifies the confidence level, as a percentage, for confidence intervals. The default is `level(95)` or as set by `set level`; see [R] **level**.

Remarks

Remarks are presented under the following headings:

> *Ordinary confidence intervals*
> *Binomial confidence intervals*
> *Poisson confidence intervals*
> *Immediate form*

Ordinary confidence intervals

▷ Example 1

Without the `binomial` or `poisson` options, `ci` produces "ordinary" confidence intervals, meaning those that are correct if the variable is distributed normally, and *asymptotically* correct for all other distributions satisfying the conditions of the Central Limit Theorem.

```
. use http://www.stata-press.com/data/r10/auto
(1978 Automobile Data)
. ci mpg price
```

Variable	Obs	Mean	Std. Err.	[95% Conf. Interval]	
mpg	74	21.2973	.6725511	19.9569	22.63769
price	74	6165.257	342.8719	5481.914	6848.6

The standard error of the mean of `mpg` is 0.67, and the 95% confidence interval is $[19.96, 22.64]$. We can obtain wider confidence intervals, 99%, by typing

```
. ci mpg price, level(99)
```

Variable	Obs	Mean	Std. Err.	[99% Conf. Interval]	
mpg	74	21.2973	.6725511	19.51849	23.07611
price	74	6165.257	342.8719	5258.405	7072.108

◁

▷ Example 2

by() breaks out the confidence intervals according to by-group; total adds an overall summary. For instance,

```
. ci mpg, by(foreign) total
```

-> foreign = Domestic

Variable	Obs	Mean	Std. Err.	[95% Conf. Interval]
mpg	52	19.82692	.657777	18.50638 21.14747

-> foreign = Foreign

Variable	Obs	Mean	Std. Err.	[95% Conf. Interval]
mpg	22	24.77273	1.40951	21.84149 27.70396

-> Total

Variable	Obs	Mean	Std. Err.	[95% Conf. Interval]
mpg	74	21.2973	.6725511	19.9569 22.63769

◁

❑ Technical Note

You can control the formatting of the numbers in the output by specifying a display format for the variable; see [U] **12.5 Formats: controlling how data are displayed**. For instance,

```
. format mpg %9.2f
. ci mpg
```

Variable	Obs	Mean	Std. Err.	[95% Conf. Interval]
mpg	74	21.30	0.67	19.96 22.64

❑

Binomial confidence intervals

▷ Example 3

We have data on employees, including a variable marking whether the employee was promoted last year.

```
. use http://www.stata-press.com/data/r10/promo
. ci promoted, binomial
```

Variable	Obs	Mean	Std. Err.	— Binomial Exact — [95% Conf. Interval]
promoted	20	.1	.067082	.0123485 .3169827

The above interval is the default for binomial data, known equivalently as both the exact binomial and the Clopper–Pearson interval.

Nominally, the interpretation of a 95% confidence interval is that under repeated samples or experiments, 95% of the resultant intervals would contain the unknown parameter in question. However, for binomial data, the actual coverage probability, regardless of method, usually differs from that interpretation. This result occurs because of the discreteness of the binomial distribution, which produces only a finite set of outcomes, meaning that coverage probabilities are subject to discrete jumps and the exact nominal level cannot always be achieved. Therefore, the term *exact confidence interval* refers to its being derived from the binomial distribution, the distribution exactly generating the data, rather than resulting in exactly the nominal coverage.

For the Clopper–Pearson interval, the actual coverage probability is guaranteed to be greater than or equal to the nominal confidence interval, here, 95%. Because of the way it is calculated—see *Methods and Formulas*—it may also be interpreted as follows: if the true probability of being promoted were 0.012, the chances of observing a result as extreme or more extreme than the result observed ($20 \cdot 0.1 = 2$ or more promotions) would be 2.5%. If the true probability of being promoted were 0.317, the chances of observing a result as extreme or more extreme than the result observed (two or fewer promotions) would be 2.5%.

◁

▷ Example 4

The Clopper–Pearson interval is desirable because it guarantees nominal coverage; however, by dropping this restriction, you may obtain accurate intervals that are not as conservative. In this vein, you might opt for the Wilson (1927) interval

```
. ci promoted, binomial wilson
```

Variable	Obs	Mean	Std. Err.	——— Wilson ———[95% Conf. Interval]	
promoted	20	.1	.067082	.0278665	.3010336

the Agresti–Coull (1998) interval

```
. ci promoted, binomial agresti
```

Variable	Obs	Mean	Std. Err.	— Agresti-Coull —[95% Conf. Interval]	
promoted	20	.1	.067082	.0156562	.3132439

or the Bayesian-derived Jeffreys interval (Brown, Cai, and DasGupta 2001)

```
. ci promoted, binomial jeffreys
```

Variable	Obs	Mean	Std. Err.	——— Jeffreys ———[95% Conf. Interval]	
promoted	20	.1	.067082	.0213725	.2838533

Picking the best interval is a matter of balancing accuracy (coverage) against precision (average interval length) and depends on sample size and success probability. Brown, Cai, and DasGupta (2001) recommend the Wilson or Jeffreys interval for small sample sizes (≤ 40) yet favor the Agresti–Coull interval for its simplicity, decent performance for sample sizes less than or equal to 40, and performance comparable to Wilson/Jeffreys for sample sizes greater than 40. They also deem the Clopper–Pearson interval to be "wastefully conservative and [...] not a good choice for practical use", unless of course one requires at a minimum the nominal coverage level.

◁

Finally, the binomial Wald confidence interval is obtained by specifying options `binomial` and `wald`. The Wald interval is the one taught in most introductory statistics courses and for the above is simply, for level $1 - \alpha$, Mean$\pm z_\alpha$(`Std. Err.`), where z_α is the $1 - \alpha/2$ quantile of the standard normal. Because its overall poor performance makes it impractical, the Wald interval is available mainly for pedagogical purposes. The binomial Wald interval is also similar to the interval produced by treating binary data as normal data and using `ci` without the `binomial` option, with two exceptions. First, when `binomial` is specified, the calculation of the standard error uses denominator n rather than $n - 1$, used for normal data. Second, confidence intervals for normal data are based on the t distribution rather than the standard normal. Of course, both discrepancies vanish as sample size increases.

❏ Technical Note

Let us repeat example 3, but this time with data in which there are no promotions over the observed period:

```
. use http://www.stata-press.com/data/r10/promonone
. ci promoted, binomial
```

Variable	Obs	Mean	Std. Err.	— Binomial Exact — [95% Conf. Interval]	
promoted	20	0	0	0	.1684335*

(*) one-sided, 97.5% confidence interval

The confidence interval is $[0, .168]$, and this is the confidence interval most books publish. It is not, however, a true 95% confidence interval because the lower tail has vanished. As Stata notes, it is a one-sided, 97.5% confidence interval. If you wanted to put 5% in the right tail, you could type `ci promoted, binomial level(90)`.

❏

❏ Technical Note

`ci` with the `binomial` option ignores any variables that do not take on the values 0 and 1 exclusively. For instance, with our automobile dataset,

```
. use http://www.stata-press.com/data/r10/auto
(1978 Automobile Data)
. ci mpg foreign, binomial
```

Variable	Obs	Mean	Std. Err.	— Binomial Exact — [95% Conf. Interval]	
foreign	74	.2972973	.0531331	.196584	.4148353

We also requested the confidence interval for `mpg`, but Stata ignored us. It does that so you can type `ci, binomial` and obtain correct confidence intervals for all the variables that are 0/1 in your data.

❏

Poisson confidence intervals

▷ Example 5

We have data on the number of bacterial colonies on a Petri dish. The dish has been divided into 36 small squares, and the number of colonies in each square has been counted. Each observation in our dataset represents a square on the dish. The variable count records the number of colonies in each square counted, which varies from 0 to 5.

```
. use http://www.stata-press.com/data/r10/petri
. ci count, poisson
```

Variable	Exposure	Mean	Std. Err.	— Poisson Exact — [95% Conf. Interval]	
count	36	2.333333	.2545875	1.861158	2.888825

ci reports that the average number of colonies per square is 2.33. If the expected number of colonies per square were as low as 1.86, the probability of observing 2.33 or more colonies per square would be 2.5%. If the expected number were as large as 2.89, the probability of observing $36 \cdot 2.33 = 84$ or fewer colonies per square would be 2.5%.

◁

❏ Technical Note

The number of "observations"—how finely the Petri dish is divided—makes no difference. The Poisson distribution is a function only of the count. In example 4, we observed a total of $2.33 \cdot 36 = 84$ colonies and a confidence interval of $[1.86 \cdot 36, 2.89 \cdot 36] = [67, 104]$. We would obtain the same $[67, 104]$ confidence interval if our dish were divided into, say, 49 squares, rather than 36.

For the counts, it is not even important that all the squares be of the same size. For *rates*, however, such differences do matter, but in an easy-to-calculate way. Rates are obtained from counts by dividing by exposure, which is typically a number multiplied by either time or an area. For our Petri dishes, we divide by an area to obtain a rate, but if our example were cast in terms of being infected by a disease, we might divide by person-years to obtain the rate. Rates are convenient because they are easier to compare: we might have 2.3 colonies per square inch or .0005 infections per person-year.

So, let us assume that we wish to obtain the number of colonies per square inch, and, moreover, that not all the "squares" on our dish are of equal size. We have a variable called area that records the area of each "square":

```
. ci count, exposure(area)
```

Variable	Exposure	Mean	Std. Err.	— Poisson Exact — [95% Conf. Interval]	
count	3	28	3.055051	22.3339	34.66591

The rates are now in more familiar terms. In our sample, there are 28 colonies per square inch and the 95% confidence interval is $[22.3, 34.7]$. When we did not specify exposure(), ci assumed that each observation contributed 1 to exposure.

❏

❏ Technical Note

As with the `binomial` option, if there were no colonies on our dish, `ci` would calculate a one-sided confidence interval:

```
. use http://www.stata-press.com/data/r10/petrinone
. ci count, poisson
```

Variable	Exposure	Mean	Std. Err.	— Poisson Exact — [95% Conf. Interval]	
count	36	0	0	0	.1024689*

(*) one-sided, 97.5% confidence interval

     ❏

Immediate form

▷ **Example 6**

We are reading a soon-to-be-published paper by a colleague. In it is a table showing the number of observations, mean, and standard deviation of 1980 median family income for the Northeast and West. We correctly think that the paper would be much improved if it included the confidence intervals. The paper claims that, for 166 cities in the Northeast, the average of median family income is $19,509 with a standard deviation of $4,379:

For the Northeast:

```
. cii 166 19509 4379
```

Variable	Obs	Mean	Std. Err.	[95% Conf. Interval]	
	166	19509	339.8763	18837.93	20180.07

For the West:

```
. cii 256 22557 5003
```

Variable	Obs	Mean	Std. Err.	[95% Conf. Interval]	
	256	22557	312.6875	21941.22	23172.78

     ◁

▷ **Example 7**

We flip a coin 10 times, and it comes up heads only once. We are shocked and decide to obtain a 99% confidence interval for this coin:

```
. cii 10 1, level(99)
```

Variable	Obs	Mean	Std. Err.	— Binomial Exact — [99% Conf. Interval]	
	10	.1	.0948683	.0005011	.5442871

     ◁

▷ Example 8

The number of reported traffic accidents in Santa Monica over a 24-hour period is 27. We need know nothing else:

```
. cii 1 27, poisson
```

Variable	Exposure	Mean	Std. Err.	— Poisson Exact — [95% Conf. Interval]	
	1	27	5.196152	17.79317	39.28358

◁

Saved Results

ci and cii saves the following in r():

Scalars

r(N)	number of observations or exposure	r(lb)	lower bound of confidence interval
r(mean)	mean	r(ub)	upper bound of confidence interval
r(se)	estimate of standard error		

Methods and Formulas

ci and cii are implemented as ado-files.

Ordinary

Define n, $\overline{x}$, and s^2 as, respectively, the number of observations, (weighted) average, and (unbiased) estimated variance of the variable in question; see [R] **summarize**.

The standard error of the mean, s_μ, is defined as $\sqrt{s^2/n}$.

Let α be $1 - l/100$, where l is the significance level specified by the user. Define t_α as the two-sided t statistic corresponding to a significance level of α with $n - 1$ degrees of freedom; t_α is obtained from Stata as invttail(n-1,0.5*α). The lower and upper confidence bounds are, respectively, $\overline{x} - s_\mu t_\alpha$ and $\overline{x} + s_\mu t_\alpha$.

Binomial

Given k successes of n trials, the estimated probability is $\widehat{p} = k/n$ with standard error $\sqrt{\widehat{p}(1 - \widehat{p})/n}$. ci calculates the exact (Clopper–Pearson) confidence interval $[p_1, p_2]$ such that

$$\Pr(K \geq k | p = p_1) = \alpha/2$$

and

$$\Pr(K \leq k | p = p_2) = \alpha/2$$

where K is distributed as binomial(n, p). The endpoints may be obtained directly by using Stata's invbinomial() function. If $k = 0$ or $k = n$, the calculation of the appropriate tail is skipped.

The Wald interval is $\widehat{p} \pm z_\alpha \sqrt{\widehat{p}(1-\widehat{p})/n}$, where z_α is the $1 - \alpha/2$ quantile of the standard normal. The interval is obtained by inverting the acceptance region of the large-sample Wald test of $H_0 : p = p_0$ versus the two-sided alternative. That is, the confidence interval is the set of all p_0 such that

$$\left| \frac{\widehat{p} - p_0}{\sqrt{n^{-1}\widehat{p}(1-\widehat{p})}} \right| \leq z_\alpha$$

The Wilson interval is a variation on the Wald interval, using the null standard error $\sqrt{n^{-1}p_0(1-p_0)}$ in place of the estimated standard error $\sqrt{n^{-1}\widehat{p}(1-\widehat{p})}$ in the above expression. Inverting this acceptance region is more complicated yet results in the closed form

$$\frac{k + z_\alpha^2/2}{n + z_\alpha^2} \pm \frac{z_\alpha n^{1/2}}{n + z_\alpha^2/2} \left\{ \widehat{p}(1-\widehat{p}) + \frac{z_\alpha^2}{4n} \right\}^{1/2}$$

The Agresti–Coull interval is basically a Wald interval that borrows its center from the Wilson interval. Defining $\widetilde{k} = k + z_\alpha^2/2$, $\widetilde{n} = n + z_\alpha^2$, and (hence) $\widetilde{p} = \widetilde{k}/\widetilde{n}$, the Agresti–Coull interval is

$$\widetilde{p} \pm z_\alpha \sqrt{\widetilde{p}(1-\widetilde{p})/\widetilde{n}}$$

When $\alpha = 0.05$, z_α is near enough to 2 that $\widetilde{p}$ can be thought of as a typical estimate of proportion where two successes and two failures have been added to the sample (Agresti and Coull, 1998). This typical estimate of proportion makes the Agresti–Coull interval an easy-to-present alternative for introductory statistics students.

The Jeffreys interval is a Bayesian interval and is based on the Jeffreys prior, which is the Beta$(1/2, 1/2)$ distribution. Assigning this prior to p results in a posterior distribution for p that is Beta with parameters $k + 1/2$ and $n - k + 1/2$. The Jeffreys interval is then taken to be the $1 - \alpha$ central posterior probability interval, namely the $\alpha/2$ and $1 - \alpha/2$ quantiles of the Beta$(k + 1/2, n - k + 1/2)$ distribution. These quantiles may be obtained directly by using Stata's `invibeta()` function.

Poisson

Given the total cases k, the estimate of the expected count λ is k, and its standard error is $\sqrt{k}$. `ci` calculates the exact confidence interval $[\lambda_1, \lambda_2]$ such that

$$\Pr(K \geq k | \lambda = \lambda_1) = \alpha/2$$

and

$$\Pr(K \leq k | \lambda = \lambda_2) = \alpha/2$$

where K is Poisson with mean λ. Solution is by Newton's method. If $k = 0$, the calculation of λ_1 is skipped. All values are then reported as rates, which are the above numbers divided by the total exposure.

Harold Jeffreys (1891–1989) was born near Durham, England, and spent more than 75 years studying and working at the University of Cambridge, principally on theoretical and observational problems in geophysics, astronomy, mathematics, and statistics. He developed a systematic Bayesian approach to inference in his monograph *Theory of Probability*.

E. B. Wilson (1879–1964) majored in mathematics at Harvard and studied and taught at Yale and MIT before returning to Harvard in 1922. He worked in mathematics, physics, and statistics. His method for binomial intervals can be considered as a precursor, for a particular problem, of Neyman's concept of confidence intervals.

Acknowledgment

We thank Nicholas J. Cox of Durham University for his assistance with the `jeffreys` and `wilson` options.

References

Agresti, A., and B. A. Coull. 1998. Approximate is better than "exact" for interval estimation of binomial proportions. *American Statistician* 52: 119–126.

Brown, L. D., T. T. Cai, and A. DasGupta. 2001. Interval estimation for a binomial proportion. *Statistical Science* 16: 101–133.

Clopper, C. J., and S. E. Pearson. 1934. The use of confidence or fiducial limits illustrated in the case of the binomial. *Biometrika* 26: 404–413.

Cook, A. H. 1990. Sir Harold Jeffreys. *Biographical Memoirs of Fellows of the Royal Society* 36: 303–333.

Feller, W. 1968. *An Introduction to Probability Theory and Its Applications*, vol. 1. 3rd ed. New York: Wiley.

Gleason, J. R. 1999. sg119: Improved confidence intervals for binomial proportions. *Stata Technical Bulletin* 52: 16–18. Reprinted in *Stata Technical Bulletin Reprints*, vol. 9, pp. 208–211.

Hamilton, L. C. 1996. *Data Analysis for Social Scientists*. Belmont, CA: Duxbury.

Hoel, P. G. 1984. *Introduction to Mathematical Statistics*. 5th ed. New York: Wiley.

Jeffreys, H. 1946. An invariant form for the prior probability in estimation problems. *Proceedings of the Royal Society, Series A* 186: 453–461.

Lindley, D. V. 2001. Harold Jeffreys. In *Statisticians of the Centuries*, ed. C. C. Heyde and E. Seneta, 402–405. New York: Springer.

Reid, C. 1982. *Neyman—from Life*. New York: Springer.

Rothman, K. J., and S. Greenland. 1998. *Modern Epidemiology*. 2nd ed. Philadelphia: Lippincott–Raven.

Seed, P. T. 2001. sg159: Confidence intervals for correlations. *Stata Technical Bulletin* 59: 27–28. Reprinted in *Stata Technical Bulletin Reprints*, vol. 10, pp. 267–269.

Stigler, S. M. 1997. Wilson, Edwin Bidwell. In *Leading Personalities in Statistical Sciences: From the Seventeenth Century to the Present*, ed. N. L. Johnson and S. Kotz, 344–346. New York: Wiley.

Utts, J. M. 2005. *Seeing Through Statistics*. 3rd ed. Belmont, CA: Brooks/Cole.

Wang, D. 2000. sg154: Confidence intervals for the ratio of two binomial proportions by Koopman's method. *Stata Technical Bulletin* 58: 16–19. Reprinted in *Stata Technical Bulletin Reprints*, vol. 10, pp. 244–247.

Wilson, E. B. 1927. Probable inference, the law of succession, and statistical inference. *Journal of the American Statistical Association* 22: 209–212.

Jerzy Neyman (1894–1981) was born in Bendery, Moldavia. He studied and then taught at Kharkov University, moving from physics to mathematics. In 1921, Neyman moved to Poland, where he worked in statistics at Bydgoszcz and then Warsaw. Receiving a Rockefeller Fellowship to work with Karl Pearson at University College London, he befriended Egon Pearson, Karl's son, and they worked together on the theory of hypothesis testing. Life in Poland became progressively more difficult, and Neyman returned to UCL to work there from 1934 to 1938. At this time, he published on the theory of confidence intervals. He then was offered a post in California at Berkeley, where he settled. Neyman established an outstanding statistics department and remained highly active in research, including applications in astronomy, meteorology, and medicine. He was one of the great statisticians of the 20th century.

Also See

[R] **bitest** — Binomial probability test

[R] **prtest** — One- and two-sample tests of proportions

[R] **ttest** — Mean-comparison tests

[R] **ameans** — Arithmetic, geometric, and harmonic means

[R] **centile** — Report centile and confidence interval

[R] **summarize** — Summary statistics

[D] **pctile** — Create variable containing percentiles

Title

> **clogit** — Conditional (fixed-effects) logistic regression

Syntax

> *clogit* *depvar* $\left[\,indepvars\,\right]$ $\left[\,if\,\right]$ $\left[\,in\,\right]$ $\left[\,weight\,\right]$, group(*varname*) $\left[\,options\,\right]$

options	description
Model	
*group(*varname*)	matched group variable
offset(*varname*)	include *varname* in model with coefficient constrained to 1
constraints(*numlist*)	apply specified linear constraints
collinear	keep collinear variables
SE/Robust	
vce(*vcetype*)	*vcetype* may be oim, robust, cluster *clustvar*, opg, bootstrap, or jackknife
nonest	do not check that panels are nested within clusters
Reporting	
level(*#*)	set confidence level; default is level(95)
or	report odds ratios
Max options	
maximize_options	control the maximization process; seldom used

*group(*varname*) is required.

bootstrap, by, jackknife, nestreg, rolling, statsby, stepwise, svy, and xi are allowed; see
  [U] **11.1.10 Prefix commands**.

Weights are not allowed with the bootstrap prefix.

vce(), nonest, and weights are not allowed with the svy prefix.

fweights, iweights, and pweights are allowed (see [U] **11.1.6 weight**), but they are interpreted to apply to groups
  as a whole, not to individual observations. See *Use of weights* below.

See [U] **20 Estimation and postestimation commands** for more capabilities of estimation commands.

Description

clogit fits what biostatisticians and epidemiologists call conditional logistic regression for matched case–control groups (see, for example, Hosmer and Lemeshow 2000, chap. 7) and what economists and other social scientists call fixed-effects logit for panel data (see, for example, Chamberlain 1980). Computationally, these models are the same.

See [R] **asclogit** if you want to fit McFadden's choice model (McFadden 1974). Also see [R] **logistic** for a list of related estimation commands.

Options

Model

group(*varname*) is required; it specifies an identifier variable (numeric or string) for the matched groups. strata(*varname*) is a synonym for group().

offset(*varname*), constraints(*numlist*), collinear; see [R] **estimation options**.

SE/Robust

vce(*vcetype*) specifies the type of standard error reported, which includes types that are derived from asymptotic theory, that are robust to some kinds of misspecification, that allow for intragroup correlation, and that use bootstrap or jackknife methods; see [R] *vce_option*.

nonest, available only with vce(cluster *clustvar*), prevents checking that matched groups are nested within clusters. It is the user's responsibility to verify that the standard errors are theoretically correct.

Reporting

level(#); see [R] **estimation options**.

or reports the estimated coefficients transformed to odds ratios, i.e., e^b rather than b. Standard errors and confidence intervals are similarly transformed. This option affects how results are displayed, not how they are estimated. or may be specified at estimation or when replaying previously estimated results.

Max options

maximize_options: <u>dif</u>ficult, <u>tech</u>nique(*algorithm_spec*), <u>iter</u>ate(#), [<u>no</u>]<u>log</u>, <u>tra</u>ce, gradient, showstep, <u>hess</u>ian, <u>shownr</u>tolerance, <u>tol</u>erance(#), <u>ltol</u>erance(#), gtolerance(#), <u>nrtol</u>erance(#), <u>nonrtol</u>erance, from(*init_specs*); see [R] **maximize**. These options are seldom used.

Setting the optimization type to technique(bhhh) resets the default *vcetype* to vce(opg).

Remarks

Remarks are presented under the following headings:

> *Matched case–control data*
> *Use of weights*
> *Fixed-effects logit*

clogit fits maximum likelihood models with a dichotomous dependent variable coded as 0/1 (more precisely, clogit interprets 0 and not 0 to indicate the dichotomy). Conditional logistic analysis differs from regular logistic regression in that the data are grouped and the likelihood is calculated relative to each group; i.e., a conditional likelihood is used. See *Methods and Formulas* at the end of this entry.

Biostatisticians and epidemiologists fit these models when analyzing matched case–control studies with $1:1$ matching, $1:k_{2i}$ matching, or $k_{1i}:k_{2i}$ matching, where i denotes the ith matched group for $i = 1, 2, \ldots, n$, where n is the total number of groups. clogit fits a model appropriate for all these matching schemes or for any mix of the schemes since the matching $k_{1i}:k_{2i}$ can vary from group to group. clogit always uses the true conditional likelihood, not an approximation. Biostatisticians and epidemiologists sometimes refer to the matched groups as "strata", but we will stick to the more generic term "group".

Economists and other social scientists fitting fixed-effects logit models have data that look exactly like the data biostatisticians and epidemiologists call $k_{1i} : k_{2i}$ matched case–control data. In terms of how the data are arranged, $k_{1i} : k_{2i}$ matching means that in the ith group the dependent variable is 1 a total of k_{1i} times and 0 a total of k_{2i} times. There are a total of $T_i = k_{1i} + k_{2i}$ observations for the ith group. This data arrangement is what economists and other social scientists call "panel data", "longitudinal data", or "cross-sectional time-series data".

So, no matter what terminology you use, the computation and the use of the clogit command is the same. The following example shows how your data should be arranged to use clogit.

▷ Example 1

Suppose that we have grouped data with the variable id containing a unique identifier for each group. Our outcome variable, y, contains 0s and 1s. If we were biostatisticians, y = 1 would indicate a case, y = 0 would be a control, and id would be an identifier variable that indicates the groups of matched case–controls subjects.

If we were economists, y = 1 might indicate that a person was unemployed at any time during a year and y = 0 that a person was employed all year, and id would be an identifier variable for persons.

If we list the first few observations of this dataset, it looks like

```
. use http://www.stata-press.com/data/r10/clogitid
. list y x1 x2 id in 1/11
```

	y	x1	x2	id
1.	0	0	4	1014
2.	0	1	4	1014
3.	0	1	6	1014
4.	1	1	8	1014
5.	0	0	1	1017
6.	0	0	7	1017
7.	1	1	10	1017
8.	0	0	1	1019
9.	0	1	7	1019
10.	1	1	7	1019
11.	1	1	9	1019

Pretending that we are biostatisticians, we describe our data as follows. The first group (id = 1014) consists of four matched persons: 1 case (y = 1) and three controls (y = 0), i.e., 1 : 3 matching. The second group has 1 : 2 matching, and the third 2 : 2.

Pretending that we are economists, we describe our data as follows. The first group consists of 4 observations (one per year) for person 1014. This person had a period of unemployment during 1 year of 4. The second person had a period of unemployment during 1 year of 3, and the third had a period of 2 years of 4.

Our independent variables are x1 and x2. To fit the conditional (fixed-effects) logistic model, we type

```
. clogit y x1 x2, group(id)
note: multiple positive outcomes within groups encountered.
Iteration 0:    log likelihood = -123.42828
Iteration 1:    log likelihood = -123.41386
Iteration 2:    log likelihood = -123.41386
Conditional (fixed-effects) logistic regression     Number of obs   =        369
                                                    LR chi2(2)      =       9.07
                                                    Prob > chi2     =     0.0107
Log likelihood = -123.41386                         Pseudo R2       =     0.0355
```

y	Coef.	Std. Err.	z	P>|z|	[95% Conf. Interval]	
x1	.653363	.2875215	2.27	0.023	.0898312	1.216895
x2	.0659169	.0449555	1.47	0.143	-.0221943	.1540281

◁

❏ Technical Note

The message "note: multiple positive outcomes within groups encountered" at the top of the clogit output for the previous example merely informs us that we have $k_{1i} : k_{2i}$ matching with $k_{1i} > 1$ for at least one group. If your data should be $1 : k_{2i}$ matched, this message tells you that there is an error in the data somewhere.

We can see the distribution of k_{1i} and $T_i = k_{1i} + k_{2i}$ for the data of the previous example by using the following steps.

```
. by id, sort: gen k1 = sum(y)
. by id: replace k1 = . if _n < _N
(303 real changes made, 303 to missing)
. by id: gen T = sum(y < .)
. by id: replace T = . if _n < _N
(303 real changes made, 303 to missing)
. tab k1
```

k1	Freq.	Percent	Cum.
1	48	72.73	72.73
2	12	18.18	90.91
3	4	6.06	96.97
4	2	3.03	100.00
Total	66	100.00	

```
. tab T
```

T	Freq.	Percent	Cum.
2	5	7.58	7.58
3	5	7.58	15.15
4	12	18.18	33.33
5	11	16.67	50.00
6	13	19.70	69.70
7	8	12.12	81.82
8	3	4.55	86.36
9	7	10.61	96.97
10	2	3.03	100.00
Total	66	100.00	

We see that k_{1i} ranges from 1 to 4 and T_i ranges from 2 to 10 for these data.

❏

❏ Technical Note

For $k_{1i} : k_{2i}$ matching (and hence in the general case of fixed-effects logit), clogit uses a recursive algorithm to compute the likelihood, which means that there are no limits on the size of T_i. However, computation time is proportional to $\sum T_i \min(k_{1i}, k_{2i})$, so clogit will take roughly 10 times longer to fit a model with 10 : 10 matching than one with 1 : 10 matching. But clogit is fast, so computation time becomes an issue only when $\min(k_{1i}, k_{2i})$ is around 100 or more. See *Methods and Formulas* for details.

❏

Matched case–control data

Here we give a more detailed example of matched case–control data.

▷ Example 2

Hosmer and Lemeshow (2000, 25) present data on matched pairs of infants, each pair having one with low birthweight and another with regular birthweight. The data are matched on age of the mother. Several possible maternal exposures are considered: race (three categories), smoking status, presence of hypertension, presence of uterine irritability, previous preterm delivery, and weight at the last menstrual period.

```
. use http://www.stata-press.com/data/r10/lowbirth, clear
(Applied Logistic Regression, Hosmer & Lemeshow, pp. 262-265)
. describe
Contains data from http://www.stata-press.com/data/r10/lowbirth.dta
  obs:           112                          Applied Logistic Regression,
                                                Hosmer & Lemeshow, pp. 262-265
  vars:           11                          21 Jan 2007 11:48
  size:         1,792 (99.8% of memory free)
```

variable name	storage type	display format	value label	variable label
pairid	byte	%8.0g		Case-control pair id
low	byte	%8.0g		Baby has low birth weight
age	byte	%8.0g		Age of mother
lwt	int	%8.0g		Mother's last menstrual weight
smoke	byte	%8.0g		Mother smoked during pregnancy
ptd	byte	%8.0g		Mother had previous preterm baby
ht	byte	%8.0g		Mother has hypertension
ui	byte	%8.0g		Uterine irritability
race1	byte	%8.0g		mother is white
race2	byte	%8.0g		mother is black
race3	byte	%8.0g		mother is other

```
Sorted by:
```

We list the case–control indicator variable, low; the match identifier variable, pairid; and two of the covariates, lwt and smoke, for the first 10 observations.

```
. list low lwt smoke pairid in 1/10
```

	low	lwt	smoke	pairid
1.	0	135	0	1
2.	1	101	1	1
3.	0	98	0	2
4.	1	115	0	2
5.	0	95	0	3
6.	1	130	0	3
7.	0	103	0	4
8.	1	130	1	4
9.	0	122	1	5
10.	1	110	1	5

We fit a conditional logistic model of low birthweight on mother's weight, race, smoking behavior, and history.

```
. clogit low lwt smoke ptd ht ui race2 race3, strata(pairid) nolog
Conditional (fixed-effects) logistic regression   Number of obs   =        112
                                                  LR chi2(7)      =      26.04
                                                  Prob > chi2     =     0.0005
Log likelihood = -25.794271                       Pseudo R2       =     0.3355
```

low	Coef.	Std. Err.	z	P>|z|	[95% Conf. Interval]	
lwt	-.0183757	.0100806	-1.82	0.068	-.0381333	.0013819
smoke	1.400656	.6278396	2.23	0.026	.1701131	2.631199
ptd	1.808009	.7886502	2.29	0.022	.2622828	3.353735
ht	2.361152	1.086128	2.17	0.030	.2323796	4.489924
ui	1.401929	.6961585	2.01	0.044	.0374836	2.766375
race2	.5713643	.689645	0.83	0.407	-.7803149	1.923044
race3	-.0253148	.6992044	-0.04	0.971	-1.39573	1.345101

We might prefer to see results presented as odds ratios. We could have specified the or option when we first fitted the model, or we can now redisplay results and specify or:

```
. clogit, or
Conditional (fixed-effects) logistic regression   Number of obs   =        112
                                                  LR chi2(7)      =      26.04
                                                  Prob > chi2     =     0.0005
Log likelihood = -25.794271                       Pseudo R2       =     0.3355
```

low	Odds Ratio	Std. Err.	z	P>|z|	[95% Conf. Interval]	
lwt	.9817921	.009897	-1.82	0.068	.9625847	1.001383
smoke	4.057862	2.547686	2.23	0.026	1.185439	13.89042
ptd	6.098293	4.80942	2.29	0.022	1.299894	28.60938
ht	10.60316	11.51639	2.17	0.030	1.261599	89.11467
ui	4.06303	2.828513	2.01	0.044	1.038195	15.90088
race2	1.770681	1.221141	0.83	0.407	.4582617	6.84175
race3	.975003	.6817263	-0.04	0.971	.2476522	3.838573

Smoking, previous preterm delivery, hypertension, uterine irritability, and possibly the mother's weight all contribute to low birthweight. race2 (mother black) and race3 (mother other) are statistically insignificant when compared with the race1 (mother white) omitted group, although the race2 effect is large. We can test the joint statistical significance of race2 and race3 by using test:

```
. test race2 race3
 ( 1)  race2 = 0
 ( 2)  race3 = 0

       chi2(  2) =    0.88
     Prob > chi2 =    0.6436
```

For a more complete description of test, see [R] **test**. test presents results in coefficients rather than odds ratios. Jointly testing that the coefficients on race2 and race3 are zero is equivalent to jointly testing that the odds ratios are 1.

Here one case was matched to one control, i.e., 1 : 1 matching. From clogit's point of view, that was not important—k_1 cases could have been matched to k_2 controls ($k_1 : k_2$ matching), and we would have fitted the model in the same way. Furthermore, the matching can change from group to group, which we have denoted as $k_{1i} : k_{2i}$ matching, where i denotes the group. clogit does not care. To fit the conditional logistic regression model, we specified the group(*varname*) option, group(pairid). The case and control are stored in separate observations. clogit knew that they were linked (in the same group) because the related observations share the same value of pairid.

◁

❏ Technical Note

clogit provides a way to extend McNemar's test to multiple controls per case (1 : k_{2i} matching) and to multiple controls matched with multiple cases ($k_{1i} : k_{2i}$ matching).

In Stata, McNemar's test is calculated by the mcc command; see [ST] **epitab**. The mcc command, however, requires that the matched case and control appear in one observation, so the data will need to be manipulated from 1 to 2 observations per stratum before using clogit. Alternatively, if you begin with clogit's 2-observations-per-group organization, you will have to change it to 1 observation per group if you wish to use mcc. In either case, reshape provides an easy way to change the organization of the data. We will demonstrate its use below, but we direct you to [D] **reshape** for a more thorough discussion.

In the example above, we used clogit to analyze the relationship between low birthweight and various characteristics of the mother. Assume that we now want to assess the relationship between low birthweight and smoking, ignoring the mother's other characteristics. Using clogit, we obtain the following results:

```
. clogit low smoke, strata(pairid) or
Iteration 0:   log likelihood = -35.425931
Iteration 1:   log likelihood = -35.419283
Iteration 2:   log likelihood = -35.419282
Conditional (fixed-effects) logistic regression     Number of obs   =        112
                                                     LR chi2(1)      =       6.79
                                                     Prob > chi2     =     0.0091
Log likelihood = -35.419282                          Pseudo R2       =     0.0875
```

low	Odds Ratio	Std. Err.	z	P>|z|	[95% Conf. Interval]	
smoke	2.75	1.135369	2.45	0.014	1.224347	6.176763

Let us compare our estimated odds ratio and 95% confidence interval with that produced by mcc. We begin by reshaping the data:

```
. keep low smoke pairid

. reshape wide smoke, i(pairid) j(low 0 1)
Data                          long   ->   wide

Number of obs.                 112   ->      56
Number of variables              3   ->       3
j variable (2 values)          low   ->   (dropped)
xij variables:
                             smoke   ->   smoke0 smoke1
```

We now have the variables smoke0 (formed from smoke and low $= 0$), recording 1 if the control mother smoked and 0 otherwise; and smoke1 (formed from smoke and low $= 1$), recording 1 if the case mother smoked and 0 otherwise. We can now use mcc:

```
. mcc smoke1 smoke0
                   |         Controls
Cases              |  Exposed   Unexposed  |      Total

         Exposed   |        8          22  |         30
       Unexposed   |        8          18  |         26

           Total   |       16          40  |         56

McNemar's chi2(1) =      6.53    Prob > chi2 = 0.0106
Exact McNemar significance probability      = 0.0161

Proportion with factor
        Cases      .5357143
        Controls   .2857143          [95% Conf. Interval]

        difference       .25     .0519726    .4480274
        ratio          1.875     1.148685    3.060565
        rel. diff.       .35     .1336258    .5663742

        odds ratio      2.75     1.179154    7.143667   (exact)
```

Both methods estimated the same odds ratio, and the 95% confidence intervals are similar. clogit produced a confidence interval of $[1.22, 6.18]$, whereas mcc produced a confidence interval of $[1.18, 7.14]$.

❑

Use of weights

With clogit, weights apply to groups as a whole, not to individual observations. For example, if there is a group in your dataset with a frequency weight of 3, there are a total of three groups in your sample with the same values of the dependent and independent variables as this one group. Weights must have the same value for all observations belonging to the same group; otherwise, an error message will be displayed.

▷ Example 3

We use the example from the above discussion of the mcc command. Here we have a total of 56 matched case–control groups, each with one case matched to one control. We had 8 matched pairs in which both the case and the control are exposed, 22 pairs in which the case is exposed and the control is unexposed, 8 pairs in which the case is unexposed and the control is exposed, and 18 pairs in which they are both unexposed.

With weights, it is easy to enter these data into Stata and run clogit.

```
. clear
. input id case exposed weight
          id       case     exposed      weight
 1. 1 1 1 8
 2. 1 0 1 8
 3. 2 1 1 22
 4. 2 0 0 22
 5. 3 1 0 8
 6. 3 0 1 8
 7. 4 1 0 18
 8. 4 0 0 18
 9. end
. clogit case exposed [w=weight], strata(id) or
(frequency weights assumed)
Iteration 0:   log likelihood = -35.425931
Iteration 1:   log likelihood = -35.419283
Iteration 2:   log likelihood = -35.419282
Conditional (fixed-effects) logistic regression    Number of obs   =        112
                                                   LR chi2(1)      =       6.79
                                                   Prob > chi2     =     0.0091
Log likelihood = -35.419282                        Pseudo R2       =     0.0875
```

case	Odds Ratio	Std. Err.	z	P>|z|	[95% Conf. Interval]	
exposed	2.75	1.135369	2.45	0.014	1.224347	6.176763

◁

Fixed-effects logit

The fixed-effects logit model can be written as

$$\Pr(y_{it} = 1 \mid \mathbf{x}_{it}) = F(\alpha_i + \mathbf{x}_{it}\boldsymbol{\beta})$$

where F is the cumulative logistic distribution

$$F(z) = \frac{\exp(z)}{1 + \exp(z)}$$

$i = 1, 2, \ldots, n$ denotes the independent units (called "groups" by clogit), and $t = 1, 2, \ldots, T_i$ denotes the observations for the ith unit (group).

Fitting this model by using a full maximum-likelihood approach leads to difficulties, however. When T_i is fixed, the maximum likelihood estimates for α_i and $\boldsymbol{\beta}$ are inconsistent (Andersen 1970 and Chamberlain 1980). This difficulty can be circumvented by looking at the probability of $\mathbf{y}_i = (y_{i1}, \ldots, y_{iT_i})$ conditional on $\sum_{t=1}^{T_i} y_{it}$. This conditional probability does not involve the α_i, so they are never estimated when the resulting conditional likelihood is used. See Hamerle and Ronning (1995) for a succinct and lucid development. See *Methods and Formulas* for the estimation equation.

▷ Example 4

We are studying unionization of women in the United States by using the `union` dataset; see [XT] **xt**. We fit the fixed-effects logit model:

```
. clear all

. set memory 4000
(4000k)

. use http://www.stata-press.com/data/r10/union
(NLS Women 14-24 in 1968)

. clogit union age grade not_smsa south black, group(idcode)
note: multiple positive outcomes within groups encountered.
note: 2744 groups (14165 obs) dropped due to all positive or
      all negative outcomes.
note: black omitted due to no within-group variance.

Iteration 0:   log likelihood = -4521.3385
Iteration 1:   log likelihood = -4516.1404
Iteration 2:   log likelihood = -4516.1385
Iteration 3:   log likelihood = -4516.1385

Conditional (fixed-effects) logistic regression   Number of obs   =     12035
                                                  LR chi2(4)      =     68.09
                                                  Prob > chi2     =    0.0000
Log likelihood = -4516.1385                       Pseudo R2       =    0.0075
```

union	Coef.	Std. Err.	z	P>|z|	[95% Conf. Interval]	
age	.0170301	.004146	4.11	0.000	.0089042	.0251561
grade	.0853572	.0418781	2.04	0.042	.0032777	.1674368
not_smsa	.0083678	.1127963	0.07	0.941	-.2127088	.2294445
south	-.748023	.1251752	-5.98	0.000	-.9933619	-.5026842

We received three messages at the top of the output. The first one, "multiple positive outcomes within groups encountered", we expected. Our data do indeed have multiple positive outcomes (union = 1) in many groups. (Here a group consists of all the observations for a particular individual.)

The second message tells us that 2744 groups were "dropped" by clogit. When either union = 0 or union = 1 for all observations for an individual, this individual's contribution to the log-likelihood is zero. Although these are perfectly valid observations in every sense, they have no effect on the estimation, so they are not included in the total "Number of obs". Hence, the reported "Number of obs" gives the effective sample size of the estimation. Here it is 12,035 observations—only 46% of the total 26,200.

We can easily check that there are indeed 2,744 groups with union either all 0 or all 1. We will generate a variable that contains the fraction of observations for each individual who has union = 1.

(*Continued on next page*)

```
. by idcode, sort: generate fraction = sum(union)/sum(union < .)

. by idcode: replace fraction = . if _n < _N
(21766 real changes made, 21766 to missing)

. tabulate fraction
```

fraction	Freq.	Percent	Cum.
0	2,481	55.95	55.95
.0833333	30	0.68	56.63
.0909091	33	0.74	57.37
.1	53	1.20	58.57
(output omitted)			
.9	10	0.23	93.59
.9090909	11	0.25	93.84
.9166667	10	0.23	94.07
1	263	5.93	100.00
Total	4,434	100.00	

Since $2481 + 263 = 2744$, we confirm what clogit did.

The third warning message from clogit said "black omitted due to no within-group variance". Obviously, race stays constant for an individual across time. Any such variables are collinear with the α_i (i.e., the fixed effects), and just as the α_i drop out of the conditional likelihood, so do all variables that are unchanging within groups. Thus they cannot be estimated with the conditional fixed-effects model.

There are several other estimators implemented in Stata that we could use with these data:

```
cloglog ... , vce(cluster idcode)
logit ... , vce(cluster idcode)
probit ... , vce(cluster idcode)
scobit ... , vce(cluster idcode)
xtcloglog ... , i(idcode)
xtgee ... , i(idcode) family(binomial) link(logit) corr(exchangeable)
xtlogit ... , i(idcode)
xtprobit ... , i(idcode)
```

See [R] **cloglog**, [R] **logit**, [R] **probit**, [R] **scobit**, [XT] **xtcloglog**, [XT] **xtgee**, [XT] **xtlogit**, and [XT] **xtprobit** for details.

◁

Saved Results

clogit saves the following in e():

Scalars

e(N)	number of observations	e(df_m)	model degrees of freedom
e(N_drop)	number of observations dropped due to all positive or negative outcomes	e(r2_p)	pseudo-R-squared
		e(ll)	log likelihood
		e(ll_0)	log likelihood, constant-only model
e(N_group_drop)	number of groups dropped due to all positive or negative outcomes	e(N_clust)	number of clusters
		e(chi2)	χ^2
		e(p)	significance
e(k)	number of parameters	e(rank)	rank of e(V)
e(k_eq)	number of equations in e(b)	e(ic)	number of iterations
e(k_eq_model)	number of equations in model Wald test	e(rc)	return code
		e(converged)	1 if converged, 0 otherwise
e(k_dv)	number of dependent variables		

Macros

e(cmd)	clogit	e(vce)	*vcetype* specified in vce()
e(cmdline)	command as typed	e(vcetype)	title used to label Std. Err.
e(depvar)	name of dependent variable	e(opt)	type of optimization
e(group)	name of group() variable	e(ml_method)	type of ml method
e(wtype)	weight type	e(user)	name of likelihood-evaluator program
e(wexp)	weight expression	e(technique)	maximization technique
e(title)	title in estimation output	e(crittype)	optimization criterion
e(clustvar)	name of cluster variable	e(properties)	b V
e(offset)	offset	e(predict)	program used to implement predict
e(chi2type)	LR; type of model χ^2 test		

Matrices

e(b)	coefficient vector	e(V)	variance–covariance matrix of the estimators
e(ilog)	iteration log (up to 20 iterations)		
e(gradient)	gradient vector		

Functions

e(sample)	marks estimation sample

Methods and Formulas

clogit is implemented as an ado-file.

Breslow and Day (1980, 247–279), Collett (2003, 251–267), and Hosmer and Lemeshow (2000, 223–259) provide a biostatistical point of view on conditional logistic regression. Hamerle and Ronning (1995) give a succinct and lucid review of fixed-effects logit; Chamberlain (1980) is a standard reference for this model. Greene (2003, chap. 21) provides a straightforward textbook description of conditional logistic regression from an economist's point of view, as well as a brief description of choice models.

Let $i = 1, 2, \ldots, n$ denote the groups and let $t = 1, 2, \ldots, T_i$ denote the observations for the ith group. Let y_{it} be the dependent variable taking on values 0 or 1. Let $\mathbf{y}_i = (y_{i1}, \ldots, y_{iT_i})$ be the outcomes for the ith group as a whole. Let $\mathbf{x}_{it}$ be a row vector of covariates. Let

$$k_{1i} = \sum_{t=1}^{T_i} y_{it}$$

be the observed number of ones for the dependent variable in the ith group. Biostatisticians would say that there are k_{1i} cases matched to $k_{2i} = T_i - k_{1i}$ controls in the ith group.

We consider the probability of a possible value of $\mathbf{y}_i$ conditional on $\sum_{t=1}^{T_i} y_{it} = k_{1i}$ (Hamerle and Ronning 1995, equation 8.33; Hosmer and Lemeshow 2000, equation 7.4),

$$\Pr\left(\mathbf{y}_i \mid \sum_{t=1}^{T_i} y_{it} = k_{1i}\right) = \frac{\exp\left(\sum_{t=1}^{T_i} y_{it}\mathbf{x}_{it}\boldsymbol{\beta}\right)}{\sum_{\mathbf{d}_i \in S_i} \exp\left(\sum_{t=1}^{T_i} d_{it}\mathbf{x}_{it}\boldsymbol{\beta}\right)}$$

where d_{it} is equal to 0 or 1 with $\sum_{t=1}^{T_i} d_{it} = k_{1i}$, and S_i is the set of all possible combinations of k_{1i} ones and k_{2i} zeros. Clearly, there are $\binom{T_i}{k_{1i}}$ such combinations, but we need not count all these combinations to compute the denominator of the above equation. It can be computed recursively.

Denote the denominator by

$$f_i(T_i, k_{1i}) = \sum_{\mathbf{d}_i \in S_i} \exp\left(\sum_{t=1}^{T_i} d_{it}\mathbf{x}_{it}\boldsymbol{\beta}\right)$$

Consider, computationally, how f_i changes as we go from a total of 1 observation in the group to 2 observations to 3, etc. Doing this, we derive the recursive formula

$$f_i(T, k) = f_i(T-1, k) + f_i(T-1, k-1)\exp(\mathbf{x}_{iT}\boldsymbol{\beta})$$

where we define $f_i(T, k) = 0$ if $T < k$ and $f_i(T, 0) = 1$.

The conditional log-likelihood is

$$\ln L = \sum_{i=1}^{n}\left\{\sum_{t=1}^{T_i} y_{it}\mathbf{x}_{it}\boldsymbol{\beta} - \log f_i(T_i, k_{1i})\right\}$$

The derivatives of the conditional log-likelihood can also be computed recursively by taking derivatives of the recursive formula for f_i.

Computation time is roughly proportional to

$$p^2 \sum_{i=1}^{n} T_i \min(k_{1i}, k_{2i})$$

where p is the number of independent variables in the model. If $\min(k_{1i}, k_{2i})$ is small, computation time is not an issue. But if it is large—say, 100 or more—patience may be required.

If T_i is large for all groups, the bias of the unconditional fixed-effects estimator is not a concern, and we can confidently use `logit` with an indicator variable for each group (provided, of course, that the number of groups does not exceed `matsize`; see [R] **matsize**).

References

Andersen, E. B. 1970. Asymptotic properties of conditional maximum likelihood estimators. *Journal of the Royal Statistical Society, Series B* 32: 283–301.

Breslow, N. E., and N. E. Day. 1980. *Statistical Methods in Cancer Research: The Analysis of Case–Control Studies, Vol. 1.* Lyon: IARC.

Chamberlain, G. 1980. Analysis of covariance with qualitative data. *Review of Economic Studies* 47: 225–238.

Collett, D. 2003. *Modelling Binary Data.* 2nd ed. London: Chapman & Hall/CRC.

Greene, W. H. 2003. *Econometric Analysis.* 5th ed. Upper Saddle River, NJ: Prentice Hall.

Hamerle, A., and G. Ronning. 1995. Panel analysis for qualitative variables. In *Handbook of Statistical Modeling for the Social and Behavioral Sciences*, ed. G. Arminger, C. C. Clogg, and M. E. Sobel, 401–451. New York: Plenum.

Hosmer, D. W., Jr., and S. Lemeshow. 2000. *Applied Logistic Regression.* 2nd ed. New York: Wiley.

Kleinbaum, D. G., and M. Klein. 2002. *Logistic Regression: A Self-Learning Text.* 2nd ed. New York: Springer.

Long, J. S., and J. Freese. 2006. *Regression Models for Categorical Dependent Variables Using Stata.* 2nd ed. College Station, TX: Stata Press.

McFadden, D. 1974. Conditional logit analysis of qualitative choice behavior. In *Frontiers in Econometrics*, ed. P. Zarembka, 105–142. New York: Academic Press.

Also See

[R] **clogit postestimation** — Postestimation tools for clogit

[R] **asclogit** — Alternative-specific conditional logit (McFadden's choice) model

[R] **logistic** — Logistic regression, reporting odds ratios

[R] **mlogit** — Multinomial (polytomous) logistic regression

[R] **nlogit** — Nested logit regression

[R] **ologit** — Ordered logistic regression

[R] **scobit** — Skewed logistic regression

[SVY] **svy estimation** — Estimation commands for survey data

[XT] **xtgee** — Fit population-averaged panel-data models by using GEE

[XT] **xtlogit** — Fixed-effects, random-effects, and population-averaged logit models

[U] **20 Estimation and postestimation commands**

Title

clogit postestimation — Postestimation tools for clogit

Description

The following postestimation commands are available for `clogit`:

command	description
adjust	adjusted predictions of $\mathbf{x}\beta$, probabilities, or $\exp(\mathbf{x}\beta)$
estat	AIC, BIC, VCE, and estimation sample summary
estat (svy)	postestimation statistics for survey data
estimates	cataloging estimation results
hausman	Hausman's specification test
lincom	point estimates, standard errors, testing, and inference for linear combinations of coefficients
linktest	link test for model specification
lrtest[1]	likelihood-ratio test
mfx	marginal effects or elasticities
nlcom	point estimates, standard errors, testing, and inference for nonlinear combinations of coefficients
predict	predictions, residuals, influence statistics, and other diagnostic measures
predictnl	point estimates, standard errors, testing, and inference for generalized predictions
suest	seemingly unrelated estimation
test	Wald tests for simple and composite linear hypotheses
testnl	Wald tests of nonlinear hypotheses

[1] `lrtest` is not appropriate with svy estimation results.

See the corresponding entries in the *Stata Base Reference Manual* for details, but see [SVY] **estat** for details about `estat` (svy).

Syntax for predict

> predict [*type*] *newvar* [*if*] [*in*] [, *statistic* <u>nooff</u>set]

statistic	description
Main	
pc1	probability of a positive outcome; the default
pu0	probability of a positive outcome, assuming fixed effect is zero
xb	linear prediction
stdp	standard error of the linear prediction
* <u>db</u>eta	Delta-β influence statistic
* <u>dx</u>2	Delta-χ^2 lack-of-fit statistic
* <u>gdb</u>eta	Delta-β influence statistic for each group
* <u>gdx</u>2	Delta-χ^2 lack-of-fit statistic for each group
* <u>h</u>at	Hosmer and Lemeshow leverage
* <u>r</u>esiduals	Pearson residuals
* <u>rsta</u>ndard	standardized Pearson residuals
<u>sc</u>ore	first derivative of the log likelihood with respect to $\mathbf{x}_j\boldsymbol{\beta}$

Unstarred statistics are available both in and out of sample; type predict ... if e(sample) ... if wanted only for the estimation sample. Starred statistics are calculated only for the estimation sample, even when if e(sample) is not specified.

Starred statistics are available for multiple controls per case-matching design only. They are not available if vce(robust), vce(cluster *clustvar*), or pweights were specified with clogit.

dbeta, dx2, gdbeta, gdx2, hat, and rstandard are not available if constraints() were specified with clogit.

Options for predict

> **Main**

pc1, the default, calculates the probability of a positive outcome conditional on one positive outcome within group.

pu0 calculates the probability of a positive outcome, assuming that the fixed effect is zero.

xb calculates the linear prediction.

stdp calculates the standard error of the linear prediction.

dbeta calculates the Delta-β influence statistic, a standardized measure of the difference in the coefficient vector that is due to deletion of the observation.

dx2 calculates the Delta-χ^2 influence statistic, reflecting the decrease in the Pearson chi-squared that is due to deletion of the observation.

gdbeta calculates the approximation to the Pregibon stratum-specific Delta-β influence statistic, a standardized measure of the difference in the coefficient vector that is due to deletion of the entire stratum.

gdx2 calculates the approximation to the Pregibon stratum-specific Delta-χ^2 influence statistic, reflecting the decrease in the Pearson chi-squared that is due to deletion of the entire stratum.

hat calculates the Hosmer and Lemeshow leverage or the diagonal element of the hat matrix.

residuals calculates the Pearson residuals.

rstandard calculates the standardized Pearson residuals.

score calculates the equation-level score, $\partial \ln L / \partial(\mathbf{x}_{it}\boldsymbol{\beta})$.

nooffset is relevant only if you specified offset(*varname*) for clogit. It modifies the calculations made by predict so that they ignore the offset variable; the linear prediction is treated as $\mathbf{x}_j\mathbf{b}$ rather than as $\mathbf{x}_j\mathbf{b} + \text{offset}_j$. This option cannot be specified with dbeta, dx2, gdbeta, gdx2, hat, and rstandard.

Remarks

predict may be used after clogit to obtain predicted values of the index $\mathbf{x}_{it}\boldsymbol{\beta}$. Predicted probabilities for conditional logistic regression must be interpreted carefully. Probabilities are estimated for each group as a whole, not for individual observations. Furthermore, the probabilities are conditional on the number of positive outcomes in the group (i.e., the number of cases and the number of controls), or it is assumed that the fixed effect is zero. predict may also be used to obtain influence and lack of fit statistics for an individual observation and for the whole group, to compute Pearson, standardized Pearson residuals and leverage values.

predict may be used for both within-sample and out-of-sample predictions.

▷ Example 1

Suppose that we have $1 : k_{2i}$ matched data and that we have previously fitted the following model:

```
. use http://www.stata-press.com/data/r10/clogitid
. clogit y x1 x2, group(id)
(output omitted)
```

To obtain the predicted values of the index, we could type predict idx, xb to create a new variable called idx. From idx, we could then calculate the predicted probabilities. Easier, however, would be to type

```
. predict phat
(option pc1 assumed; conditional probability for single outcome within group)
```

phat would then contain the predicted probabilities.

Again the predicted probabilities are really predicted probabilities for the group as a whole (i.e., they are the predicted probability of observing $y_{it} = 1$ *and* $y_{it'} = 0$ for all $t' \neq t$). Thus, if we want to obtain the predicted probabilities for the estimation sample, it is important that, when we make the calculation, predictions be restricted to the same sample on which we estimated the data. We cannot predict the probabilities and then just keep the relevant ones because the entire sample determines each probability. Thus, assuming that we are not attempting to make out-of-sample predictions, we type

```
. predict phat2 if e(sample)
(option pc1 assumed; conditional probability for single outcome within group)
```
◁

Methods and Formulas

All postestimation commands listed above are implemented as ado-files.

Recall that $i = 1, \dots, n$ denote the groups and $t = 1, \dots, T_i$ denote the observations for the ith group.

`predict` produces probabilities of a positive outcome within group conditional on there being one positive outcome (`pc1`),

$$
\Pr\left(y_{it} = 1 \,\middle|\, \sum_{t=1}^{T_i} y_{it} = 1\right) = \frac{\exp(\mathbf{x}_{it}\boldsymbol{\beta})}{\sum_{t=1}^{T_i} \exp(\mathbf{x}_{it}\boldsymbol{\beta})}
$$

or `predict` calculates the unconditional `pu0`:

$$
\Pr(y_{it} = 1) = \frac{\exp(\mathbf{x}_{it}\boldsymbol{\beta})}{1 + \exp(\mathbf{x}_{it}\boldsymbol{\beta})}
$$

Let $N = \sum_{j=1}^{n} T_j$ denote the total number of observations, p denote the number of covariates, and $\widehat{\theta}_{it}$ denote the conditional predicted probabilities of a positive outcome (`pc1`).

For the multiple control per case $(1 : k_{2i})$ matching, Hosmer and Lemeshow (2000, 248–251) propose the following diagnostics:

The Pearson residual is

$$
r_{it} = \frac{(y_{it} - \widehat{\theta}_{it})}{\sqrt{\widehat{\theta}_{it}}}
$$

The leverage (hat) value is defined as

$$
h_{it} = \widehat{\theta}_{it}\widetilde{\mathbf{x}}_{it}^{T}(\widetilde{\mathbf{X}}^{T}\mathbf{U}\widetilde{\mathbf{X}})^{-1}\widetilde{\mathbf{x}}_{it}
$$

where $\widetilde{\mathbf{x}}_{it} = \mathbf{x}_{it} - \sum_{j=1}^{T_i} \mathbf{x}_{ij}\widehat{\theta}_{ij}$ is the $1 \times p$ row vector of centered by a weighted stratum-specific mean covariate values, $\mathbf{U}_N = \text{diag}\{\widehat{\theta}_{it}\}$, and the rows of $\widetilde{\mathbf{X}}_{N \times p}$ are comprised of $\widetilde{\mathbf{x}}_{it}$ values.

The standardized Pearson residual is

$$
r_{sit} = \frac{r_{it}}{\sqrt{1 - h_{it}}}
$$

The lack of fit and influence diagnostics for an individual observation are (respectively) computed as

$$
\Delta\chi_{it}^{2} = r_{sit}^{2}
$$

and

$$
\Delta\widehat{\beta}_{it} = \Delta\chi_{it}^{2}\frac{h_{it}}{1 - h_{it}}
$$

The lack of fit and influence diagnostics for the groups are the group-specific totals of the respective individual diagnostics shown above.

Also See

[R] **clogit** — Conditional (fixed-effects) logistic regression

[U] **20 Estimation and postestimation commands**

Title

---
cloglog — Complementary log-log regression

Syntax

cloglog *depvar* [*indepvars*] [*if*] [*in*] [*weight*] [, *options*]

options	description
Model	
<u>noconst</u>ant	suppress constant term
<u>off</u>set(*varname*)	include *varname* in model with coefficient constrained to 1
asis	retain perfect predictor variables
<u>constraints</u>(*constraints*)	apply specified linear constraints
<u>coll</u>inear	keep collinear variables
SE/Robust	
vce(*vcetype*)	*vcetype* may be oim, <u>r</u>obust, <u>c</u>luster *clustvar*, opg, <u>boot</u>strap, or <u>jack</u>knife
Reporting	
<u>level</u>(#)	set confidence level; default is level(95)
<u>ef</u>orm	report exponentiated coefficients
Max options	
maximize_options	control the maximization process; seldom used

depvar and *indepvars* may contain time-series operators; see [U] **11.4.3 Time-series varlists**.

bootstrap, by, jackknife, nestreg, rolling, statsby, stepwise, svy, and xi are allowed; see
[U] **11.1.10 Prefix commands**.

Weights are not allowed with the bootstrap prefix.

vce() and weights are not allowed with the svy prefix.

fweights, iweights, and pweights are allowed; see [U] **11.1.6 weight**.

See [U] **20 Estimation and postestimation commands** for more capabilities of estimation commands.

Description

cloglog fits maximum-likelihood complementary log-log models.

See [R] **logistic** for a list of related estimation commands.

Options

  Model

noconstant, offset(*varname*); see [R] **estimation options**.

asis forces retention of perfect predictor variables and their associated perfectly predicted observations
and may produce instabilities in maximization; see [R] **probit**.

constraints(*constraints*), collinear; see [R] **estimation options**.

⌐ SE/Robust ⌐

vce(*vcetype*) specifies the type of standard error reported, which includes types that are derived from asymptotic theory, that are robust to some kinds of misspecification, that allow for intragroup correlation, and that use bootstrap or jackknife methods; see [R] *vce_option*.

⌐ Reporting ⌐

level(*#*); see [R] **estimation options**.

eform displays the exponentiated coefficients and corresponding standard errors and confidence intervals.

⌐ Max options ⌐

maximize_options: difficult, technique(*algorithm_spec*), iterate(*#*), [no]log, trace, gradient, showstep, hessian, shownrtolerance, tolerance(*#*), ltolerance(*#*), gtolerance(*#*), nrtolerance(*#*), nonrtolerance, from(*init_specs*); see [R] **maximize**. These options are seldom used.

Setting the optimization type to technique(bhhh) resets the default *vcetype* to vce(opg).

Remarks

Remarks are presented under the following headings:

> *Introduction to complementary log-log regression*
> *Robust standard errors*

Introduction to complementary log-log regression

cloglog fits maximum likelihood models with dichotomous dependent variables coded as 0/1 (or, more precisely, coded as 0 and not 0).

▷ Example 1

We have data on the make, weight, and mileage rating of 22 foreign and 52 domestic automobiles. We wish to fit a model explaining whether a car is foreign based on its weight and mileage. Here is an overview of our data:

(*Continued on next page*)

```
. use http://www.stata-press.com/data/r10/auto
(1978 Automobile Data)

. keep make mpg weight foreign

. describe

Contains data from http://www.stata-press.com/data/r10/auto.dta
  obs:            74                          1978 Automobile Data
  vars:            4                          13 Apr 2007 17:45
  size:         1,998 (99.7% of memory free)  (_dta has notes)
```

variable name	storage type	display format	value label	variable label
make	str18	%-18s		Make and Model
mpg	int	%8.0g		Mileage (mpg)
weight	int	%8.0gc		Weight (lbs.)
foreign	byte	%8.0g	origin	Car type

```
Sorted by:  foreign
     Note:  dataset has changed since last saved
```

```
. inspect foreign
```

foreign:	Car type		Number of Observations		
			Total	Integers	Nonintegers
#		Negative	–	–	–
#		Zero	52	52	–
#		Positive	22	22	–
#			────	────	────
#	#	Total	74	74	–
#	#	Missing	–		
0	1		74		

```
    (2 unique values)
        foreign is labeled and all values are documented in the label.
```

The variable foreign takes on two unique values, 0 and 1. The value 0 denotes a domestic car, and 1 denotes a foreign car.

The model that we wish to fit is

$$\Pr(\text{foreign} = 1) = F(\beta_0 + \beta_1\text{weight} + \beta_2\text{mpg})$$

where $F(z) = 1 - \exp\{-\exp(z)\}$.

To fit this model, we type

```
. cloglog foreign weight mpg
Iteration 0:   log likelihood = -33.526894
Iteration 1:   log likelihood = -27.868343
Iteration 2:   log likelihood =  -27.74293
Iteration 3:   log likelihood = -27.742769
Iteration 4:   log likelihood = -27.742769
```

```
Complementary log-log regression            Number of obs     =        74
                                            Zero outcomes     =        52
                                            Nonzero outcomes  =        22

                                            LR chi2(2)        =     34.58
Log likelihood = -27.742769                 Prob > chi2       =    0.0000
```

foreign	Coef.	Std. Err.	z	P>|z|	[95% Conf. Interval]	
weight	-.0029153	.0006974	-4.18	0.000	-.0042823	-.0015483
mpg	-.1422911	.076387	-1.86	0.062	-.2920069	.0074247
_cons	10.09694	3.351841	3.01	0.003	3.527448	16.66642

We find that heavier cars are less likely to be foreign and that cars yielding better gas mileage are also less likely to be foreign, at least when holding the weight of the car constant.

See [R] **maximize** for an explanation of the output.

◁

❏ Technical Note

Stata interprets a value of 0 as a negative outcome (failure) and treats all other values (except missing) as positive outcomes (successes). Thus, if your dependent variable takes on the values 0 and 1, 0 is interpreted as failure and 1 as success. If your dependent variable takes on the values 0, 1, and 2, 0 is still interpreted as failure, but both 1 and 2 are treated as successes.

If you prefer a more formal mathematical statement, when you type cloglog y x, Stata fits the model

$$\Pr(y_j \neq 0 \mid \mathbf{x}_j) = 1 - \exp\left\{-\exp(\mathbf{x}_j\boldsymbol{\beta})\right\}$$

❏

Robust standard errors

If you specify the vce(robust) option, cloglog reports robust standard errors, as described in [U] **20.15 Obtaining robust variance estimates**. For the model of foreign on weight and mpg, the robust calculation increases the standard error of the coefficient on mpg by 44%:

(Continued on next page)

```
. cloglog foreign weight mpg, vce(robust)
Iteration 0:   log pseudolikelihood = -33.526894
Iteration 1:   log pseudolikelihood = -27.868343
Iteration 2:   log pseudolikelihood =  -27.74293
Iteration 3:   log pseudolikelihood = -27.742769
Iteration 4:   log pseudolikelihood = -27.742769
```

```
Complementary log-log regression            Number of obs    =         74
                                            Zero outcomes    =         52
                                            Nonzero outcomes =         22

                                            Wald chi2(2)     =      29.74
Log pseudolikelihood = -27.742769           Prob > chi2      =     0.0000
```

		Robust				
foreign	Coef.	Std. Err.	z	P>|z|	[95% Conf. Interval]	
weight	-.0029153	.0007484	-3.90	0.000	-.0043822	-.0014484
mpg	-.1422911	.1102466	-1.29	0.197	-.3583704	.0737882
_cons	10.09694	4.317305	2.34	0.019	1.635174	18.5587

Without vce(robust), the standard error for the coefficient on mpg was reported to be .076, with a resulting confidence interval of $[-.29, .01]$.

The vce(cluster *clustvar*) option can relax the independence assumption required by the complementary log-log estimator to being just independence between clusters. To demonstrate this ability, we will switch to a different dataset.

We are studying unionization of women in the United States by using the union dataset; see [XT] **xt**. We fit the following model, ignoring that women are observed an average of 5.9 times each in this dataset:

```
. clear
. set memory 3m
. use http://www.stata-press.com/data/r10/union
(NLS Women 14-24 in 1968)
. cloglog union age grade not_smsa south southXt
Iteration 0:   log likelihood = -13609.987
Iteration 1:   log likelihood = -13544.321
Iteration 2:   log likelihood = -13544.202
Iteration 3:   log likelihood = -13544.202
```

```
Complementary log-log regression            Number of obs    =      26200
                                            Zero outcomes    =      20389
                                            Nonzero outcomes =       5811

                                            LR chi2(5)       =     640.06
Log likelihood = -13544.202                 Prob > chi2      =     0.0000
```

union	Coef.	Std. Err.	z	P>|z|	[95% Conf. Interval]	
age	.0085897	.0023034	3.73	0.000	.0040752	.0131043
grade	.0447187	.0057069	7.84	0.000	.0335334	.0559039
not_smsa	-.1906552	.0317663	-6.00	0.000	-.252916	-.1283943
south	-.6446647	.0557644	-11.56	0.000	-.7539609	-.5353684
southXt	.0068271	.0047297	1.44	0.149	-.0024428	.0160971
_cons	-1.966755	.0991779	-19.83	0.000	-2.16114	-1.77237

The reported standard errors in this model are probably meaningless. Women are observed repeatedly, and so the observations are not independent. Looking at the coefficients, we find a large southern effect against unionization and little time trend. The vce(cluster *clustvar*) option provides a way to fit this model and obtains correct standard errors:

```
. cloglog union age grade not_smsa south southXt, vce(cluster id) nolog
Complementary log-log regression                Number of obs    =      26200
                                                Zero outcomes    =      20389
                                                Nonzero outcomes =       5811
                                                Wald chi2(5)     =     160.78
Log pseudolikelihood = -13544.202               Prob > chi2      =     0.0000
                            (Std. Err. adjusted for 4434 clusters in idcode)
```

| | | Robust | | | | |
union	Coef.	Std. Err.	z	P>|z|	[95% Conf.	Interval]
age	.0085897	.0033831	2.54	0.011	.0019591	.0152204
grade	.0447187	.0125927	3.55	0.000	.0200375	.0693999
not_smsa	-.1906552	.064189	-2.97	0.003	-.3164632	-.0648471
south	-.6446647	.0833807	-7.73	0.000	-.8080878	-.4812415
southXt	.0068271	.0063038	1.08	0.279	-.0055282	.0191824
_cons	-1.966755	.186019	-10.57	0.000	-2.331346	-1.602165

These standard errors are roughly 50% larger than those reported by the inappropriate conventional calculation. By comparison, another way we could fit this model is with an equal-correlation population-averaged complementary log-log model:

```
. xtcloglog union age grade not_smsa south southXt, pa nolog
GEE population-averaged model                   Number of obs    =      26200
Group variable:                       idcode    Number of groups =       4434
Link:                                 cloglog   Obs per group: min =         1
Family:                               binomial                  avg =       5.9
Correlation:                      exchangeable                  max =        12
                                                Wald chi2(5)     =     232.44
Scale parameter:                           1    Prob > chi2      =     0.0000
```

union	Coef.	Std. Err.	z	P>|z|	[95% Conf.	Interval]
age	.0045777	.0021754	2.10	0.035	.0003139	.0088415
grade	.0544267	.0095097	5.72	0.000	.035788	.0730654
not_smsa	-.1051731	.0430512	-2.44	0.015	-.189552	-.0207943
south	-.6578891	.061857	-10.64	0.000	-.7791266	-.5366515
southXt	.0142329	.004133	3.44	0.001	.0061325	.0223334
_cons	-2.074687	.1358008	-15.28	0.000	-2.340851	-1.808522

The coefficient estimates are similar, but these standard errors are smaller than those produced by cloglog, vce(cluster *clustvar*). This finding is as we would expect. If the within-panel correlation assumptions are valid, the population-averaged estimator should be more efficient.

In addition to this estimator, we may use the xtgee command to fit a panel estimator (with complementary log-log link) and any number of assumptions on the within-idcode correlation.

cloglog, vce(cluster *clustvar*) is robust to assumptions about within-cluster correlation. That is, it inefficiently sums within cluster for the standard error calculation rather than attempting to exploit what might be assumed about the within-cluster correlation (as do the xtgee population-averaged models).

Saved Results

cloglog saves the following in e():

Scalars

e(N)	number of observations	e(ll)	log likelihood
e(k)	number of parameters	e(ll_0)	log likelihood, constant-only model
e(k_eq)	number of equations in e(b)	e(N_clust)	number of clusters
e(k_eq_model)	number of equations in model	e(chi2)	χ^2
	Wald test	e(p)	significance
e(k_dv)	number of dependent variables	e(rank)	rank of e(V)
e(N_f)	number of zero outcomes	e(ic)	number of iterations
e(N_s)	number of nonzero outcomes	e(rc)	return code
e(df_m)	model degrees of freedom	e(converged)	1 if converged, 0 otherwise

Macros

e(cmd)	cloglog	e(vce)	*vcetype* specified in vce()
e(cmdline)	command as typed	e(vcetype)	title used to label Std. Err.
e(depvar)	name of dependent variable	e(opt)	type of optimization
e(wtype)	weight type	e(ml_method)	type of ml method
e(wexp)	weight expression	e(user)	name of likelihood-evaluator program
e(title)	title in estimation output	e(technique)	maximization technique
e(clustvar)	name of cluster variable	e(crittype)	optimization criterion
e(offset)	offset	e(properties)	b V
e(chi2type)	Wald or LR; type of model χ^2 test	e(predict)	program used to implement predict

Matrices

e(b)	coefficient vector	e(V)	variance–covariance matrix of
e(ilog)	iteration log (up to 20 iterations)		the estimators
e(gradient)	gradient vector		

Functions

e(sample)	marks estimation sample

Methods and Formulas

cloglog is implemented as an ado-file.

Complementary log-log analysis (related to the gompit model, so named because of its relationship to the Gompertz distribution) is an alternative to logit and probit analysis, but it is unlike these other estimators in that the transformation is not symmetric. Typically this model is used when the positive (or negative) outcome is rare.

The log-likelihood function for complementary log-log is

$$\ln L = \sum_{j \in S} w_j \ln F(\mathbf{x}_j \mathbf{b}) + \sum_{j \notin S} w_j \ln \left\{ 1 - F(\mathbf{x}_j \mathbf{b}) \right\}$$

where S is the set of all observations j such that $y_j \neq 0$, $F(z) = 1 - \exp\left\{-\exp(z)\right\}$, and w_j denotes the optional weights. $\ln L$ is maximized as described in [R] **maximize**.

We can fit a gompit model by reversing the success–failure sense of the dependent variable and using cloglog.

If robust standard errors are requested, the calculation described in *Methods and Formulas* of [R] **regress** is carried forward with $\mathbf{u}_j = [\exp(\mathbf{x}_j\mathbf{b})\exp\left\{-\exp(\mathbf{x}_j\mathbf{b})\right\}/F(\mathbf{x}_j\mathbf{b})]\mathbf{x}_j$ for the positive outcomes and $\left\{-\exp(\mathbf{x}_j\mathbf{b})\right\}\mathbf{x}_j$ for the negative outcomes.

Acknowledgment

We thank Joseph Hilbe of Arizona State University for providing the inspiration for the `cloglog` command (Hilbe 1996, 1998).

References

Clayton, D., and M. Hills. 1993. *Statistical Models in Epidemiology*. Oxford: Oxford University Press.

Hilbe, J. 1996. sg53: Maximum-likelihood complementary log-log regression. *Stata Technical Bulletin* 32: 19–20. Reprinted in *Stata Technical Bulletin Reprints*, vol. 6, pp. 129–131.

——. 1998. sg53.2: Stata-like commands for complementary log-log regression. *Stata Technical Bulletin* 41: 23. Reprinted in *Stata Technical Bulletin Reprints*, vol. 7, pp. 166–167.

Long, J. S. 1997. *Regression Models for Categorical and Limited Dependent Variables*. Thousand Oaks, CA: Sage.

Long, J. S., and J. Freese. 2006. *Regression Models for Categorical Dependent Variables Using Stata*. 2nd ed. College Station, TX: Stata Press.

Xu, J., and J. S. Long. 2005. Confidence intervals for predicted outcomes in regression models for categorical outcomes. *Stata Journal* 5: 537–559.

Also See

[R] **cloglog postestimation** — Postestimation tools for cloglog

[R] **clogit** — Conditional (fixed-effects) logistic regression

[R] **glm** — Generalized linear models

[R] **logistic** — Logistic regression, reporting odds ratios

[R] **scobit** — Skewed logistic regression

[SVY] **svy estimation** — Estimation commands for survey data

[XT] **xtcloglog** — Random-effects and population-averaged cloglog models

[U] **20 Estimation and postestimation commands**

Title

> **cloglog postestimation** — Postestimation tools for cloglog

Description

The following postestimation commands are available for `cloglog`:

command	description
adjust[1]	adjusted predictions of $\mathbf{x}\beta$ or $\exp(\mathbf{x}\beta)$
estat	AIC, BIC, VCE, and estimation sample summary
estat (svy)	postestimation statistics for survey data
estimates	cataloging estimation results
lincom	point estimates, standard errors, testing, and inference for linear combinations of coefficients
linktest	link test for model specification
lrtest[2]	likelihood-ratio test
mfx	marginal effects or elasticities
nlcom	point estimates, standard errors, testing, and inference for nonlinear combinations of coefficients
predict	predictions, residuals, influence statistics, and other diagnostic measures
predictnl	point estimates, standard errors, testing, and inference for generalized predictions
suest	seemingly unrelated estimation
test	Wald tests for simple and composite linear hypotheses
testnl	Wald tests of nonlinear hypotheses

[1] `adjust` is not appropriate with time-series operators.

[2] `lrtest` is not appropriate with svy estimation results.

See the corresponding entries in the *Stata Base Reference Manual* for details, but see [SVY] **estat** for details about `estat` (svy).

Syntax for predict

> predict [*type*] *newvar* [*if*] [*in*] [, *statistic* <u>nooff</u>set]

statistic	description
Main	
<u>pr</u>	probability of a positive outcome; the default
xb	linear prediction
stdp	standard error of the linear prediction
<u>sco</u>re	first derivative of the log likelihood with respect to $\mathbf{x}_j\beta$

These statistics are available both in and out of sample; type `predict ... if e(sample) ...` if wanted only for the estimation sample.

Options for predict

___ Main ___

pr, the default, calculates the probability of a positive outcome.

xb calculates the linear prediction.

stdp calculates the standard error of the linear prediction.

score calculates the equation-level score, $\partial \ln L / \partial(\mathbf{x}_j \boldsymbol{\beta})$.

nooffset is relevant only if you specified offset(*varname*) for cloglog. It modifies the calculations made by predict so that they ignore the offset variable; the linear prediction is treated as $\mathbf{x}_j \mathbf{b}$ rather than as $\mathbf{x}_j \mathbf{b} + \text{offset}_j$.

Remarks

Once you have fitted a model, you can obtain the predicted probabilities by using the predict command for both the estimation sample and other samples; see [U] **20 Estimation and postestimation commands** and [R] **predict**. Here we will make only a few comments.

predict without arguments calculates the predicted probability of a positive outcome. With the xb option, it calculates the linear combination $\mathbf{x}_j \mathbf{b}$, where $\mathbf{x}_j$ are the independent variables in the jth observation and $\mathbf{b}$ is the estimated parameter vector.

With the stdp option, predict calculates the standard error of the linear prediction, which is not adjusted for replicated covariate patterns in the data.

▷ Example 1

In example 1 in [R] **cloglog**, we fitted the complementary log-log model cloglog foreign weight mpg. To obtain predicted probabilities,

```
. use http://www.stata-press.com/data/r10/auto
(1978 Automobile Data)
. cloglog foreign weight mpg
 (output omitted )
. predict p
(option pr assumed; Pr(foreign))
. summarize foreign p
```

Variable	Obs	Mean	Std. Dev.	Min	Max
foreign	74	.2972973	.4601885	0	1
p	74	.2928348	.29732	.0032726	.9446067

◁

Methods and Formulas

All postestimation commands listed above are implemented as ado-files.

Also See

[R] **cloglog** — Complementary log-log regression

[U] **20 Estimation and postestimation commands**

Title

cnreg — Censored-normal regression

Syntax

<u>cn</u>reg *depvar* [*indepvars*] [*if*] [*in*] [*weight*], <u>cen</u>sored(*varname*) [*options*]

options	description
Model	
* <u>cen</u>sored(*varname*)	variable indicating whether *depvar* is not censored (0), left censored (-1), or right censored (1)
<u>off</u>set(*varname*)	include *varname* in model with coefficient constrained to 1
SE/Robust	
vce(*vcetype*)	*vcetype* may be oim, <u>r</u>obust, <u>cl</u>uster *clustvar*, <u>boot</u>strap, or <u>jack</u>knife
Reporting	
<u>level</u>(#)	set confidence level; default is level(95)
Max options	
maximize_options	control the maximization process; seldom used

* censored(*varname*) is required.

bootstrap, by, jackknife, nestreg, rolling, statsby, stepwise, svy, and xi are allowed;
  see [U] **11.1.10 Prefix commands**.

Weights are not allowed with the bootstrap prefix.

aweights are not allowed with the jackknife prefix.

vce() and weights are not allowed with the svy prefix.

aweights, fweights, pweights, and iweights are allowed; see [U] **11.1.6 weight**.

See [U] **20 Estimation and postestimation commands** for more capabilities of estimation commands.

Description

cnreg fits a model of *depvar* on *indepvars*, where *depvar* contains both observations and censored observations on the process. Censoring values may vary from observation to observation.

Options

┌── Model └──

censored(*varname*) is required. *varname* is a variable indicating if *depvar* is censored and, if so, whether the censoring is left or right. 0 indicates that *depvar* is not censored. -1 indicates left censoring; the true value is known only to be less than or equal to the value recorded in *depvar*. $+1$ indicates right censoring; the true value is known only to be greater than or equal to the value recorded in *depvar*.

offset(*varname*); see [R] **estimation options**.

302

⌐ SE/Robust ⌐

vce(*vcetype*) specifies the type of standard error reported, which includes types that are derived from asymptotic theory, that are robust to some kinds of misspecification, that allow for intragroup correlation, and that use bootstrap or jackknife methods; see [R] *vce_option*.

⌐ Reporting ⌐

level(*#*); see [R] **estimation options**.

⌐ Max options ⌐

maximize_options: <u>iter</u>ate(*#*), [<u>no</u>]log, <u>tr</u>ace, <u>tol</u>erance(*#*), <u>ltol</u>erance(*#*); see [R] **maximize**. These options are seldom used.

Unlike most maximum likelihood commands, cnreg defaults to nolog—it suppresses the iteration log. log will display the iteration log.

Remarks

cnreg is a generalization of tobit; see [R] **tobit**. Rather than there being constant lower- and upper-censoring points, each observation can be censored at a different point. With cnreg, you specify censored(*varname*) rather than the ll(*#*) and/or ul(*#*) options. *varname* records a 0 if the observation is not censored, a −1 if it is left censored, and a +1 if it is right censored.

▷ Example 1

We are fitting an earnings model. The data are censored from above in that, if the respondent's income was $60,000 or more, the interviewer recorded simply "in excess of $60,000". We might fit the model by using tobit, specifying ul(60000) (or, if we are working in natural logs, ul(11)). Later, we get another dataset to merge with our current data. In the second survey, earnings were again censored, but this time at $100,000. We can no longer use tobit since some observations are censored at $60,000 and others at $100,000. We could, however, use

```
. generate cens=0
. replace cens=1 if survey==1 & earnings==60000
. replace cens=1 if survey==2 & earnings==100000
. cnreg earnings etc., censored(cens)
```
◁

▷ Example 2

Censored-normal regression has been used to fit duration-time models, but such use has now declined in favor of models with more reasonable assumptions (e.g., exponential or Weibull; see [ST] **streg**) or no assumption except proportionality among groups (see [ST] **stcox**). There are, however, occasions when a normally distributed time variable may be reasonable.

We have (fictional) data on the date various newspapers adopted VDT (video display terminal) editing, stored in date as days since 1/1/60. We also know the newspaper's circulation (lncltn) and whether the newspaper is a subsidiary or family owned (famown). Our data begin in 1982 and end in 1991. We have left censoring if the paper adopted VDT editing before 1982. We have right censoring if the paper had not adopted VDT editing by 1991.

In our data, `date` is recorded as missing if the paper is still not using VDTs or if it began doing so before 1982. The variable `before82` is 1 if the paper started using VDTs before 1982:

```
. use http://www.stata-press.com/data/r10/news

. generate cnsrd=0                                    (nobody censored yet)

. replace cnsrd=-1 if before82
(24 real changes made)

. replace date=mdy(1,1,1982) if before82
(24 real changes made)

. replace cnsrd=1 if date>=.
(11 real changes made)

. replace date=mdy(1,1,1991) if date>=.
(11 real changes made)

. cnreg date lncltn famown, censored(cnsrd)
```

```
Censored-normal regression                    Number of obs   =        100
                                              LR chi2(2)      =     201.09
                                              Prob > chi2     =     0.0000
Log likelihood = -519.74678                   Pseudo R2       =     0.1621
```

date	Coef.	Std. Err.	t	P>|t|	[95% Conf. Interval]	
lncltn	-1377.138	76.92723	-17.90	0.000	-1529.797	-1224.478
famown	576.8444	185.5287	3.11	0.002	208.6686	945.0201
_cons	24439.46	815.7107	29.96	0.000	22820.71	26058.21
/sigma	607.9846	53.19381			502.4232	713.546

```
Observation summary:        24   left-censored observations
                            65       uncensored observations
                            11 right-censored observations
```

We estimate that family-owned newspapers adopted VDT editing 577 days later than non–family-owned newspapers (but remember, the data are fictional).

◁

❑ Technical Note

Censored-normal regression and tobit models achieve their results by assuming that the distribution of the error term is normal. This is a rather standard assumption, but for these two models, it is vital. Results from censored-normal regression and tobit are not robust to other distributions of the errors; see Goldberger (1983). Moreover, heteroskedastic error terms can wreak havoc on such models; see Hurd (1979).

◁

Saved Results

cnreg saves the following in e():

Scalars

e(N)	number of observations	e(df_r)	residual degrees of freedom
e(N_unc)	number of uncensored observations	e(r2_p)	pseudo-R-squared
e(N_lc)	number of left-censored observations	e(chi2)	χ^2 statistic
e(N_rc)	number of right-censored observations	e(ll)	log likelihood
e(llopt)	contents of ll(), if specified	e(ll_0)	log likelihood, constant-only model
e(ulopt)	contents of ul(), if specified	e(N_clust)	number of clusters
e(k_aux)	number of auxiliary parameters	e(converged)	1 if converged, 0 otherwise
e(df_m)	model degrees of freedom		

Macros

e(cmd)	cnreg	e(chi2type)	Wald or LR; type of model χ^2 test
e(cmdline)	command as typed	e(vce)	*vcetype* specified in vce()
e(depvar)	name of dependent variable	e(vcetype)	title used to label Std. Err.
e(censored)	variable specified in censored()	e(crittype)	optimization criterion
e(wtype)	weight type	e(properties)	b V
e(wexp)	weight expression	e(footnote)	program used to implement the footnote display
e(title)	title in estimation output		
e(clustvar)	name of cluster variable	e(predict)	program used to implement predict
e(offset)	offset		

Matrices

e(b)	coefficient vector	e(V)	variance–covariance matrix of the estimators

Functions

e(sample)	marks estimation sample

Methods and Formulas

cnreg is implemented as an ado-file.

See the *Methods and Formulas* section of [R] **intreg**.

References

Amemiya, T. 1973. Regression analysis when the dependent variable is truncated normal. *Econometrica* 41: 997–1016.

Goldberger, A. S. 1983. Abnormal selection bias. In *Studies in Econometrics, Time Series, and Multivariate Statistics*, ed. S. Karlin, T. Amemiya, and L. A. Goodman, 67–84. New York: Academic Press.

Hurd, M. 1979. Estimation in truncated samples when there is heteroscedasticity. *Journal of Econometrics* 11: 247–258.

Johnston, J., and J. DiNardo. 1997. *Econometric Methods*. 4th ed. New York: McGraw–Hill.

Kendall, M. G., and A. Stuart. 1973. *The Advanced Theory of Statistics: Inference and Relationship, Vol. 2*. New York: Hafner.

Kmenta, J. 1997. *Elements of Econometrics*. 2nd ed. Ann Arbor: University of Michigan Press.

Long, J. S. 1997. *Regression Models for Categorical and Limited Dependent Variables*. Thousand Oaks, CA: Sage.

Maddala, G. S., and K. Lahiri. 2006. *Introduction to Econometrics*. 4th ed. New York: Wiley.

Stewart, M. B. 1983. On least-squares estimation when the dependent variable is grouped. *Review of Economic Studies* 50: 737–753.

Tobin, J. 1958. Estimation of relationships for limited dependent variables. *Econometrica* 26: 24–36.

Also See

[R] **cnreg postestimation** — Postestimation tools for cnreg

[R] **intreg** — Interval regression

[R] **tobit** — Tobit regression

[R] **regress** — Linear regression

[SVY] **svy estimation** — Estimation commands for survey data

[XT] **xtintreg** — Random-effects interval-data regression models

[XT] **xttobit** — Random-effects tobit models

[U] **20 Estimation and postestimation commands**

Title

> **cnreg postestimation** — Postestimation tools for cnreg

Description

The following postestimation commands are available for `cnreg`:

command	description
adjust	adjusted predictions of $\mathbf{x}\boldsymbol{\beta}$
estat	AIC, BIC, VCE, and estimation sample summary
estat (svy)	postestimation statistics for survey data
estimates	cataloging estimation results
lincom	point estimates, standard errors, testing, and inference for linear combinations of coefficients
linktest	link test for model specification
lrtest[1]	likelihood-ratio test
mfx	marginal effects or elasticities
nlcom	point estimates, standard errors, testing, and inference for nonlinear combinations of coefficients
predict	predictions, residuals, influence statistics, and other diagnostic measures
predictnl	point estimates, standard errors, testing, and inference for generalized predictions
test	Wald tests for simple and composite linear hypotheses
testnl	Wald tests of nonlinear hypotheses

[1] `lrtest` is not appropriate with `svy` estimation results.

See the corresponding entries in the *Stata Base Reference Manual* for details, but see [SVY] **estat** for details about estat (svy).

Syntax for predict

> predict [*type*] *newvar* [*if*] [*in*] [, *statistic* <u>nooff</u>set]

> predict [*type*] { *stub** | *newvar*$_{reg}$ *newvar*$_{sigma}$ } [*if*] [*in*], <u>sc</u>ores

statistic	description
Main	
xb	linear prediction; the default
stdp	standard error of the prediction
stdf	standard error of the forecast
<u>pr</u>(*a*,*b*)	$\Pr(a < y_j < b)$
e(*a*,*b*)	$E(y_j \mid a < y_j < b)$
<u>ystar</u>(*a*,*b*)	$E(y_j^*)$, $y_j^* = \max\{a, \min(y_j, b)\}$

These statistics are available both in and out of sample; type `predict ... if e(sample) ...` if wanted only for the estimation sample.

`stdf` is not allowed with `svy` estimation results.

where a and b may be numbers or variables; a missing ($a \geq .$) means $-\infty$, and b missing ($b \geq .$) means $+\infty$; see [U] **12.2.1 Missing values**.

Options for predict

⌐ Main ⌐

xb, the default, calculates the linear prediction.

stdp calculates the standard error of the prediction, which can be thought of as the standard error of the predicted expected value or mean for the observation's covariate pattern. The standard error of the prediction is also referred to as the standard error of the fitted value.

stdf calculates the standard error of the forecast, which is the standard error of the point prediction for 1 observation. It is commonly referred to as the standard error of the future or forecast value. By construction, the standard errors produced by stdf are always larger than those produced by stdp; see [R] **regress** *Methods and Formulas*.

pr(a,b) calculates $\Pr(a < \mathbf{x}_j\mathbf{b} + u_j < b)$, the probability that $y_j|\mathbf{x}_j$ would be observed in the interval (a, b).

a and b may be specified as numbers or variable names; *lb* and *ub* are variable names;
pr(20,30) calculates $\Pr(20 < \mathbf{x}_j\mathbf{b} + u_j < 30)$;
pr(*lb,ub*) calculates $\Pr(lb < \mathbf{x}_j\mathbf{b} + u_j < ub)$; and
pr(20,*ub*) calculates $\Pr(20 < \mathbf{x}_j\mathbf{b} + u_j < ub)$.

a missing ($a \geq .$) means $-\infty$; pr(.,30) calculates $\Pr(-\infty < \mathbf{x}_j\mathbf{b} + u_j < 30)$;
pr(*lb*,30) calculates $\Pr(-\infty < \mathbf{x}_j\mathbf{b} + u_j < 30)$ in observations for which $lb \geq .$
and calculates $\Pr(lb < \mathbf{x}_j\mathbf{b} + u_j < 30)$ elsewhere.

b missing ($b \geq .$) means $+\infty$; pr(20,.) calculates $\Pr(+\infty > \mathbf{x}_j\mathbf{b} + u_j > 20)$;
pr(20,*ub*) calculates $\Pr(+\infty > \mathbf{x}_j\mathbf{b} + u_j > 20)$ in observations for which $ub \geq .$
and calculates $\Pr(20 < \mathbf{x}_j\mathbf{b} + u_j < ub)$ elsewhere.

e(a,b) calculates $E(\mathbf{x}_j\mathbf{b} + u_j \mid a < \mathbf{x}_j\mathbf{b} + u_j < b)$, the expected value of $y_j|\mathbf{x}_j$ conditional on $y_j|\mathbf{x}_j$ being in the interval (a, b), meaning that $y_j|\mathbf{x}_j$ is censored. a and b are specified as they are for pr().

ystar(a,b) calculates $E(y_j^*)$, where $y_j^* = a$ if $\mathbf{x}_j\mathbf{b} + u_j \leq a$, $y_j^* = b$ if $\mathbf{x}_j\mathbf{b} + u_j \geq b$, and $y_j^* = \mathbf{x}_j\mathbf{b} + u_j$ otherwise, meaning that y_j^* is truncated. a and b are specified as they are for pr().

nooffset is relevant only if you specified offset(*varname*). It modifies the calculations made by predict so that they ignore the offset variable; the linear prediction is treated as $\mathbf{x}_j\mathbf{b}$ rather than as $\mathbf{x}_j\mathbf{b} + \text{offset}_j$.

scores calculates equation-level score variables.

The first new variable will contain $\partial \ln L/\partial(\mathbf{x}_j\boldsymbol{\beta})$.

The second new variable will contain $\partial \ln L/\partial\sigma$.

Methods and Formulas

All postestimation commands listed above are implemented as ado-files.

Also See

[R] **cnreg** — Censored-normal regression

[U] **20 Estimation and postestimation commands**

Title

cnsreg — Constrained linear regression

Syntax

cnsreg *depvar indepvars* [*if*] [*in*] [*weight*], c̲onstraints(*constraints*) [*options*]

options	description
Model	
* c̲onstraints(*constraints*)	apply specified linear constraints
noc̲onstant	suppress constant term
SE/Robust	
vce(*vcetype*)	*vcetype* may be ols, r̲obust, c̲luster *clustvar*, boot̲strap, or jackknife
Reporting	
l̲evel(*#*)	set confidence level; default is level(95)
† mse1	force MSE to be 1

* c̲onstraints(*constraints*) is required.

depvar and *indepvars* may contain time-series operators; see [U] **11.4.3 Time-series varlists**.

bootstrap, by, jackknife, rolling, statsby, svy, and xi are allowed; see [U] **11.1.10 Prefix commands**.

Weights are not allowed with the bootstrap prefix.

aweights are not allowed with the jackknife prefix.

vce(), mse1, and weights are not allowed with the svy prefix.

† mse1 does not appear in the dialog.

aweights, fweights, pweights, and iweights are allowed; see [U] **11.1.6 weight**.

See [U] **20 Estimation and postestimation commands** for more capabilities of estimation commands.

Description

cnsreg fits constrained linear regression models. cnsreg typed without arguments redisplays the previous cnsreg results.

Options

 ⌐ Model ⌐

constraints(*constraints*), noconstant; see [R] **estimation options**.

 ⌐ SE/Robust ⌐

vce(*vcetype*) specifies the type of standard error reported, which includes types that are derived from asymptotic theory, that are robust to some kinds of misspecification, that allow for intragroup correlation, and that use bootstrap or jackknife methods; see [R] *vce_option*.

vce(ols), the default, uses the standard variance estimator for ordinary least-squares regression.

level(#); see [R] **estimation options**.

The following option is available with cnsreg but is not shown in the dialog box:

mse1 is used only in programs and ado-files that use cnsreg to fit models other than constrained linear regression. mse1 sets the mean squared error to 1, thus forcing the variance–covariance matrix of the estimators to be $(\mathbf{X'DX})^{-1}$ (see *Methods and Formulas* in [R] **matsize**) and affecting calculated standard errors. Degrees of freedom for t statistics are calculated as n rather than $n - p + c$, where p is the total number of parameters (prior to restrictions and including the constant) and c is the number of constraints.

mse1 is not allowed with the svy prefix.

Remarks

For a discussion of constrained linear regression, see Greene (2003, 100) or Johnston and DiNardo (1997, 97).

▷ Example 1

In principle, we can obtain constrained linear regression estimates by modifying the list of independent variables. For instance, if we wanted to fit the model

$$\texttt{mpg} = \beta_0 + \beta_1\,\texttt{price} + \beta_2\,\texttt{weight} + u$$

and constrain $\beta_1 = \beta_2$, we could write

$$\texttt{mpg} = \beta_0 + \beta_1(\texttt{price} + \texttt{weight}) + u$$

and run a regression of mpg on price + weight. The estimated coefficient on the sum would be the constrained estimate of β_1 and β_2. Using cnsreg, however, is easier:

```
. use http://www.stata-press.com/data/r10/auto
(1978 Automobile Data)
. constraint 1 price = weight
. cnsreg mpg price weight, constraint(1)
```

Constrained linear regression	Number of obs =	74
	F(1, 72) =	37.59
	Prob > F =	0.0000
	Root MSE =	4.722

(1) price - weight = 0

mpg	Coef.	Std. Err.	t	P>|t|	[95% Conf. Interval]
price	-.0009875	.0001611	-6.13	0.000	-.0013086	-.0006664
weight	-.0009875	.0001611	-6.13	0.000	-.0013086	-.0006664
_cons	30.36718	1.577958	19.24	0.000	27.22158	33.51278

We define constraints by using the constraint command; see [R] **constraint**. We fit the model with cnsreg and specify the constraint number or numbers in the constraints() option.

Just to show that the results above are correct, here is the result of applying the constraint by hand:

```
. generate x = price + weight
. regress mpg x
```

Source	SS	df	MS		
Model	838.065767	1	838.065767		
Residual	1605.39369	72	22.2971346		
Total	2443.45946	73	33.4720474		

```
                                                Number of obs =      74
                                                F(  1,     72) =   37.59
                                                Prob > F       =  0.0000
                                                R-squared      =  0.3430
                                                Adj R-squared  =  0.3339
                                                Root MSE       =   4.722
```

mpg	Coef.	Std. Err.	t	P>|t|	[95% Conf. Interval]
x	-.0009875	.0001611	-6.13	0.000	-.0013086 -.0006664
_cons	30.36718	1.577958	19.24	0.000	27.22158 33.51278

◁

▷ Example 2

Models can be fitted subject to multiple simultaneous constraints. We simply define the constraints and then include the constraint numbers in the constraints() option. For instance, say that we wish to fit the model

$$\text{mpg} = \beta_0 + \beta_1 \, \text{price} + \beta_2 \, \text{weight} + \beta_3 \, \text{displ} + \beta_4 \, \text{gear\_ratio} + \beta_5 \, \text{foreign} +$$
$$\beta_6 \, \text{length} + u$$

subject to the constraints

$$\beta_1 = \beta_2 = \beta_3 = \beta_6$$
$$\beta_4 = -\beta_5 = \beta_0/20$$

(This model, like the one in example 1, is admittedly senseless.) We fit the model by typing

```
. constraint 1 price=weight
. constraint 2 displ=weight
. constraint 3 length=weight
. constraint 5 gear_ratio = -foreign
. constraint 6 gear_ratio = _cons/20
```

```
. cnsreg mpg price weight displ gear_ratio foreign length, c(1-3,5-6)
```

Constrained linear regression

```
Number of obs =      74
F(  2,    72) =  785.20
Prob > F      =  0.0000
Root MSE      =  4.6823
```

```
( 1)  price - weight = 0
( 2)  - weight + displacement = 0
( 3)  - weight + length = 0
( 4)  gear_ratio + foreign = 0
( 5)  gear_ratio - .05 _cons = 0
```

mpg	Coef.	Std. Err.	t	P>|t|	[95% Conf. Interval]	
price	-.000923	.0001534	-6.02	0.000	-.0012288	-.0006172
weight	-.000923	.0001534	-6.02	0.000	-.0012288	-.0006172
displacement	-.000923	.0001534	-6.02	0.000	-.0012288	-.0006172
gear_ratio	1.326114	.0687589	19.29	0.000	1.189046	1.463183
foreign	-1.326114	.0687589	-19.29	0.000	-1.463183	-1.189046
length	-.000923	.0001534	-6.02	0.000	-.0012288	-.0006172
_cons	26.52229	1.375178	19.29	0.000	23.78092	29.26365

There are many ways we could have specified the constraints() option (which we abbreviated c() above). We typed c(1-3,5-6), meaning that we want constraints 1 through 3 and 5 and 6; those numbers correspond to the constraints we defined. The only reason we did not use the number 4 was to emphasize that constraints do not have to be consecutively numbered. We typed c(1-3,5-6), but we could have typed c(1,2,3,5,6) or c(1-3,5,6) or c(1-2,3,5,6) or even c(1-6), which would have worked as long as constraint 4 was not defined. If we had previously defined a constraint 4, c(1-6) would have included it.

◁

Saved Results

cnsreg saves the following in e():

Scalars

e(N)	number of observations	e(rmse)	root mean squared error
e(df_m)	model degrees of freedom	e(ll)	log likelihood
e(df_r)	residual degrees of freedom	e(N_clust)	number of clusters
e(F)	F statistic		

Macros

e(cmd)	cnsreg	e(clustvar)	name of cluster variable
e(cmdline)	command as typed	e(vce)	*vcetype* specified in vce()
e(depvar)	name of dependent variable	e(vcetype)	title used to label Std. Err.
e(wtype)	weight type	e(properties)	b V
e(wexp)	weight expression	e(predict)	program used to implement predict
e(title)	title in estimation output		

Matrices

e(b)	coefficient vector	e(V)	variance–covariance matrix of the estimators

Functions

e(sample)	marks estimation sample

Methods and Formulas

cnsreg is implemented as an ado-file.

Let n be the number of observations, p be the total number of parameters (prior to restrictions and including the constant), and c be the number of constraints. The coefficients are calculated as $\mathbf{b}' = \mathbf{T}\{(\mathbf{T}'\mathbf{X}'\mathbf{W}\mathbf{X}\mathbf{T})^{-1}(\mathbf{T}'\mathbf{X}'\mathbf{W}\mathbf{y} - \mathbf{T}'\mathbf{X}'\mathbf{W}\mathbf{X}\mathbf{a}')\} + \mathbf{a}'$, where $\mathbf{T}$ and $\mathbf{a}$ are as defined in [P] **makecns**. $\mathbf{W} = \mathbf{I}$ if no weights are specified. If weights are specified, let $\mathbf{v}: 1 \times n$ be the specified weights. If fweight frequency weights are specified, $\mathbf{W} = \mathrm{diag}(\mathbf{v})$. If aweight analytic weights are specified, then $\mathbf{W} = \mathrm{diag}[\mathbf{v}/(\mathbf{1}'\mathbf{v})(\mathbf{1}'\mathbf{1})]$, meaning that the weights are normalized to sum to the number of observations.

The mean squared error is $s^2 = (\mathbf{y}'\mathbf{W}\mathbf{y} - 2\mathbf{b}'\mathbf{X}'\mathbf{W}\mathbf{y} + \mathbf{b}'\mathbf{X}'\mathbf{W}\mathbf{X}\mathbf{b})/(n - p + c)$. The variance–covariance matrix is $s^2\mathbf{T}(\mathbf{T}'\mathbf{X}'\mathbf{W}\mathbf{X}\mathbf{T})^{-1}\mathbf{T}'$.

References

Greene, W. H. 2003. *Econometric Analysis*. 5th ed. Upper Saddle River, NJ: Prentice Hall.

Johnston, J., and J. DiNardo. 1997. *Econometric Methods*. 4th ed. New York: McGraw–Hill.

Also See

[R] **cnsreg postestimation** — Postestimation tools for cnsreg

[R] **regress** — Linear regression

[SVY] **svy estimation** — Estimation commands for survey data

[U] **20 Estimation and postestimation commands**

Title

> **cnsreg postestimation** — Postestimation tools for cnsreg

Description

The following postestimation commands are available for cnsreg:

command	description
adjust[1]	adjusted predictions of $\mathbf{x}\beta$
estat	AIC, BIC, VCE, and estimation sample summary
estat (svy)	postestimation statistics for survey data
estimates	cataloging estimation results
lincom	point estimates, standard errors, testing, and inference for linear combinations of coefficients
linktest	link test for model specification
lrtest[2]	likelihood-ratio test
mfx	marginal effects or elasticities
nlcom	point estimates, standard errors, testing, and inference for nonlinear combinations of coefficients
predict	predictions, residuals, influence statistics, and other diagnostic measures
predictnl	point estimates, standard errors, testing, and inference for generalized predictions
suest	seemingly unrelated estimation
test	Wald tests for simple and composite linear hypotheses
testnl	Wald tests of nonlinear hypotheses

[1] adjust is not appropriate with time-series operators.

[2] lrtest is not appropriate with svy estimation results.

See the corresponding entries in the *Stata Base Reference Manual* for details, but see
[SVY] **estat** for details about estat (svy).

Syntax for predict

> predict [*type*] *newvar* [*if*] [*in*] [, *statistic*]

statistic	description
Main	
xb	linear prediction; the default
residuals	residuals
stdp	standard error of the prediction
stdf	standard error of the forecast
pr(a,b)	$\Pr(a < y_j < b)$
e(a,b)	$E(y_j \mid a < y_j < b)$
ystar(a,b)	$E(y_j^*)$, $y_j^* = \max\{a, \min(y_j, b)\}$
score	equivalent to residuals

These statistics are available both in and out of sample; type predict ... if e(sample) ... if wanted only for the estimation sample.

stdf is not allowed with svy estimation results.

where a and b may be numbers or variables; a missing ($a \geq .$) means $-\infty$, and b missing ($b \geq .$) means $+\infty$; see [U] **12.2.1 Missing values**.

Options for predict

    ⌐ Main ⌐

xb, the default, calculates the linear prediction.

residuals calculates the residuals, that is, $y_j - \mathbf{x}_j\mathbf{b}$.

stdp calculates the standard error of the prediction, which can be thought of as the standard error of the predicted expected value or mean for the observation's covariate pattern. The standard error of the prediction is also referred to as the standard error of the fitted value.

stdf calculates the standard error of the forecast, which is the standard error of the point prediction for 1 observation. It is commonly referred to as the standard error of the future or forecast value. By construction, the standard errors produced by stdf are always larger than those produced by stdp; see [R] **regress** *Methods and Formulas*.

pr(a,b) calculates $\Pr(a < \mathbf{x}_j\mathbf{b} + u_j < b)$, the probability that $y_j|\mathbf{x}_j$ would be observed in the interval (a, b).

> a and b may be specified as numbers or variable names; *lb* and *ub* are variable names;
> pr(20,30) calculates $\Pr(20 < \mathbf{x}_j\mathbf{b} + u_j < 30)$;
> pr(*lb,ub*) calculates $\Pr(lb < \mathbf{x}_j\mathbf{b} + u_j < ub)$; and
> pr(20,*ub*) calculates $\Pr(20 < \mathbf{x}_j\mathbf{b} + u_j < ub)$.

> a missing ($a \geq .$) means $-\infty$; pr(.,30) calculates $\Pr(-\infty < \mathbf{x}_j\mathbf{b} + u_j < 30)$;
> pr(*lb*,30) calculates $\Pr(-\infty < \mathbf{x}_j\mathbf{b} + u_j < 30)$ in observations for which $lb \geq .$
> and calculates $\Pr(lb < \mathbf{x}_j\mathbf{b} + u_j < 30)$ elsewhere.

> b missing ($b \geq .$) means $+\infty$; pr(20,.) calculates $\Pr(+\infty > \mathbf{x}_j\mathbf{b} + u_j > 20)$;
> pr(20,*ub*) calculates $\Pr(+\infty > \mathbf{x}_j\mathbf{b} + u_j > 20)$ in observations for which $ub \geq .$
> and calculates $\Pr(20 < \mathbf{x}_j\mathbf{b} + u_j < ub)$ elsewhere.

e(a,b) calculates $E(\mathbf{x}_j\mathbf{b} + u_j \mid a < \mathbf{x}_j\mathbf{b} + u_j < b)$, the expected value of $y_j|\mathbf{x}_j$ conditional on $y_j|\mathbf{x}_j$ being in the interval (a, b), meaning that $y_j|\mathbf{x}_j$ is censored. a and b are specified as they are for pr().

ystar(a,b) calculates $E(y_j^*)$, where $y_j^* = a$ if $\mathbf{x}_j\mathbf{b} + u_j \leq a$, $y_j^* = b$ if $\mathbf{x}_j\mathbf{b} + u_j \geq b$, and $y_j^* = \mathbf{x}_j\mathbf{b} + u_j$ otherwise, meaning that y_j^* is truncated. a and b are specified as they are for pr().

score is equivalent to residuals for linear regression models.

Methods and Formulas

All postestimation commands listed above are implemented as ado-files.

Also See

[R] **cnsreg** — Constrained linear regression

[U] **20 Estimation and postestimation commands**

Title

constraint — Define and list constraints

Syntax

Define constraints

constraint [define] # [exp=exp | coeflist]

List constraints

constraint dir [numlist | _all]

constraint list [numlist | _all]

Drop constraints

constraint drop { numlist | _all }

Programmer's commands

constraint get #

constraint free

where *coeflist* is as defined in [R] **test** and # is restricted to the range 1–1,999, inclusive.

Description

constraint defines, lists, and drops linear constraints. Constraints are for use by models that allow constrained estimation.

Constraints are defined by the constraint command. The currently defined constraints can be listed by either constraint list or constraint dir; both do the same thing. Existing constraints can be eliminated by constraint drop.

constraint get and constraint free are programmer's commands. constraint get returns the contents of the specified constraint in macro r(contents) and returns in scalar r(defined) 0 or 1—1 being returned if the constraint was defined. constraint free returns the number of a free (unused) constraint in macro r(free).

Remarks

Using constraints is discussed in [R] **cnsreg**, [R] **mlogit**, and [R] **reg3**; this entry is concerned only with practical aspects of defining and manipulating constraints.

▷ Example 1

Constraints are numbered from 1 to 1,999, and we assign the number when we define the constraint:

```
. use http://www.stata-press.com/data/r10/sysdsn
. constraint 2 [Indemnity]site2 = 0
```

The currently defined constraints can be listed by `constraint list`:

```
. constraint list
      2:  [Indemnity]site2 = 0
```

`constraint drop` drops constraints:

```
. constraint drop 2
. constraint list
```

The empty list after `constraint list` indicates that no constraints are defined. Below we demonstrate the various syntaxes allowed by `constraint`:

```
. constraint 1 [Indemnity]
. constraint 10 [Indemnity]: site1 site2
. constraint 11 [Indemnity]: site3
. constraint 20 [Prepaid=Uninsure]: site1 site2 site3
. constraint 21 [Prepaid=Uninsure]: nonwhite
. constraint 30 [Prepaid]
. constraint 31 [Insure]
. constraint list
      1:  [Indemnity]
     10:  [Indemnity]: site1 site2
     11:  [Indemnity]: site3
     20:  [Prepaid=Uninsure]: site1 site2 site3
     21:  [Prepaid=Uninsure]: nonwhite
     30:  [Prepaid]
     31:  [Insure]
. constraint drop 20-25, 31
. constraint list
      1:  [Indemnity]
     10:  [Indemnity]: site1 site2
     11:  [Indemnity]: site3
     30:  [Prepaid]
. constraint drop _all
. constraint list
```

◁

❑ Technical Note

The `constraint` command does not check the syntax of the constraint itself because a constraint can be interpreted only in the context of a model. Thus `constraint` is willing to define constraints that later will not make sense. Any errors in the constraints will be detected and mentioned at the time of estimation.

❑

Reference

Weesie, J. 1999. sg100: Two-stage linear constrained estimation. *Stata Technical Bulletin* 47: 24–30. Reprinted in *Stata Technical Bulletin Reprints*, vol. 8, pp. 217–225.

Also See

[R] **cnsreg** — Constrained linear regression

Title

> **copyright** — Display copyright information

Syntax

```
copyright
```

Description

`copyright` presents copyright notifications concerning tools, libraries, etc., used in the construction of Stata.

Remarks

The correct form for a copyright notice is

Copyright *dates* by *author/owner*

The word "Copyright" is spelled out. You can use the © symbol, but "(C)" has never been given legal recognition. The phrase "All Rights Reserved" was historically required but is no longer needed.

Currently, most works are copyrighted from the moment they are written, and no copyright notice is required. Copyright concerns the protection of the expression and structure of facts and ideas, not the facts and ideas themselves. Copyright concerns the ownership of the expression and not the name given to the expression, which is covered under trademark law.

Copyright law as it exists today began in England in 1710 with the Statute of Anne, *An Act for the Encouragement of Learning by Vesting the Copies of Printed Books in the Authors or Purchases of Such Copies, during the Times therein mentioned*. In 1672, Massachusetts introduced the first copyright law in what was to become the United States. After the Revolutionary War, copyright was introduced into the U.S. Constitution in 1787 and went into effect on May 31, 1790. On June 9, 1790, the first copyright in the United States was registered for *The Philadelphia Spelling Book* by John Barry.

There are significant differences in the understanding of copyright in the English- and non-English-speaking world. The Napoleonic or Civil Code, the dominant legal system in the non-English-speaking world, splits the rights into two classes: the author's economic rights and the author's moral rights. Moral rights are available only to "natural persons". Legal persons (corporations) have economic rights but not moral rights.

Also See

Copyright page of this book

Title

> **correlate** — Correlations (covariances) of variables or coefficients

Syntax

Display correlation matrix or covariance matrix

> <u>cor</u>relate [*varlist*] [*if*] [*in*] [*weight*] [, *correlate_options*]

Display all pairwise correlation coefficients

> pwcorr [*varlist*] [*if*] [*in*] [*weight*] [, *pwcorr_options*]

correlate_options	description
Options	
<u>_coef</u>	display correlation or covariance matrix of the coefficients
<u>means</u>	display means, standard deviations, minimums, and maximums with matrix
<u>nof</u>ormat	ignore display format associated with variables
<u>covariance</u>	display covariances
<u>wrap</u>	allow wide matrices to wrap

pwcorr_options	description
Main	
<u>o</u>bs	print number of observations for each entry
sig	print significance level for each entry
<u>print</u>(#)	significance level for displaying coefficients
<u>star</u>(#)	significance level for displaying with a star
<u>bonf</u>erroni	use Bonferroni-adjusted significance level
<u>sidak</u>	use Šidák-adjusted significance level

varlist may contain time-series operators; see [U] **11.4.3 Time-series varlists**.
by is allowed with correlate and pwcorr; see [D] **by**.
aweights and fweights are allowed; see [U] **11.1.6 weight**.

Description

The correlate command displays the correlation matrix or covariance matrix for a group of variables or for the coefficients of the most recent estimation. If *varlist* and _coef are not specified, the matrix is displayed for all variables in the data. Also see [R] **estat**.

pwcorr displays all the pairwise correlation coefficients between the variables in *varlist* or, if *varlist* is not specified, all the variables in the dataset.

Options for correlate

⌐ Options ⌐

_coef displays the correlations (or covariances if covariance is also specified) between the coefficients of the last estimation.

means displays summary statistics (means, standard deviations, minimums, and maximums) with the matrix.

noformat displays the summary statistics requested by the means option in g format, regardless of the display formats associated with the variables.

covariance displays the covariances rather than the correlation coefficients.

wrap requests that no action be taken on wide correlation matrices to make them readable. It prevents Stata from breaking wide matrices into pieces to enhance readability. You might want to specify this option if you are displaying results in a window wider than 80 characters. Then you may need to set linesize to however many characters you can display across a line; see [R] **log**.

Options for pwcorr

⌐ Main ⌐

obs adds a line to each row of the matrix reporting the number of observations used to calculate the correlation coefficient.

sig adds a line to each row of the matrix reporting the significance level of each correlation coefficient.

print(#) specifies the significance level of correlation coefficients to be printed. Correlation coefficients with larger significance levels are left blank in the matrix. Typing pwcorr, print(.10) would list only correlation coefficients significant at the 10% level or better.

star(#) specifies the significance level of correlation coefficients to be starred. Typing pwcorr, star(.05) would star all correlation coefficients significant at the 5% level or better.

bonferroni makes the Bonferroni adjustment to calculated significance levels. This option affects printed significance levels and the print() and star() options. Thus pwcorr, print(.05) bonferroni prints coefficients with Bonferroni-adjusted significance levels of .05 or less.

sidak makes the Šidák adjustment to calculated significance levels. This option affects printed significance levels and the print() and star() options. Thus pwcorr, print(.05) sidak prints coefficients with Šidák-adjusted significance levels of .05 or less.

Remarks

Remarks are presented under the following headings:

> *correlate*
> *pwcorr*

correlate

Typing correlate by itself produces a correlation matrix for all variables in the dataset. If you specify the *varlist*, a correlation matrix for just those variables is displayed.

▷ Example 1

We have state data on demographic characteristics of the population. To obtain a correlation matrix, we type correlate:

```
. use http://www.stata-press.com/data/r10/census3
(1980 Census data by state)

. correlate
(obs=50)
```

	state	brate	pop	medage	division	region	mrgrate
state	1.0000						
brate	0.0208	1.0000					
pop	-0.0540	-0.2830	1.0000				
medage	-0.0624	-0.8800	0.3294	1.0000			
division	-0.1345	0.6356	-0.1081	-0.5207	1.0000		
region	-0.1339	0.6086	-0.1515	-0.5292	0.9688	1.0000	
mrgrate	0.0509	0.0677	-0.1502	-0.0177	0.2280	0.2490	1.0000
dvcrate	-0.0655	0.3508	-0.2064	-0.2229	0.5522	0.5682	0.7700
medage2	-0.0621	-0.8609	0.3324	0.9984	-0.5162	-0.5239	-0.0202

	dvcrate	medage2
dvcrate	1.0000	
medage2	-0.2192	1.0000

Since we did not specify the wrap option, Stata did its best to make the result readable by breaking the table into two parts.

To obtain the correlations between mrgrate, dvcrate, and medage, we type correlate mrgrate dvcrate medage:

```
. correlate mrgrate dvcrate medage
(obs=50)
```

	mrgrate	dvcrate	medage
mrgrate	1.0000		
dvcrate	0.7700	1.0000	
medage	-0.0177	-0.2229	1.0000

◁

▷ Example 2

The variable pop in our previous example represents the total population of the state. Thus, to obtain population-weighted correlations among mrgrate, dvcrate, and medage, we type

```
. correlate mrgrate dvcrate medage [w=pop]
(analytic weights assumed)
(sum of wgt is   2.2591e+08)
(obs=50)
```

	mrgrate	dvcrate	medage
mrgrate	1.0000		
dvcrate	0.5854	1.0000	
medage	-0.1316	-0.2833	1.0000

◁

With the covariance option, correlate can be used to obtain covariance matrices, as well as correlation matrices, for both weighted and unweighted data.

▷ Example 3

To obtain the matrix of covariances between mrgrate, dvcrate, and medage, we type correlate mrgrate dvcrate medage, covariance:

```
. correlate mrgrate dvcrate medage, covariance
(obs=50)

             |  mrgrate   dvcrate    medage
-------------+---------------------------------
     mrgrate |  .000662
      dvcrate |  .000063   1.0e-05
       medage | -.000769  -.001191   2.86775
```

We could have obtained the pop-weighted covariance matrix by typing correlate mrgrate dvcrate medage [w=pop], covariance.

◁

By specifying the _coef option, you can use correlate to display the correlation or covariance matrix for the coefficients from the last fitted model. You may not specify a *varlist*, in *range*, or if *exp* since the sample was determined by the most recent estimation command. You may not specify a weight, either. If the most recent estimates were weighted, the correlation (covariance) matrix displayed is correctly weighted already.

▷ Example 4

The only possible forms of correlate with the _coef option are correlate, _coef and correlate, _coef covariance. Before you can use either, you must fit a model—here we will use regression—but you can use correlate after any estimation procedure, even maximum likelihood procedures such as clogit, logistic, logit, ologit, and mlogit. We run a regression of mrgrate on medage and medage2, excluding Nevada from the data:

```
. use http://www.stata-press.com/data/r10/census6
(1980 Census data by state)

. regress mrgrate medage medage2 if state !="Nevada"
```

Source	SS	df	MS		
Model	.000179759	2	.00008988		
Residual	.00034728	46	7.5496e-06		
Total	.000527039	48	.00001098		

Number of obs = 49
F(2,   46) = 11.91
Prob > F   = 0.0001
R-squared   = 0.3411
Adj R-squared = 0.3124
Root MSE   = .00275

mrgrate	Coef.	Std. Err.	t	P>|t|	[95% Conf. Interval]
medage	-.0041675	.0040848	-1.02	0.313	-.0123899 .0040548
medage2	.0000518	.0000692	0.75	0.459	-.0000876 .0001911
_cons	.0928522	.0602392	1.54	0.130	-.0284031 .2141075

To obtain the correlation matrix of the coefficients, we now type correlate, _coef:

```
. correlate, _coef
             |  medage  medage2    _cons
-------------+---------------------------
      medage |  1.0000
     medage2 | -0.9984   1.0000
       _cons | -0.9983   0.9935   1.0000
```

The results show much collinearity among the explanatory variables.

We did not have to type the `correlate` command immediately after running the regression. We could have gone on to do other things—retrieving the residuals, plotting them, and so on—before requesting to see the correlation matrix. Stata never forgets the most recent estimates (unless you explicitly eliminate them by typing `discard`).

We can obtain the covariance matrix of the estimators by typing `correlate, _coef covariance`:

```
. correlate, _coef covariance
             |   medage  medage2    _cons
-------------+----------------------------
      medage |  .000017
     medage2 | -2.8e-07  4.8e-09
       _cons | -.000246  4.1e-06  .003629
```

◁

pwcorr

`correlate` calculates correlation coefficients by using casewise deletion: when you request correlations of variables $x_1, x_2, \ldots, x_k$, any observation for which any of $x_1, x_2, \ldots, x_k$ is missing is not used. Thus if x_3 and x_4 have no missing values, but x_2 is missing for half the data, the correlation between x_3 and x_4 is calculated using only the half of the data for which x_2 is not missing. Of course, you can obtain the correlation between x_3 and x_4 by using all the data by typing `correlate $x_3$ $x_4$`.

`pwcorr` makes obtaining such pairwise correlation coefficients easier.

▷ Example 5

Using the `auto.dta`, we investigate the correlation between several of the variables.

```
. use http://www.stata-press.com/data/r10/auto1
(Automobile Models)
. pwcorr mpg price rep78 foreign, obs sig
             |      mpg    price    rep78  foreign
-------------+------------------------------------
         mpg |   1.0000
             |
             |       74
             |
       price |  -0.4594   1.0000
             |   0.0000
             |       74       74
             |
       rep78 |   0.3739   0.0066   1.0000
             |   0.0016   0.9574
             |       69       69       69
             |
     foreign |   0.3613   0.0487   0.5922   1.0000
             |   0.0016   0.6802   0.0000
             |       74       74       69       74
```

```
. pwcorr mpg price headroom rear_seat trunk rep78 foreign, print(.05) star(.01)
                    mpg     price headroom rear_s~t    trunk    rep78  foreign

       mpg |    1.0000
     price |   -0.4594*  1.0000
  headroom |   -0.4220*            1.0000
 rear_seat |   -0.5213*  0.4194*   0.5238*  1.0000
     trunk |   -0.5703*  0.3143*   0.6620*  0.6480*  1.0000
     rep78 |    0.3739*                                       1.0000
   foreign |    0.3613*           -0.2939  -0.2409  -0.3594*  0.5922*  1.0000

. pwcorr mpg price headroom rear_seat trunk rep78 foreign, print(.05) bon
                    mpg     price headroom rear_s~t    trunk    rep78  foreign

       mpg |    1.0000
     price |   -0.4594   1.0000
  headroom |   -0.4220             1.0000
 rear_seat |   -0.5213   0.4194    0.5238   1.0000
     trunk |   -0.5703             0.6620   0.6480   1.0000
     rep78 |    0.3739                                        1.0000
   foreign |    0.3613                               -0.3594  0.5922   1.0000
```
◁

❏ Technical Note

The `correlate` command will report the correlation matrix of the data, but there are occasions when you need the matrix stored as a Stata matrix so that you can further manipulate it. You can obtain the matrix by typing

```
. matrix accum R = varlist, nocons dev
. matrix R = corr(R)
```

The first line places the cross-product matrix of the data in matrix R. The second line converts that to a correlation matrix. Also see [P] **matrix define** and [P] **matrix accum**.

❏

Saved Results

`correlate` saves the following in `r()`:

Scalars

r(N)	number of observations	r(Var_1)	variance of first variable (covariance only)
r(rho)	ρ (first and second variables)	r(Var_2)	variance of second variable (covariance only)
r(cov_12)	covariance (covariance only)		

`pwcorr` will leave in its wake only the results of the last call that it makes internally to `correlate` for the correlation between the last variable and itself. Only rarely is this feature useful.

Methods and Formulas

`pwcorr` is implemented as an ado-file.

For a discussion of correlation, see, for instance, Snedecor and Cochran (1989, 177–195); for a more introductory explanation, see Edwards (1984).

According to Snedecor and Cochran (1989, 180), the term "co-relation" was first proposed by Galton (1888). The product-moment correlation coefficient is often called the Pearson product-moment correlation coefficient because Pearson (1896, 1898) was partially responsible for popularizing its use. See Stigler (1986) for information on the history of correlation.

The estimate of the product-moment correlation coefficient, ρ, is

$$\widehat{\rho} = \frac{\sum_{i=1}^{n} w_i (x_i - \overline{x})(y_i - \overline{y})}{\sqrt{\sum_{i=1}^{n} w_i (x_i - \overline{x})^2} \sqrt{\sum_{i=1}^{n} w_i (y_i - \overline{y})^2}}$$

where w_i are the weights, if specified, or $w_i = 1$ if weights are not specified. $\overline{x} = (\sum w_i x_i)/(\sum w_i)$ is the mean of x, and $\overline{y}$ is similarly defined.

The unadjusted significance level is calculated by `pwcorr` as

$$p = 2 * \mathtt{ttail}(n - 2, |\widehat{\rho}| \sqrt{n - 2} / \sqrt{1 - \widehat{\rho}^2})$$

Let v be the number of variables specified so that $k = v(v-1)/2$ correlation coefficients are to be estimated. If `bonferroni` is specified, the adjusted significance level is $p' = \min(1, kp)$. If `sidak` is specified, $p' = \min\left\{1, 1 - (1 - p)^k\right\}$. In both cases, see *Methods and Formulas* in [R] **oneway** for a more complete description of the logic behind these adjustments.

As for the correlation or covariance matrix of estimators, see the appropriate estimation command. When `correlate` is used after a maximum likelihood procedure, it obtains the covariance matrix from the inverse of the information matrix.

Carlo Emilio Bonferroni (1892–1960) studied in Turin and taught there and in Bari and Florence. He published on actuarial mathematics, probability, statistics, analysis, geometry, and mechanics. His work on probability inequalities has been applied to simultaneous statistical inference, although the method known as Bonferroni adjustment usually relies only on an inequality established earlier by Boole.

Karl Pearson (1857–1936) studied mathematics at Cambridge. He was professor of applied mathematics (1884–1911) and eugenics (1911–1933) at University College London. His publications include literary, historical, philosophical, and religious topics. Statistics became his main interest in the early 1890s after he learned about its application to biological problems. His work centered on distribution theory, the method of moments, correlation, and regression. Pearson introduced the chi-squared test and the terms coefficient of variation, contingency table, heteroskedastic, histogram, homoskedastic, kurtosis, mode, random sampling, random walk, skewness, standard deviation, and truncation. Despite many strong qualities he also fell into prolonged disagreements with others, most notably William Bateson and R. A. Fisher.

Zbyněk Šidák (1933–1999) was an outstanding Czech statistician and probabilist who worked on Markov chains, rank tests, multivariate distribution theory and multiple comparison methods.

References

Dewey, M. E., and E. Seneta. 2001. Carlo Emilio Bonferroni. In *Statisticians of the Centuries*, ed. C. C. Heyde and E. Seneta, 411–414. New York: Springer.

Edwards, A. L. 1984. *An Introduction to Linear Regression and Correlation*. 2nd ed. New York: Freeman.

Eisenhart, C. 1974. Pearson, Karl. In *Dictionary of Scientific Biography*, ed. C. C. Gillispie, 10: 447–473. New York: Charles Scribner's Sons.

Galton, F. 1888. Co-relations and their measurement, chiefly from anthropometric data. *Proceedings of the Royal Society of London* 45: 135–145.

Gleason, J. R. 1996. sg51: Inference about correlations using the Fisher z-transform. *Stata Technical Bulletin* 32: 13–18. Reprinted in *Stata Technical Bulletin Reprints*, vol. 6, pp. 121–128.

Goldstein, R. 1996. sg52: Testing dependent correlation coefficients. *Stata Technical Bulletin* 32: 18. Reprinted in *Stata Technical Bulletin Reprints*, vol. 6, pp. 128–129.

Pearson, K. 1896. Mathematical contributions to the theory of evolution. III. Regression, heredity, and panmixia. *Philosophical Transactions of the Royal Society of London, Series A* 187: 253–318.

Pearson, K., and L. N. G. Filon. 1898. Mathematical contributions to the theory of evolution. IV. On the probable errors of frequency constants and on the influence of random selection on variation and correlation. *Philosophical Transactions of the Royal Society of London, Series A* 191: 229–311.

Porter, T. M. 2004. *Karl Pearson: The Scientific Life in a Statistical Age*. Princeton, NJ: Princeton University Press.

Rodgers, J. L., and W. A. Nicewander. 1988. Thirteen ways to look at the correlation coefficient. *American Statistician* 42: 59–66.

Rovine, M. J., and A. von Eye. 1997. A 14th way to look at the correlation coefficient: Correlation as the proportion of matches. *American Statistician* 51: 42–46.

Seed, P. T. 2001. sg159: Confidence intervals for correlations. *Stata Technical Bulletin* 59: 27–28. Reprinted in *Stata Technical Bulletin Reprints*, vol. 10, pp. 267–269.

Seidler, J., J. Vondráček, and I. Saxl. 2000. The life and work of Zbyněk Šidák (1933–1999). *Applications of Mathematics* 45: 321–336.

Snedecor, G. W., and W. G. Cochran. 1989. *Statistical Methods*. 8th ed. Ames, IA: Iowa State University Press.

Stigler, S. M. 1986. *The History of Statistics: The Measurement of Uncertainty before 1900*. Cambridge, MA: Belknap Press.

Wolfe, F. 1997. sg64: pwcorrs: An enhanced correlation display. *Stata Technical Bulletin* 35: 22–25. Reprinted in *Stata Technical Bulletin Reprints*, vol. 6, pp. 163–167.

——. 1999. sg64.1: Update to pwcorrs. *Stata Technical Bulletin* 49: 17. Reprinted in *Stata Technical Bulletin Reprints*, vol. 9, p. 159.

Also See

[R] **pcorr** — Partial correlation coefficients

[R] **spearman** — Spearman's and Kendall's correlations

[R] **summarize** — Summary statistics

[R] **tetrachoric** — Tetrachoric correlations for binary variables

Title

> **cumul** — Cumulative distribution

Syntax

cumul *varname* $\big[$ *if* $\big]$ $\big[$ *in* $\big]$ $\big[$ *weight* $\big]$, <u>g</u>enerate(*newvar*) $\big[$ *options* $\big]$

options	description
Main	
*<u>g</u>enerate(*newvar*)	create variable *newvar*
<u>f</u>req	use frequency units for cumulative
equal	generate equal cumulatives for tied values

*generate(*newvar*) is required.

by is allowed; see [D] **by**.

fweights and aweights are allowed; see [U] **11.1.6 weight**.

Description

cumul creates *newvar*, defined as the empirical cumulative distribution function of *varname*.

Options

> ┌ Main ┐

generate(*newvar*) is required. It specifies the name of the new variable to be created.

freq specifies that the cumulative be in frequency units; otherwise, it is normalized so that *newvar* is 1 for the largest value of *varname*.

equal requests that observations with equal values in *varname* get the same cumulative value in *newvar*.

Jean Baptiste Joseph Fourier (1768–1830) was born in Auxerre in France. He got caught up in the Revolution and its aftermath and was twice arrested and imprisoned, between periods of studying and teaching mathematics. Fourier joined Napoleon's army in its invasion of Egypt in 1798 as scientific adviser, returning to France in 1801, when he was appointed Prefect of the Department of Isère. While Prefect, Fourier did his important mathematical work on the theory of heat, based on what are now called Fourier series. This work was published in 1822, despite the skepticism of Lagrange, Laplace, Legendre, and others—who found the work lacking in generality and even rigor—and disagreements of both priority and substance with Biot and Poisson.

Remarks

▷ Example 1

cumul is most often used with graph to graph the empirical cumulative distribution. For instance, we have data on the median family income of 957 U.S. cities:

```
. use http://www.stata-press.com/data/r10/hsng
(1980 Census housing data)

. cumul faminc, gen(cum)

. sort cum

. line cum faminc, ylab(, grid) ytitle("") xlab(, grid)
> title("Cumulative of median family income")
> subtitle("1980 Census, 957 U.S. Cities")
```

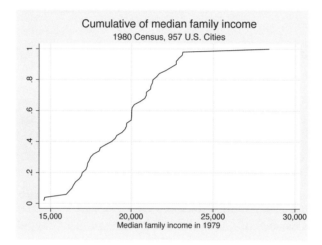

It would have been enough to type line cum faminc, but we wanted to make the graph look better; see [G] **graph twoway line**.

If we had wanted a weighted cumulative, we would have typed cumul faminc [w=pop] at the first step.   ◁

▷ Example 2

To graph two (or more) cumulatives on the same graph, use cumul and stack; see [D] **stack**. For instance, we have data on the average January and July temperatures of 956 U.S. cities:

```
. use http://www.stata-press.com/data/r10/citytemp
(City Temperature Data)

. cumul tempjan, gen(cjan)

. cumul tempjuly, gen(cjuly)

. stack  cjan tempjan  cjuly tempjuly, into(c temp) wide clear

. line cjan cjuly temp, sort ylab(, grid) ytitle("") xlab(, grid)
> xtitle("Temperature (F)")
> title("Cumulatives:" "Average January and July Temperatures")
> subtitle("956 U.S. Cities") clstyle(. dot)
```

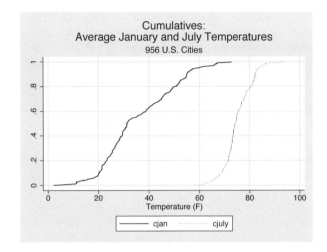

As before, it would have been enough to type line cjan cjuly temp, sort. See [D] **stack** for an explanation of how the stack command works.

◁

❏ Technical Note

According to Beniger and Robyn (1978), Fourier (1821) published the first graph of a cumulative frequency distribution, which was later given the name "ogive" by Galton (1875).

❏

Methods and Formulas

cumul is implemented as an ado-file.

Acknowledgment

The equal option was added by Nicholas J. Cox, Durham University, Durham, UK.

References

Beniger, J. R., and D. L. Robyn. 1978. Quantitative graphics in statistics: A brief history. *American Statistician* 32: 1–11.

Clayton, D., and M. Hills. 1999. gr37: Cumulative distribution function plots. *Stata Technical Bulletin* 49: 10–12. Reprinted in *Stata Technical Bulletin Reprints*, vol. 9, pp. 96–98.

Cox, N. J. 1999. gr41: Distribution function plots. *Stata Technical Bulletin* 51: 12–16. Reprinted in *Stata Technical Bulletin Reprints*, vol. 9, pp. 108–112.

Fourier, J. B. J. 1821. Notions générales, sur la population. *Recherches Statistiques sur la Ville de Paris et le Département de la Seine* 1: 1–70.

Galton, F. 1875. Statistics by intercomparison, with remarks on the law of frequency of error. *Philosophical Magazine* 49: 33–46.

Wilk, M. B., and R. Gnanadesikan. 1968. Probability plotting methods for the analysis of data. *Biometrika* 55: 1–17.

Also See

[D] **stack** — Stack data

[R] **diagnostic plots** — Distributional diagnostic plots

[R] **kdensity** — Univariate kernel density estimation

Title

cusum — Graph cumulative spectral distribution

Syntax

cusum *yvar xvar* [*if*] [*in*] [, *options*]

options	description
Main	
generate(*newvar*)	save cumulative sum in *newvar*
yfit(*fitvar*)	calculate cumulative sum against *fitvar*
nograph	suppress the plot
nocalc	suppress cusum test statistics
Cusum plot	
connect_options	affect the rendition of the plotted line
Add plots	
addplot(*plot*)	add plots to the generated graph
Y axis, X axis, Titles, Legend, Overall	
twoway_options	any options other than by() documented in [G] ***twoway_options***

Description

cusum graphs the cumulative sum (cusum) of a binary (0/1) variable *yvar* against a (usually) continuous variable *xvar*.

Options

⌐ Main ⌐

generate(*newvar*) saves the cusum in *newvar*.

yfit(*fitvar*) calculates a cusum against *fitvar*, that is, the running sums of the "residuals" *fitvar* minus *yvar*. Typically, *fitvar* is the predicted probability of a positive outcome obtained from a logistic regression analysis.

nograph suppresses the plot.

nocalc suppresses calculation of the cusum test statistics.

⌐ Cusum plot ⌐

connect_options affect the rendition of the plotted line; see [G] ***connect_options***.

⌐ Add plots ⌐

addplot(*plot*) provides a way to add other plots to the generated graph. See [G] ***addplot_option***.

> Y axis, X axis, Titles, Legend, Overall

twoway_options are any of the options documented in [G] ***twoway_options***, excluding by(). These include options for titling the graph (see [G] ***title_options***) and for saving the graph to disk (see [G] ***saving_option***).

Remarks

The cusum is the running sum of the proportion of ones in the sample, a constant number, minus *yvar*,

$$c_j = \sum_{k=1}^{j} f - yvar_{(k)}, \qquad 1 \le j \le N$$

where $f = (\sum yvar)/N$ and $yvar_{(k)}$ refers to the corresponding value of *yvar* when *xvar* is placed in ascending order: $xvar_{(k+1)} \ge xvar_{(k)}$. Tied values of *xvar* are broken at random. If you want them broken the same way in two runs, you must set the random-number seed to the same value before giving the cusum command; see [D] **generate**.

A U-shaped or inverted U-shaped cusum indicates, respectively, a negative or a positive trend of *yvar* with *xvar*. A sinusoidal shape is evidence of a nonmonotonic (for example, quadratic) trend. cusum displays the maximum absolute cusum for monotonic and nonmonotonic trends of *yvar* on *xvar*. These are nonparametric tests of departure from randomness of *yvar* with respect to *xvar*. Approximate values for the tests are given.

▷ Example 1

For the automobile dataset, auto.dta, we wish to investigate the relationship between foreign (0 = domestic, 1 = foreign) and car weight as follows:

```
. use http://www.stata-press.com/data/r10/auto
(1978 Automobile Data)

. cusum foreign weight
```

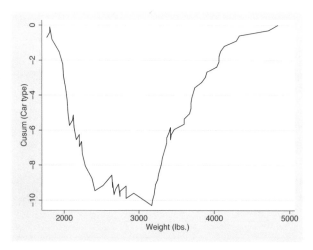

Variable	Obs	Pr(1)	CusumL	zL	Pr>zL	CusumQ	zQ	Pr>zQ
foreign	74	0.2973	10.30	3.963	0.000	3.32	0.469	0.320

The resulting plot, which is U-shaped, suggests a negative monotonic relationship. The trend is confirmed by a highly significant linear cusum statistic, labeled CusumL in the output above.

Some 29.73% of the cars are foreign (coded 1). The proportion of foreign cars diminishes with increasing weight. The domestic cars are crudely heavier than the foreign ones. We could have discovered that by typing `table foreign, stats(mean weight)`, but such an approach does not give the full picture of the relationship. The quadratic cusum (CusumQ) is not significant, so we do not suspect any tendency for the very heavy cars to be foreign rather than domestic. A slightly enhanced version of the plot shows the preponderance of domestic (coded 0) cars at the heavy end of the weight axis:

```
. label values foreign
. cusum foreign weight, s(none) recast(scatter) mlabel(foreign) mlabp(0)
```

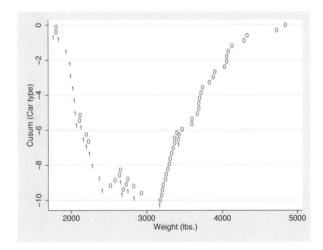

Variable	Obs	Pr(1)	CusumL	zL	Pr>zL	CusumQ	zQ	Pr>zQ
foreign	74	0.2973	10.30	3.963	0.000	2.92	0.064	0.475

The example is, of course, artificial, as we would not really try to model the probability of a car being foreign given its weight.

◁

Saved Results

cusum saves the following in r():

Scalars

r(N)	number of observations	r(P_zl)	p-value for test (linear)
r(prop1)	proportion of positive outcomes	r(cusumq)	quadratic cusum
r(cusuml)	cusum	r(zq)	test (quadratic)
r(zl)	test (linear)	r(P_zq)	p-value for test (quadratic)

Methods and Formulas

cusum is implemented as an ado-file.

Acknowledgment

cusum was written by Patrick Royston of the MRC Clinical Trials Unit, London.

References

Royston, P. 1992. The use of cusums and other techniques in modelling continuous covariates in logistic regression. *Statistics in Medicine* 11: 1115–1129.

——. 1993. sqv7: Cusum plots and tests for binary variables. *Stata Technical Bulletin* 12: 16–17. Reprinted in *Stata Technical Bulletin Reprints*, vol. 2, pp. 175–177.

Also See

[R] **logistic** — Logistic regression, reporting odds ratios

[R] **logit** — Logistic regression, reporting coefficients

[R] **probit** — Probit regression

Title

db — Launch dialog

Syntax

Syntax for db

   db *commandname*

For programmers

   db *commandname* [, message(*string*) debug dryrun]

Set system parameter

   set maxdb *#* [, <u>permanently</u>]

where *#* must be between 5 and 1,000.

Description

db is the command-line way to launch a dialog for a Stata command.

The second syntax (which is the same but includes options) is for use by programmers.

If you wish to allow the launching of dialogs from a help file, see [P] **smcl** for information on the dialog SMCL directive.

set maxdb sets the maximum number of dialog boxes whose contents are remembered from one invocation to the next during a session. The default value of maxdb is 50.

Options

message(*string*) specifies that *string* be passed to the dialog box where it can be referred to from the __MESSAGE STRING property.

debug specifies that the underlying dialog box be loaded with debug messaging turned on.

dryrun specifies that, rather than launching the dialog, db show the commands it would issue to launch the dialog.

permanently specifies that, in addition to making the change right now, the maxdb setting be remembered and become the default setting when you invoke Stata.

Remarks

The usual way to launch a dialog is to open the **Data**, **Graphics**, or **Statistics** menu and to make your selection from there. When you know the name of the command that you want to run, however, db provides a way to invoke the dialog from the command line.

db follows the same abbreviation rules that Stata's command-line interface follows. So, to launch the dialog for regress, you can type

```
. db regress
```

or

```
. db reg
```

Say that you use the dialog box for `regress`, either by selecting

Statistics > Linear regression and related > Linear regression

or by typing

```
db regress
```

You fit a regression.

Much later during the session, you return to the `regress` dialog box. It will have the contents as you left them if (1) you have not typed `clear all` between the first and second invocations; (2) you have not typed `discard` between the two invocations; and (3) you have not used more than 50 different dialog boxes—regardless of how many times you have used each—between the first and second invocations of `regress`. If you use 51 or more, the contents of the `regress` dialog box will be forgotten.

`set maxdb` determines how many different dialog boxes are remembered. A dialog box takes, on average, about 20 KB of memory, so the 50 default corresponds to allowing dialog boxes to consume about 1 MB of memory.

Methods and Formulas

`db` is implemented as an ado-file.

Also See

[R] **query** — Display system parameters

Title

> **diagnostic plots** — Distributional diagnostic plots

Syntax

Symmetry plot

    symplot *varname* [*if*] [*in*] [, *options*₁]

Ordered values of varname against quantiles of uniform distribution

    quantile *varname* [*if*] [*in*] [, *options*₁]

Quantiles of varname₁ against quantiles of varname₂

    qqplot *varname*₁ *varname*₂ [*if*] [*in*] [, *options*₁]

Quantiles of varname against quantiles of normal distribution

    qnorm *varname* [*if*] [*in*] [, *options*₂]

Standardized normal probability plot

    pnorm *varname* [*if*] [*in*] [, *options*₂]

Quantiles of varname against quantiles of χ^2 distribution

    qchi *varname* [*if*] [*in*] [, *options*₃]

χ^2 probability plot

    pchi *varname* [*if*] [*in*] [, *options*₃]

*options*₁	description
Plot	
marker_options	change look of markers (color, size, etc.)
marker_label_options	add marker labels; change look or position
Reference line	
rlopts(*cline_options*)	affect rendition of the reference line
Add plots	
addplot(*plot*)	add other plots to the generated graph
Y axis, X axis, Titles, Legend, Overall	
twoway_options	any options other than by() documented in [G] ***twoway_options***

options$_2$	description
Main	
<u>grid</u>	add grid lines
Plot	
marker_options	change look of markers (color, size, etc.)
marker_label_options	add marker labels; change look or position
Reference line	
<u>rlopts</u>(*cline_options*)	affect rendition of the reference line
Add plots	
addplot(*plot*)	add other plots to the generated graph
Y axis, X axis, Titles, Legend, Overall	
twoway_options	any options other than by() documented in [G] *twoway_options*

options$_3$	description
Main	
<u>grid</u>	add grid lines
df(#)	degrees of freedom of χ^2 distribution; default is df(1)
Plot	
marker_options	change look of markers (color, size, etc.)
marker_label_options	add marker labels; change look or position
Reference line	
<u>rlopts</u>(*cline_options*)	affect rendition of the reference line
Add plots	
addplot(*plot*)	add other plots to the generated graph
Y axis, X axis, Titles, Legend, Overall	
twoway_options	any options other than by() documented in [G] *twoway_options*

Description

symplot graphs a symmetry plot of *varname*.

quantile plots the ordered values of *varname* against the quantiles of a uniform distribution.

qqplot plots the quantiles of *varname*$_1$ against the quantiles of *varname*$_2$ (Q–Q plot).

qnorm plots the quantiles of *varname* against the quantiles of the normal distribution (Q–Q plot).

pnorm graphs a standardized normal probability plot (P–P plot).

qchi plots the quantiles of *varname* against the quantiles of a χ^2 distribution (Q–Q plot).

pchi graphs a χ^2 probability plot (P–P plot).

See [R] **regress postestimation** for regression diagnostic plots and [R] **logistic postestimation** for logistic regression diagnostic plots.

Options for symplot, quantile, and qqplot

⌐ Plot ⌐

marker_options affect the rendition of markers drawn at the plotted points, including their shape, size, color, and outline; see [G] *marker_options*.

marker_label_options specify if and how the markers are to be labeled; see [G] *marker_label_options*.

⌐ Reference line ⌐

rlopts(*cline_options*) affect the rendition of the reference line; see [G] *cline_options*.

⌐ Add plots ⌐

addplot(*plot*) provides a way to add other plots to the generated graph; see [G] *addplot_option*.

⌐ Y axis, X axis, Titles, Legend, Overall ⌐

twoway_options are any of the options documented in [G] *twoway_options*, excluding by(). These include options for titling the graph (see [G] *title_options*) and for saving the graph to disk (see [G] *saving_option*).

Options for qnorm and pnorm

⌐ Main ⌐

grid adds grid lines at the .05, .10, .25, .50, .75, .90, and .95 quantiles when specified with qnorm. With pnorm, grid is equivalent to yline(.25,.5,.75) xline(.25,.5,.75).

⌐ Plot ⌐

marker_options affect the rendition of markers drawn at the plotted points, including their shape, size, color, and outline; see [G] *marker_options*.

marker_label_options specify if and how the markers are to be labeled; see [G] *marker_label_options*.

⌐ Reference line ⌐

rlopts(*cline_options*) affect the rendition of the reference line; see [G] *cline_options*.

⌐ Add plots ⌐

addplot(*plot*) provides a way to add other plots to the generated graph; see [G] *addplot_option*.

⌐ Y axis, X axis, Titles, Legend, Overall ⌐

twoway_options are any of the options documented in [G] *twoway_options*, excluding by(). These include options for titling the graph (see [G] *title_options*) and for saving the graph to disk (see [G] *saving_option*).

Options for qchi and pchi

> ### Main

grid adds grid lines at the .05, .10, .25, .50, .75, .90, and .95 quantiles when specified with qchi. With pchi, grid is equivalent to yline(.25,.5,.75) xline(.25,.5,.75).

df(#) specifies the degrees of freedom of the χ^2 distribution. The default is 1.

> ### Plot

marker_options affect the rendition of markers drawn at the plotted points, including their shape, size, color, and outline; see [G] *marker_options*.

marker_label_options specify if and how the markers are to be labeled; see [G] *marker_label_options*.

> ### Reference line

rlopts(*cline_options*) affect the rendition of the reference line; see [G] *cline_options*.

> ### Add plots

addplot(*plot*) provides a way to add other plots to the generated graph; see [G] *addplot_option*.

> ### Y axis, X axis, Titles, Legend, Overall

twoway_options are any of the options documented in [G] *twoway_options*, excluding by(). These include options for titling the graph (see [G] *title_options*) and for saving the graph to disk (see [G] *saving_option*).

Remarks

Remarks are presented under the following headings:

> *symplot*
> *quantile*
> *qqplot*
> *qnorm*
> *pnorm*
> *qchi*
> *pchi*

symplot

▷ Example 1

We have data on 74 automobiles. To make a symmetry plot of the variable `price`, we type

```
. use http://www.stata-press.com/data/r10/auto
(1978 Automobile Data)
. symplot price
```

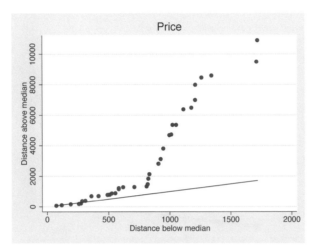

All points would lie along the reference line (defined as $y = x$) if car prices were symmetrically distributed. The points in this plot lie above the reference line, indicating that the distribution of car prices is skewed to the right—the most expensive cars are far more expensive than the least expensive cars are inexpensive.

The logic works as follows: a variable z is distributed symmetrically if

$$\text{median} - z_{(i)} = z_{(N+1-i)} - \text{median}$$

where $z_{(i)}$ indicates the ith-order statistic of z. `symplot` graphs $y_i = \text{median} - z_{(i)}$ versus $x_i = z_{(N+1-i)} - \text{median}$.

For instance, consider the largest and smallest values of `price` in the example above. The most expensive car costs $15,906, and the least expensive, $3,291. Let's compare these two cars with the typical car in the data and see how much more it costs to buy the most expensive car, and compare that with how much less it costs to buy the least expensive car. If the automobile price distribution is symmetric, the price differences would be the same.

Before we can make this comparison, we must agree on a definition for the word "typical". Let's agree that "typical" means median. The price of the median car is $5,006.50, so the most expensive car costs $10,899.50 more than the median car, and the least expensive car costs $1,715.50 less than the median car. We now have one piece of evidence that the car price distribution is not symmetric. We can repeat the experiment for the second-most-expensive car and the second-least-expensive car. We find that the second-most-expensive car costs $9,494.50 more than the median car, and the second-least-expensive car costs $1,707.50 less than the median car. We now have more evidence. We can continue doing this with the third most expensive and the third least expensive, and so on.

Once we have all these numbers, we want to compare each pair and ask how similar, on average, they are. The easiest way to do that is to plot all the pairs.

◁

quantile

▷ Example 2

We have data on the prices of 74 automobiles. To make a quantile plot of `price`, we type

```
. use http://www.stata-press.com/data/r10/auto
(1978 Automobile Data)

. quantile price, rlopts(clpattern(dash))
```

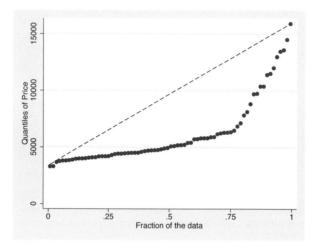

We changed the pattern of the reference line by specifying `rlopts(clpattern(dash))`.

In a quantile plot, each value of the variable is plotted against the fraction of the data that have values less than that fraction. The diagonal line is a reference line. If automobile prices were rectangularly distributed, all the data would be plotted along the line. Since all the points are below the reference line, we know that the price distribution is skewed right.

◁

qqplot

▷ Example 3

We have data on the weight and country of manufacture of 74 automobiles. We wish to compare the distributions of weights for domestic and foreign automobiles:

```
. use http://www.stata-press.com/data/r10/auto
(1978 Automobile Data)

. generate weightd=weight if !foreign
(22 missing values generated)

. generate weightf=weight if foreign
(52 missing values generated)
```

. qqplot weightd weightf

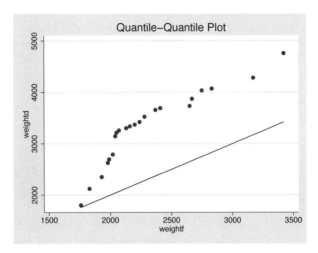

qnorm

▷ Example 4

Continuing with our price data on 74 automobiles, we now wish to compare the distribution of price with the normal distribution:

```
. qnorm price, grid ylabel(, angle(horizontal) axis(1))
> ylabel(, angle(horizontal) axis(2))
```

The result shows that the distributions are different.

◁

❏ Technical Note

The idea behind `qnorm` is recommended strongly by Miller (1997): he calls it probit plotting. His recommendations from much practical experience should interest many users. "My recommendation for detecting nonnormality is *probit plotting*" (Miller 1997, 10). "If a deviation from normality cannot be spotted by eye on probit paper, it is not worth worrying about. I never use the Kolmogorov–Smirnov test (or one of its cousins) or the χ^2 test as a preliminary test of normality. They do not tell you how the sample is differing from normality, and I have a feeling they are more likely to detect irregularities in the middle of the distribution than in the tails" (Miller 1997, 13–14).

❏

pnorm

▷ Example 5

Quantile–normal plots emphasize the tails of the distribution. Normal probability plots put the focus on the center of the distribution:

```
. pnorm price, grid
```

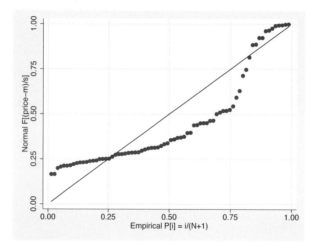

◁

qchi

▷ Example 6

Suppose that we want to examine the distribution of the sum of squares of `price` and `mpg`, standardized for their variances.

```
. egen c1 = std(price)
. egen c2 = std(mpg)
. generate ch = c1^2 + c2^2
```

```
. qchi ch, df(2) grid ylabel(, alt axis(2)) xlabel(, alt axis(2))
```

The quadratic form is clearly not χ^2 with 2 degrees of freedom.

◁

pchi

▷ Example 7

We can focus on the center of the distribution by doing a probability plot:

```
. pchi ch, df(2) grid
```

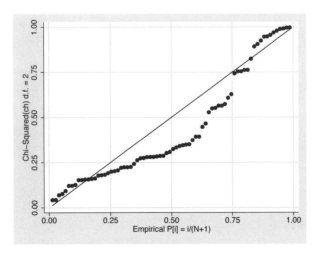

◁

Methods and Formulas

`symplot`, `quantile`, `qnorm`, `pnorm`, `qchi`, `pchi`, and `qqplot` are implemented as ado-files. Let $x_{(1)}$, $x_{(2)}$, ..., $x_{(N)}$ be the data sorted in ascending order.

If a continuous variable x has a cumulative distribution function $F(x) = P(X \leq x) = p$, the quantiles x_{p_i} are such that $F(x_{p_i}) = p_i$. For example, if $p_i = 0.5$, $x_{0.5}$ is the median. When we plot data, the probabilities p_i are often referred to as plotting positions. There are many different conventions for choice of plotting positions, given $x_{(1)} \leq \ldots \leq x_{(N)}$. Most belong to the family $(i - a)/(N - 2a + 1)$. $a = 0.5$ (suggested by Hazen) and $a = 0$ (suggested by Weibull) are popular choices.

For a wider discussion of the calculation of plotting positions, see Cox (2002).

`symplot` plots median $- x_{(i)}$ versus $x_{(N+1-i)} -$ median.

`quantile` plots $x_{(i)}$ versus $(i - 0.5)/N$ (the Hazen position).

`qnorm` plots $x_{(i)}$ against q_i, where $q_i = \Phi^{-1}(p_i)$, Φ is the cumulative normal distribution, and $p_i = i/(N + 1)$ (the Weibull position).

`pnorm` plots $\Phi\{(x_i - \widehat{\mu})/\widehat{\sigma}\}$ versus $p_i = i/(N + 1)$, where $\widehat{\mu}$ is the mean of the data and $\widehat{\sigma}$ is the standard deviation.

`qchi` and `pchi` are similar to `qnorm` and `pnorm`; the cumulative χ^2 distribution is used in place of the cumulative normal distribution.

`qqplot` is just a two-way scatterplot of one variable against the other after both variables have been sorted into ascending order, and both variables have the same number of nonmissing observations. If the variables have unequal numbers of nonmissing observations, interpolated values of the variable with more data are plotted against the variable with fewer data.

Ramanathan Gnanadesikan (1932–) was born in Madras. He obtained degrees from the Universities of Madras and North Carolina. He worked in industry at Procter and Gamble, Bell Labs, and Bellcore, as well as in universities, retiring from Rutgers in 1998. Among many contributions to statistics he is especially well known for work on probability plotting, robustness, outlier detection, clustering, classification, and pattern recognition.

Martin Bradbury Wilk (1922–) was born in Montreal. He obtained degrees in chemical engineering and statistics from McGill and Iowa State Universities. After several posts in statistics in industry and universities (including periods at Princeton, Bell Labs, and Rutgers), Wilk was appointed Chief Statistician at Statistics Canada (1980–1986). He is especially well known for his work with Gnanadesikan on probability plotting and with Shapiro on tests for normality.

Acknowledgments

We thank Peter A. Lachenbruch of the Department of Public Health, Oregon State University, for writing the original version of `qchi` and `pchi`. Patrick Royston of the MRC Clinical Trials Unit in London also published a similar command in the *Stata Technical Bulletin* (Royston 1996).

References

Chambers, J. M., W. S. Cleveland, B. Kleiner, and P. A. Tukey. 1983. *Graphical Methods for Data Analysis.* Belmont, CA: Wadsworth.

Cox, N. J. 1999. gr42: Quantile plots, generalized. *Stata Technical Bulletin* 51: 16–18. Reprinted in *Stata Technical Bulletin Reprints*, vol. 9, pp. 113–116.

——. 2001. gr42.1: Quantile plots, generalized: Update to Stata 7. *Stata Technical Bulletin* 61: 10. Reprinted in *Stata Technical Bulletin Reprints*, vol. 10, pp. 55–56.

——. 2002. Speaking Stata: On getting functions to do the work. *Stata Journal* 2: 411–427.

——. 2004a. Speaking Stata: Graphing distributions. *Stata Journal* 4: 66–88.

——. 2004b. gr42.2: Quantile plots, generalized. *Stata Journal* 4: 97.

——. 2005a. Speaking Stata: Density probability plots. *Stata Journal* 5: 259–273.

——. 2005b. Speaking Stata: The protean quantile plot. *Stata Journal* 5: 442–460.

——. 2005c. Speaking Stata: Smoothing in various directions. *Stata Journal* 5: 574–593.

Daniel, C., and F. S. Wood. 1980. *Fitting Equations to Data: Computer Analysis of Multifactor Data.* 2nd ed. New York: Wiley.

Gan, F. F., K. J. Koehler, and J. C. Thompson. 1991. Probability plots and distribution curves for assessing the fit of probability models. *American Statistician* 45: 14–21.

Hamilton, L. C. 1992. *Regression with Graphics: A Second Course in Applied Statistics*, 10–17. Belmont, CA: Duxbury.

——. 2006. *Statistics with Stata (Updated for Version 9).* Belmont, CA: Duxbury.

Hoaglin, D. C. 1985. Using quantiles to study shape. In *Exploring Data Tables, Trends, and Shapes*, ed. D. C. Hoaglin, F. Mosteller, and J. W. Tukey, 417–460. New York: Wiley.

Kettenring, J. R. 2001. A conversation with Ramanathan Gnanadesikan. *Statistical Science* 16: 295–309.

Miller, R. G., Jr. 1997. *Beyond ANOVA: Basics of Applied Statistics.* London: Chapman & Hall.

Nolan, D., and T. Speed. 2000. *Stat Labs: Mathematical Statistics Through Applications.* New York: Springer.

Royston, J. P. 1996. sg47: A plot and a test for the χ^2 distribution. *Stata Technical Bulletin* 29: 26–27. Reprinted in *Stata Technical Bulletin Reprints*, vol. 5, pp. 142–144.

Scotto, M. G. 2000. sg140: The Gumbel quantile plot and a test for choice of extreme models. *Stata Technical Bulletin* 55: 23–25. Reprinted in *Stata Technical Bulletin Reprints*, vol. 10, pp. 156–159.

Wilk, M. B., and R. Gnanadesikan. 1968. Probability plotting methods for the analysis of data. *Biometrika* 55: 1–17.

Also See

[R] **cumul** — Cumulative distribution

[R] **kdensity** — Univariate kernel density estimation

[R] **logistic postestimation** — Postestimation tools for logistic

[R] **lv** — Letter-value displays

[R] **regress postestimation** — Postestimation tools for regress

Title

> **display** — Substitute for a hand calculator

Syntax

<u>di</u>splay *exp*

Description

display displays strings and values of scalar expressions.

display really has many more features and a more complex syntax diagram, but the diagram shown above is adequate for interactive use. For a full discussion of display's capabilities, see [P] **display**.

Remarks

display can be used as a substitute for a hand calculator.

▷ Example 1

display 2+2 produces the output 4. Stata variables may also appear in the expression, such as in display myvar/2. Since display works only with scalars, the resulting calculation is performed for only the first observation. You could type display myvar[10]/2 to display the calculation for the 10th observation. Here are more examples:

```
. display sqrt(2)/2
.70710678
. display normal(-1.1)
.13566606
. di (57.2-3)/(12-2)
5.42
. display myvar/10
7
. display myvar[10]/2
3.5
```

◁

Also See

[P] **display** — Display strings and values of scalar expressions

[U] **13 Functions and expressions**

Title

> **do** — Execute commands from a file

Syntax

$\{do \mid \underline{run}\}$ *filename* $\left[\ arguments\ \right]$ $\left[\ ,\ nostop\ \right]$

Description

do and run cause Stata to execute the commands stored in *filename* just as if they were entered from the keyboard. do echoes the commands as it executes them, whereas run is silent. If *filename* is specified without an extension, .do is assumed.

Option

nostop allows the do-file to continue executing even if an error occurs. Normally, Stata stops executing the do-file when it detects an error (nonzero return code).

Remarks

You can create *filename* (called a *do-file*) using Stata's Do-file Editor; see [R] **doedit**. This file will be a standard ASCII (text) file. A complete discussion of do-files can be found in [U] **16 Do-files**.

You can also create *filename* by using a non-Stata text editor; see [D] **shell** for a way to invoke your favorite editor from inside Stata. Make sure that you save the file in ASCII format.

If the path or *filename* contains spaces, it should be enclosed in double quotes.

Reference

Jenkins, S. P. 2006 Stata tip 32: Do not stop. *Stata Journal* 6: 281.

Also See

[R] **doedit** — Edit do-files and other text files

[P] **include** — Include commands from file

[GSM] **14 Using the Do-file Editor—automating Stata**

[GSU] **14 Using the Do-file Editor—automating Stata**

[GSW] **14 Using the Do-file Editor—automating Stata**

[U] **15 Printing and preserving output**

[U] **16 Do-files**

Title

> **doedit** — Edit do-files and other text files

Syntax

<u>doed</u>it [*filename*]

Description

doedit opens a text editor that lets you edit do-files and other text files.

The Do-file Editor lets you submit several commands to Stata at once.

Remarks

Clicking the **Do-file Editor** button is equivalent to typing doedit.

doedit, typed by itself, invokes the editor with an empty document. If you specify *filename*, that file is displayed in the editor.

You may have more than one Do-file Editor open at once. Each time you submit the doedit command, a new window will be opened.

A tutorial discussion of doedit is found in the *Getting Started with Stata* manual. Read [U] **16 Do-files** for an explanation of do-files, and then read [GS] **14 Using the Do-file Editor—automating Stata** to learn how to use the Do-file Editor to create and execute do-files.

Also See

[GSM] **14 Using the Do-file Editor—automating Stata**

[GSU] **14 Using the Do-file Editor—automating Stata**

[GSW] **14 Using the Do-file Editor—automating Stata**

[U] **16 Do-files**

Title

| **dotplot** — Comparative scatterplots |

Syntax

Dotplot of varname, with one column per value of groupvar

    dotplot *varname* $\big[$ *if* $\big]$ $\big[$ *in* $\big]$ $\big[$, *options* $\big]$

Dotplot for each variable in varlist, with one column per variable

    dotplot *varlist* $\big[$ *if* $\big]$ $\big[$ *in* $\big]$ $\big[$, *options* $\big]$

options	description
Options	
<u>over</u>(*groupvar*)	display one columnar dotplot for each value of *groupvar*
nx(*#*)	horizontal dot density; default is nx(0)
ny(*#*)	vertical dot density; default is ny(35)
<u>incr</u>(*#*)	label every *#* group; default is incr(1)
mean | <u>med</u>ian	plot a horizontal line of pluses at the mean or median
<u>b</u>ounded	use minimum and maximum as boundaries
<u>bar</u>	plot horizontal dashed lines at shoulders of each group
<u>nog</u>roup	use the actual values of *yvar*
<u>cen</u>ter	center the dot for each column
Plot	
marker_options	change look of markers (color, size, etc.)
marker_label_options	add marker labels; change look or position
Y axis, X axis, Titles, Legend, Overall	
twoway_options	any options other than by() documented in [G] ***twoway_options***

Description

A dotplot is a scatterplot with values grouped together vertically ("binning", as in a histogram) and with plotted points separated horizontally. The aim is to display all the data for several variables or groups in one compact graphic.

In the first syntax, dotplot produces a columnar dotplot of *varname*, with one column per value of *groupvar*. In the second syntax, dotplot produces a columnar dotplot for each variable in *varlist*, with one column per variable; over(*groupvar*) is not allowed. In each case, the "dots" are plotted as small circles to increase readability.

Options

___Options___

over(*groupvar*) identifies the variable for which dotplot will display one columnar dotplot for each value of *groupvar*.

nx(#) sets the horizontal dot density. A larger value of # will increase the dot density, reducing the horizontal separation between dots. This option will increase the separation between columns if two or more groups or variables are used.

ny(#) sets the vertical dot density (number of "bins" on the y axis). A larger value of # will result in more bins and a plot that is less spread out horizontally. # should be determined in conjunction with nx() to give the most pleasing appearance.

incr(#) specifies how the x axis is to be labeled. incr(1), the default, labels all groups. incr(2) labels every second group.

[mean | median] plots a horizontal line of pluses at the mean or median of each group.

bounded forces the minimum and maximum of the variable to be used as boundaries of the smallest and largest bins. It should be used with one variable whose support is not the whole of the real line and whose density does not tend to zero at the ends of its support, e.g., a uniform random variable or an exponential random variable.

bar plots horizontal dashed lines at the shoulders of each group. The "shoulders" are taken to be the upper and lower quartiles unless mean has been specified; here they will be the mean plus or minus the standard deviation.

nogroup uses the actual values of *yvar* rather than grouping them (the default). This option may be useful if *yvar* takes on only a few values.

center centers the dots for each column on a hidden vertical line.

___Plot___

marker_options affect the rendition of markers drawn at the plotted points, including their shape, size, color, and outline; see [G] ***marker_options***.

marker_label_options specify if and how the markers are to be labeled; see [G] ***marker_label_options***.

___Y axis, X axis, Titles, Legend, Overall___

twoway_options are any of the options documented in [G] ***twoway_options***, excluding by(). These include options for titling the graph (see [G] ***title_options***) and for saving the graph to disk (see [G] ***saving_option***).

Remarks

dotplot produces a figure that has elements of a boxplot, a histogram, and a scatterplot. Like a boxplot, it is most useful for comparing the distributions of several variables or the distribution of 1 variable in several groups. Like a histogram, the figure provides a crude estimate of the density, and, as with a scatterplot, each symbol (dot) represents 1 observation.

▷ Example 1

dotplot may be used as an alternative to Stata's histogram graph for displaying the distribution of 1 variable.

```
. set seed 123456789
. set obs 1000
. generate norm = invnormal(uniform())
. dotplot norm, title("Normal distribution, sample size 1000")
```

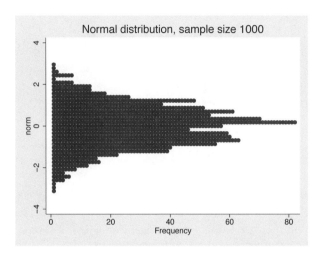

◁

▷ Example 2

The over() option lets us use dotplot to compare the distribution of 1 variable within different levels of a grouping variable. The options center, median, and bar create a graph that may be compared with Stata's boxplot; see [G] **graph box**. The next graph illustrates this option using Stata's automobile dataset.

(Continued on next page)

```
. use http://www.stata-press.com/data/r10/auto
(1978 Automobile Data)
. dotplot mpg, over(foreign) nx(25) ny(10) center median bar
```

◁

▷ Example 3

The second version of dotplot lets us compare the distribution of several variables. In the next graph, all 10 variables contain measurements on tumor volume.

```
. use http://www.stata-press.com/data/r10/dotgr
. dotplot g1r1-g1r10, ytitle("Tumor volume, cu mm")
```

◁

▷ Example 4

When using the first form with the over() option, we can encode a third dimension in a dotplot by using a different plotting symbol for different groups. The third dimension cannot be encoded with a *varlist*. The example is of a hypothetical matched case–control study. The next graph shows the exposure of each individual in each matched stratum. Cases are marked by the letter 'x', and controls are marked by the letter 'o'.

```
. use http://www.stata-press.com/data/r10/dotdose
. label define symbol 0 "o" 1 "x"
. label values case symbol
. dotplot dose, over(strata) m(none) mlabel(case) mlabp(0) center
```

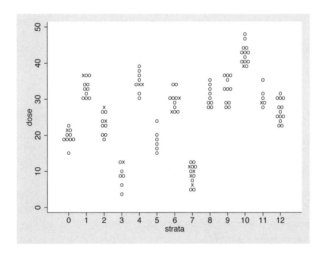

◁

▷ Example 5

dotplot can also be used with two virtually continuous variables as an alternative to jittering the data to distinguish ties. We must use the xlab option, since otherwise dotplot will attempt to label too many points on the *x* axis. It is often useful in such instances to use a value of nx that is smaller than the default. That was not necessary in this example, partly because of our choice of symbols.

```
. use http://www.stata-press.com/data/r10/auto
(1978 Automobile Data)
. generate byte hi_price = (price>10000) if price < .
. label define symbol 0 "|" 1 "o"
. label values hi_price symbol
```

(Continued on next page)

```
. dotplot weight, over(gear_ratio) m(none) mlabel(hi_price) mlabp(0) center xlabel(#5)
```

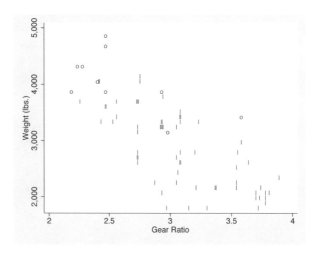

▷ Example 6

The following figure is included mostly for aesthetic reasons. It also demonstrates dotplot's ability to cope with even very large datasets. The sample size for each variable is 10,000, so it may take a long time to print!

```
. clear all
. set seed 123456789
. set obs 10000
. gen norm0 = invnormal(uniform())
. gen norm1 = invnormal(uniform()) + 1
. gen norm2 = invnormal(uniform()) + 2
. label variable norm0 "N(0,1)"
. label variable norm1 "N(1,1)"
. label variable norm2 "N(2,1)"
. dotplot norm0 norm1 norm2
```

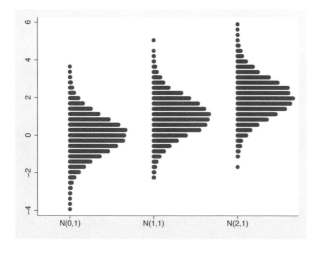

Saved Results

dotplot saves the following in r():

Scalars
r(nx)   horizontal dot density
r(ny)   vertical dot density

Methods and Formulas

dotplot is implemented as an ado-file.

Acknowledgments

dotplot was written by Peter Sasieni of the Wolfson Institute of Preventive Medicine, London, and Patrick Royston of the MRC Clinical Trials Unit, London.

References

Sasieni, P., and J. P. Royston. 1994. gr14: dotplot: Comparative scatterplots. *Stata Technical Bulletin* 19: 8–10. Reprinted in *Stata Technical Bulletin Reprints*, vol. 4, pp. 50–54.

——. 1996. Dotplots. *Applied Statistics* 45: 219–234.

Title

> **dstdize** — Direct and indirect standardization

Syntax

Direct standardization

> dstdize *charvar popvar stratavars* $\left[\,if\,\right]$ $\left[\,in\,\right]$, by(*groupvars*) $\left[\,dstdize\_options\,\right]$

Indirect standardization

> istdize *casevar$_s$ popvar$_s$ stratavars* $\left[\,if\,\right]$ $\left[\,in\,\right]$ using *filename*,
>
> $\left\{\,\underline{\text{pop}}\text{vars}(casevar_p\ popvar_p)\,|\,\text{rate}(ratevar_p\,\{\#\,|\,crudevar_p\})\,\right\}$
>
> $\left[\,istdize\_options\,\right]$

dstdize_options	description
Main	
* by(*groupvars*)	study populations
<u>us</u>ing(*filename*)	use standard population from Stata dataset
<u>base</u>(#|*string*)	use standard population from a value of grouping variable
<u>l</u>evel(#)	set confidence level; default is level(95)
Options	
<u>sav</u>ing(*filename*)	save computed standard population distribution as a Stata dataset
<u>f</u>ormat(%*fmt*)	final summary table display format; default is %10.0g
<u>print</u>	include table summary of standard population in output
nores	suppress saving results in r()

* by(*groupvars*) is required.

istdize_options	description
Main	
* <u>pop</u>vars(*casevar$_p$ popvar$_p$*)	for standard population, *casevar$_p$* is number of cases and *popvar$_p$* is number of individuals
* rate(*ratevar$_p$* {#|*crudevar$_p$*})	*ratevar$_p$* is stratum-specific rates and # or *crudevar$_p$* is the crude case rate value or variable
<u>l</u>evel(#)	set confidence level; default is level(95)
Options	
by(*groupvars*)	variables identifying study populations
<u>f</u>ormat(%*fmt*)	final summary table display format; default is %10.0g
<u>print</u>	include table summary of standard population in output

*Either popvars(*casevar$_p$ popvar$_p$*) or rate(*ratevar$_p$* {#\|*crudevar$_p$*}) must be specified.

Description

dstdize produces standardized rates for *charvar*, which are defined as a weighted average of the stratum-specific rates. These rates can be used to compare the characteristic *charvar* across different populations identified by *groupvars*. Weights used in the standardization are given by *popvar*; the strata across which the weights are to be averaged are defined by *stratavars*.

istdize produces indirectly standardized rates for a study population based on a standard population. This standardization method is appropriate when the stratum-specific rates for the population being studied are either unavailable or based on small samples and thus are unreliable. The standardization uses the stratum-specific rates of a standard population to calculate the expected number of cases in the study population(s), sums them, and then compares them with the actual number of cases observed. The standard population is in another Stata data file specified by using *filename*, and it must contain *popvar* and *stratavars*.

In addition to calculating rates, the indirect standardization command produces point estimates and exact confidence intervals of the study population's standardized mortality ratio (SMR), if death is the event of interest, or the standardized incidence ratio (SIR) for studies of incidence. Here we refer to both ratios as SMR.

casevar$_s$ is the variable name for the study population's number of cases (usually deaths). It must contain integers, and for each group, defined by *groupvar*, each subpopulation identified by *stratavars* must have the same values or missing.

popvar$_s$ identifies the number of subjects represented by each observation in the study population.

stratavars define the strata.

Options for dstdize

___ Main ___

by(*groupvars*) is required for the dstdize command; it specifies the variables identifying the study populations. If base() is also specified, there must be only one variable in the by() group. If you do not have a variable for this option, you can generate one by using something like gen newvar=1 and then use newvar as the argument to this option.

using(*filename*) or base(*# | string*) may be used to specify the standard population. You may not specify both options. using(*filename*) supplies the name of a .dta file containing the standard population. The standard population must contain the *popvar* and the *stratavars*. If using() is not specified, the standard population distribution will be obtained from the data. base(*# | string*) lets you specify one of the values of *groupvar*—either a numeric value or a string—to be used as the standard population. If neither base() nor using() is specified, the entire dataset is used to determine an estimate of the standard population.

level(*#*) specifies the confidence level, as a percentage, for a confidence interval of the adjusted rate. The default is level(95) or as set by set level; see [U] **20.7 Specifying the width of confidence intervals**.

___ Options ___

saving(*filename*) saves the computed standard population distribution as a Stata dataset that can be used in further analyses.

format(*%fmt*) specifies the format in which to display the final summary table. The default is %10.0g.

print includes a table summary of the standard population before displaying the study population results.

nores suppresses saving results in r(). This option is seldom specified. Some saved results are stored in matrices. If there are more groups than matsize, dstdize will report "matsize too small". Then you can either increase matsize or specify nores. The nores option does not change how results are calculated but specifies that results need not be left behind for use by other programs.

Options for istdize

_____| Main |_____

popvars(*casevar$_p$ popvar$_p$*) or rate(*ratevar$_p$ #|ratevar$_p$ crudevar$_p$*) must be specified with the istdize command. Only one of these two options is allowed. These options are used to describe the standard population's data.

With popvars(*casevar$_p$ popvar$_p$*), *casevar$_p$* records the number of cases (deaths) for each stratum in the standard population, and *popvar$_p$* records the total number of individuals in each stratum (individuals at risk).

With rate(*ratevar$_p$* {*#|crudevar$_p$*}), *ratevar$_p$* contains the stratum-specific rates. *#|crudevar$_p$* specifies the crude case rate either by a variable name or optionally by the crude case rate value. If a crude rate variable is used, it must be the same for all observations, although it could be missing for some.

level(*#*) specifies the confidence level, as a percentage, for a confidence interval of the adjusted rate. The default is level(95) or as set by set level; see [U] **20.7 Specifying the width of confidence intervals**.

_____| Options |_____

by(*groupvars*) specifies variables identifying study populations when more than one exists in the data. If this option is not specified, the entire study population is treated as one group.

format(%*fmt*) specifies the format in which to display the final summary table. The default is %10.0g.

print outputs a table summary of the standard population before displaying the study population results.

Remarks

Remarks are presented under the following headings:

> *Direct standardization*
> *Indirect standardization*

In epidemiology and other fields, you will often need to compare rates for some characteristic across different populations. These populations often differ on factors associated with the characteristic under study; thus, directly comparing overall rates may be misleading.

See van Belle et al. (2004, 642–684), Fleiss, Levin, and Paik (2003, chap. 19), or Kirkwood and Sterne (2003, chap. 25) for a discussion of direct and indirect standardization.

Direct standardization

The direct method of adjusting for differences among populations involves computing the overall rates that would result if, instead of having different distributions, all populations had the same standard distribution. The standardized rate is defined as a weighted average of the stratum-specific rates, with the weights taken from the standard distribution. Direct standardization may be applied only when the specific rates for a given population are available.

dstdize generates adjusted summary measures of occurrence, which can be used to compare prevalence, incidence, or mortality rates between populations that may differ on certain characteristics (e.g., age, gender, race). These underlying differences may affect the crude prevalence, mortality, or incidence rates.

▷ Example 1

We have data (Rothman 1986, 42) on mortality rates for Sweden and Panama for 1962, and we wish to compare mortality in these two countries:

```
. use http://www.stata-press.com/data/r10/mortality
(1962 Mortality, Sweden & Panama)

. describe

Contains data from http://www.stata-press.com/data/r10/mortality.dta
  obs:            6                         1962 Mortality, Sweden & Panama
  vars:           4                         14 Apr 2007 16:18
  size:         114 (99.8% of memory free)
```

variable name	storage type	display format	value label	variable label
nation	str6	%9s		Nation
age_category	byte	%9.0g	age_lbl	Age Category
population	float	%10.0gc		Population in Age Category
deaths	float	%9.0gc		Deaths in Age Category

```
Sorted by:
. list, sepby(nation) abbrev(12) divider
```

	nation	age_category	population	deaths
1.	Sweden	0 - 29	3145000	3,523
2.	Sweden	30 - 59	3057000	10,928
3.	Sweden	60+	1294000	59,104
4.	Panama	0 - 29	741,000	3,904
5.	Panama	30 - 59	275,000	1,421
6.	Panama	60+	59,000	2,456

We divide the total number of cases in the population by the population to obtain the *crude rate*:

```
. collapse (sum) pop deaths, by(nation)

. list, abbrev(10) divider
```

	nation	population	deaths
1.	Panama	1075000	7,781
2.	Sweden	7496000	73,555

```
. generate crude = deaths/pop
```

```
. list, abbrev(10) divider
```

	nation	population	deaths	crude
1.	Panama	1075000	7,781	.0072381
2.	Sweden	7496000	73,555	.0098126

If we examine the total number of deaths in the two nations, the total crude mortality rate in Sweden is higher than that in Panama. From the original data, we see one possible explanation: Swedes are older than Panamanians, making direct comparison of the mortality rates difficult.

Direct standardization lets us remove the distortion caused by the different age distributions. The adjusted rate is defined as the weighted sum of the crude rates, where the weights are given by the standard distribution. Suppose that we wish to standardize these mortality rates to the following age distribution:

```
. use http://www.stata-press.com/data/r10/1962, clear
(Standard Population Distribution)
. list, abbrev(12) divider
```

	age_category	population
1.	0 - 29	.35
2.	30 - 59	.35
3.	60+	.3

```
. sort age_cat
. save 1962
file 1962.dta saved
```

If we multiply the above weights for the age strata by the crude rate for the corresponding age category, the sum gives us the standardized rate.

```
. use http://www.stata-press.com/data/r10/mortality
(1962 Mortality, Sweden & Panama)
. gen crude=deaths/pop
. drop pop
. sort age_cat
. merge age_cat using 1962
variable age_category does not uniquely identify observations in the master data
age_category was byte now float
. list, sepby(age_category) abbrev(12)
```

	nation	age_category	deaths	crude	population	_merge
1.	Sweden	0 - 29	3,523	.0011202	.35	3
2.	Panama	0 - 29	3,904	.0052686	.35	3
3.	Sweden	30 - 59	10,928	.0035747	.35	3
4.	Panama	30 - 59	1,421	.0051673	.35	3
5.	Sweden	60+	59,104	.0456754	.3	3
6.	Panama	60+	2,456	.0416271	.3	3

```
. generate product = crude*pop
. by nation, sort: egen adj_rate = total(product)
```

```
. drop _merge

. sort nation age_category

. list, sepby(nation)
```

	nation	age_ca~y	deaths	crude	popula~n	product	adj_rate
1.	Panama	0 - 29	3,904	.0052686	.35	.001844	.0161407
2.	Panama	30 - 59	1,421	.0051673	.35	.0018085	.0161407
3.	Panama	60+	2,456	.0416271	.3	.0124881	.0161407
4.	Sweden	0 - 29	3,523	.0011202	.35	.0003921	.0153459
5.	Sweden	30 - 59	10,928	.0035747	.35	.0012512	.0153459
6.	Sweden	60+	59,104	.0456754	.3	.0137026	.0153459

Comparing the standardized rates indicates that the Swedes have a slightly lower mortality rate.

To perform the above analysis with dstdize,

```
. use http://www.stata-press.com/data/r10/mortality, clear
(1962 Mortality, Sweden & Panama)

. dstdize deaths pop age_cat, by(nation) using(1962)
```

```
-> nation= Panama
                       ──────Unadjusted───── Std.
                            Pop.   Stratum   Pop.
    Stratum      Pop.   Cases Dist. Rate[s] Dst[P]  s*P

     0 - 29    741000    3904  0.689 0.0053  0.350 0.0018
    30 - 59    275000    1421  0.256 0.0052  0.350 0.0018
        60+     59000    2456  0.055 0.0416  0.300 0.0125

Totals:      1075000    7781     Adjusted Cases:   17351.2
                                    Crude Rate:     0.0072
                                 Adjusted Rate:     0.0161
                   95% Conf. Interval: [0.0156, 0.0166]
```

```
-> nation= Sweden
                       ──────Unadjusted───── Std.
                            Pop.   Stratum   Pop.
    Stratum      Pop.   Cases Dist. Rate[s] Dst[P]  s*P

     0 - 29   3145000    3523  0.420 0.0011  0.350 0.0004
    30 - 59   3057000   10928  0.408 0.0036  0.350 0.0013
        60+   1294000   59104  0.173 0.0457  0.300 0.0137

Totals:      7496000   73555     Adjusted Cases: 115032.5
                                    Crude Rate:     0.0098
                                 Adjusted Rate:     0.0153
                   95% Conf. Interval: [0.0152, 0.0155]
```

```
Summary of Study Populations:
    nation          N     Crude    Adj_Rate      Confidence Interval

    Panama    1075000  0.007238    0.016141    [  0.015645,    0.016637]
    Sweden    7496000  0.009813    0.015346    [  0.015235,    0.015457]
```

The summary table above lets us make a quick inspection of the results within the study populations, and the detail tables give the behavior among the strata within the study populations.

◁

▷ Example 2

We have individual-level data on persons in four cities over several years. Included in the data is a variable indicating whether the person has high blood pressure, together with information on the person's age, sex, and race. We wish to obtain standardized high blood pressure rates for each city for 1990 and 1992, using, as the standard, the age, sex, and race distribution of the four cities and two years combined.

Our dataset contains

```
. use http://www.stata-press.com/data/r10/hbp
. describe
Contains data from http://www.stata-press.com/data/r10/hbp.dta
  obs:         1,130
  vars:            7                          21 Feb 2007 06:42
  size:       23,730 (96.2% of memory free)
```

variable name	storage type	display format	value label	variable label
id	str10	%10s		Record identification number
city	byte	%8.0g		
year	int	%8.0g		
sex	byte	%8.0g	sexfmt	
age_group	byte	%8.0g	agefmt	
race	byte	%8.0g	racefmt	
hbp	byte	%8.0g	yn	high blood pressure

```
Sorted by:
```

The `dstdize` command is designed to work with aggregate data but will work with individual-level data only if we create a variable recording the population represented by each observation. For individual-level data, this is one:

```
. gen pop = 1
```

On the next page, we specify `print` to obtain a listing of the standard population and `level(90)` to request 90% rather than 95% confidence intervals. Typing `if year==1990 | year==1992` restricts the data to the two years for both summary tables and the standard population.

. dstdize hbp pop age race sex if year==1990 | year==1992, by(city year) print level(90)

```
─────────────Standard Population──────────────
                  Stratum      Pop.    Dist.
─────────────────────────────────────────────
15 - 19    Black   Female      35     0.077
15 - 19    Black     Male      44     0.097
15 - 19 Hispanic   Female       5     0.011
15 - 19 Hispanic     Male      10     0.022
15 - 19    White   Female       7     0.015
15 - 19    White     Male       5     0.011
20 - 24    Black   Female      43     0.095
20 - 24    Black     Male      67     0.147
20 - 24 Hispanic   Female      14     0.031
20 - 24 Hispanic     Male      13     0.029
20 - 24    White   Female       4     0.009
20 - 24    White     Male      21     0.046
25 - 29    Black   Female      17     0.037
25 - 29    Black     Male      44     0.097
25 - 29 Hispanic   Female       7     0.015
25 - 29 Hispanic     Male      13     0.029
25 - 29    White   Female       9     0.020
25 - 29    White     Male      16     0.035
30 - 34    Black   Female      16     0.035
30 - 34    Black     Male      32     0.070
30 - 34 Hispanic   Female       2     0.004
30 - 34 Hispanic     Male       3     0.007
30 - 34    White   Female       5     0.011
30 - 34    White     Male      23     0.051
─────────────────────────────────────────────
```

Total: 455
(6 observations excluded due to missing values)

-> city year= 1 1990

	Stratum	Pop.	Cases	Unadjusted Pop. Dist.	Stratum Rate[s]	Std. Pop. Dst[P]	s*P
15 - 19	Black Female	6	2	0.128	0.3333	0.077	0.0256
15 - 19	Black Male	6	0	0.128	0.0000	0.097	0.0000
15 - 19	Hispanic Male	1	0	0.021	0.0000	0.022	0.0000
20 - 24	Black Female	3	0	0.064	0.0000	0.095	0.0000
20 - 24	Black Male	11	0	0.234	0.0000	0.147	0.0000
25 - 29	Black Female	4	0	0.085	0.0000	0.037	0.0000
25 - 29	Black Male	6	1	0.128	0.1667	0.097	0.0161
25 - 29	Hispanic Female	2	0	0.043	0.0000	0.015	0.0000
25 - 29	White Female	1	0	0.021	0.0000	0.020	0.0000
30 - 34	Black Female	1	0	0.021	0.0000	0.035	0.0000
30 - 34	Black Male	6	0	0.128	0.0000	0.070	0.0000

Totals: 47   3   Adjusted Cases: 2.0
                                                  Crude Rate: 0.0638
                                               Adjusted Rate: 0.0418
                              90% Conf. Interval: [0.0074, 0.0761]

(output omitted)

```
-> city year= 5 1992
```

		Stratum	Pop.	Cases	Unadjusted Pop. Dist.	Stratum Rate[s]	Std. Pop. Dst[P]	s*P
15 - 19	Black	Female	6	0	0.087	0.0000	0.077	0.0000
15 - 19	Black	Male	9	0	0.130	0.0000	0.097	0.0000
15 - 19	Hispanic	Female	1	0	0.014	0.0000	0.011	0.0000
15 - 19	Hispanic	Male	2	0	0.029	0.0000	0.022	0.0000
15 - 19	White	Female	2	0	0.029	0.0000	0.015	0.0000
15 - 19	White	Male	1	0	0.014	0.0000	0.011	0.0000
20 - 24	Black	Female	13	0	0.188	0.0000	0.095	0.0000
20 - 24	Black	Male	10	0	0.145	0.0000	0.147	0.0000
20 - 24	Hispanic	Male	1	0	0.014	0.0000	0.029	0.0000
20 - 24	White	Male	3	0	0.043	0.0000	0.046	0.0000
25 - 29	Black	Female	2	0	0.029	0.0000	0.037	0.0000
25 - 29	Black	Male	2	0	0.029	0.0000	0.097	0.0000
25 - 29	Hispanic	Male	3	0	0.043	0.0000	0.029	0.0000
25 - 29	White	Male	1	0	0.014	0.0000	0.035	0.0000
30 - 34	Black	Female	4	0	0.058	0.0000	0.035	0.0000
30 - 34	Black	Male	5	0	0.072	0.0000	0.070	0.0000
30 - 34	Hispanic	Male	2	0	0.029	0.0000	0.007	0.0000
30 - 34	White	Female	1	0	0.014	0.0000	0.011	0.0000
30 - 34	White	Male	1	1	0.014	1.0000	0.051	0.0505

```
Totals:                        69        1      Adjusted Cases:      3.5
                                                   Crude Rate:      0.0145
                                                Adjusted Rate:      0.0505
                          90% Conf. Interval: [0.0505, 0.0505]
```

Summary of Study Populations:

city year	N	Crude	Adj_Rate	Confidence Interval	
1 1990	47	0.063830	0.041758	[0.007427,	0.076089]
1 1992	56	0.017857	0.008791	[0.000000,	0.022579]
2 1990	64	0.046875	0.044898	[0.009072,	0.080724]
2 1992	67	0.029851	0.014286	[0.002537,	0.026035]
3 1990	69	0.159420	0.088453	[0.050093,	0.126813]
3 1992	37	0.189189	0.046319	[0.025271,	0.067366]
5 1990	46	0.043478	0.022344	[0.002044,	0.042644]
5 1992	69	0.014493	0.050549	[0.050549,	0.050549]

Indirect standardization

Standardization of rates can be performed via the indirect method whenever the stratum-specific rates are either unknown or unreliable. If the stratum-specific rates are known, the direct standardization method is preferred.

To apply the indirect method, you must have the following information:

1. The observed number of cases in each population to be standardized, O. For example, if death rates in two states are being standardized using the U.S. death rate for the same period, you must know the total number of deaths in each state.

2. The distribution across the various strata for the population being studied, $n_1, \ldots, n_k$. If you are standardizing the death rate in the two states adjusting for age, you must know the number of individuals in each of the k age groups.

3. The stratum-specific rates for the standard population, $p_1, \ldots, p_k$. For example, you must have the U.S. death rate for each stratum (age group).

4. The crude rate of the standard population, C. For example, you must have the U.S. mortality rate for the year.

The indirect adjusted rate is then

$$R_{\text{indirect}} = C\frac{O}{E}$$

where E is the expected number of cases (deaths) in each population. See *Methods and Formulas* for a more detailed description of calculations.

▷ Example 3

This example is borrowed from Kahn and Sempos (1989, 95–105). We want to compare 1970 mortality rates in California and Maine, adjusting for age. Although we have age-specific population counts for the two states, we lack age-specific death rates. Direct standardization is not feasible here. We can use the U.S. population census data for the same year to produce indirectly standardized rates for these two states.

From the U.S. census, the standard population for this example was entered into Stata and saved in popkahn.dta.

```
. use http://www.stata-press.com/data/r10/popkahn, clear
. list age pop deaths rate, sep(4)
```

	age	population	deaths	rate
1.	<15	57,900,000	103,062	.00178
2.	15-24	35,441,000	45,261	.00128
3.	25-34	24,907,000	39,193	.00157
4.	35-44	23,088,000	72,617	.00315
5.	45-54	23,220,000	169,517	.0073
6.	55-64	18,590,000	308,373	.01659
7.	65-74	12,436,000	445,531	.03583
8.	75+	7,630,000	736,758	.09656

The standard population contains for each age stratum the total number of individuals (pop) and both the age-specific mortality rate (rate) and the number of deaths. The standard population need not contain all three. If we have only the age-specific mortality rate, we can use the $\mathtt{rate}(ratevar_p\ crudevar_p)$ or $\mathtt{rate}(ratevar_p\ \#)$ option, where $crudevar_p$ refers to the variable containing the total population's crude death rate or $\#$ is the total population's crude death rate.

Now let's look at the states' data (study population):

```
. use http://www.stata-press.com/data/r10/kahn
. list, sep(4)
```

	state	age	populat~n	death	st	death_~e
1.	California	<15	5,524,000	166,285	1	.0016
2.	California	15-24	3,558,000	166,285	1	.0013
3.	California	25-34	2,677,000	166,285	1	.0015
4.	California	35-44	2,359,000	166,285	1	.0028
5.	California	45-54	2,330,000	166,285	1	.0067
6.	California	55-64	1,704,000	166,285	1	.0154
7.	California	65-74	1,105,000	166,285	1	.0328
8.	California	75+	696,000	166,285	1	.0917
9.	Maine	<15	286,000	11,051	2	.0019
10.	Maine	15-24	168,000	.	2	.0011
11.	Maine	25-34	110,000	.	2	.0014
12.	Maine	35-44	109,000	.	2	.0029
13.	Maine	45-54	110,000	.	2	.0069
14.	Maine	55-64	94,000	.	2	.0173
15.	Maine	65-74	69,000	.	2	.039
16.	Maine	75+	46,000	.	2	.1041

For each state the number of individuals in each stratum (age group) is contained in the variable pop. The death variable is the total number of deaths observed in the state during the year. It must have the same value for all observations in the group, as for California, or it could be missing in all but one observation per group, as for Maine.

To match these two datasets, the strata variables must have the same name in both datasets and ideally the same levels. If a level is missing from either dataset, that level will not be included in the standardization.

With kahn.dta in memory, we now execute the command. We will use the print option to obtain the standard population's summary table, and since we have both the standard population's age-specific count and deaths, we will specify the popvars($casevar_p\ popvar_p$) option. Or, we could specify the rate(rate 0.00945) option since we know that 0.00945 is the U.S. crude death rate for 1970.

```
. istdize death pop age using http://www.stata-press.com/data/r10/popkahn,
> by(state) pop(deaths pop) print
```

```
————Standard Population————
Stratum                  Rate

  <15               0.00178
15-24               0.00128
25-34               0.00157
35-44               0.00315
45-54               0.00730
55-64               0.01659
65-74               0.03583
  75+               0.09656
```

Standard population's crude rate: 0.00945

```
-> state= California
                Indirect Standardization
                Standard
                Population     Observed       Cases
Stratum           Rate       Population     Expected

  <15           0.0018        5524000        9832.72
15-24           0.0013        3558000        4543.85
25-34           0.0016        2677000        4212.46
35-44           0.0031        2359000        7419.59
45-54           0.0073        2330000       17010.10
55-64           0.0166        1704000       28266.14
65-74           0.0358        1105000       39587.63
  75+           0.0966         696000       67206.23
```

Totals: 19953000   178078.73

```
                              Observed Cases:    166285
                              SMR (Obs/Exp):        0.93
         SMR exact 95% Conf. Interval: [0.9293, 0.9383]
                                 Crude Rate:      0.0083
                              Adjusted Rate:      0.0088
                    95% Conf. Interval: [0.0088, 0.0089]
```

(Continued on next page)

```
-> state= Maine
                    Indirect Standardization
                    Standard
                    Population    Observed      Cases
      Stratum         Rate       Population    Expected
```

Stratum	Standard Population Rate	Observed Population	Cases Expected
<15	0.0018	286000	509.08
15-24	0.0013	168000	214.55
25-34	0.0016	110000	173.09
35-44	0.0031	109000	342.83
45-54	0.0073	110000	803.05
55-64	0.0166	94000	1559.28
65-74	0.0358	69000	2471.99
75+	0.0966	46000	4441.79

```
Totals:                          992000      10515.67
                                       Observed Cases:      11051
                                       SMR (Obs/Exp):        1.05
                    SMR exact 95% Conf. Interval: [1.0314, 1.0707]
                                           Crude Rate:      0.0111
                                        Adjusted Rate:      0.0099
                             95% Conf. Interval: [0.0097, 0.0101]
```

Summary of Study Populations (Rates):

state	Cases Observed	Crude	Adj_Rate	Confidence Interval
California	166285	0.008334	0.008824	[0.008782, 0.008866]
Maine	11051	0.011140	0.009931	[0.009747, 0.010118]

Summary of Study Populations (SMR):

state	Cases Observed	Cases Expected	SMR	Exact Confidence Interval
California	166285	178078.73	0.934	[0.929290, 0.938271]
Maine	11051	10515.67	1.051	[1.031405, 1.070687]

◁

Saved Results

dstdize saves the following in r():

Scalars

    r(k)   number of populations

Macros

    r(by)   variable names specified in by()

    r(c#)   values of r(by) for #th group

    r(se)   standard errors of adjusted rates

    r(ub)   upper bounds of confidence intervals for adjusted rates

    r(lb)   lower bounds of confidence intervals for adjusted rates

Matrices

    r(Nobs) $1 \times k$ vector of number of observations

    r(crude) $1 \times k$ vector of crude rates (*)

    r(adj) $1 \times k$ vector of adjusted rates (*)

                 (*) If, in a group, the number of observations is 0, then 9
                   is stored for the corresponding crude and adjusted rates.

Methods and Formulas

dstdize and istdize are implemented as ado-files.

The directly standardized rate, S_{R}, is defined by

$$S_{\mathrm{R}} = \frac{\sum_{i=1}^{k} w_i R_i}{\sum_{i=1}^{k} w_i}$$

(Rothman 1986, 44), where R_i is the stratum-specific rate in stratum i, and w_i is the weight for stratum i derived from the standard population.

If n_i is the population of stratum i, the standard error, $\mathrm{se}(S_{\mathrm{R}})$, in stratified sampling for proportions (ignoring the finite population correction) is

$$\mathrm{se}(S_{\mathrm{R}}) = \frac{1}{\sum w_i} \sqrt{\sum_{i=1}^{k} \frac{w_i{}^2 R_i (1 - R_i)}{n_i}}$$

(Cochran 1977, 108), from which the confidence intervals are calculated.

For indirect standardization, define O as the observed number of cases in each population to be standardized; $n_1, \ldots, n_k$ as the distribution across the various strata for the population being studied; $R_1, \ldots, R_k$ as the stratum-specific rates for the standard population; and C as the crude rate of the standard population. The expected number of cases (deaths), E, in each population is obtained by applying the standard population stratum-specific rates, $R_1, \ldots, R_k$, to the study populations:

$$E = \sum_{i=1}^{k} n_i R_i$$

The indirectly adjusted rate is then

$$R_{\mathrm{indirect}} = C \frac{O}{E}$$

and O/E is the study population's standardized mortality ratio (SMR) if death is the event of interest or the standardized incidence ratio (SIR) for studies of disease (or other) incidence.

The exact confidence interval is calculated for each estimated SMR by assuming a Poisson process as described in Breslow and Day (1987, 69–71). These intervals are obtained by first calculating the upper and lower bounds for the confidence interval of the Poisson-distributed observed events, O—say, L and U, respectively—and then computing $\mathrm{SMR}_L = L/E$ and $\mathrm{SMR}_U = U/E$.

Acknowledgments

We gratefully acknowledge the collaboration of Dr. Joel A. Harrison, Cigna Healthcare of Texas; Dr. José Maria Pacheco from the Departamento de Epidemiologia, Faculdade de Saúde Pública/USP, Sao Paulo, Brazil; and Dr John L. Moran from The Queen Elizabeth Hospital, Woodville, Australia.

References

Breslow, N. E., and N. E. Day. 1987. *Statistical Methods in Cancer Research: The Design and Analysis of Cohort Studies, Vol. 2.* Lyon: IARC.

Cleves, M. A. 1998. sg80: Indirect standardization. *Stata Technical Bulletin* 42: 43–47. Reprinted in *Stata Technical Bulletin Reprints*, vol. 7, pp. 224–228.

Cochran, W. G. 1977. *Sampling Techniques.* 3rd ed. New York: Wiley.

Fleiss, J. L., B. Levin, and M. C. Paik. 2003. *Statistical Methods for Rates and Proportions.* 3rd ed. New York: Wiley.

Forthofer, R. N., and E. S. Lee. 1995. *Introduction to Biostatistics: A Guide to Design, Analysis, and Discovery.* New York: Academic Press.

Juul, S. 2006. *An Introduction to Stata for Health Researchers.* College Station, TX: Stata Press.

Kahn, H. A., and C. T. Sempos. 1989. *Statistical Methods in Epidemiology.* New York: Oxford University Press.

Kirkwood, B. R., and J. A. C. Sterne. 2003. *Essential Medical Statistics.* 2nd ed. Malden, MA: Blackwell.

McGuire, T. J., and J. A. Harrison. 1994. sbe11: Direct standardization. *Stata Technical Bulletin* 21: 5–9. Reprinted in *Stata Technical Bulletin Reprints*, vol. 4, pp. 88–94.

Pagano, M., and K. Gauvreau. 2000. *Principles of Biostatistics.* 2nd ed. Belmont, CA: Duxbury.

Rothman, K. J. 1986. *Modern Epidemiology.* Boston: Little, Brown.

van Belle, G., L. D. Fisher, P. J. Heagerty, and T. Lumley. 2004. *Biostatistics: A Methodology for the Health Sciences.* 2nd ed. New York: Wiley.

Wang, D. 2000. sbe40: Modeling mortality data using the Lee–Carter model. *Stata Technical Bulletin* 57: 15–17. Reprinted in *Stata Technical Bulletin Reprints*, vol. 10, pp. 118–121.

Also See

[ST] **epitab** — Tables for epidemiologists

[SVY] **direct standardization** — Direct standardization of means, proportions, and ratios

Title

> **dydx** — Calculate numeric derivatives and integrals

Syntax

Derivatives of numeric functions

> dydx *yvar* *xvar* $\left[\textit{if}\right]$ $\left[\textit{in}\right]$, <u>g</u>enerate(*newvar*) $\left[\textit{dydx\_options}\right]$

Integrals of numeric functions

> integ *yvar* *xvar* $\left[\textit{if}\right]$ $\left[\textit{in}\right]$ $\left[\text{, }\textit{integ\_options}\right]$

dydx_options	description
Main	
* <u>g</u>enerate(*newvar*)	create variable named *newvar*
replace	overwrite the existing variable

> *generate(*newvar*) is required.

integ_options	description
Main	
<u>g</u>enerate(*newvar*)	create variable named *newvar*
<u>t</u>rapezoid	use trapezoidal rule to compute integrals; default is cubic splines
<u>i</u>nitial(#)	initial value of integral; default is 0
replace	overwrite the existing variable

by is allowed with dydx and integ; see [D] **by**.

Description

dydx and integ calculate derivatives and integrals of numeric "functions".

Options

> ⌐ Main ⌐

generate(*newvar*) specifies the name of the new variable to be created. It must be specified with dydx.

trapezoid requests that the trapezoidal rule $\big($the sum of $(x_i - x_{i-1})(y_i + y_{i-1})/2\big)$ be used to compute integrals. The default is cubic splines, which give superior results for most smooth functions; for irregular functions, trapezoid may give better results.

initial(#) specifies the initial condition for calculating definite integrals; see *Methods and Formulas* below. If not specified, the initial condition is taken as 0.

replace specifies that if an existing variable is specified for generate(), it should be overwritten.

Remarks

dydx and integ lets you to extend Stata's graphics capabilities beyond data analysis and into mathematics. (See Gould 1993 for another command that draws functions.)

▷ Example 1

We graph $y = e^{-x/6}\sin(x)$ over the interval $[0, 12.56]$:

```
. range x 0 12.56 100
obs was 0, now 100
. generate y = exp(-x/6)*sin(x)
. label variable y "exp(-x/6)*sin(x)"
. twoway connected y x, connect(i) yline(0)
```

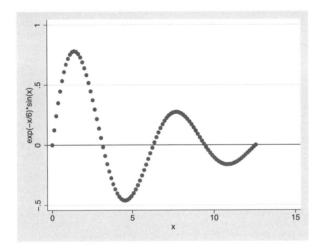

We estimate the derivative by using dydx and compute the relative difference between this estimate and the true derivative.

```
. dydx y x, gen(dy)
. generate dytrue = exp(-x/6)*(cos(x) - sin(x)/6)
. generate error = abs(dy - dytrue)/dytrue
```

The error is greatest at the endpoints, as we would expect. The error is approximately 0.5% at each endpoint, but the error quickly falls to less than 0.01%.

```
. label variable error "Error in derivative estimate"
. twoway line error x, ylabel(0(.002).006)
```

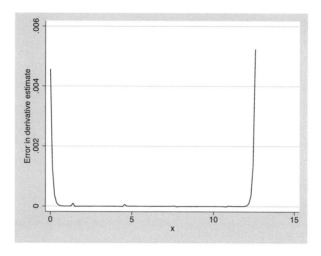

We now estimate the integral by using `integ`:

```
. integ y x, gen(iy)
number of points = 100
integral        = .85316396
. generate iytrue = (36/37)*(1 - exp(-x/6)*(cos(x) + sin(x)/6))
. display iytrue[_N]
.85315901
. display abs(r(integral) - iytrue[_N])/iytrue[_N]
5.799e-06
. generate diff = iy - iytrue
```

The relative difference between the estimate $\big($stored in $r(\texttt{integral})\big)$ and the true value of the integral is about 6×10^{-6}. A graph of the absolute difference (`diff`) is shown below. Here error is cumulative. Again most of the error is due to a relatively poorer fit near the endpoints.

```
. label variable diff "Error in integral estimate"
. twoway line diff x, ylabel(0(5.00e-06).00001)
```

◁

Saved Results

dydx saves the following in r():

Macros

r(y)   name of *yvar*

integ saves the following in r():

Scalars

r(N_points)   number of unique x points
r(integral)   estimate of the integral

Methods and Formulas

dydx and integ are implemented as ado-files.

Consider a set of data points, $(x_1, y_1), \ldots, (x_n, y_n)$, generated by a function $y = f(x)$. dydx and integ first fit these points with a cubic spline, which is then analytically differentiated (integrated) to give an approximation for the derivative (integral) of f.

The cubic spline (see, for example, Press et al. [1992]) consists of $n - 1$ cubic polynomials $P_i(x)$, with the ith one defined on the interval $[x_i, x_{i+1}]$,

$$P_i(x) = y_i a_i(x) + y_{i+1} b_i(x) + y_i'' c_i(x) + y_{i+1}'' d_i(x)$$

where

$$a_i(x) = \frac{x_{i+1} - x}{x_{i+1} - x_i} \qquad\qquad b_i(x) = \frac{x - x_i}{x_{i+1} - x_i}$$

$$c_i(x) = \frac{1}{6}(x_{i+1} - x_i)^2 a_i(x)[\{a_i(x)\}^2 - 1] \qquad d_i(x) = \frac{1}{6}(x_{i+1} - x_i)^2 b_i(x)[\{b_i(x)\}^2 - 1]$$

and y_i'' and y_{i+1}'' are constants whose values will be determined as described below. The notation for these constants is justified because $P_i''(x_i) = y_i''$ and $P_i''(x_{i+1}) = y_{i+1}''$.

Since $a_i(x_i) = 1$, $a_i(x_{i+1}) = 0$, $b_i(x_i) = 0$, and $b_i(x_{i+1}) = 1$. Therefore, $P_i(x_i) = y_i$, and $P_i(x_{i+1}) = y_{i+1}$. Thus the P_i jointly define a function that is continuous at the interval boundaries. The first derivative should be continuous at the interval boundaries; that is,

$$P_i'(x_{i+1}) = P_{i+1}'(x_{i+1})$$

The above $n - 2$ equations (one equation for each point except the two endpoints) and the values of the first derivative at the endpoints, $P_1'(x_1)$ and $P_{n-1}'(x_n)$, determine the n constants y_i''.

The value of the first derivative at an endpoint is set to the value of the derivative obtained by fitting a quadratic to the endpoint and the two adjacent points; namely, we use

$$P_1'(x_1) = \frac{y_1 - y_2}{x_1 - x_2} + \frac{y_1 - y_3}{x_1 - x_3} - \frac{y_2 - y_3}{x_2 - x_3}$$

and a similar formula for the upper endpoint.

dydx approximates $f'(x_i)$ using $P_i'(x_i)$.

`integ` approximates $F(x_i) = F(x_1) + \int_{x_1}^{x_i} f(x)\,dx$ using

$$I_0 + \sum_{k=1}^{i-1} \int_{x_k}^{x_{k+1}} P_k(x)\,dx$$

where I_0 (an estimate of $F(x_1)$) is the value specified by the `initial(#)` option. If the `trapezoid` option is specified, `integ` approximates the integral by using the trapezoidal rule:

$$I_0 + \sum_{k=1}^{i-1} \frac{1}{2}(x_{k+1} - x_k)(y_{k+1} + y_k)$$

If there are ties among the x_i, the mean of y_i is computed at each set of ties and the cubic spline is fitted to these values.

Acknowledgment

The present versions of `dydx` and `integ` were inspired by the `dydx2` command written by Patrick Royston of the MRC Clinical Trials Unit, London.

References

Gould, W. W. 1993. ssi5.1: Graphing functions. *Stata Technical Bulletin* 16: 23–26. Reprinted in *Stata Technical Bulletin Reprints*, vol. 3, pp. 188–193.

——. 1997. crc46: Better numerical derivatives and integrals. *Stata Technical Bulletin* 35: 3–5. Reprinted in *Stata Technical Bulletin Reprints*, vol. 6, pp. 8–12.

Press, W. H., S. A. Teukolsky, W. T. Vetterling, and B. P. Flannery. 1992. *Numerical Recipes in C, The Art of Scientific Computing*. 2nd ed. Cambridge: Cambridge University Press.

Also See

[D] **obs** — Increase the number of observations in a dataset

[D] **range** — Generate numerical range

Title

> **eivreg** — Errors-in-variables regression

Syntax

eivreg *depvar* [*indepvars*] [*if*] [*in*] [*weight*]

　　[, <u>r</u>eliab(*indepvar # [indepvar # [. . .]]*) <u>l</u>evel(*#*)]

bootstrap, by, jackknife, rolling, statsby, and xi are allowed; see [U] **11.1.10 Prefix commands**.
Weights are not allowed with the bootstrap prefix.
aweights are not allowed with the jackknife prefix.
aweights and fweights are allowed; see [U] **11.1.6 weight**.
See [U] **20 Estimation and postestimation commands** for more capabilities of estimation commands.

Description

eivreg fits errors-in-variables regression models.

Options

> Model

reliab(*indepvar # [indepvar # [. . .]]*) specifies the measurement reliability for each independent
variable measured with error. Reliabilities are specified as pairs consisting of an independent
variable name (a name that appears in *indepvars*) and the corresponding reliability r, $0 < r \le 1$.
Independent variables for which no reliability is specified are assumed to have reliability 1. If the
option is not specified, all variables are assumed to have reliability 1, and the result is thus the
same as that produced by regress (the ordinary least-squares results).

> Reporting

level(*#*); see [R] **estimation options**.

Remarks

For an introduction to errors-in-variables regression, see Kmenta (1997, 352–357).

Errors-in-variables regression models are useful when one or more of the independent variables are
measured with additive noise. Standard regression (as performed by regress) would underestimate
the effect of the variable, and the other coefficients in the model can be biased to the extent that
they are correlated with the poorly measured variable. You can adjust for the biases if you know the
reliability:

$$r = 1 - \frac{\text{noise variance}}{\text{total variance}}$$

That is, given the model $\mathbf{y} = \mathbf{X}\mathcal{B} + \mathbf{u}$, for some variable $\mathbf{x}_i$ in $\mathbf{X}$, the $\mathbf{x}_i$ is observed with error,
$\mathbf{x}_i = \mathbf{x}_i^* + \mathbf{e}$, and the noise variance is the variance of $\mathbf{e}$. The total variance is the variance of $\mathbf{x}_i$.

▷ Example 1

Say that in our automobile data, the weight of cars was measured with error, and the reliability of our measured weight is 0.85. The result of this would be to underestimate the effect of `weight` in a regression of, say, `price` on `weight` and `foreign`, and it would also bias the estimate of the coefficient on `foreign` (because being of foreign manufacture is correlated with the weight of cars). We would ignore all this if we fitted the model with `regress`:

```
. use http://www.stata-press.com/data/r10/auto
(1978 Automobile Data)

. regress price weight foreign
```

Source	SS	df	MS
Model	316859273	2	158429637
Residual	318206123	71	4481776.38
Total	635065396	73	8699525.97

Number of obs = 74
F(2,   71) = 35.35
Prob > F = 0.0000
R-squared = 0.4989
Adj R-squared = 0.4848
Root MSE = 2117

price	Coef.	Std. Err.	t	P>|t|	[95% Conf. Interval]	
weight	3.320737	.3958784	8.39	0.000	2.531378	4.110096
foreign	3637.001	668.583	5.44	0.000	2303.885	4970.118
_cons	-4942.844	1345.591	-3.67	0.000	-7625.876	-2259.812

With `eivreg`, we can account for our measurement error:

```
. eivreg price weight foreign, r(weight .85)
```

variable	assumed reliability
weight	0.8500
*	1.0000

Errors-in-variables regression

Number of obs = 74
F(2,   71) = 50.37
Prob > F = 0.0000
R-squared = 0.6483
Root MSE = 1773.54

price	Coef.	Std. Err.	t	P>|t|	[95% Conf. Interval]	
weight	4.31985	.431431	10.01	0.000	3.459601	5.180099
foreign	4637.32	624.5362	7.43	0.000	3392.03	5882.609
_cons	-8257.017	1452.086	-5.69	0.000	-11152.39	-5361.639

The effect of weight is increased — as we knew it would be — and here the effect of foreign manufacture is also increased. A priori, we knew only that the estimate of `foreign` might be biased; we did not know the direction.

◁

❑ Technical Note

Swept under the rug in our example is how we would determine the reliability, r. We can easily see that a variable is measured with error, but we may not know the reliability since the ingredients for calculating r depend on the unobserved noise.

For our example, we made up a value for r, and in fact we do not believe that weight is measured with error at all, so the reported `eivreg` results have no validity. The `regress` results were the statistically correct results here.

But let's say that we do suspect that weight is measured with error and that we do not know r. We could then experiment with various values of r to describe the sensitivity of our estimates to possible error levels. We may not know r, but r does have a simple interpretation, and we could probably produce a sensible range for r by thinking about how the data were collected.

If the reliability, r, is less than the R^2 from a regression of the poorly measured variable on all the other variables, including the dependent variable, the information might as well not have been collected; no adjustment to the final results is possible. For our automobile data, running a regression of weight on foreign and price would result in an R^2 of .6743. Thus the reliability must be at least .6743 here. If we specify a reliability that is too small, eivreg will inform us and refuse to fit the model:

```
. eivreg price weight foreign, r(weight .6742)
reliability r() too small
r(399);
```

Returning to our problem of how to estimate r, too small or not, if the measurements are summaries of scaled items, the reliability may be estimated using the alpha command; see [R] **alpha**. If the score is computed from factor analysis and the data are scored using predict's default options (see [MV] **factor postestimation**), the square of the standard deviation of the score is an estimate of the reliability.

❏

❏ Technical Note

Consider a model with more than one variable measured with error. For instance, say that our model is that price is a function of weight, foreign, and mileage rating and that both weight and mileage rating are measured with error.

```
. eivreg price weight foreign mpg, r(weight .85 mpg .9)
```

variable	assumed reliability		Errors-in-variables regression	
			Number of obs =	74
weight	0.8500		F(3, 70) =	429.14
mpg	0.9000		Prob > F =	0.0000
*	1.0000		R-squared =	0.9728
			Root MSE =	496.41

price	Coef.	Std. Err.	t	P>|t|	[95% Conf. Interval]	
weight	12.88302	.6820532	18.89	0.000	11.52271	14.24333
foreign	8268.951	352.8719	23.43	0.000	7565.17	8972.732
mpg	999.2043	73.60037	13.58	0.000	852.413	1145.996
_cons	-56473.19	3710.015	-15.22	0.000	-63872.58	-49073.8

❏

Saved Results

eivreg saves the following in e():

Scalars

e(N)	number of observations	e(r2)	R-squared
e(df_m)	model degrees of freedom	e(F)	F statistic
e(df_r)	residual degrees of freedom	e(rmse)	root mean squared error

Macros

e(cmd)	eivreg	e(wtype)	weight type
e(cmdline)	command as typed	e(wexp)	weight expression
e(depvar)	name of dependent variable	e(properties)	b V
e(rellist)	*indepvars* and associated reliabilities	e(predict)	program used to implement predict

Matrices

e(b)	coefficient vector	e(V)	variance–covariance matrix of the estimators

Functions

e(sample)	marks estimation sample

Methods and Formulas

eivreg is implemented as an ado-file.

Let the model to be fitted be

$$\mathbf{y} = \mathbf{X}^* \mathcal{B} + \mathbf{e}$$
$$\mathbf{X} = \mathbf{X}^* + \mathbf{U}$$

where $\mathbf{X}^*$ are the true values and $\mathbf{X}$ the observed values. Let $\mathbf{W}$ be the user-specified weights. If no weights are specified, $\mathbf{W} = \mathbf{I}$. If weights are specified, let $\mathbf{v}$ be the specified weights. If fweight frequency weights are specified, then $\mathbf{W} = \mathrm{diag}(\mathbf{v})$. If aweight analytic weights are specified, then $\mathbf{W} = \mathrm{diag}\{\mathbf{v}/(\mathbf{1}'\mathbf{v})(\mathbf{1}'\mathbf{1})\}$, meaning that the weights are normalized to sum to the number of observations.

The estimates $\mathbf{b}$ of $\mathcal{B}$ are obtained as $\mathbf{A}^{-1}\mathbf{X}'\mathbf{W}\mathbf{y}$, where $\mathbf{A} = \mathbf{X}'\mathbf{W}\mathbf{X} - \mathbf{S}$. $\mathbf{S}$ is a diagonal matrix with elements $N(1 - r_i)s_i^2$. N is the number of observations, r_i is the user-specified reliability coefficient for the ith explanatory variable or 1 if not specified, and s_i^2 is the (appropriately weighted) variance of the variable.

The variance–covariance matrix of the estimators is obtained as $s^2\mathbf{A}^{-1}\mathbf{X}'\mathbf{W}\mathbf{X}\mathbf{A}^{-1}$, where the root mean squared error $s^2 = (\mathbf{y}'\mathbf{W}\mathbf{y} - \mathbf{b}\mathbf{A}\mathbf{b}')/(N - p)$, where p is the number of estimated parameters.

Reference

Kmenta, J. 1997. *Elements of Econometrics*. 2nd ed. Ann Arbor, MI: University of Michigan Press.

Also See

[R] **eivreg postestimation** — Postestimation tools for eivreg

[R] **regress** — Linear regression

[U] **20 Estimation and postestimation commands**

Title

eivreg postestimation — Postestimation tools for eivreg

Description

The following postestimation commands are available for `eivreg`:

command	description
adjust	adjusted predictions of $\mathbf{x}\beta$
estat	VCE and estimation sample summary
estimates	cataloging estimation results
lincom	point estimates, standard errors, testing, and inference for linear combinations of coefficients
linktest	link test for model specification
mfx	marginal effects or elasticities
nlcom	point estimates, standard errors, testing, and inference for nonlinear combinations of coefficients
predict	predictions, residuals, influence statistics, and other diagnostic measures
predictnl	point estimates, standard errors, testing, and inference for generalized predictions
test	Wald tests for simple and composite linear hypotheses
testnl	Wald tests of nonlinear hypotheses

See the corresponding entries in the *Stata Base Reference Manual* for details.

Syntax for predict

predict [*type*] *newvar* [*if*] [*in*] [, *statistic*]

statistic	description
Main	
xb	linear prediction; the default
residuals	residuals
stdp	standard error of the prediction
stdf	standard error of the forecast
pr(*a*,*b*)	$\Pr(a < y_j < b)$
e(*a*,*b*)	$E(y_j \mid a < y_j < b)$
ystar(*a*,*b*)	$E(y_j^*)$, $y_j^* = \max\{a, \min(y_j, b)\}$

These statistics are available both in and out of sample; type predict ... if e(sample) ... if wanted only for the estimation sample.

where a and b may be numbers or variables; a missing ($a \geq .$) means $-\infty$, and b missing ($b \geq .$) means $+\infty$; see [U] **12.2.1 Missing values**.

Options for predict

⌐ Main ⌐

xb, the default, calculates the linear prediction.

residuals calculates the residuals, that is, $y_j - \mathbf{x}_j\mathbf{b}$.

stdp calculates the standard error of the prediction, which can be thought of as the standard error of the predicted expected value or mean for the observation's covariate pattern. The standard error of the prediction is also referred to as the standard error of the fitted value.

stdf calculates the standard error of the forecast, which is the standard error of the point prediction for 1 observation and is commonly referred to as the standard error of the future or forecast value. By construction, the standard errors produced by stdf are always larger than those produced by stdp; see [R] **regress** *Methods and Formulas*.

pr(*a*,*b*) calculates $\Pr(a < \mathbf{x}_j\mathbf{b} + u_j < b)$, the probability that $y_j|\mathbf{x}_j$ would be observed in the interval (a, b).

  a and *b* may be specified as numbers or variable names; *lb* and *ub* are variable names;
  pr(20,30) calculates $\Pr(20 < \mathbf{x}_j\mathbf{b} + u_j < 30)$;
  pr(*lb*,*ub*) calculates $\Pr(lb < \mathbf{x}_j\mathbf{b} + u_j < ub)$; and
  pr(20,*ub*) calculates $\Pr(20 < \mathbf{x}_j\mathbf{b} + u_j < ub)$.

  a missing ($a \geq .$) means $-\infty$; pr(.,30) calculates $\Pr(-\infty < \mathbf{x}_j\mathbf{b} + u_j < 30)$;
  pr(*lb*,30) calculates $\Pr(-\infty < \mathbf{x}_j\mathbf{b} + u_j < 30)$ in observations for which *lb* $\geq$.
  and calculates $\Pr(lb < \mathbf{x}_j\mathbf{b} + u_j < 30)$ elsewhere.

  b missing ($b \geq .$) means $+\infty$; pr(20,.) calculates $\Pr(+\infty > \mathbf{x}_j\mathbf{b} + u_j > 20)$;
  pr(20,*ub*) calculates $\Pr(+\infty > \mathbf{x}_j\mathbf{b} + u_j > 20)$ in observations for which *ub* $\geq$.
  and calculates $\Pr(20 < \mathbf{x}_j\mathbf{b} + u_j < ub)$ elsewhere.

e(*a*,*b*) calculates $E(\mathbf{x}_j\mathbf{b} + u_j \mid a < \mathbf{x}_j\mathbf{b} + u_j < b)$, the expected value of $y_j|\mathbf{x}_j$ conditional on $y_j|\mathbf{x}_j$ being in the interval (a, b), meaning that $y_j|\mathbf{x}_j$ is censored. *a* and *b* are specified as they are for pr().

ystar(*a*,*b*) calculates $E(y_j^*)$, where $y_j^* = a$ if $\mathbf{x}_j\mathbf{b} + u_j \leq a$, $y_j^* = b$ if $\mathbf{x}_j\mathbf{b} + u_j \geq b$, and $y_j^* = \mathbf{x}_j\mathbf{b} + u_j$ otherwise, meaning that y_j^* is truncated. *a* and *b* are specified as they are for pr().

Methods and Formulas

All postestimation commands listed above are implemented as ado-files.

Also See

[R] **eivreg** — Errors-in-variables regression

[U] **20 Estimation and postestimation commands**

Title

error messages — Error messages and return codes

Description

Whenever Stata detects that something is wrong — that what you typed is uninterpretable, that you are trying to do something you should not be trying to do, or that you requested the impossible — Stata responds by typing a message describing the problem, together with a *return code*. For instance,

```
. lsit
unrecognized command:  lsit
r(199);
. list myvar
variable myvar not found
r(111);
. test a=b
last estimates not found
r(301);
```

In each case, the message is probably sufficient to guide you to a solution. When we typed `lsit`, Stata responded with "unrecognized command". We meant to type `list`. When we typed `list myvar`, Stata responded with "variable myvar not found". There is no variable named `myvar` in our data. When we typed `test a=b`, Stata responded with "last estimates not found". `test` tests hypotheses about previously fitted models, and we have not yet fitted a model.

The numbers in parentheses in the `r(199)`, `r(111)`, and `r(301)` messages are called the *return codes*. To find out more about these messages, type `search rc #`, where # is the number returned in the parentheses.

▷ Example 1

```
. search rc 301

[P]     error messages . . . . . . . . . . . . . . . . . . . . Return code 301
        last estimates not found;
        You typed an estimation command such as regress without arguments
        or attempted to perform a test or typed predict, but there were no
        previous estimation results.
```
◁

Programmers should see [P] **error** for details on programming error messages.

Also See

[R] **search** — Search Stata documentation

Title

> **estat** — Postestimation statistics

Syntax

Common subcommands

Obtain information criteria

> estat ic $\left[\, , \, \text{n}(\#) \right]$

Summarize estimation sample

> estat <u>summ</u>arize $\left[\, eqlist \, \right] \left[\, , \, estat\_summ\_options \, \right]$

Display covariance matrix estimates

> estat vce $\left[\, , \, estat\_vce\_options \, \right]$

Command-specific subcommands

> estat $subcommand_1 \left[\, , \, options_1 \, \right]$

> estat $subcommand_2 \left[\, , \, options_2 \, \right]$

> $\dots$

estat_summ_options	description
<u>eq</u>uation	display summary by equation
<u>l</u>abels	display variable labels
<u>noh</u>eader	suppress the header
<u>now</u>eights	ignore weights

eqlist is rarely used and specifies the variables, with optional equation name, to be summarized. *eqlist* may be *varlist* or ($eqname_1$: *varlist*) ($eqname_2$: *varlist*).... *varlist* may contain time-series operators; see [U] **11.4.3 Time-series varlists**.

estat_vce_options	description
<u>cov</u>ariance	display as covariance matrix; the default
<u>c</u>orrelation	display as correlation matrix
<u>eq</u>uation(*spec*)	display only specified equations
<u>b</u>lock	display submatrices by equation
<u>d</u>iag	display submatrices by equation; diagonal blocks only
<u>f</u>ormat(%*fmt*)	display format for covariances and correlations
<u>nolin</u>es	suppress lines between equations

Description

estat displays scalar- and matrix-valued statistics after estimation; it complements predict, which calculates variables after estimation. Exactly what statistics estat can calculate depends on the previous estimation command.

Three sets of statistics are so commonly used that they are available after all estimation commands that store the model log-likelihood. estat ic displays Akaike's and Schwarz's Bayesian information criteria. estat summarize summarizes the variables used by the command and automatically restricts the sample to e(sample); it also summarizes the weight variable and cluster structure, if specified. estat vce displays the covariance or correlation matrix of the parameter estimates of the previous model.

Option for estat ic

n(#) specifies the N to be used in calculating BIC; see [R] **BIC note**.

Options for estat summarize

equation requests that the dependent variables and the independent variables in the equations be displayed in the equation-style format of estimation commands, repeating the summary information about variables entered in more than one equation.

labels displays variable labels.

noheader suppresses the header.

noweights ignores the weights, if any, from the previous estimation command. The default when weights are present is to perform a weighted summarize on all variables except the weight variable itself. An unweighted summarize is performed on the weight variable.

Options for estat vce

covariance displays the matrix as a variance–covariance matrix; this is the default.

correlation displays the matrix as a correlation matrix rather than a variance–covariance matrix. rho is a synonym.

equation(spec) selects part of the VCE to be displayed. If spec is eqlist, the VCE for the listed equations is displayed. If spec is eqlist1 \ eqlist2, the part of the VCE associated with the equations in eqlist1 (rowwise) and eqlist2 (columnwise) is displayed. If spec is *, all equations are displayed. equation() implies block if diag is not specified.

block displays the submatrices pertaining to distinct equations separately.

diag displays the diagonal submatrices pertaining to distinct equations separately.

format(%fmt) specifies the number format for displaying the elements of the matrix. The default is format(%10.0g) for covariances and format(%8.4f) for correlations. See [U] **12.5 Formats: controlling how data are displayed** for more information.

nolines suppresses lines between equations.

Remarks

estat displays a variety of scalar- and matrix-valued statistics after you have estimated the parameters of a model. Exactly what statistics estat can calculate depends on the estimation command used, and command-specific statistics are detailed in that command's postestimation manual entry. The rest of this entry discusses three sets of statistics that are available after all estimation commands.

Remarks are presented under the following headings:

> estat ic
> estat summarize
> estat vce

estat ic

estat ic calculates two information criteria used to compare models. Unlike likelihood-ratio, Wald, and similar testing procedures, the models need not be nested to compare the information criteria. Because they are based on the log-likelihood function, information criteria are available only after commands that report the log likelihood.

In general, "smaller is better": given two models, the one with the smaller AIC fits the data better than the one with the larger AIC. As with the AIC, a smaller BIC indicates a better-fitting model. For AIC and BIC formulas, see *Methods and Formulas*.

▷ Example 1

In [R] **mlogit**, we fit a model explaining the type of insurance a person has on the basis of age, gender, race, and site of study. Here we refit the model with and without the site dummies and compare the models.

```
. use http://www.stata-press.com/data/r10/sysdsn3
(Health insurance data)

. mlogit insure age male nonwhite
(output omitted )

. estat ic
```

Model	Obs	ll(null)	ll(model)	df	AIC	BIC
.	615	-555.8545	-545.5833	8	1107.167	1142.54

Note: N=Obs used in calculating BIC; see **[R] BIC note**

```
. mlogit insure age male nonwhite site2 site3
(output omitted )

. estat ic
```

Model	Obs	ll(null)	ll(model)	df	AIC	BIC
.	615	-555.8545	-534.3616	12	1092.723	1145.783

Note: N=Obs used in calculating BIC; see **[R] BIC note**

The AIC indicates that the model including the site dummies fits the data better, whereas the BIC indicates the opposite. As is often the case, different model-selection criteria have led to conflicting conclusions.

◁

❑ Technical Note

glm and binreg, ml report a slightly different version of AIC and BIC; see [R] **glm** for the formulas used. That version is commonly used within the GLM literature; see, for example, Hardin and Hilbe (2007). The literature on information criteria is vast; see among others Akaike (1973), Sawa (1978), and Raftery (1995). Judge et al. (1985) contains a discussion of using information criteria in econometrics.

❑

estat summarize

Often when fitting a model, you will also be interested in obtaining summary statistics, such as the sample means and standard deviations of the variables in the model. estat summarize makes this process simple. The output displayed is similar to that obtained by typing

    . summarize *varlist* if e(sample)

without the need to type the *varlist* containing the dependent and independent variables.

▷ Example 2

Continuing with the previous multinomial logit model, here we summarize the variables using estat summarize

    . estat summarize
    Estimation sample mlogit   Number of obs =   615

Variable	Mean	Std. Dev.	Min	Max
insure	1.596748	.6225846	1	3
age	44.46832	14.18523	18.1109	86.0725
male	.2504065	.4335998	0	1
nonwhite	.196748	.3978638	0	1
site2	.3707317	.4833939	0	1
site3	.3138211	.4644224	0	1

◁

The output in the previous example contains all the variables in one table, though mlogit presents its results in a multiple-equation format. For models in which the same variables appear in all equations, that is fine; but for other multiple-equation models, we may prefer to have the variables separated by the equation in which they appear. The equation option makes this possible.

▷ Example 3

Systems of simultaneous equations typically have different variables in each equation, and the equation option of estat summarize is helpful in such situations. In [R] **reg3** we have a model of supply and demand. We first refit the model and then call estat summarize.

    . use http://www.stata-press.com/data/r10/supDem
    . reg3 (Demand:quantity price pcompete income) (Supply:quantity price praw),
    > endog(price)
      (*output omitted*)

```
. estat summarize, equation
Estimation sample reg3                    Number of obs =    49
```

Variable	Mean	Std. Dev.	Min	Max
depvar				
quantity	12.61818	2.774952	7.71069	20.0477
quantity	12.61818	2.774952	7.71069	20.0477
Demand				
price	32.70944	2.882684	26.3819	38.4769
pcompete	5.929975	3.508264	.207647	11.5549
income	7.811735	4.18859	.570417	14.0077
Supply				
price	32.70944	2.882684	26.3819	38.4769
praw	4.740891	2.962565	.151028	9.79881

The first block of the table contains statistics on the dependent (or, more accurately, left-hand side) variables, and since we specified quantity as the left-hand-side variable in both equations, it is listed twice. The second block refers to the variables in the first equation we specified, which we labeled "Demand" in our call to reg3; and the final block refers to the supply equation.

◁

estat vce

estat vce allows you to display the variance–covariance matrix (VCE) of the parameters of the previously fitted model, as either a covariance matrix or a correlation matrix.

▷ Example 4

Returning to the mlogit example, we type

```
. use http://www.stata-press.com/data/r10/sysdsn3
(Health insurance data)

. mlogit insure age male nonwhite, nolog
Multinomial logistic regression                Number of obs   =       615
                                               LR chi2(6)      =     20.54
                                               Prob > chi2     =    0.0022
Log likelihood = -545.58328                    Pseudo R2       =    0.0185
```

insure	Coef.	Std. Err.	z	P>|z|	[95% Conf. Interval]	
Prepaid						
age	-.0111915	.0060915	-1.84	0.066	-.0231305	.0007475
male	.5739825	.2005221	2.86	0.004	.1809665	.9669985
nonwhite	.7312659	.218978	3.34	0.001	.302077	1.160455
_cons	.1567003	.2828509	0.55	0.580	-.3976773	.7110778
Uninsure						
age	-.0058414	.0114114	-0.51	0.609	-.0282073	.0165245
male	.5102237	.3639793	1.40	0.161	-.2031626	1.22361
nonwhite	.433314	.4106255	1.06	0.291	-.3714972	1.238125
_cons	-1.811165	.5348606	-3.39	0.001	-2.859473	-.7628577

(insure==Indemnity is the base outcome)

```
. estat vce, block
```

Covariance matrix of coefficients of mlogit model

covariances of equation Prepaid

	age	male	nonwhite	_cons
age	.00003711			
male	-.00015303	.0402091		
nonwhite	-.00008948	.00470608	.04795135	
_cons	-.00159095	-.00398961	-.00628886	.08000462

covariances of equation Uninsure (row) by equation Prepaid (column)

	age	male	nonwhite	_cons
age	.00001753	-.00007926	-.00004564	-.00076886
male	-.00007544	.02188398	.0023186	-.00145923
nonwhite	-.00004577	.00250588	.02813553	-.00263872
_cons	-.00077045	-.00130535	-.00257593	.03888032

covariances of equation Uninsure

	age	male	nonwhite	_cons
age	.00013022			
male	-.00050406	.13248095		
nonwhite	-.00026145	.01505449	.16861327	
_cons	-.00562159	-.01686629	-.02474852	.28607591

The block option is particularly useful for multiple-equation estimators. The first block of output here corresponds to the VCE of the estimated parameters for the first equation — the square roots of the diagonal elements of this matrix are equal to the standard errors of the first equation's parameters. Similarly, the final block corresponds to the VCE of the parameters for the second equation. The middle block shows the covariances between the estimated parameters of the first and second equations.

◁

Saved Results

estat ic saves the following in r():

  Matrices
   r(S) 1 × 6 matrix of results:
     1. sample size   4. degrees of freedom
     2. log likelihood of null model 5. AIC
     3. log likelihood of full model 6. BIC

estat summarize saves the following in r():

  Matrices
   r(stats) k × 4 matrix of means, standard
          deviations, minimums, and maximums

estat vce saves the following in r():

  Matrices
   r(V) VCE or correlation matrix

Methods and Formulas

`estat` is implemented as an ado-file.

Akaike's (1974) information criterion is defined as

$$\text{AIC} = -2\ln L + 2k$$

where $\ln L$ is the maximized log-likelihood of the model and k is the number of parameters estimated. Some authors define the AIC as the expression above divided by the sample size.

Schwarz's (1978) Bayesian information criterion is another measure of fit defined as

$$\text{BIC} = -2\ln L + k\ln N$$

where N is the sample size. See [R] **BIC note** for additional information on calculating and interpreting BIC.

Hirotugu Akaike (1927–) was born in Fujinomiya City, Shizuoka Prefecture, Japan. He gained B.A. and Ph.D. degrees from the University of Tokyo. Akaike's career from 1952 at the Institute of Statistical Mathematics in Japan culminated in service as Director General; since 1994 he has been Professor Emeritus. His best known work in a prolific career is on what is now known as the Akaike information criterion (AIC), which was formulated to help selection of the most appropriate model from a number of candidates.

Gideon E. Schwarz (1933–) is Professor Emeritus of Statistics at the Hebrew University, Jerusalem. He was born in Salzburg, Austria and obtained an M.Sc. in 1956 from the Hebrew University and a Ph.D. in 1961 from Columbia University. He is interested in stochastic processes, sequential analysis, probability, and geometry and is responsible for the Bayesian information criterion (BIC).

References

Akaike, H. 1973. Information theory and an extension of the maximum likelihood principle. In *Second International Symposium on Information Theory*, ed. B. N. Petrov and F. Csaki, 267–281. Budapest: Akailseoniai–Kiudo.

———. 1974. A new look at the statistical model identification. *IEEE Transactions on Automatic Control* 19: 716–723.

Belsley, D. A., E. Kuh, and R. E. Welsch. 1980. *Regression Diagnostics: Identifying Influential Data and Sources of Collinearity*. New York: Wiley.

Findley, D. F., and E. Parzen. 1995. A conversation with Hirotugu Akaike. *Statistical Science* 10: 104–117.

Hardin, J. W., and J. Hilbe. 2007. *Generalized Linear Models and Extensions*. 2nd ed. College Station, TX: Stata Press.

Judge, G. G., W. E. Griffiths, R. C. Hill, H. Lütkepohl, and T.-C. Lee. 1985. *The Theory and Practice of Econometrics*. 2nd ed. New York: Wiley.

Neter, J., W. Wasserman, and M. H. Kutner. 2004. *Applied Linear Regression Models*. 4th ed. Homewood, IL: Irwin.

Raftery, A. 1995. Bayesian model selection in social research. In *Sociological Methodology*, vol. 25, ed. P. V. Marsden, 111–163. Oxford: Blackwell.

Sawa, T. 1978. Information criteria for discriminating among alternative regression models. *Econometrica* 46: 1273–1291.

Schwarz, G. 1978. Estimating the dimension of a model. *Annals of Statistics* 6: 461–464.

Also See

[R] **estimates** — Save and manipulate estimation results

[R] **summarize** — Summary statistics

[P] **estat programming** — Controlling estat after user-written commands

[U] **20 Estimation and postestimation commands**

Title

estimates — Save and manipulate estimation results

Syntax

command	reference
Save and use results from disk	
estimates save *filename*	[R] **estimates save**
estimates use *filename*	[R] **estimates save**
estimates describe using *filename*	[R] **estimates describe**
estimates esample: ...	[R] **estimates save**
Store and restore estimates in memory	
estimates store *name*	[R] **estimates store**
estimates restore *name*	[R] **estimates store**
estimates query	[R] **estimates store**
estimates dir	[R] **estimates store**
estimates drop *namelist*	[R] **estimates store**
estimates clear	[R] **estimates store**
Set titles and notes	
estimates title: *text*	[R] **estimates title**
estimates title	[R] **estimates title**
estimates notes: *text*	[R] **estimates notes**
estimates notes	[R] **estimates notes**
estimates notes list ...	[R] **estimates notes**
estimates notes drop ...	[R] **estimates notes**
Report	
estimates describe [*name*]	[R] **estimates describe**
estimates replay [*namelist*]	[R] **estimates replay**
Tables and statistics	
estimates table [*namelist*]	[R] **estimates table**
estimates stats [*namelist*]	[R] **estimates stats**
estimates for *namelist*: ...	[R] **estimates for**

Description

estimates allows you to store and manipulate estimation results:

1. You can save estimation results in a file for use in later sessions.

2. You can store estimation results in memory so that you can

    a. switch among separate estimation results and

    b. form tables combining separate estimation results.

Remarks

estimates is for use after you have fitted a model, be it with regress, logistic, etc. You can use estimates after any estimation command, whether it be an official estimation command of Stata or a user-written one.

estimates has three separate but related capabilities:

1. You can save estimation results in a file on disk, so you can use them later, even in a different Stata session.

2. You can store up to 300 estimation results in memory so that they are at your fingertips.

3. You can make tables comparing any results you have stored in memory.

Remarks are presented under the following headings:

> *Saving and using estimation results*
> *Storing and restoring estimation results*
> *Comparing estimation results*
> *Jargon*

Saving and using estimation results

After you have fitted a model, say, with regress,

```
. use http://www.stata-press.com/data/r10/auto
(1978 Automobile Data)
. regress mpg weight displ foreign
  (output omitted )
```

You can save the results in a file:

```
. estimates save basemodel
(file basemodel.ster saved)
```

Later, say, in a different session, you can reload those results:

```
. estimates use basemodel
```

The situation is now nearly identical to what it was immediately after you fitted the model. You can replay estimation results,

```
. regress
  (output omitted )
```

you can perform tests,

```
. test foreign==0
  (output omitted )
```

and you can use any postestimation command or postestimation capability of Stata. The only difference is that Stata no longer knows what was the estimation sample, e(sample) in Stata jargon. When you reload the estimation results, you might not even have the original data in memory. That is okay. Stata will know to refuse to calculate anything that can be calculated only on the original estimation sample.

If it is important that you use a postestimation command that can be used only on the original estimation sample, there is a way you can do that. You use the original data and then use estimates esample to tell Stata what was the original sample.

See [R] **estimates save** for details.

Storing and restoring estimation results

Storing and restoring estimation results in memory is much like saving them to disk. You type

```
. estimates store base
```

to save the current estimation results under the name base, and you type

```
. estimates restore base
```

to get them back later. You can find out what you have stored by typing

```
. estimates dir
```

Saving estimation results to disk is more permanent than storing them in memory, so why would you want merely to store them? The answer is that, once they are stored, you can use other estimates commands to produce tables and reports from them.

See [R] **estimates store** for details about the estimates store and restore commands.

Comparing estimation results

Let's say that you have done the following:

```
. use http://www.stata-press.com/data/r10/auto
(1978 Automobile Data)
. regress mpg weight displ
(output omitted)
. estimates store base
. regress mpg weight displ foreign
(output omitted)
. estimates store alt
```

You can now get a table comparing the coefficients:

```
. estimates table base alt
```

Variable	base	alt
weight	-.00656711	-.00677449
displacement	.00528078	.00192865
foreign		-1.6006312
_cons	40.084522	41.847949

estimates table can do much more; see [R] **estimates table**. Also see [R] **estimates stats**. It works similarly to table but produces model comparisons in terms of BIC and AIC.

Jargon

You know that if you fit a model, say, by typing

```
. regress mpg weight displacement
```

you can later replay the results by typing

```
. regress
```

and you can do tests and calculate other postestimation statistics by typing

```
. test displacement==0
. estat vif
. predict mpghat
```

As a result, we often refer to *the estimation results* or the *current estimation results* or the *most recent estimation results* or the *last estimation results* or the *estimation results in memory*.

With `estimates store` and `estimates restore`, you can have many estimation results in memory. One set of those, the set most recently estimated, or the set most recently restored, are the *current* or *active* estimation results, which you can replay, which you can test, or from which you can calculate postestimation statistics.

Current and *active* are the two words we will use interchangeably from now on.

Also See

[R] **estimates save** — Save and use estimation results

[R] **estimates describe** — Describe estimation results

[R] **estimates store** — Store and restore estimation results

[R] **estimates title** — Set title for estimation results

[R] **estimates notes** — Add notes to estimation results

[R] **estimates replay** — Redisplay estimation results

[R] **estimates table** — Compare estimation results

[R] **estimates stats** — Model statistics

[R] **estimates for** — Repeat postestimation command across models

Title

> **estimates describe** — Describe estimation results

Syntax

> estimates <u>desc</u>ribe
>
> estimates <u>desc</u>ribe *name*
>
> estimates <u>desc</u>ribe using *filename*

Description

estimates describe describes the current (active) estimates. Reported are the command line that produced the estimates, any title that was set by [R] **estimates title**, and any notes that were added by [R] **estimates notes**.

estimates describe *name* does the same but reports results for estimates stored by [R] **estimates store**.

estimates describe using *filename* does the same but reports results for estimates saved by [R] **estimates save**. If *filename* is specified without an extension, .ster is assumed.

Remarks

estimates describe can be used to describe the estimation results currently in memory,

```
. estimates describe
Estimation results produced by
    . regress mpg weight displ if foreign
```

or to describe results saved by estimates save in a .ster file:

```
. estimates describe using final
Estimation results "Final results" saved on 12apr2007 14:20, produced by
    . logistic myopic age sex drug1 drug2 if complete==1
Notes:
  1.  Used file patient.dta
  2.  "datasignature myopic age sex drug1 drug2 if complete==1"
      reports 148:5(58763):2252897466:3722318443
  3.  must be reviewed by rgg
```

▷ Example 1

```
. use http://www.stata-press.com/data/r10/auto
(1978 Automobile Data)
. regress mpg weight displ if foreign
(output omitted)
. estimates notes: file 'c(filename)'
. datasignature
74:12(71728):3831085005:1395876116
. estimates notes: datasignature report 'r(datasignature)'
```

```
. estimates save foreign
file foreign.ster saved
. regress mpg weight displ if !foreign
(output omitted)
. estimates describe using foreign
Estimation results saved on 02may2007 10:33, produced by
    . regress mpg weight displ if foreign
Notes:
  1.  file http://www.stata-press.com/data/r10/auto.dta
  2.  datasignature report 74:12(71728):3831085005:1395876116
```

◁

Saved Results

estimates describe and estimates describe *name* save the following in r():

Macros
    r(title)   title
    r(cmdline)   original command line

estimates describe using *filename* saves the above in r() and saves the following:

Scalars
    r(datetime)   %tc value of date/time file saved

Methods and Formulas

estimates describe is implemented as an ado-file.

Also See

[R] **estimates** — Save and manipulate estimation results

Title

estimates for — Repeat postestimation command across models

Syntax

estimates for *namelist* [, *options*] : *postestimation_command*

where *namelist* is a name, a list of names, _all, or *. A name may be ., meaning the current (active) estimates. _all and * mean the same thing.

options	description
noheader	do not display title
nostop	do not stop if command fails

Description

estimates for performs *postestimation_command* on each estimation result specified.

Options

noheader suppresses the display of the header as *postestimation_command* is executed each time.

nostop specifies that execution of *postestimation_command* is to be performed on the remaining models even if it fails on some.

Remarks

In the example that follows, we fit a model two different ways, store the results, and then use estimates for to perform the same test on both of them:

▷ Example 1

```
. use http://www.stata-press.com/data/r10/auto
(1978 Automobile Data)
. gen fwgt = foreign*weight
. gen dwgt = !foreign*weight
. gen gpm = 1/mpg
. regress gpm fwgt dwgt displ foreign
(output omitted)
. estimates store reg
. qreg gpm fwgt dwgt displ foreign
(output omitted)
. estimates store qreg
```

```
. estimates for reg qreg: test fwgt==dwgt
```

Model **reg**

```
( 1)   fwgt - dwgt = 0
       F(  1,    69) =    4.87
            Prob > F =    0.0307
```

Model **qreg**

```
( 1)   fwgt - dwgt = 0
       F(  1,    69) =    0.07
            Prob > F =    0.7937
```

◁

Methods and Formulas

estimates for is implemented as an ado-file.

Also See

[R] **estimates** — Save and manipulate estimation results

Title

> **estimates notes** — Add notes to estimation results

Syntax

estimates notes: *text*

estimates notes

estimates notes list [in *noterange*]

estimates notes drop in *noterange*

where *noterange* is # or #/# and where # may be a number, the letter f (meaning first), or the letter l (meaning last).

Description

estimates notes: *text* adds a note to the current (active) estimation results.

estimates notes and estimates notes list list the current notes.

estimates notes drop in *noterange* eliminates the specified notes.

Remarks

After adding or removing notes, if estimates have been stored, do not forget to store them again. If estimates have been saved, do not forget to save them again.

Notes are most useful when you intend to save estimation results in a file; see [R] **estimates save**. For instance, after fitting a model, you might type

```
. estimates note:  I think these are final
. estimates save lock2
```

and then, later when going through your files, you could type

```
. estimates use lock2
. estimates notes
  1.  I think these are final
```

Up to 9,999 notes can be attached to estimation results. If estimation results are important, we recommend that you add a note identifying the .dta dataset you used. The best way to do that is to type

```
. estimates notes:  file 'c(filename)'
```

because 'c(filename)' will expand to include not just the name of the file but also its full path; see [P] **creturn**.

If estimation results took a long time to estimate—say, they were produced by asmprobit or gllamm—see [R] **asmprobit** and http://www.gllamm.org—it is also a good idea to add a data signature. A data signature takes less time to compute than reestimation when you need proof that you really have the right dataset. The easy way to do that is to type

```
. datasignature
74:12(71728):3831085005:1395876116

. estimates notes:  datasignature reports 'r(datasignature)'
```

Now when you ask to see the notes, you will see

```
. estimates notes
  1.  I think these are final
  2.  file C:\project\one\pat4.dta
  3.  datasignature reports 74:12(71728):3831085005:1395876116
```

See [D] **datasignature**. Notes need not be positive. You might set a note to be, "I need to check that age is defined correctly."

▷ Example 1

```
. use http://www.stata-press.com/data/r10/auto
(1978 Automobile Data)

. regress mpg weight displ if foreign
  (output omitted)

. estimates notes: file 'c(filename)'

. datasignature
74:12(71728):3831085005:1395876116

. estimates notes: datasignature report 'r(datasignature)'

. estimates save foreign
file foreign.ster saved

. estimates notes list in 1/2
    1.   file http://www.stata-press.com/data/r10/auto.dta
    2.   datasignature report 74:12(71728):3831085005:1395876116

. estimates notes drop in 2
  (1 note dropped)

. estimates notes
    1.   file http://www.stata-press.com/data/r10/auto.dta
```

◁

Methods and Formulas

estimates notes is implemented as an ado-file.

Also See

[R] **estimates** — Save and manipulate estimation results

Title

> **estimates replay** — Redisplay estimation results

Syntax

<u>est</u>imates <u>r</u>eplay

<u>est</u>imates <u>r</u>eplay *namelist*

where *namelist* is a name, a list of names, _all, or *. A name may be ., meaning the current (active) estimates. _all and * mean the same thing.

Description

estimates replay redisplays the current (active) estimation results, just as typing the name of the estimation command would do.

estimates replay *namelist* redisplays each specified estimation result. The active estimation results are left unchanged.

Remarks

In the example that follows, we fit a model two different ways, store the results, use estimates for to perform the same test on both of them, and then replay the results:

▷ Example 1

```
. use http://www.stata-press.com/data/r10/auto
(1978 Automobile Data)
. gen fwgt = foreign*weight
. gen dwgt = !foreign*weight
. gen gpm = 1/mpg
. regress gpm fwgt dwgt displ foreign
(output omitted)
. estimates store reg
. qreg gpm fwgt dwgt displ foreign
(output omitted)
. estimates store qreg
. estimates for reg qreg: test fwgt==dwgt
```

Model **reg**

```
( 1)  fwgt - dwgt = 0

     F(  1,    69) =     4.87
          Prob > F =    0.0307
```

```
Model qreg
```

```
( 1)  fwgt - dwgt = 0
      F(  1,    69) =     0.07
           Prob > F =     0.7937
. estimates replay
```

```
Model qreg
```

Median regression					Number of obs =	74
Raw sum of deviations .7555689 (about .05)						
Min sum of deviations .3201479					Pseudo R2 =	0.5763

gpm	Coef.	Std. Err.	t	P>|t|	[95% Conf. Interval]	
fwgt	.0000155	2.87e-06	5.40	0.000	9.76e-06	.0000212
dwgt	.0000147	1.88e-06	7.81	0.000	.0000109	.0000184
displacement	.0000179	.0000147	1.22	0.226	-.0000113	.0000471
foreign	.0065352	.0078098	0.84	0.406	-.009045	.0221153
_cons	.0003134	.0042851	0.07	0.942	-.0082351	.0088618

```
. estimates replay reg
```

```
Model reg
```

Source	SS	df	MS		Number of obs =	74
					F(4, 69) =	61.62
Model	.009342436	4	.002335609		Prob > F =	0.0000
Residual	.002615192	69	.000037901		R-squared =	0.7813
					Adj R-squared =	0.7686
Total	.011957628	73	.000163803		Root MSE =	.00616

gpm	Coef.	Std. Err.	t	P>|t|	[95% Conf. Interval]	
fwgt	.00002	3.27e-06	6.12	0.000	.0000135	.0000265
dwgt	.0000123	2.30e-06	5.36	0.000	7.75e-06	.0000169
displacement	.0000296	.0000187	1.58	0.119	-7.81e-06	.000067
foreign	-.0117756	.0086088	-1.37	0.176	-.0289497	.0053986
_cons	.0053352	.0046748	1.14	0.258	-.0039909	.0146612

◁

Methods and Formulas

estimates replay is implemented as an ado-file.

Also See

[R] **estimates** — Save and manipulate estimation results

Title

> **estimates save** — Save and use estimation results

Syntax

estimates save *filename* [, replace]

estimates use *filename*

estimates esample: [*varlist*] [*if*] [*in*] [*weight*]
    [, replace stringvars(*varlist*) zeroweight]

estimates esample

Description

estimates save *filename* saves the current (active) estimation results in *filename*.

estimates use *filename* loads the results saved in *filename* into the current (active) estimation results.

In both cases, if *filename* is specified without an extension, .ster is assumed.

estimates esample: (note the colon) resets e(sample). After estimates use *filename*, e(sample) is set to contain 0, meaning that none of the observations currently in memory were used in obtaining the estimates.

estimates esample (without a colon) displays how e(sample) is currently set.

Options

replace, used with estimates save, specifies that *filename* can be replaced if it already exists.

replace, used with estimates esample:, specifies that e(sample) can be replaced even if it is already set.

stringvars(*varlist*), used with estimates esample:, specifies string variables. Observations containing variables that contain "" will be omitted from e(sample).

zeroweight, used with estimates esample:, specifies that observations with zero weights are to be included in e(sample).

Remarks

See [R] **estimates** for an overview of the estimates commands.

For a description of estimates save and use, see *Saving and using estimation results* in [R] **estimates**.

The rest of this entry concerns e(sample).

Remarks are presented under the following headings:

> *Setting e(sample)*
> *Resetting e(sample)*
> *Determining who set e(sample)*

Setting e(sample)

After `estimates use` *filename*, the situation is nearly identical to what it was immediately after you fitted the model. The one difference is that `e(sample)` is set to 0.

`e(sample)` is Stata's function to mark which observations among those currently in memory were used in producing the estimates. For instance, you might type

```
. use http://www.stata-press.com/data/r10/auto
(1978 Automobile Data)
. regress mpg weight displ if foreign
(output omitted )
. summarize mpg if e(sample)
(output omitted )
```

and `summarize` would report the summary statistics for the observations `regress` in fact used, which would exclude not only observations for which `foreign = 0` but also any observations for which `mpg`, `weight`, or `displ` was missing.

If you saved the above estimation results and then reloaded them, however, `summarize mpg if e(sample)` would produce

```
. summarize mpg if e(sample)
```

Variable	Obs	Mean	Std. Dev.	Min	Max
mpg	0				

Stata thinks that none of these observations were used in producing the estimates currently loaded.

What else could Stata think? When you `estimates use` *filename*, you do not have to have the original data in memory. Even if you do have data in memory that look like the original data, it might not be. Setting `e(sample)` to 0 is the safe thing to do. There are some postestimation statistics, for instance, that are appropriate only when calculated on the estimation sample. Setting `e(sample)` to 0 ensures that, should you ask for one of them, you will get back a null result.

We recommend that you leave `e(sample)` set to 0. But what if you really need to calculate that postestimation statistic? Well, you can get it, but you are going to take responsibility for setting `e(sample)` correctly. Here we just happen to know that all the foreign observations were used, so we can type

```
. estimates esample:  if foreign
```

If all the observations had been used, we could simply type

```
. estimates esample:
```

The safe thing to do, however, is to look at the estimation command—`estimates describe` will show it to you—and then type

```
. estimates esample:  mpg weight displ if foreign
```

Resetting e(sample)

estimates esample: will allow you to not only set but also reset e(sample). If e(sample) has already been set (say that you just fitted the model), and you try to set it, you will see

```
. estimates esample:  mpg weight displ if foreign
no; e(sample) already set
r(322);
```

Here you can specify the `replace` option:

```
. estimates esample:  mpg weight displ if foreign, replace
```

We do not recommend resetting e(sample), but the situation can arise where you need to. Imagine that you estimates use *filename*, you set e(sample), and then realize you set it wrong. Here you would want to reset it.

Determining who set e(sample)

estimates esample without a colon will report whether and how e(sample) was set. You might see

```
. estimates esample
  e(sample) set by estimation command
```

or

```
. estimates esample
  e(sample) set by user
```

or

```
. estimates esample
  e(sample) not set (0 assumed)
```

Saved Results

estimates esample without the colon saves macro r(who), which will contain cmd, user, or zero'd.

Methods and Formulas

estimates save, estimates use, estimates esample:, and estimates esample are implemented as ado-files.

Also See

[R] **estimates** — Save and manipulate estimation results

Title

Syntax

estimates stats [*namelist*] [, n(*#*)]

where *namelist* is a name, a list of names, _all, or *. A name may be ., meaning the current (active) estimates. _all and * mean the same thing.

Description

estimates stats reports model-selection statistics, including the Akaike information criterion (AIC) and the Bayesian information criterion (BIC). These measures are appropriate for maximum likelihood models.

If estimates stats is used for a non–likelihood based model such as qreg, missing values are reported.

Option

n(*#*) specifies the N to be used in calculating BIC; see [R] **BIC note**.

Remarks

If you type estimates stats without arguments, a table for the most recent estimation results will be shown:

```
. logistic foreign mpg weight displ
(output omitted )

. estimates stats
```

Model	Obs	ll(null)	ll(model)	df	AIC	BIC
.	74	-45.03321	-20.59083	4	49.18167	58.39793

Note: N=Obs used in calculating BIC; see **[R] BIC note**

About the note at the bottom of the table: N is an ingredient in the calculation of BIC; see the reference [R] **BIC note**. The note changes if you specify option n(), which tells estimates stats what N to use. $N = $ Obs is the default.

About the table itself: ll(null) is the log likelihood for the constant-only model, ll(model) is the log likelihood for the model, df is the number of degrees of freedom, and AIC and BIC are the Akaike and Bayesian information criteria.

Models with smaller values of an information criterion are considered preferable.

estimates stats can compare estimation results:

```
. use http://www.stata-press.com/data/r10/auto
(1978 Automobile Data)
. logistic foreign mpg weight displ
  (output omitted )
. estimates store full
. logistic foreign mpg weight
  (output omitted )
. estimates store sub
. estimates stats full sub
```

Model	Obs	ll(null)	ll(model)	df	AIC	BIC
full	74	-45.03321	-20.59083	4	49.18167	58.39793
sub	74	-45.03321	-27.17516	3	60.35031	67.26251

Note: N=Obs used in calculating BIC; see **[R] BIC note**

Saved Results

estimates stats saves the following in r():

Matrices
  r(S)   matrix with 6 columns (N, ll0, ll, df, AIC, and BIC) and rows corresponding to models in table

Methods and Formulas

estimates stats is implemented as an ado-file.

See [R] **BIC note**.

Also See

[R] **estimates** — Save and manipulate estimation results

Title

> **estimates store** — Store and restore estimation results

Syntax

estimates store *name* [, nocopy]

estimates restore *name*

estimates query

estimates dir [*namelist*]

estimates drop *namelist*

estimates clear

where *namelist* is a name, a list of names, _all, or *. _all and * mean the same thing.

Description

estimates store *name* saves the current (active) estimation results under the name *name*.

estimates restore *name* loads the results saved under *name* into the current (active) estimation results.

estimates query tells you whether the current (active) estimates have been stored and, if so, the name.

estimates dir displays a list of the stored estimates.

estimates drop *namelist* drops the specified stored estimation results.

estimates clear drops all stored estimation results.

estimates clear, estimates drop _all, and estimates drop * do the same thing. estimates drop and estimates clear do not eliminate the current (active) estimation results.

Option

nocopy, used with estimates store, specifies that the current (active) estimation results are to be moved into *name* rather than copied. Typing

    . estimates store hold, nocopy

is the same as typing

    . estimates store hold
    . ereturn clear

except that the former is faster. Option nocopy is sometimes used by programmers.

Remarks

`estimates store` stores estimation results in memory so that you can access them later.

```
. use http://www.stata-press.com/data/r10/auto
(1978 Automobile Data)
. regress mpg weight displ
  (output omitted )
. estimates store myreg
. ... you do other things, including fitting other models ...
. estimates restore myreg
. regress
  (same output shown again)
```

After `estimates restore myreg`, things are once again just as they were, estimationwise, just after you typed `regress mpg weight displ`.

`estimates store` stores results in memory. When you exit Stata, those stored results vanish. If you wish to make a permanent copy of your estimation results, see [R] **estimates save**.

The purpose of making copies in memory is (1) so that you can quickly switch between them and (2) so that you make tables comparing estimation results. Concerning the latter, see [R] **estimates table** and [R] **estimates stats**.

Saved Results

`estimates dir` saves the following in `r()`:

Macros
  r(names)   names of stored results

Methods and Formulas

`estimates store`, `estimates restore`, `estimates query`, `estimates dir`, `estimates drop`, and `estimates clear` are implemented as ado-files.

Also See

[R] **estimates** — Save and manipulate estimation results

Title

> **estimates table** — Compare estimation results

Syntax

estimates table $\big[$ *namelist* $\big]$ $\big[$, *options* $\big]$

where *namelist* is a name, a list of names, _all, or *. A name may be ., meaning the current (active) estimates. _all and * mean the same thing.

options	description
Main	
stats(*scalarlist*)	report *scalarlist* in table
star$\big[$ (*#1 #2 #3*) $\big]$	use stars to denote significance levels
Options	
keep(*coeflist*)	report coefficients in order specified
drop(*coeflist*)	omit specified coefficients from table
equations(*matchlist*)	match equations of models as specified
Numerical formats	
b$\big[$ (%*fmt*) $\big]$	how to format coefficients, which are always reported
se$\big[$ (%*fmt*) $\big]$	report standard errors and use optional format
t$\big[$ (%*fmt*) $\big]$	report t or z and use optional format
p$\big[$ (%*fmt*) $\big]$	report p-values and use optional format
stfmt(%*fmt*)	how to format scalar statistics
General format	
varwidth(*#*)	use *#* characters to display variable names and statistics
modelwidth(*#*)	use *#* characters to display model names
eform	display coefficients in exponentiated form
label	display variable labels rather than variable names
newpanel	display statistics in separate table from coefficients
style(oneline)	put vertical line after variable names; the default
style(columns)	put vertical line separating every column
style(noline)	suppress all vertical lines
coded	display compact table
†title(*string*)	title for table

†title() does not appear in the dialog box.

414

where

1. A *scalarlist* is a list of any or all of the names of scalars stored in e(), plus aic, bic, and rank.

2. *#1 #2 #3* are three numbers such as .05 .01 .001.

3. A *coeflist* is a list of coefficient names, each name of which may be simple (e.g., price), an equation name followed by a colon (e.g., mean:), or a full name (e.g., mean:price). Names are separated by blanks.

4. A *matchlist* specifies how equations from different estimation results are to be matched. If you need to specify a *matchlist*, the solution is usually 1, as in equations(1). The full syntax is

$$matchlist := term \left[, \ term \ \dots \ \right]$$

$$term := \left[eqname = \right] \#{:}\#\dots{:}\#$$

$$\left[eqname = \right] \#$$

See equations() under *Options* below.

5. *%fmt* is any valid Stata numerical display format.

Description

estimates table displays a table of coefficients and statistics for one or more sets of estimation results.

Options

▢ Main

stats(*scalarlist*) specifies one or more scalar statistics to be displayed in the table. *scalarlist* may contain

aic	Akaike's information criterion
bic	Schwarz's information criterion
rank	rank of e(V) (# of free parameters in model)

along with the names of any scalars stored in e(). The specified statistics do not have to be available for all estimation results being displayed.

Example: stats(N ll chi2 aic) specifies that e(N), e(ll), e(chi2), and Akaike's information criterion be included. In Stata, e(N) records the number of observations; e(ll), the log likelihood; and e(chi2), the chi-squared test that all coefficients in the first equation of the model are equal to zero.

star and star(*#1 #2 #3*) specifies that stars (asterisks) are to be used to mark significance. The second syntax specifies the significance levels for one, two, and three stars. If you specify simply star, that is equivalent to specifying star(.05 .01 .001), which means one star (*) if $p < .05$, two stars (**) if $p < .01$, and three stars (***) if $p < .001$.

Options star and star() may not be combined with option se, t, or p.

keep(*coeflist*) and drop(*coeflist*) are alternatives; they specify coefficients to be included or omitted from the table. The default is to display all coefficients.

If keep() is specified, it specifies not only the coefficients to be included but also the order in which they appear.

A *coeflist* is a list of coefficient names, each name of which may be simple (e.g., price), an equation name followed by a colon (e.g., mean:), or a full name (e.g., mean:price). Names are separated from each other by blanks.

When full names are not specified, all coefficients that match the partial specification are included. For instance, drop(_cons) would omit _cons for all equations.

equations(*matchlist*) specifies how the equations of the models in *namelist* are to be matched. The default is to match equations by name. Matching by name usually works well when all results were fitted by the same estimation command. When you are comparing results from different estimation commands, however, specifying equations() may be necessary.

The most common usage is equations(1), which indicates that all first equations are to be matched into one equation named #1.

matchlist has the syntax

> *term* $\big[$, *term* ... $\big]$

where *term* is

> $\big[$ *eqname* = $\big]$ #:#...:#   (syntax 1)

> $\big[$ *eqname* = $\big]$ #   (syntax 2)

In syntax 1, each # is a number or a period (.). If a number, it specifies the position of the equation in the corresponding model; 1:3:1 would indicate that equation 1 in the first model matches equation 3 in the second, which matches equation 1 in the third. A period indicates that there is no corresponding equation in the model; 1:.:1 indicates that equation 1 in the first matches equation 1 in the third.

In syntax 2, you specify just one number, say, 1 or 2, and that is shorthand for 1:1...:1 or 2:2...:2, meaning that equation 1 matches across all models specified or that equation 2 matches across all models specified.

Now that you can specify a *term*, you can put that together into a *matchlist* by separating one term from the other by commas. In what follows, we will assume that three names were specified,

    . estimates table alpha beta gamma, ...

equations(1) is equivalent to equations(1:1:1). We would be saying that the first equations match across the board.

equations(1:.:1) would specify that equation 1 matches in models alpha and gamma but that there is nothing corresponding in model beta.

equations(1,2) is equivalent to equations(1:1:1, 2:2:2). We would be saying that the first equations match across the board, and so do the second equations.

equations(1, 2:.:2) would specify that the first equation matches across the board, that the second equations match for models alpha and gamma, and that there is nothing equivalent to equation 2 in model beta.

If equations() is specified, equations not matched by position are matched by name.

b(%*fmt*) specifies how the coefficients are to be displayed. You might specify b(%9.2f) to make decimal points line up. There is also an option b, which specifies that coefficients are to be displayed, but that is just included for consistency with options se, t, and p. Coefficients are always displayed.

se, t, and p specify that standard errors, t or z statistics, and significance levels are to be displayed. The default is not to display them. se(%*fmt*), t(%*fmt*), and p(%*fmt*) specify that each is to be displayed and specifies the display format to be used to format them.

stfmt(%*fmt*) specifies the format for displaying the scalar statistics included by options stats().

varwidth(#) specifies the number of character positions used to display the names of the variables and statistics. The default is 12.

modelwidth(#) specifies the number of character positions used to display the names of the models. The default is 12.

eform displays coefficients in exponentiated form. For each coefficient, $\exp(\beta)$ rather than β is displayed, and standard errors are transformed appropriately. Display of the intercept, if any, is suppressed.

label specifies that variable labels be displayed instead of variable names.

newpanel specifies that the statistics be displayed in a table separated by a blank line from the table with coefficients rather than in the style of another equation in the table of coefficients.

style(*stylespec*) specifies the style of the coefficient table.

style(oneline) specifies that a vertical line be displayed after the variables but not between the models. This is the default.

style(columns) specifies that vertical lines be displayed after each column.

style(noline) specifies that no vertical lines be displayed.

coded specifies that a compact table be displayed. This format is especially useful for comparing variables that are included in a large collection of models.

The following option is available with estimates table but is not shown in the dialog box:

title(*string*) specifies the title to appear above the table.

Remarks

If you type estimates table without arguments, a table of the most recent estimation results will be shown:

```
. use http://www.stata-press.com/data/r10/auto
(1978 Automobile Data)

. regress mpg weight displ
  (output omitted)
```

```
. estimates table
```

Variable	active
weight	-.00656711
displacement	.00528078
_cons	40.084522

The real use of estimates table, however, is for comparing estimation results, and that requires using it after estimates store:

```
. regress mpg weight displ
(output omitted )
. estimates store base
. regress mpg weight displ foreign
(output omitted )
. estimates store alt
. qreg mpg weight displ foreign
(output omitted )
. estimates store qreg
. estimates table base alt qreg, stats(r2)
```

Variable	base	alt	qreg
weight	-.00656711	-.00677449	-.00595056
displacement	.00528078	.00192865	.00018552
foreign		-1.6006312	-2.1326004
_cons	40.084522	41.847949	39.213348
r2	.6529307	.66287957	

Saved Results

estimates table saves the following in r():

Macros
    r(names)   names of results used

Matrices
    r(coef)   matrix M: $n \times 2*m$
                $M[i, 2j-1] = i$th parameter estimate for model j;
                $M[i, 2j] = $ variance of $M[i, 2j-1]$; $i=1,...,n$; $j=1,...,m$
    r(stats)   matrix S: $k \times m$ (if option stats() specified)
                $S[i, j] = i$th statistic for model j; $i=1,...,k$; $j=1,...,m$

Methods and Formulas

estimates table is implemented as an ado-file.

Also See

[R] **estimates** — Save and manipulate estimation results

Title

estimates title — Set title for estimation results

Syntax

estimates title: [*text*]

estimates title

Description

estimates title: (note the colon) sets or clears the title for the current estimation results. The title is used by [R] **estimates table** and [R] **estimates stats**.

estimates title without the colon displays the current title.

Remarks

After setting the title, if estimates have been stored, do not forget to store them again:

```
. use http://www.stata-press.com/data/r10/auto
(1978 Automobile Data)
. regress mpg gear turn
(output omitted )
. estimates store reg
```

Now lets add a title:

```
. estimates title: "My regression"
. estimates store reg
```

Methods and Formulas

estimates title: and estimates title are implemented as ado-files.

Also See

[R] **estimates** — Save and manipulate estimation results

Title

---
estimation options — Estimation options

Description

This entry describes the options common to many estimation commands. Not all the options documented below work with all estimation commands; see the documentation for the particular estimation command. If an option is listed there, it is applicable.

Options

⌐ Model ¬

noconstant suppresses the constant term (intercept) in the model.

offset(*varname*) specifies that *varname* be included in the model with the coefficient constrained to be 1.

exposure(*varname*) specifies a variable that reflects the amount of exposure over which the *depvar* events were observed for each observation; ln(*varname*) with coefficient constrained to be 1 is entered into the log-link function.

constraints(*numlist* | *matname*) specifies the linear constraints to be applied during estimation. The default is to perform unconstrained estimation. See [R] **reg3** for the use of constraints in multiple-equation contexts.

> constraints(*numlist*) specifies the constraints by number after they have been defined by using the constraint command; see [R] **constraint**. Some commands (for example, slogit) allow only constraints(*numlist*).
>
> constraints(*matname*) specifies a matrix containing the constraints; see [P] **makecns**.
>
> constraints(*clist*) is used by some estimation commands, such as mlogit, where *clist* has the form $\#\left[-\#\right]\left[,\#\left[-\#\right]\ \ldots\right]$.

collinear specifies that the estimation command not remove collinear variables. Usually there is no reason to leave collinear variables in place, and in fact doing so usually causes the estimation to fail because of the matrix singularity caused by the collinearity. However, with certain models, the variables may be collinear, yet the model is fully identified because of constraints or other features of the model. In such cases, using option collinear allows the estimation to take place, leaving the equations with collinear variables intact.

Suppose that your model contains indicator variables generated from a factor variable that has three levels. Assume also that your model includes the intercept (constant) term. If the three indicator variables are included in the model, they will be collinear with the constant term and the model would not be identifiable. However, including a constraint so that the three regression coefficients associated with the indicator variables sum to zero would reduce the rank of the model so that it is identifiable. Applying this constraint changes the interpretation of the regression coefficients and the constant term.

⌐ Reporting ¬

level(*#*) specifies the confidence level, as a percentage, for confidence intervals. The default is level(95) or as set by set level; see [U] **20.7 Specifying the width of confidence intervals**.

Also See

[U] **20 Estimation and postestimation commands**

Title

> **exit** — Exit Stata

Syntax

exit [, clear]

Description

Typing exit causes Stata to stop processing and return control to the operating system. If the dataset in memory has changed since the last save command, you must specify the clear option before Stata will let you exit.

exit may also be used for exiting do-files or programs; see [P] **exit**.

Stata for Windows users may also exit Stata by clicking the close button or by pressing *Alt-F4*.

Stata for Macintosh users may also exit Stata by clicking the close button.

Stata(GUI) users may also exit Stata by clicking the close button.

Option

clear permits you to exit, even if the current dataset has not been saved.

Remarks

Type exit to leave Stata and return to the operating system. If the dataset in memory has changed since the last time it was saved, however, Stata will refuse. At that point, you can either save the dataset and then type exit or type exit, clear:

```
. exit
no; data in memory would be lost
r(4);
. exit, clear
```

Also See

[P] **exit** — Exit from a program or do-file

Title

exlogistic — Exact logistic regression

Syntax

exlogistic *depvar* *indepvars* [*if*] [*in*] [*weight*] [, *options*]

options	description
Model	
<u>condvars</u>(*varlist*)	condition on variables in *varlist*
<u>group</u>(*varname*)	groups/strata are stratified by unique values of *varname*
<u>binomial</u>(*varname* | #)	data are in binomial form and the number of trials is contained in *varname* or in #
<u>estconstant</u>	estimate constant term; do not condition on the number of successes
<u>noconstant</u>	suppress constant term
Terms	
<u>terms</u>(*termsdef*)	terms definition
Options	
<u>memory</u>(#[b|k|m|g])	set limit on memory usage; default is memory(10m)
<u>saving</u>(*filename*)	save the joint conditional distribution to *filename*
Reporting	
<u>level</u>(#)	set confidence level; default is level(95)
coef	report estimated coefficients
<u>test</u>(*testopt*)	report significance of observed sufficient statistic, conditional scores test, or conditional probabilities test
mue(*varlist*)	compute the median unbiased estimates for *varlist*
midp	use the mid-p-value rule
<u>nolog</u>	do not display the enumeration log

by, statsby, and xi are allowed; see [U] **11.1.10 Prefix commands**.

fweights are allowed; see [U] **11.1.6 weight**.

See [U] **20 Estimation and postestimation commands** for more capabilities of estimation commands.

Description

exlogistic fits an exact logistic regression model of *depvar* on *indepvars*.

exlogistic is an alternative to logistic, the standard maximum likelihood–based logistic regression; see [R] **logistic**. exlogistic produces more accurate inference in small samples because it does not depend on asymptotic results and exlogistic can better deal with one-way causation, such as the case where all females are observed to have a positive outcome.

exlogistic with option group(*varname*) is an alternative to clogit, the conditional logistic regression estimator; see [R] **clogit**. Like clogit, exlogistic conditions on the number of positive outcomes within stratum.

depvar can be specified in two ways. It can be zero/nonzero, with zero indicating failure and nonzero representing positive outcomes (successes), or if you specify the `binomial(`*varname* | #`)` option, *depvar* may contain the number of positive outcomes within each trial.

`exlogistic` is computationally intensive. Unlike the usual estimators, rather than calculating coefficients for all independent variables in one combined calculation, results for each independent variable are calculated separately with the other independent variables temporarily conditioned out. You can save considerable computer time by skipping the parameter calculations for variables that are not of direct interest. Specify such variables in the `condvars()` option rather than among the *indepvars*; see `condvars()` below.

Unlike Stata's other estimation commands, you may not use `test`, `lincom`, or other postestimation commands after `exlogistic`. Given the method used to calculate estimates, hypothesis tests must be performed during estimation by using `exlogistic`'s option `terms()`; see `terms()` below.

Options

   ⌐ Model ⌐

`condvars(`*varlist*`)` specifies variables whose parameter estimates are not of interest to you. You can save substantial computer time and memory moving such variables from *indepvars* to `condvars()`. Understand that you will get the same results for `x1` and `x3` whether you type

    . `exlogistic y x1 x2 x3 x4`

or

    . `exlogistic y x1 x3, condvars(x2 x4)`

`group(`*varname*`)` specifies the variable defining the strata, if any. A constant term is assumed for each stratum identified in *varname*, and the sufficient statistics for *indepvars* are conditioned on the observed number of successes within each group. This makes the model estimated equivalent to that estimated by `clogit`, Stata's conditional logistic regression command (see [R] **clogit**). `group()` may not be specified with `noconstant` or `estconstant`.

`binomial(`*varname* | #`)` indicates that the data are in binomial form and *depvar* contains the number of successes. *varname* contains the number of trials for each observation. If all observations have the same number of trials, you can instead specify the number as an integer. The number of trials must be a positive integer at least as great as the number of successes. If `binomial()` is not specified, the data are assumed to be Bernoulli, meaning that *depvar* equaling zero/nonzero records a single failure or success.

`estconstant` estimates the constant term. By default, the models are assumed to have an intercept (constant), but the value of the intercept is not calculated. That is, the conditional distribution of the sufficient statistics for the *indepvars* is computed given the number of successes in *depvar*, thus conditioning out the constant term of the model. Use `estconstant` if you want the estimate of the intercept reported. `estconstant` may not be specified with `group()`.

`noconstant`; see [R] **estimation options**. `noconstant` may not be specified with `group()`.

   ⌐ Terms ⌐

`terms(`*termname* = *variable* ... *variable* [, *termname* = *variable* ... *variable* ...]`)` defines additional terms of the model on which you want `exlogistic` to perform joint-significance hypothesis tests. By default, `exlogistic` reports tests individually on each variable in *indepvars*. For instance, if variables `x1` and `x3` are in *indepvars*, and you want to jointly test their significance, specify terms(t1=x1 x3). To also test the joint significance of x2 and x4, specify terms(t1=x1 x3, t2=x2 x4). Each variable can be assigned to one term only.

⌐ Options ⌐

memory(#[b|k|m|g]) sets a limit on the amount of memory exlogistic can use when computing the conditional distribution of the parameter sufficient statistics. The default is memory(10m), where m stands for megabyte, or 1,048,576 bytes. The following are also available: b stands for byte; k for kilobyte, which is equal to 1,024 bytes; and g for gigabyte, which is equal to 1,024 megabytes. The minimum setting allowed is 1m and the maximum is 512m or 0.5g, but do not attempt to use more memory than is available on your computer.

saving(*filename* [, replace]) saves the joint conditional distribution to *filename*. This distribution is conditioned on those variables specified in condvars(). Use replace to replace an existing file with *filename*. A Stata data file is created containing all the feasible values of the parameter sufficient statistics. The variable names are the same as those in *indepvars*, in addition to a variable named _f_ containing the feasible value frequencies (sometimes referred to as the condition numbers).

⌐ Reporting ⌐

level(*#*); [R] **estimation options**. The level(*#*) option will not work on replay because confidence intervals are based on estimator-specific enumerations. To change the confidence level, you must refit the model.

coef reports the estimated coefficients rather than odds ratios (exponentiated coefficients). coef may be specified when the model is fitted or upon replay. coef affects only how results are displayed and not how they are estimated.

test(sufficient|score|probability) reports the significance level of the observed sufficient statistics, the conditional scores tests, or the conditional probabilities tests, respectively. The default is test(sufficient). If terms() is included in the specification, the conditional scores test and the conditional probabilities test are applied to each term providing conditional inference for several parameters simultaneously. All the statistics are computed at estimation time regardless of which is specified. Each statistic may thus also be displayed postestimation without having to refit the model; see [R] **exlogistic postestimation**.

mue(*varlist*) specifies that median unbiased estimates (MUEs) be reported for the variables in *varlist*. By default, the conditional maximum likelihood estimates (CMLEs) are reported, except for those parameters for which the CMLEs are infinite. Specify mue(_all) if you want MUEs for all the *indepvars*.

midp instructs exlogistic to use the mid-p-value rule when computing the MUEs, significance levels, and confidence intervals. This adjustment is for the discreteness of the distribution and halves the value of the discrete probability of the observed statistic before adding it to the p-value. The mid-p-value rule cannot be applied to MUEs whose corresponding parameter CMLE is infinite.

nolog prevents the display of the enumeration log. By default, the enumeration log is displayed, showing the progress of computing the conditional distribution of the sufficient statistics.

Remarks

Exact logistic regression is the estimation of the logistic model parameters by using the conditional distribution of the parameter sufficient statistics. The estimates are referred to as the conditional maximum likelihood estimates (CMLEs). This technique was first introduced by Cox and Snell (1989) as an alternative to using maximum likelihood estimation, which can perform poorly for small sample sizes. For stratified data, exact logistic regression is a small-sample alternative to conditional logistic regression. See [R] **logit**, [R] **logistic**, and [R] **clogit** to obtain maximum likelihood estimates (MLEs) for the logistic model and the conditional logistic model. For a comprehensive overview of exact logistic regression, see Mehta and Patel (1995).

Let Y_i denote a Bernoulli random variable where we observe the outcome $Y_i = y_i$, $i = 1, \ldots, n$. Associated with each independent observation is a $1 \times p$ vector of covariates $\mathbf{x}_i$. We will denote $\pi_i = \Pr(Y_i \mid \mathbf{x}_i)$ and let the logit function model the relationship between Y_i and $\mathbf{x}_i$,

$$\log\left(\frac{\pi_i}{1 - \pi_i}\right) = \theta + \mathbf{x}_i \boldsymbol{\beta}$$

where the constant term θ and the $p \times 1$ vector of regression parameters $\boldsymbol{\beta}$ are unknown. The probability of observing $Y_i = y_i$, $i = 1, \ldots, n$ is

$$\Pr(\mathbf{Y} = \mathbf{y}) = \prod_{i=1}^{n} \mu_i^{y_i} (1 - \mu_i)^{1 - y_i}$$

where $\mathbf{Y} = (Y_1, \ldots, Y_n)$ and $\mathbf{y} = (y_1, \ldots, y_n)$. The MLEs for θ and $\boldsymbol{\beta}$ maximize the log of this function.

The sufficient statistics for θ and β_j, $j = 1, \ldots, p$ are $M = \sum_{i=1}^{n} Y_i$ and $T_j = \sum_{i=1}^{n} Y_i x_{ij}$, respectively, and we observe $M = m$ and $T_j = t_j$. By default, `exlogistic` tallies the conditional distribution of $\mathbf{T} = (T_1, \ldots, T_p)$ given $M = m$. This distribution will have a size of $\binom{n}{m}$. (It would have a size of 2^n without conditioning on $M = m$.) Denote one of these vectors $\mathbf{T}^{(k)} = (t_1^{(k)}, \ldots, t_p^{(k)})$, $k = 1, \ldots, N$, with combinatorial coefficient (frequency) c_k, $\sum_{k=1}^{N} c_k = \binom{n}{m}$. For each independent variable x_j, $j = 1, \ldots, p$, we reduce the conditional distribution further by conditioning on all other observed sufficient statistics $T_l = t_l$, $l \neq j$. The conditional probability of observing $T_j = t_j$ has the form

$$\Pr(T_j = t_j \mid T_l = t_l, l \neq j, M = m) = \frac{c \, e^{t_j \beta_j}}{\sum_k c_k e^{t_j^{(k)} \beta_j}}$$

where the sum is over the subset of $\mathbf{T}$ vectors such that $(T_1^{(k)} = t_1, \ldots, T_j^{(k)} = t_j^{(k)}, \ldots, T_p^{(k)} = t_p)$ and c is the combinatorial coefficient associated with the observed $\mathbf{t}$. The CMLE for β_j maximizes the log of this function.

Specifying nuisance variables in `condvars()` will reduce the size of the conditional distribution by conditioning on their observed sufficient statistics as well as conditioning on $M = m$. This reduces the amount of memory consumed at the cost of not obtaining regression estimates for those variables specified in `condvars()`.

Inferences from MLEs rely on asymptotics and if your sample size is small these inferences may not be valid. On the other hand, inferences from the CMLEs are exact in the sense that they use the conditional distribution of the sufficient statistics outlined above.

For small datasets, it is common for the dependent variable to be completely determined by the data. Here the MLEs and the CMLEs are unbounded. `exlogistic` will instead compute the median unbiased estimate (MUE), the regression estimate that places the observed sufficient statistic at the median of the conditional distribution.

▷ Example 1

One example presented by Mehta and Patel (1995) is data from a prospective study of perinatal infection and human immunodeficiency virus type 1 (HIV-1). We use a variation of this dataset. There was an investigation (Hutto, Parks, and Lai 1991) into whether the blood serum levels of glycoproteins CD4 and CD8 measured in infants at 6 months of age might predict their development of HIV infection. The blood serum levels are coded as ordinal values 0, 1, and 2.

```
. use http://www.stata-press.com/data/r10/hiv1
(prospective study of perinatal infection of HIV-1)
. list
```

	hiv	cd4	cd8
1.	1	0	0
2.	0	0	0
3.	1	0	2
4.	1	1	0
5.	0	1	0
	(output omitted)		
46.	0	2	1
47.	0	2	2

We first obtain the MLEs from `logistic` so that we can compare the estimates and associated statistics with the CMLEs from `exlogistic`.

```
. logistic hiv cd4 cd8, coef
```

Logistic regression

```
                                      Number of obs   =        47
                                      LR chi2(2)      =     15.75
                                      Prob > chi2     =    0.0004
Log likelihood = -20.751687           Pseudo R2       =    0.2751
```

hiv	Coef.	Std. Err.	z	P>|z|	[95% Conf. Interval]	
cd4	-2.541669	.8391893	-3.03	0.002	-4.18645	-.8968886
cd8	1.658586	.8210821	2.02	0.043	.049295	3.267878
_cons	.5132389	.6808967	0.75	0.451	-.8212941	1.847772

```
. exlogistic hiv cd4 cd8, coef
Enumerating sample-space combinations:
observation  1:    enumerations =          2
observation  2:    enumerations =          3
 (output omitted)
observation 46:    enumerations =        601
observation 47:    enumerations =        326
```

Exact logistic regression

```
                                      Number of obs =          47
                                      Model score   =    13.34655
                                      Pr >= score   =      0.0006
```

hiv	Coef.	Suff.	2*Pr(Suff.)	[95% Conf. Interval]	
cd4	-2.387632	10	0.0004	-4.699633	-.8221807
cd8	1.592366	12	0.0528	-.0137905	3.907876

`exlogistic` produced a log showing how many records are generated as it processes each observation. The primary purpose of the log is to provide feedback since generating the distribution can be time consuming, but we also see from the last entry that the joint distribution for the sufficient statistics for `cd4` and `cd8` conditioned on the total number of successes has 326 unique values (but a size of $\binom{47}{14} = 341{,}643{,}774{,}795$).

The statistics for `logistic` are based on asymptotics: for a large sample size, each Z statistic will be approximately normally distributed (with a mean of zero and a standard deviation of one) if the associated regression parameter is zero. The question is whether a sample size of 47 is large enough.

On the other hand, the p-values computed by `exlogistic` are from the conditional distributions of the sufficient statistics for each parameter given the sufficient statistics for all other parameters. In this sense, these p-values are exact. By default, `exlogistic` reports the sufficient statistics for the regression parameters and the probability of observing a more extreme value. These are single-parameter tests for $H_0: \beta_{\text{cd4}} = 0$ and $H_0: \beta_{\text{cd8}} = 0$ versus the two-sided alternatives. The conditional scores test, located in the coefficient table header, is testing that both $H_0: \beta_{\text{cd4}} = 0$ and $H_0: \beta_{\text{cd8}} = 0$. We find these p-values to be in fair agreement with the Wald and likelihood-ratio tests from `logistic`.

The confidence intervals for `exlogistic` are computed from the exact conditional distributions. The exact confidence intervals are asymmetrical about the estimate and are wider than the normal-based confidence intervals from `logistic`.

Both estimation techniques indicate that the incidence of HIV infection decreases with increasing CD4 blood serum levels and increases with increasing CD8 blood serum levels. The constant term is missing from the exact logistic coefficient table because we conditioned out its observed sufficient statistic when tallying the joint distribution of the sufficient statistics for the `cd4` and `cd8` parameters.

The `test()` option provides two other test statistics used in exact logistic: the conditional scores test, `test(score)`, and the conditional probabilities test, `test(probability)`. For comparison, we display the individual parameter conditional scores test.

```
. exlogistic, test(score) coef
Exact logistic regression                    Number of obs =          47
                                             Model score   =  13.34655
                                             Pr >= score   =    0.0006
```

hiv	Coef.	Score	Pr>=Score	[95% Conf. Interval]
cd4	-2.387632	12.88022	0.0003	-4.699633 -.8221807
cd8	1.592366	4.604816	0.0410	-.0137905 3.907876

For the probabilities test, the probability statistic is computed from (1) in *Methods and Formulas* with $\beta = 0$. For this example, the significance of the probabilities tests matches the scores tests so they are not displayed here.

◁

❑ Technical Note

Typically, the value of θ, the constant term, is of little interest, as well as perhaps some of the parameters in β, but we need to include all parameters in the model to correctly specify it. By conditioning out the nuisance parameters, we can reduce the size of joint conditional distribution that is used to estimate the regression parameters of interest. The option `condvars()` allows you to specify a *varlist* of nuisance variables. By default, `exlogistic` conditions on the sufficient statistic of θ, which is the number of successes. You can save computation time and computer memory by using the `condvars()` option since infeasible values of the sufficient statistics associated with the variables in `condvars()` can be dropped from consideration before all n observations are processed.

Specifying some of your independent variables in condvar() will not change the estimated regression coefficients of the remaining independent variables. For instance, in example 1 if we instead type

```
. exlogistic hiv cd4, condvars(cd8) coef
```

the regression coefficient for cd4 (as well as all associated inference) will be identical.

One reason to have multiple variables in *indepvars* is to make conditional inference of several parameters simultaneously by using the terms() option. If you do not wish to test several parameters simultaneously, it may be more efficient to obtain estimates for individual variables by calling exlogistic multiple times with one variable in *indepvars* and all other variables listed in condvars(). The estimates will be the same as those with all variables in *indepvars*.

❏

❏ Technical Note

If you fit a clogit model to the HIV data from example 1, you will find that the estimates differ from those from exlogistic. (To fit the clogit model, you will have to create a group variable that includes all observations.) The regression estimates will be different because clogit conditions on the constant term only, whereas the estimates from exlogistic condition on the sufficient statistic of the other regression parameter as well as the constant term.

❏

▷ Example 2

The HIV data presented in table IV of Mehta and Patel (1995) are in a binomial form, where the variable hiv contains the HIV cases that tested positive and the variable n contains the number of individuals with the same CD4 and CD8 levels, the binomial number-of-trials parameter. Here *depvar* is hiv and we use the binomial(n) option to identify the number-of-trials variable.

```
. use http://www.stata-press.com/data/r10/hiv_n, clear
(prospective study of perinatal infection of HIV-1; binomial form)
. list
```

	cd4	cd8	hiv	n
1.	0	2	1	1
2.	1	2	2	2
3.	0	0	4	7
4.	1	1	4	12
5.	2	2	1	3
6.	1	0	2	7
7.	2	0	0	2
8.	2	1	0	13

Further, the cd4 and cd8 variables of the hiv dataset are actually factor variables, where each has the ordered levels of $(0, 1, 2)$. Another approach to the analysis is to use indicator variables, and per Mehta and Patel (1995), we use a 0–1 coding scheme that will give us the odds ratio of level 0 versus 2 and level 1 versus 2.

```
. gen byte cd4_0 = (cd4==0)
. gen byte cd4_1 = (cd4==1)
```

```
. gen byte cd8_0 = (cd8==0)

. gen byte cd8_1 = (cd8==1)

. exlogistic hiv cd4_0 cd4_1 cd8_0 cd8_1, terms(cd4=cd4_0 cd4_1,
> cd8=cd8_0 cd8_1) binomial(n) test(probability) saving(dist) nolog
note: saving distribution to file dist
note: CMLE estimate for cd4_0 is +inf; computing MUE
note: CMLE estimate for cd4_1 is +inf; computing MUE
note: CMLE estimate for cd8_0 is -inf; computing MUE
note: CMLE estimate for cd8_1 is -inf; computing MUE
Exact logistic regression                    Number of obs  =          47
Binomial variable: n                         Model prob.    =    3.19e-06
                                             Pr <= prob.    =     0.0011
```

hiv	Odds Ratio	Prob.	Pr<=Prob.	[95% Conf. Interval]	
cd4		.0007183	0.0055		
cd4_0	18.82831*	.007238	0.0072	1.714079	+Inf
cd4_1	11.53732*	.0063701	0.0105	1.575285	+Inf
cd8		.0053212	0.0323		
cd8_0	.1056887*	.0289948	0.0290	0	1.072531
cd8_1	.0983388*	.0241503	0.0242	0	.9837203

```
(*) median unbiased estimates (MUE)

. matrix list e(sufficient)

e(sufficient)[1,4]
    cd4_0  cd4_1  cd8_0  cd8_1
r1      5      8      6      4

. display e(n_possible)
1091475
```

Here we used `terms()` to specify two terms in the model, `cd4` and `cd8`, that make up the two `cd4` and `cd8` indicator variables. By doing so, we obtained a conditional probabilities test for `cd4`, simultaneously testing both `cd4_0` and `cd4_1`, and `cd8`, simultaneously testing both `cd8_0` and `cd8_1`. The significance levels for the two terms are 0.0055 and 0.0323, respectively.

We also have an example where the dependent variable is completely determined by the independent variables and CMLEs are infinite. If we try to obtain MLEs, `logistic` will drop each variable and then terminate with a no-data error, error number 2000.

```
. use http://www.stata-press.com/data/r10/hiv_n, clear
(prospective study of perinatal infection of HIV-1; binomial form)

. gen byte cd4_0 = (cd4==0)

. gen byte cd4_1 = (cd4==1)

. gen byte cd8_0 = (cd8==0)

. gen byte cd8_1 = (cd8==1)

. expand n
(39 observations created)

. logistic hiv cd4_0 cd4_1 cd8_0 cd8_1
note: cd4_0 != 0 predicts success perfectly
      cd4_0 dropped and 8 obs not used

note: cd4_1 != 0 predicts success perfectly
      cd4_1 dropped and 21 obs not used

note: cd8_0 != 0 predicts failure perfectly
      cd8_0 dropped and 2 obs not used

outcome = cd8_1 <= 0 predicts data perfectly
r(2000);
```

In the example, `exlogistic` generates the joint conditional distribution of T_{cd4_0}, T_{cd4_1}, T_{cd8_0}, and T_{cd8_1} given $M = 14$ (the number of individuals that tested positive) and for reference we list the observed sufficient statistics that are stored in the matrix `e(sufficient)`. Below we take that distribution and further condition on $T_{cd4\_1} = 8$, $T_{cd8\_0} = 6$, $T_{cd8\_1} = 4$, giving the conditional distribution of T_{cd4_0}. Here we see that the observed sufficient statistic $T_{cd4\_0} = 5$ is last in the sorted listing or, equivalently, T_{cd4_0} is at the domain boundary of the conditional probability distribution. When this occurs, the conditional probability distribution is monotonically increasing in β_{cd4_0} and a maximum does not exist.

```
. use dist, clear
. keep if cd4_1==8 & cd8_0==6 & cd8_1==4
(4139 observations deleted)
. list, sep(0)
```

	_f_	cd4_0	cd4_1	cd8_0	cd8_1
1.	1668667	0	8	6	4
2.	18945542	1	8	6	4
3.	55801053	2	8	6	4
4.	55867350	3	8	6	4
5.	17423175	4	8	6	4
6.	1091475	5	8	6	4

When the CMLEs are infinite, the MUEs are computed (Hirji, Tsiatis, and Mehta 1989). For the `cd4_0` estimate, we compute the value $\overline{\beta}_{cd4\_0}$ such that

$$\Pr(T_{cd4\_0} \geq 5 \mid \beta_{cd4\_0} = \overline{\beta}_{cd4\_0}, T_{cd4\_1} = 8, T_{cd8\_0} = 6, T_{cd8\_1} = 4, M = 14) = 1/2$$

using (1) in *Methods and Formulas*.

The output is in agreement with example 1: there is an increase in risk of HIV infection for a CD4 blood serum level of 0 relative to a level of 2 and for a level of 1 relative to a level of 2; there is a decrease in risk of HIV infection for a CD8 blood serum level of 0 relative to a level of 2 and for a level of 1 relative to a level of 2.

We also displayed `e(n_possible)`. This is the combinatorial coefficient associated with the observed sufficient statistics. The same value is found in the `_f_` variable of the conditional distribution dataset listed above. The size of the distribution is $\binom{47}{14} = 341{,}643{,}774{,}795$. This can be verified by summing the `_f_` variable of the generated conditional distribution dataset.

```
. use dist, clear
. summarize _f_, meanonly
. di %15.1f r(sum)
341643774795.0
```

◁

▷ Example 3

One can think of exact logistic regression as a covariate-adjusted exact binomial. To demonstrate this point we will use `exlogistic` to compute a binomial confidence interval for m successes of n trials, by fitting the constant-only model, and compare it with the confidence interval computed by [R] **ci**. We will use the `saving()` option to retain the dataset containing the feasible values for the constant term sufficient statistic, namely, the number of successes, m, given n trials and their associated combinatorial coefficients $\binom{n}{m}$, $m = 0, 1, \ldots, n$.

```
. input y

            y
  1. 1
  2. 0
  3. 1
  4. 0
  5. 1
  6. 1
  7. end
. ci y, binomial
```

| | Obs | Mean | Std. Err. | — Binomial Exact — [95% Conf. Interval] | |
Variable					
y	6	.6666667	.1924501	.2227781	.9567281

```
. exlogistic y, estconstant nolog coef saving(binom)
note: saving distribution to file binom

Exact logistic regression
```
                                       Number of obs =   6

y	Coef.	Suff.	2*Pr(Suff.)	[95% Conf. Interval]	
_cons	.6931472	4	0.6875	-1.24955	3.096017

We use the postestimation program `estat predict` to transform the estimated constant term and its confidence bounds by using the inverse logit function, `invlogit()` (see [D] **functions**), The standard error for the estimated probability is computed using the delta method.

```
. estat predict
```

y	Predicted	Std. Err.	[95% Conf. Interval]	
Probability	0.6667	0.1925	0.2228	0.9567

```
. use binom, replace
. list, sep(0)
```

	_f_	_cons_
1.	1	0
2.	6	1
3.	15	2
4.	20	3
5.	15	4
6.	6	5
7.	1	6

Examining the listing of the generated data, the values contained in the variable _cons_ are the feasible values of M and the values contained in the variable _f_ are the binomial coefficients $\binom{6}{m}$ with total $\sum_{m=0}^{6} \binom{6}{m} = 2^6 = 64$. In the coefficient table, the sufficient statistic for the constant term, labeled Suff., is $m = 4$. This value is located at record 5 of the dataset. Therefore, the two-tailed probability of the sufficient statistic is computed as $0.6875 = 2(15 + 6 + 1)/64$.

The constant term is the value of θ that maximizes the probability of observing $M = 4$; see (1) of *Methods and Formulas*:

$$\Pr(M = 4|\theta) = \frac{15e^{4\alpha}}{1 + 6e^{\alpha} + 15e^{2\alpha} + 20e^{3\alpha} + 15e^{4\alpha} + 6e^{5\alpha} + e^{6\alpha}}$$

The maximum is at the value $\theta = \log 2$, which is demonstrated in the figure below.

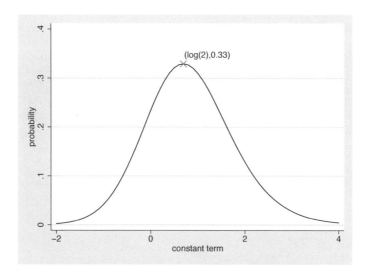

The lower and upper confidence bounds are the values of θ such that $\Pr(M \geq 4|\theta) = 0.025$ and $\Pr(M \leq 4|\theta) = 0.025$, respectively. These probabilities are plotted in the figure below for $\theta \in [-2, 4]$.

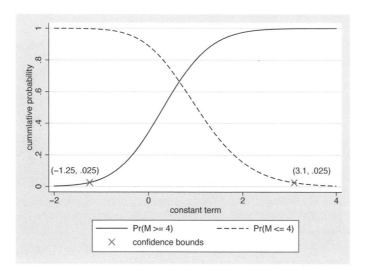

▷ Example 4

This example demonstrates the group() option, which allows the analysis of stratified data. Here the logistic model is

$$\log\left(\frac{\pi_{ik}}{1 - \pi_{ik}}\right) = \theta_k + \mathbf{x}_{ki}\boldsymbol{\beta}$$

where k indexes the s strata, $k = 1, \ldots, s$, and θ_k is the strata-specific constant term whose sufficient statistic is $M_k = \sum_{i=1}^{n_k} Y_{ki}$.

Mehta and Patel (1995) use a case–control study to demonstrate this model, which is useful in comparing the estimates from exlogistic and clogit. This study was intended to determine the role of birth complications in people with schizophrenia (Garsd 1988). Siblings from seven families took part in the study, and each individual was classified as normal or schizophrenic. A birth complication index is recorded for each individual that ranges from 0, an uncomplicated birth, to 15, a very complicated birth. Some of the frequencies contained in variable f are greater than 1 and these count different births at different times where the individual has the same birth complications index, found in variable BCindex.

```
. use http://www.stata-press.com/data/r10/schizophrenia, clear
(case-control study on birth complications for people with schizophrenia)
. list, sepby(family)
```

	family	BCindex	schizo	f
1.	1	6	0	1
2.	1	7	0	1
3.	1	3	0	2
4.	1	2	0	3
5.	1	5	0	1
6.	1	0	0	1
7.	1	15	1	1
8.	2	2	1	1
9.	2	0	0	1
10.	3	2	0	1
11.	3	9	1	1
12.	3	1	0	1
13.	4	2	1	1
14.	4	0	0	4
15.	5	3	1	1
16.	5	6	0	1
17.	5	0	1	1
18.	6	3	0	1
19.	6	0	1	1
20.	6	0	0	2
21.	7	2	0	1
22.	7	6	1	1

```
. exlogistic schizo BCindex [fw=f], group(family) test(score) coef
Enumerating sample-space combinations:
observation 1:    enumerations =        2
observation 2:    enumerations =        3
observation 3:    enumerations =        4
observation 4:    enumerations =        5
observation 5:    enumerations =        6
   (output omitted )
observation 21:   enumerations =       72
observation 22:   enumerations =       40

Exact logistic regression              Number of obs      =         29
Group variable: family                 Number of groups   =          7

                                       Obs per group: min =          2
                                                      avg =        4.1
                                                      max =         10

                                       Model score        =    6.32803
                                       Pr >= score        =     0.0167
```

schizo	Coef.	Score	Pr>=Score	[95% Conf. Interval]	
BCindex	.3251178	6.328033	0.0167	.0223423	.7408832

The asymptotic alternative for this model can be estimated using clogit, (equivalently, xtlogit, fe) and is listed below for comparison. We must expand the data since clogit will not accept frequency weights if they are not constant within the groups.

```
. expand f
(7 observations created)

. clogit schizo BCindex, group(family) nolog
note: multiple positive outcomes within groups encountered.
Conditional (fixed-effects) logistic regression   Number of obs   =         29
                                                  LR chi2(1)      =       5.20
                                                  Prob > chi2     =     0.0226
Log likelihood = -6.2819819                       Pseudo R2       =     0.2927
```

schizo	Coef.	Std. Err.	z	P>|z|	[95% Conf. Interval]	
BCindex	.3251178	.1678981	1.94	0.053	-.0039565	.654192

Both techniques compute the same regression estimate for the BCindex, which might not be too surprising since both estimation techniques condition on the total number of successes in each group. The difference lies in the p-values and confidence intervals. The p-value testing $H_0: \beta_{\text{BCindex}} = 0$ is approximately 0.0167 for the exact conditional scores test and 0.053 for the asymptotic Wald test. Moreover, the exact confidence interval is asymmetric about the estimate and does not contain zero.

◁

❑ Technical Note

The option memory(#) limits the amount of memory that exlogistic will consume when computing the conditional distribution of the parameter sufficient statistics. memory() is independent of the system setting c(memory) (see set memory in [D] **memory**), and it is possible for exlogistic to exceed the memory limit specified in c(memory) without terminating. By default, a log is provided that displays the number of enumerations (the size of the conditional distribution) after processing each observation. Typically, you will see the number of enumerations increase, and then at some point they will decrease as the multivariate shift algorithm (Hirji, Mehta, and Patel 1987) determines that some of the enumerations cannot achieve the observed sufficient statistics of the conditioning variables. When the algorithm is complete, however, it is necessary to store the conditional distribution of the parameter sufficient statistics as a dataset. It is possible, therefore, to get a memory error when the algorithm has completed and c(memory) is not large enough to store the conditional distribution.

❏

❏ Technical Note

Computing the conditional distributions and reported statistics requires data sorting and numerical comparisons. If there is at least one single-precision variable specified in the model, exlogistic will make comparisons with a relative precision of 2^{-5}. Otherwise a relative precision of 2^{-11} is used. Be careful if you use [D] **recast** to promote a single-precision variable to double precision. You might try listing the data in full precision (maybe %20.15g, see [D] **format**) to make sure that this is really what you want. See [D] **data types** for information on precision of numeric storage types.

❏

Saved Results

exlogistic saves the following in e():

Scalars

e(N)	number of observations
e(k_groups)	number of groups
e(n_possible)	number of distinct possible outcomes where sum(sufficient) equals observed e(sufficient)
e(n_trials)	binomial #-of-trials parameter
e(sum_y)	sum of *depvar*
e(k_indvars)	number of independent variables
e(k_terms)	number of model terms
e(k_condvars)	number of conditioning variables
e(condcons)	conditioned on the constant(s) indicator
e(midp)	mid-p-value rule indicator
e(eps)	relative difference tolerance

Macros

e(cmd)	exlogistic
e(cmdline)	command as typed
e(title)	Exact logistic regression
e(depvar)	dependent variable
e(indvars)	independent variables
e(condvars)	conditional variables
e(groupvar)	group variable
e(binomial)	binomial # of trials variable
e(level)	confidence level
e(wtype)	weight type
e(wexp)	weight expression
e(properties)	b
e(datasignature)	the checksum
e(datasignaturevars)	variables used in calculation of checksum
e(predict)	program used to implement predict
e(estat_cmd)	program used to implement estat

Matrices

e(b)	coefficient vector
e(mue_indicators)	indicator for elements of e(b) estimated using MUE instead of CMLE
e(se)	e(b) standard errors (CMLEs only)
e(ci)	matrix of e(level) confidence intervals for e(b)
e(sum_y_groups)	sum of e(depvar) for each group
e(N_g)	number of observations in each group
e(sufficient)	sufficient statistics for e(b)
e(p_sufficient)	p-value for e(sufficient)
e(scoretest)	conditional scores tests for *indepvars*
e(p_scoretest)	p-values for e(scoretest)
e(probtest)	conditional probabilities tests for *indepvars*
e(p_probtest)	p-value for e(probtest)
e(scoretest_m)	conditional scores tests for model terms
e(p_scoretest_m)	p-value for e(scoretest_m)
e(probtest_m)	conditional probabilities tests for model terms
e(p_probtest_m)	p-value for e(probtest_m)

Functions

e(sample)	marks estimation sample

Methods and Formulas

exlogistic is implemented as an ado-file.

Let $\{Y_1, Y_2, \ldots, Y_n\}$ be a set of n independent Bernoulli random variables, each of which can realize two outcomes $\{0, 1\}$. For each $i = 1, \ldots, n$, we observe $Y_i = y_i$ and associated with each observation is the covariate row vector of length p, $\mathbf{x}_i = (x_{i1}, \ldots, x_{ip})$. Denote $\boldsymbol{\beta} = (\beta_1, \ldots, \beta_p)^T$ to be the column vector of regression parameters and θ the constant. The sufficient statistic for β_j is $T_j = \sum_{i=1}^{n} Y_i x_{ij}$, $j = 1, \ldots, p$, and for θ is $M = \sum_{i=1}^{n} Y_i$. We observe $T_j = t_j$, $t_j = \sum_{i=1}^{n} y_i x_{ij}$ and $M = m$, $m = \sum_{i=1}^{n} y_i$. The probability of observing $(Y_1 = y_1, Y_2 = y_2, \ldots, Y_n = y_n)$ is

$$\Pr(Y_1 = y_1, \ldots, Y_n = y_n \mid \boldsymbol{\beta}, \mathbf{X}) = \frac{\exp(m\theta + \mathbf{t}\boldsymbol{\beta})}{\prod_{i=1}^{n}\{1 + \exp(\theta + \mathbf{x}_i\boldsymbol{\beta})\}}$$

where $\mathbf{t} = (t_1, \ldots, t_p)$ and $\mathbf{X} = (\mathbf{x}_1^T, \ldots, \mathbf{x}_n^T)^T$.

The joint distribution of the sufficient statistics $\mathbf{T}$ is obtained by summing over all possible binary sequences $Y_1, \ldots, Y_n$ such that $\mathbf{T} = \mathbf{t}$ and $M = m$. This probability function is

$$\Pr(T_1 = t_1, \ldots, T_p = t_p, M = m \mid \boldsymbol{\beta}, \mathbf{X}) = \frac{c(\mathbf{t}, m) \exp(m\theta + \mathbf{t}\boldsymbol{\beta})}{\prod_{i=1}^{n}\{1 + \exp(\theta + \mathbf{x}_i\boldsymbol{\beta})\}}$$

where $c(\mathbf{t}, m)$ is the combinatorial coefficient of $(\mathbf{t}, m)$, or the number of distinct binary sequences $Y_1, \ldots, Y_n$ such that $\mathbf{T} = \mathbf{t}$ and $M = m$ (Cox and Snell 1989).

Conditional distribution and CMLE

Without loss of generality, we will restrict our discussion to computing the CMLE of β_1. If we condition on observing $M = m$ and $T_2 = t_2, \ldots, T_p = t_p$, the probability function of $(T_1 \mid \beta_1, T_2 = t_2, \ldots, T_p = t_p, M = m)$ is

$$\Pr(T_1 = t_1 \mid \beta_1, T_2 = t_2, \ldots, T_p = t_p, M = m) = \frac{c(\mathbf{t}, m)e^{t_1\beta_1}}{\sum_u c(u, t_2, \ldots, t_p, m)e^{u\beta_1}} \quad (1)$$

where the sum in the denominator is over all possible values of T_1 such that $M = m$ and $T_2 = t_2, \ldots, T_p = t_p$ and $c(u, t_2, \ldots, t_p, m)$ is the combinatorial coefficient of $(u, t_2, \ldots, t_p, m)$ (Cox and Snell 1989). The CMLE for β_1 is the value $\widehat{\beta}_1$ that maximizes the log of (1). This optimization task is carried out by ml, using the conditional frequency distribution of $(T_1 \mid T_2 = t_2, \ldots, T_p = t_p, M = m)$ as a dataset. Generating the joint conditional distribution is efficiently computed using the multivariate shift algorithm described by Hirji, Mehta, and Patel (1987).

Difficulties in computing $\widehat{\beta}_1$ arise if the observed $(T_1 = t_1, \ldots, T_p = t_p, M = m)$ lies on the boundaries of the distribution of $(T_1 \mid T_2 = t_2, \ldots, T_p = t_p, M = m)$, where the conditional probability function is monotonically increasing (or decreasing) in β_1. Here the CMLE is plus infinity if it is on the upper boundary, $\Pr(T_1 \leq t_1 \mid T_2 = t_2, \ldots, T_p = t_p, M = m) = 1$, and is minus infinity if it is on the lower boundary of the distribution, $\Pr(T_1 \geq t_1 \mid T_2 = t_2, \ldots, T_p = t_p, M = m) = 1$. This concept is demonstrated in example 2. When infinite CMLEs occur, the MUE is computed.

Median unbiased estimates and exact CI

The MUE is computed using the technique outlined by Hirji, Tsiatis, and Mehta (1989). First, we find the values of $\beta_1^{(u)}$ and $\beta_1^{(l)}$ such that

$$\begin{aligned} \Pr(T_1 \leq t_1 \mid \beta_1 = \beta_1^{(u)}, T_2 = t_2, \ldots, T_p = t_p, M = m) = \\ \Pr(T_1 \geq t_1 \mid \beta_1 = \beta_1^{(l)}, T_2 = t_2, \ldots, T_p = t_p, M = m) = 1/2 \end{aligned} \quad (2)$$

The MUE is then $\overline{\beta}_1 = \left(\beta_1^{(l)} + \beta_1^{(u)}\right)/2$. However, if T_1 is equal to the minimum of the domain of the conditional distribution, $\beta^{(l)}$ does not exist and $\overline{\beta}_1 = \beta^{(u)}$. If T_1 is equal to the maximum of the domain of the conditional distribution, $\beta^{(u)}$ does not exist and $\overline{\beta}_1 = \beta^{(l)}$.

Confidence bounds for β are computed similarly, except that we substitute $\alpha/2$ for $1/2$ in (2), where $1 - \alpha$ is the confidence level. Here $\beta_1^{(l)}$ would then be the lower confidence bound and $\beta_1^{(u)}$ would be the upper confidence bound (see example 3).

Conditional hypothesis tests

To test $H_0: \beta_1 = 0$ versus $H_1: \beta_1 \neq 0$ we obtain the exact p-value from $\sum_{u \in E} f_1(u) - f_1(t_1)/2$, if the mid-$p$-value rule is used and $\sum_{u \in E} f_1(u)$ otherwise. Here E is a critical region and we define $f_1(u) = \Pr(T_1 = u \mid \beta_1 = 0, T_2 = t_2, \ldots, T_p = t_p, M = m)$ for ease of notation. There are two popular ways to define the critical region: the conditional probabilities test and the conditional scores test (Mehta and Patel 1995). The critical region when using the conditional probabilities test is all values of the sufficient statistic for β_1 that have a probability less than or equal to that of the observed t_1, $E_p = \{u : f_1(u) \leq f_1(t_1)\}$. The critical region of the conditional scores test is defined as all values of the sufficient statistic for β_1 such that its score is greater than or equal to that of t_1,

$$E_s = \left\{u : (u - \mu_1)^2/\sigma_1^2 \geq (t_1 - \mu_1)^2/\sigma_1^2)\right\}$$

Here μ_1 and σ_1^2 are the mean and variance of $(T_1 \mid \beta_1 = 0, T_2 = t_2, \ldots, T_p = t_p, M = m)$.

The score statistic is defined as

$$
\left(\frac{\partial \ell(\beta)}{\partial \beta} \right)^2 \left(-\mathbf{E} \left[\frac{\partial^2 \ell(\beta)}{\partial \beta^2} \right] \right)^{-1}
$$

evaluated at H_0: $\beta = 0$, where ℓ is the log of (1). The score test simplifies to $(t - \mathbf{E}\,[T|\beta])^2 / \mathrm{var}(T|\beta)$ (Hirji 2006), where the mean and variance are computed from the conditional distribution of the sufficient statistic with $\beta = 0$ and t is the observed sufficient statistic.

Sufficient-statistic p-value

The p-value for testing H_0: $\beta_1 = 0$ versus the two-sided alternative when $(T_1 = t_1 | T_2 = t_2, \ldots, T_p = t_p)$ is computed as $2 \cdot \min(p_l, p_u)$, where

$$
p_l = \frac{\sum_{u \leq t_1} c(u, t_2, \ldots, t_p, m)}{\sum_u c(u, t_2, \ldots, t_p, m)}
$$

$$
p_u = \frac{\sum_{u \geq t_1} c(u, t_2, \ldots, t_p, m)}{\sum_u c(u, t_2, \ldots, t_p, m)}
$$

It is the probability of observing a more extreme T_1.

References

Cox, D. R., and E. J. Snell. 1989. *Analysis of Binary Data*. 2nd ed. London: Chapman & Hall.

Garsd, A. 1988. Schizophrenia and birth complications. Unpublished manuscript.

Hirji, K. F.. 2006. *Exact Analysis of Discrete Data*. Boca Raton: Chapman & Hall.

Hirji, K. F., C. R. Mehta, and N. R. Patel. 1987. Computing distributions for exact logistic regression. *Journal of the American Statistical Association* 82: 1110–1117.

Hirji, K. F., A. A. Tsiatis, and C. R. Mehta. 1989. Median unbiased estimation for binary data. *American Statistician* 43: 7–11.

Hutto, C., W. P. Parks, and S. Lai. 1991. A hospital based prospective study of perinatal infection with HIV-1. *Journal of Pediatrics* 118: 347–353.

Mehta, C. R., and N. R. Patel. 1995. Exact logistic regression: Theory and examples. *Statistics in Medicine* 14: 2143–2160.

Also See

[R] **exlogistic postestimation** — Postestimation tools for exlogistic

[R] **binreg** — Generalized linear models: Extensions to the binomial family

[R] **clogit** — Conditional (fixed-effects) logistic regression

[R] **expoisson** — Exact Poisson regression

[R] **logit** — Logistic regression, reporting coefficients

[R] **logistic** — Logistic regression, reporting odds ratios

[U] **20 Estimation and postestimation commands**

Title

---
exlogistic postestimation — Postestimation tools for exlogistic
---

Description

The following postestimation commands are of special interest after `exlogistic`:

command	description
estat predict	single-observation prediction
estat se	report odds ratio or coefficient asymptotic standard errors

For information about these commands, see below.

The following standard postestimation command is also available:

command	description
estat summarize	estimation sample summary

`estat summarize` is not allowed if the `binomial()` option was specified in `exlogistic`.

Special-interest postestimation commands

`estat predict` computes a predicted probability (or linear predictor), its asymptotic standard error, and its exact confidence interval for 1 observation. Predictions are carried out by estimating the constant coefficient after shifting the independent variables and conditioned variables by the values specified in the `at()` option or by their medians. Therefore, predictions must be done with the estimation sample in memory. If a different dataset is used or if the dataset is modified, then an error will result.

`estat se` reports odds ratio or coefficient asymptotic standard errors. The estimates are stored in the matrix `r(estimates)`.

Syntax for estat predict

estat predict $\left[\, ,\ \textit{options}\,\right]$

options	description
pr	probability; the default
xb	linear effect
at(*atspec*)	use the specified values for the *indepvars* and condvars()
level(#)	set confidence level for the predicted value; default is level(95)
memory(#$\left[\,$b$\,|\,$k$\,|\,$m$\,|\,$g$\,\right]$)	set limit on memory usage; default is memory(10m)
nolog	do not display the enumeration log

These statistics are available only for the estimation sample.

Options for estat predict

pr, the default, calculates the probability.

xb calculates the linear effect.

at(*varname* = # [[*varname* = #] [...]]) specifies values to use in computing the predicted value. Here *varname* is one of the independent variables, *indepvars*, or the conditioned variables, condvars(). The default is to use the median of each independent and conditioned variable.

level(#) specifies the confidence level, as a percentage, for confidence intervals. The default is level(95) or as set by set level; see [U] **20.7 Specifying the width of confidence intervals**.

memory(#[b|k|m|g]) sets a limit on the amount of memory estat predict can use when generating the conditional distribution of the constant parameter sufficient statistic. The default is memory(10m), where m stands for megabyte, or 1,048,576 bytes. The following are also available: b stands for byte; k for kilobyte, which is equal to 1,024 bytes; and g for gigabyte, which is equal to 1,024 megabytes. The minimum setting allowed is 1m and the maximum is 512m or 0.5g, but do not attempt to use more memory than is available on your computer. Also see *Remarks* on enumerating the conditional distribution in [R] **exlogistic**.

nolog prevents the display of the enumeration log. By default, the enumeration log is displayed showing the progress of enumerating the distribution of the observed successes conditioned on the independent variables shifted by the values specified in at() (or by their medians). See *Methods and Formulas* in [R] **exlogistic** for details of the computations.

Syntax for estat se

estat se [, coef]

Option for estat se

coef requests that the estimated coefficients and their asymptotic standard errors be reported. The default is to report the odds ratios and their asymptotic standard errors.

Remarks

Predictions must be done using the estimation sample. This is because the prediction is really an estimated constant coefficient (the intercept) after shifting the independent variables and conditioned variables by the values specified in at() or by their medians. The justification for this approach can be seen by rewriting the model as

$$\log\left(\frac{\pi_i}{1-\pi_i}\right) = (\alpha + \mathbf{x}_0\boldsymbol{\beta}) + (\mathbf{x}_i - \mathbf{x}_0)\boldsymbol{\beta}$$

where $\mathbf{x}_0$ are the specified values for the *indepvars* (Metha and Patel 1995). Since the estimation of the constant term is required, this technique is not appropriate for stratified models that used the group() option.

▷ Example 1

To demonstrate, we return to the second example in [R] **exlogistic** using data from a prospective study of perinatal infection and human immunodeficiency virus (HIV-1). Here there was an investigation into whether the blood serum levels of CD4 and CD8 measured in infants at 6 months of age might predict their development of HIV infection. The blood serum levels are coded as ordinal values 0, 1, and 2. These data are used by Mehta and Patel (1995) as an exposition of exact logistic.

```
. use http://www.stata-press.com/data/r10/hiv_n
(prospective study of perinatal infection of HIV-1; binomial form)

. gen byte cd4_0 = (cd4==0)

. gen byte cd4_1 = (cd4==1)

. gen byte cd8_0 = (cd8==0)

. gen byte cd8_1 = (cd8==1)

. exlogistic hiv cd4_0 cd4_1 cd8_0 cd8_1, terms(cd4=cd4_0 cd4_1,
> cd8=cd8_0 cd8_1) binomial(n) test(probability) saving(dist, replace)
  (output omitted )

. estat predict
Enumerating sample-space combinations:
observation 1:    enumerations =        3
observation 2:    enumerations =       12
observation 3:    enumerations =        5
observation 4:    enumerations =        5
observation 5:    enumerations =        5
observation 6:    enumerations =       35
observation 7:    enumerations =       15
observation 8:    enumerations =       15
observation 9:    enumerations =        9
observation 10:   enumerations =        9
observation 11:   enumerations =        5
observation 12:   enumerations =       18
note: CMLE estimate for _cons is -inf; computing MUE

Predicted value at cd4_0 = 0, cd4_1 = 0, cd8_0 = 0, cd8_1 = 1
```

hiv	Predicted	Std. Err.	[95% Conf. Interval]	
Probability	0.0390*	N/A	0.0000	0.1962

```
(*) identifies median unbiased estimates (MUE); because an MUE
    is computed, there is no SE estimate
```

Since we did not specify values using the at() option, the median values of the *indepvars* are used for the prediction. By default, medians are used instead of means because we want to use values that are observed in the dataset. If the means of the binary variables cd4_0–cd8_1 were used, we would have created floating point variables in $(0, 1)$ that not only do not properly represent the indicator variables but would be a source of computational inefficiency in generating the conditional distribution. Because the MUE is computed for the predicted value, there is no standard error estimate.

From the example discussions in [R] **exlogistic**, the infants at highest risk are those with a CD4 level of 0 and a CD8 level of 2. Below we use the at() option to make a prediction at these blood serum levels.

```
. estat predict, at(cd4_0=1 cd4_1=0 cd8_0=0 cd8_1=0) nolog
note: CMLE estimate for _cons is +inf; computing MUE

Predicted value at cd4_0 = 1, cd4_1 = 0, cd8_0 = 0, cd8_1 = 0
```

hiv	Predicted	Std. Err.	[95% Conf. Interval]	
Probability	0.9063*	N/A	0.4637	1.0000

(*) identifies median unbiased estimates (MUE); because an MUE
    is computed, there is no SE estimate

◁

Saved Results

estat predict saves the following in r():

Scalars
  r(imue)   1 if r(pred) is an MUE and 0 if a CMLE
  r(pred)   estimated probability or the linear effect
  r(se)   asymptotic standard error of r(pred)

Macros
  r(estimate) prediction type: pr or xb
  r(level)   confidence level

Matrices
  r(ci)   confidence interval
  r(x)   *indepvars* and condvars() values

Methods and Formulas

All postestimation commands listed above are implemented as ado-files and using Mata.

Reference

Mehta, C. R., and N. R. Patel. 1995. Exact logistic regression: Theory and examples. *Statistics in Medicine* 14: 2143–2160.

Also See

[R] **exlogistic** — Exact logistic regression

[U] **20 Estimation and postestimation commands**

Title

expoisson — Exact Poisson regression

Syntax

expoisson *depvar indepvars* $\begin{bmatrix} if \end{bmatrix}$ $\begin{bmatrix} in \end{bmatrix}$ $\begin{bmatrix} weight \end{bmatrix}$ $\begin{bmatrix} , options \end{bmatrix}$

options	description
Model	
<u>cond</u>vars(*varlist*)	condition on variables in *varlist*
<u>group</u>(*varname*)	groups/strata are stratified by unique values of *varname*
<u>exp</u>osure(*varname$_e$*)	include ln(*varname$_e$*) in model with coefficient constrained to 1
<u>off</u>set(*varname$_o$*)	include *varname$_o$* in model with coefficient constrained to 1
Options	
<u>mem</u>ory(#$\begin{bmatrix} b \mid k \mid m \mid g \end{bmatrix}$)	set limit on memory usage; default is memory(25m)
<u>sav</u>ing(*filename*)	save the joint conditional distribution to *filename*
Reporting	
<u>level</u>(#)	set confidence level; default is level(95)
irr	report incidence-rate ratios
<u>test</u>(*testopt*)	report significance of observed sufficient statistic, conditional scores test, or conditional probabilities test
mue(*varlist*)	compute the median unbiased estimates for *varlist*
midp	use the mid-p-value rule
<u>nolog</u>	do not display the enumeration log

by, statsby, and xi are allowed; see [U] **11.1.10 Prefix commands**.

fweights are allowed; see [U] **11.1.6 weight**.

See [U] **20 Estimation and postestimation commands** for more capabilities of estimation commands.

Description

expoisson fits an exact Poisson regression model of *depvar* on *indepvars*. Exact Poisson regression is an alternative to standard maximum likelihood (ML)–based Poisson regression (see [R] **poisson**) that offers more accurate inference in small samples because it does not depend on asymptotic results. For stratified data, expoisson is an alternative to fixed-effects Poisson regression (see xtpoisson, fe in [XT] **xtpoisson**); like fixed-effects Poisson regression, exact Poisson regression conditions on the number of events in each stratum.

Exact Poisson regression is computationally intensive, so if you have regressors whose parameter estimates are not of interest (i.e., nuisance parameters), you should specify those variables in option condvars() instead of in *indepvars*.

Options

_____ Model _____

condvars(*varlist*) specifies variables whose parameter estimates are not of interest to you. You can save substantial computer time and memory moving such variables from *indepvars* to condvars(). Understand that you will get the same results for x1 and x3 whether you type

> . expoisson y x1 x2 x3 x4

or

> . expoisson y x1 x3, condvars(x2 x4)

group(*varname*) specifies the variable defining the strata, if any. A constant term is assumed for each stratum identified in *varname*, and the sufficient statistics for *indepvars* are conditioned on the observed number of successes within each group (as well as other variables in the model). The group variable must be integer valued.

exposure(*varname$_e$*), offset(*varname$_o$*); see [R] **estimation options**.

_____ Options _____

memory(#$\left[\,b\,|\,k\,|\,m\,|\,g\,\right]$) sets a limit on the amount of memory expoisson can use when computing the conditional distribution of the parameter sufficient statistics. The default is memory(25m), where m stands for megabyte, or 1,048,576 bytes. The following are also available: b stands for byte; k for kilobyte, which is equal to 1,024 bytes; and g for gigabyte, which is equal to 1,024 megabytes. The minimum setting allowed is 1m and the maximum is 512m or 0.5g, but do not attempt to use more memory than is available on your computer.

saving(*filename* $\left[\,,\,$ replace $\right]$) saves the joint conditional distribution for each independent variable specified in *indepvars*. There is one file for each variable and it is named using the prefix *filename* with the variable name appended. For example, saving(mydata) with an independent variable named X would generate a data file named mydata_X.dta. Use replace to replace an existing file. Each file contains the conditional distribution for one of the independent variables specified in *indepvars* conditioned on all other *indepvars* and those variables specified in condvars(). There are two variables in each data file: the feasible sufficient statistics for the variable's parameter and their associated weights. The weights variable is named _w_.

_____ Reporting _____

level(#); see [R] **estimation options**. The level(#) option will not work on replay because confidence intervals are based on estimator-specific enumerations. To change the confidence level, you must refit the model.

irr reports estimated coefficients transformed to incidence-rate ratios, that is, $\exp(\beta)$ rather than β. Standard errors and confidence intervals are similarly transformed. This option affects how results are displayed, not how they are estimated or stored. irr may be specified at estimation or when replaying previously estimated results.

test(sufficient | score | probability) reports the significance level of the observed sufficient statistic, the conditional scores tests, or the conditional probabilities tests. The default is test(sufficient). All the statistics are computed at estimation time, and each statistic may be displayed postestimation; see [R] **expoisson postestimation**.

mue(*varlist*) specifies that median unbiased estimates (MUEs) be reported for the variables in *varlist*. By default, the conditional maximum likelihood estimates (CMLEs) are reported, except for those parameters for which the CMLEs are infinite. Specify mue(_all) if you want MUEs for all the *indepvars*.

midp instructs expoisson to use the mid-p-value rule when computing the MUEs, significance levels, and confidence intervals. This adjustment is for the discreteness of the distribution by halving the value of the discrete probability of the observed statistic before adding it to the p-value. The mid-p-value rule cannot be MUEs whose corresponding parameter CMLE is infinite.

nolog prevents the display of the enumeration log. By default, the enumeration log is displayed, showing the progress of computing the conditional distribution of the sufficient statistics.

Remarks

Exact Poisson regression estimates the model parameters by using the conditional distributions of the parameters' sufficient statistics, and the resulting parameter estimates are known as CMLEs. Exact Poisson regression is a small-sample alternative to the ML Poisson model. See [R] **poisson** and [XT] **xtpoisson** to obtain maximum likelihood estimates (MLEs) for the Poisson model and the fixed-effects Poisson model.

Let Y_i denote a Poisson random variable where we observe the outcome $Y_i = y_i$, $i = 1, \ldots, n$. Associated with each independent observation is a $1 \times p$ vector of covariates $\mathbf{x}_i$. We will denote $\mu_i = E[Y_i \mid \mathbf{x}_i]$ and use the log linear model to model the relationship between Y_i and $\mathbf{x}_i$,

$$\log(\mu_i) = \theta + \mathbf{x}_i \boldsymbol{\beta}$$

where the constant term θ and the $p \times 1$ vector of regression parameters $\boldsymbol{\beta}$ are unknown. The probability of observing $Y_i = y_i$, $i = 1, \ldots, n$ is

$$\Pr(\mathbf{Y} = \mathbf{y}) = \prod_{i=1}^{n} \frac{\mu_i^{y_i} e^{-\mu_i}}{y_i!}$$

where $\mathbf{Y} = (Y_1, \ldots, Y_n)$ and $\mathbf{y} = (y_1, \ldots, y_n)$. The MLEs for θ and $\boldsymbol{\beta}$ maximize the log of this function.

The sufficient statistics for θ and β_j, $j = 1, \ldots, p$ are $M = \sum_{i=1}^{n} Y_i$ and $T_j = \sum_{i=1}^{n} Y_i x_{ij}$, respectively, and we observe $M = m$ and $T_j = t_j$. expoisson tallies the conditional distribution for each T_j, given the other sufficient statistics $T_l = t_l$, $l \neq j$ and $M = m$. Denote one of these values to be $t_j^{(k)}$, $k = 1, \ldots, N$, with weight w_k that accounts for all of the generated $\mathbf{Y}$ vectors that give rise to $t_j^{(k)}$. [See (2) in *Methods and Formulas* for a more concise definition of the weights.] The conditional probability of observing $T_j = t_j$ has the form

$$\Pr(T_j = t_j \mid T_l = t_l, l \neq j, M = m) = \frac{w \, e^{t_j \beta_j}}{\sum_k w_k e^{t_j^{(k)} \beta_j}} \tag{1}$$

where the sum is over the subset of $\mathbf{T}$ vectors such that $(T_1^{(k)} = t_1, \ldots, T_j^{(k)} = t_j^{(k)}, \ldots, T_p^{(k)} = t_p)$ and w is the weight associated with the observed $\mathbf{t}$. The CMLE for β_j maximizes the log of this function.

Specifying nuisance variables in condvars() prevents expoisson from estimating their associated regression coefficients. These variables are still conditional variables when tallying the conditional distribution for the variables in *indepvars*.

Inferences from MLEs rely on asymptotics, and if your sample size is small these inferences may not be valid. On the other hand, inferences from the CMLEs are exact in that they use the conditional distribution of the sufficient statistics outlined above.

For small datasets, the dependent variable can be completely determined by the data. Here the MLEs and the CMLEs are unbounded. When this occurs, expoisson will compute the median unbiased estimate (MUE), the regression estimate that places the observed sufficient statistic at the median of the conditional distribution.

See [R] **exlogistic** for a more thorough discussion of exact estimation and related statistics.

▷ Example 1

Armitage and Berry (1994) fitted a log-linear model to data containing the number of cerebrovascular accidents experienced by 41 men during a fixed period, each of whom had recovered from a previous cerebrovascular accident and was hypertensive. Sixteen men received treatment, and in the original data there are three age groups (40–49, 50–59, ≥60), but we pool the first two age groups in order to simplify the example. Armitage and Berry point out that this was not a controlled trial, but the data are useful to inquire whether there is evidence of fewer accidents for the treatment group and if age may be an important factor. The dependent variable count contains the number of accidents, the variable treat is an indicator for the treatment group (1 = treatment, 0 = control), and age is an indicator for the age group (0 = 40−59; 1 = ≥60).

First, we load the data, list it, and tabulate the cerebrovascular accident counts by treatment and age group.

```
. use http://www.stata-press.com/data/r10/cerebacc
(cerebrovascular accidents in hypotensive-treated and control groups)

. list
```

	treat	count	age
1.	control	0	40/59
2.	control	0	>=60
3.	control	1	40/59
4.	control	1	>=60
5.	control	2	40/59
6.	control	2	>=60
7.	control	3	40/59
8.	treatment	0	40/59
	(output omitted)		
35.	treatment	0	40/59
36.	treatment	0	40/59
37.	treatment	0	40/59
38.	treatment	0	40/59
39.	treatment	1	40/59
40.	treatment	1	40/59
41.	treatment	1	40/59

```
. tabulate treat age [fw=count]
```

hypotensive drug treatment	age group 40/59	>=60	Total
control	15	10	25
treatment	4	0	4
Total	19	10	29

Next we estimate the CMLE with expoisson and, for comparison, the MLE with poisson.

```
. expoisson count treat age

Estimating: treat
Enumerating sample-space combinations:
observation 1:    enumerations =        11
observation 2:    enumerations =        11
observation 3:    enumerations =        11
      (output omitted )
observation 39:   enumerations =       410
observation 40:   enumerations =       410
observation 41:   enumerations =        30

Estimating: age
Enumerating sample-space combinations:
observation 1:    enumerations =         5
observation 2:    enumerations =        15
observation 3:    enumerations =        15
      (output omitted )
observation 39:   enumerations =       455
observation 40:   enumerations =       455
observation 41:   enumerations =        30

Exact Poisson regression
                                    Number of obs =          41
```

count	Coef.	Suff.	2*Pr(Suff.)	[95% Conf. Interval]	
treat	-1.594306	4	0.0026	-3.005089	-.4701708
age	-.5112067	10	0.2794	-1.416179	.3429232

```
. poisson count treat age, nolog
Poisson regression                          Number of obs   =          41
                                            LR chi2(2)      =       10.64
                                            Prob > chi2     =      0.0049
Log likelihood =  -38.97981                 Pseudo R2       =      0.1201
```

count	Coef.	Std. Err.	z	P>|z|	[95% Conf. Interval]	
treat	-1.594306	.5573614	-2.86	0.004	-2.686714	-.5018975
age	-.5112067	.4043525	-1.26	0.206	-1.303723	.2813096
_cons	.233344	.2556594	0.91	0.361	-.2677391	.7344271

expoisson generates an enumeration log for each independent variable in *indepvars*. The conditional distribution of the parameter sufficient statistic is tallied for each independent variable. The conditional distribution for treat, for example, has 30 records containing the weights, w_k, and feasible sufficient statistics, $t_{\text{treat}}^{(k)}$. In essence, the set of points $(w_k, t_{\text{treat}}^{(k)})$, $k = 1, \ldots, 30$, tallied by expoisson now become the data to estimate the regression coefficient for treat, using (1) as the likelihood. Remember that one of the 30 $(w_k, t_{\text{treat}}^{(k)})$ must contain the observed sufficient statistic, $t_{\text{treat}} = \sum_{i=1}^{41} \text{treat}_i \times \text{count}_i = 4$, and its relative position in the sorted set of points (sorted by $t_{\text{treat}}^{(k)}$) is how the sufficient statistic significance is computed. This algorithm is repeated for variable age.

The regression coefficients for treat and age are numerically identical for both Poisson models. Both models indicate that the treatment is significant at reducing the rate of cerebrovascular accidents, $\approx e^{-1.59} \approx 0.204$, or a reduction of about 80%. There is no significant age effect.

The *p*-value for the treatment regression coefficient sufficient statistic indicates that the treatment effect is a bit more significant than the corresponding asymptotic Z statistic from poisson. Moreover, the exact confidence intervals are wider than their asymptotic counterparts.

◁

▷ Example 2

Agresti (2002) used the data from Laird and Oliver (1981) to demonstrate the Poisson model for modeling rates. The data consist of patient survival after heart valve replacement operations. The sample consists of 109 patients that are classified by type of heart valve (aortic, mitral) and by age (<55, ≥ 55). Follow-up observations cover lengths from 3 to 97 months, and the time at risk, or exposure, is stored in the variable TAR. The response is whether the subject died. First, we take a look at the data and then estimate the incidence rates (IRs) with expoisson and poisson.

```
. use http://www.stata-press.com/data/r10/heartvalve
(heart valve replacement data)
. list
```

	age	valve	deaths	TAR
1.	< 55	aortic	4	1259
2.	< 55	mitral	1	2082
3.	>= 55	aortic	7	1417
4.	>= 55	mitral	9	1647

The age variable is coded 0 for age <55 and 1 for age ≥ 55, and the valve variable is coded 0 for the aortic valve and 1 for the mitral valve. The total number of deaths, $M = 21$, is small enough that enumerating the conditional distributions for age and valve type is feasible and asymptotic inferences associated with standard ML Poisson regression may be questionable.

```
. expoisson deaths age valve, exposure(TAR) irr
Estimating: age
Enumerating sample-space combinations:
observation 1:    enumerations =          11
observation 2:    enumerations =          11
observation 3:    enumerations =         132
observation 4:    enumerations =          22

Estimating: valve
Enumerating sample-space combinations:
observation 1:    enumerations =          17
observation 2:    enumerations =          17
observation 3:    enumerations =         102
observation 4:    enumerations =          22
Exact Poisson regression
                                            Number of obs =          4
```

deaths	IRR	Suff.	2*Pr(Suff.)	[95% Conf. Interval]	
age	3.390401	16	0.0194	1.182297	11.86935
valve	.7190197	10	0.5889	.2729881	1.870068
TAR	(exposure)				

(Continued on next page)

```
. poisson deaths age valve, exposure(TAR) irr nolog
```

Poisson regression

Number of obs	=	4
LR chi2(2)	=	7.62
Prob > chi2	=	0.0222
Pseudo R2	=	0.3178

Log likelihood = -8.1747285

deaths	IRR	Std. Err.	z	P>|z|	[95% Conf. Interval]	
age	3.390401	1.741967	2.38	0.017	1.238537	9.280965
valve	.7190197	.3150492	-0.75	0.452	.3046311	1.6971
TAR	(exposure)					

The CMLE and the MLE are numerically identical. The death rate for the older age group is about 3.4 times higher than the younger age group, and this difference is significant at the 5% level. This means that for every death in the younger group each month, we would expect about three deaths from the older group. The IR estimate for valve type is approximately 0.72, but it is not significant. The exact Poisson confidence intervals are a bit wider than the asymptotic CIs.

You can use ir (see [ST] **epitab**) to estimate IRs and exact CIs for one covariate, and we compare these CIs with those from expoisson, where we estimate the incidence rate using age only.

```
. ir deaths age TAR
```

	age of patient		
	Exposed	Unexposed	Total
number of deaths	16	5	21
time at risk	3064	3341	6405
Incidence Rate	.0052219	.0014966	.0032787
	Point estimate		[95% Conf. Interval]
Inc. rate diff.	.0037254		.00085 .0066007
Inc. rate ratio	3.489295		1.221441 12.17875 (exact)
Attr. frac. ex.	.7134092		.1812948 .9178898 (exact)
Attr. frac. pop	.5435498		

(midp) Pr(k>=16) =		0.0049 (exact)
(midp) 2*Pr(k>=16) =		0.0099 (exact)

```
. expoisson deaths age, exposure(TAR) irr midp nolog
```

Exact Poisson regression

Number of obs = 4

deaths	IRR	Suff.	2*Pr(Suff.)	[95% Conf. Interval]	
age	3.489295	16	0.0099	1.324926	10.64922
TAR	(exposure)				

mid-p-value computed for the probabilities and CIs

Both ir and expoisson give identical IRs and p-values. Both report the two-sided exact significance by using the mid-p-value rule that accounts for the discreteness in the distribution by subtracting $p_{1/2} = \Pr(T = t)/2$ from $p_l = \Pr(T \le t)$ and $p_g = \Pr(T \ge t)$, computing $2 \cdot \min(p_l - p_{1/2}, p_g - p_{1/2})$. By default, expoisson will not use the mid-p-value rule (when you exclude the midp option) and here the two-sided exact significance is $2 \cdot \min(p_l, p_g) = 0.0158$. The confidence intervals differ since expoisson uses the mid-p-value rule when computing the confidence intervals, yet ir does not. You can verify this by executing `expoisson` without the `midp` option for this example and you will get the same CIs as `ir`.

You can replay `expoisson` to view the conditional scores test or the conditional probabilities test by using the `test()` option.

```
. expoisson, test(score) irr
Exact Poisson regression
                                          Number of obs =           4
```

deaths	IRR	Score	Pr>=Score	[95% Conf. Interval]
age	3.489295	6.76528	0.0113	1.324926	10.64922
TAR	(exposure)				

```
mid-p-value computed for the probabilities and CIs
```

All the statistics for `expoisson` are defined in *Methods and Formulas* for [R] **exlogistic**. Apart from enumerating the conditional distributions for the logistic and Poisson sufficient statistics, computationally, the primary difference between `exlogistic` and `expoisson` are the weighting values in the likelihood for the parameter sufficient statistics.   ◁

▷ Example 3

In this example, we fabricate data that will demonstrate the difference between the CMLE and the MUE when the CMLE is not infinite. A difference in these estimates will be more pronounced when the probability of the coefficient sufficient statistic is skewed when plotted as a function of the regression coefficient.

```
. input y x
                y          x
  1. 0 2
  2. 1 1
  3. `1 0
  4. 0 0
  5. 0 .5
  6. 1 .5
  7. 2 .01
  8. 3 .001
  9. 4 .0001
 10. end
. expoisson y x, test(score)
Enumerating sample-space combinations:
observation 1:     enumerations =          13
observation 2:     enumerations =          91
observation 3:     enumerations =         169
observation 4:     enumerations =         169
observation 5:     enumerations =         313
observation 6:     enumerations =         313
observation 7:     enumerations =        1469
observation 8:     enumerations =        5525
observation 9:     enumerations =        5287
Exact Poisson regression
                                          Number of obs =           9
```

y	Coef.	Score	Pr>=Score	[95% Conf. Interval]
x	-1.534468	2.955318	0.0810	-3.761718	.0485548

```
. expoisson y x, test(score) mue(x) nolog
```
Exact Poisson regression

Number of obs =   9

y	Coef.	Score	Pr>=Score	[95% Conf. Interval]
x	-1.309268*	2.955318	0.0810	-3.761718 .0485548

(*) median unbiased estimates (MUE)

We observe (x_i, y_i), $i = 1, \ldots, 9$. If we condition on $m = \sum_{i=1}^{9} y_i = 12$, the conditional distribution of $T_x = \sum_i Y_i x_i$ has a size of 5,287 elements. For each entry in this enumeration a realization of $Y_i = y_i^{(k)}$, $k = 1, \ldots, 5,287$, is generated such that $\sum_i y_i^{(k)} = 12$. One of these realizations produces the observed $t_x = \sum_i y_i x_i \approx 1.5234$.

Below is a graphical display comparing the CMLE with the MUE. We plot $\Pr(T_x = t_x \mid M = 12, \beta_x)$ versus β_x, $-6 \leq \beta_x \leq 1$, in the upper panel and the cumulative probabilities, $\Pr(T_x \leq t_x \mid M = 12, \beta_x)$ and $\Pr(T_x \geq t_x \mid M = 12, \beta_x)$, in the lower panel.

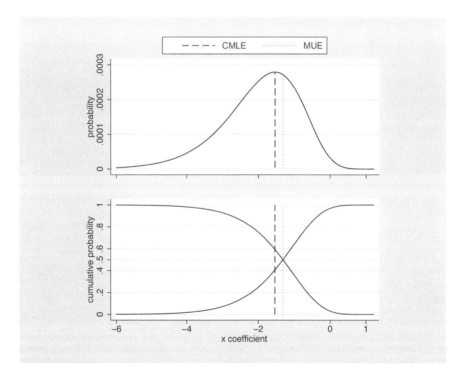

The location of the CMLE, indicated by the dashed line, is at the mode of the probability profile, and the MUE, indicated by the dotted line, is to the right of the mode. If we solve for the $\beta_x^{(u)}$ and $\beta_x^{(l)}$ such that $\Pr(T_x \leq t_x \mid M = 12, \beta_x^{(u)}) = 1/2$ and $\Pr(T_x \geq t_x \mid M = 12, \beta_x^{(l)}) = 1/2$, the MUE is $(\beta_x^{(u)} + \beta_x^{(l)})/2$. As you can see in the lower panel, the MUE cuts through the intersection of these cumulative probability profiles.

◁

❑ Technical Note

The option memory(*#*) limits the amount of memory that expoisson will consume when computing the conditional distribution of the parameter sufficient statistics. memory() is independent of the system setting c(memory) (see **set memory** in [D] **memory**), and expoisson can exceed the memory limit specified in c(memory) without terminating. By default, a log is provided that displays the number of enumerations (the size of the conditional distribution) after processing each observation. Typically, you will see the number of enumerations increase, and then at some point they will decrease as the multivariate shift algorithm (Hirji, Mehta, and Patel 1987) determines that some of the enumerations cannot achieve the observed sufficient statistics of the conditioning variables. When the algorithm is complete, however, it is necessary to store the conditional distribution of the parameter sufficient statistics as a dataset. It is possible, therefore, to get a memory error when the algorithm has completed and c(memory) is not large enough to store the conditional distribution.   ❑

❑ Technical Note

Computing the conditional distributions and reported statistics requires data sorting and numerical comparisons. If there is at least one single-precision variable specified in the model, expoisson will make comparisons with a relative precision of 2^{-5}. Otherwise a relative precision of 2^{-11} is used. Be careful if you use [D] **recast** to promote a single-precision variable to double precision. You might try listing the data in full precision (maybe %20.15g; see [D] **format**) to make sure that this is really what you want. See [D] **data types** for information on precision of numeric storage types.   ❑

Saved Results

expoisson saves the following in e():

Scalars

e(N)	number of observations
e(k_groups)	number of groups
e(relative_weight)	relative weight for the observed e(sufficient) and e(condvars)
e(sum_y)	sum of *depvar*
e(k_indvars)	number of independent variables
e(k_condvars)	number of conditioning variables
e(midp)	mid-*p*-value rule indicator
e(eps)	relative difference tolerance

Macros

e(cmd)	expoisson
e(cmdline)	command as typed
e(title)	Exact Poisson regression
e(depvar)	dependent variable
e(indvars)	independent variables
e(condvars)	conditional variables
e(groupvar)	group variable
e(level)	confidence level
e(wtype)	weight type
e(wexp)	weight expression
e(properties)	b V
e(datasignature)	the checksum
e(datasignaturevars)	variables used in calculation of checksum
e(predict)	program used to implement predict
e(estat_cmd)	program used to implement estat

Matrices

e(b)	coefficient vector
e(mue_indicators)	indicator for elements of e(b) estimated using MUE instead of CMLE
e(se)	e(b) standard errors (CMLEs only)
e(ci)	matrix of e(level) confidence intervals for e(b)
e(sum_y_groups)	sum of e(depvar) for each group
e(N_g)	number of observations in each group
e(sufficient)	sufficient statistics for e(b)
e(p_sufficient)	*p*-value for e(sufficient)
e(scoretest)	conditional scores tests for *indepvars*
e(p_scoretest)	*p*-values for e(scoretest)
e(probtest)	conditional probability tests for *indepvars*
e(p_probtest)	*p*-value for e(probtest)

Functions

e(sample)	marks estimation sample

Methods and Formulas

expoisson is implemented as an ado-file.

Let $\{Y_1, Y_2, \ldots, Y_n\}$ be a set of n independent Poisson random variables. For each $i = 1, \ldots, n$, we observe $Y_i = y_i \geq 0$ and associated with each observation is the covariate row vector of length p, $\mathbf{x}_i = (x_{i1}, \ldots, x_{ip})$. Denote $\boldsymbol{\beta} = (\beta_1, \ldots, \beta_p)^T$ to be the column vector of regression parameters and θ the constant. The sufficient statistic for β_j is $T_j = \sum_{i=1}^{n} Y_i x_{ij}$, $j = 1, \ldots, p$, and for θ is $M = \sum_{i=1}^{n} Y_i$. We observe $T_j = t_j$, $t_j = \sum_{i=1}^{n} y_i x_{ij}$ and $M = m$, $m = \sum_{i=1}^{n} y_i$. Let κ_i be the exposure for the ith observation. Then the probability of observing $(Y_1 = y_1, Y_2 = y_2, \ldots, Y_n = y_n)$ is

$$\Pr(Y_1 = y_1, \ldots, Y_n = y_n \mid \boldsymbol{\beta}, \mathbf{X}, \boldsymbol{\kappa}) = \frac{\exp(m\theta + \mathbf{t}\boldsymbol{\beta})}{\exp\{\sum_{i=1}^{n} \kappa_i \exp(\theta + \mathbf{x}_i\boldsymbol{\beta})\}} \prod_{i=1}^{n} \frac{\kappa_i^{y_i}}{y_i!}$$

where $\mathbf{t} = (t_1, \ldots, t_p)$, $\mathbf{X} = (\mathbf{x}_1^T, \ldots, \mathbf{x}_n^T)^T$, and $\boldsymbol{\kappa} = (\kappa_1, \ldots, \kappa_n)^T$.

The joint distribution of the sufficient statistics $(\mathbf{T}, M)$ is obtained by summing over all possible sequences $Y_1 \geq 0, \ldots, Y_n \geq 0$ such that $\mathbf{T} = \mathbf{t}$ and $M = m$. This probability function is

$$\Pr(T_1 = t_1, \ldots, T_p = t_p, M = m \mid \boldsymbol{\beta}, \mathbf{X}, \boldsymbol{\kappa}) = \frac{\exp(m\theta + \mathbf{t}\boldsymbol{\beta})}{\exp\{\sum_{i=1}^{n} \kappa_i \exp(\theta + \mathbf{x}_i\boldsymbol{\beta})\}} \left(\sum_{\mathbf{u}} \prod_{i=1}^{n} \frac{\kappa_i^{u_i}}{u_i!} \right)$$

where the sum $\sum_{\mathbf{u}}$ is over all nonnegative vectors $\mathbf{u}$ of length n such that $\sum_{i=1}^{n} u_i = m$ and $\sum_{i=1}^{n} u_i \mathbf{x}_i = \mathbf{t}$.

Conditional distribution

Without loss of generality, we will restrict our discussion to the conditional distribution of the sufficient statistic for β_1, T_1. If we condition on observing $M = m$ and $T_2 = t_2, \ldots, T_p = t_p$, the probability function of $(T_1 \mid \beta_1, T_2 = t_2, \ldots, T_p = t_p, M = m)$ is

$$\Pr(T_1 = t_1 \mid \beta_1, T_2 = t_2, \ldots, T_p = t_p, M = m) = \frac{\left(\sum_{\mathbf{u}} \prod_{i=1}^{n} \frac{\kappa_i^{u_i}}{u_i!} \right) e^{t_1\beta_1}}{\sum_{\mathbf{v}} \left(\prod_{i=1}^{n} \frac{\kappa_i^{v_i}}{v_i!} \right) e^{\beta_1 \sum_i v_i x_{i1}}} \tag{2}$$

where the sum $\sum_{\mathbf{u}}$ is over all nonnegative vectors $\mathbf{u}$ of length n such that $\sum_{i=1}^{n} u_i = m$ and $\sum_{i=1}^{n} u_i \mathbf{x}_i = \mathbf{t}$, and the sum $\sum_{\mathbf{v}}$ is over all nonnegative vectors $\mathbf{v}$ of length n such that $\sum_{i=1}^{n} v_i = m$, $\sum_{i=1}^{n} v_i x_{i2} = t_2$, ..., and $\sum_{i=1}^{n} v_i x_{ip} = t_p$. The CMLE for β_1 is the value that maximizes the log of (1). This optimization task is carried out by [R] **ml**, using the conditional distribution of $(T_1 \mid T_2 = t_2, \ldots, T_p = t_p, M = m)$ as a dataset. This dataset consists of the feasible values and weights for T_1,

$$\left\{ \left(s_1, \ \prod_{i=1}^{n} \frac{\kappa_i^{v_i}}{v_i!} \right) : \sum_{i=1}^{n} v_i = m, \sum_{i=1}^{n} v_i x_{i1} = s_1, \sum_{i=1}^{n} v_i x_{i2} = t_2, \ldots, \sum_{i=1}^{n} v_i x_{ip} = t_p \right\}$$

Computing the CMLE, MUE, confidence intervals, conditional hypothesis tests, and sufficient statistic p-values are discussed in *Methods and Formulas* of [R] **exlogistic**. The only difference between the two techniques is the use of the weights; that is, the weights for exact logistic are the combinatorial coefficients, $c(\mathbf{t}, m)$, in (1) of the *Methods and Formulas* section of [R] **exlogistic**. expoisson and exlogistic use the same ml likelihood evaluator to compute the CMLEs as well as the same ado-programs and Mata functions to compute the MUEs and estimate statistics.

References

Armitage, P., and G. Berry. 1994. *Statistical Methods in Medical Research*. 3rd ed. Oxford: Blackwell.

Agresti, A. 2002. *Categorical Data Analysis*. 2nd ed. Hoboken, NJ: Wiley.

Cox, D. R., and E. J. Snell. 1989. *Analysis of Binary Data*. 2nd ed. London: Chapman & Hall.

Hirji, K. F., C. R. Mehta, and N. R. Patel. 1987. Computing distributions for exact logistic regression. *Journal of the American Statistical Association* 82: 1110–1117.

Laird, N. M., D. Olivier. 1981. Covariance analysis of censored survival data using log-linear analysis techniques. *Journal of the American Statistical Association* 76: 231–240.

Also See

[R] **expoisson postestimation** — Postestimation tools for expoisson

[R] **poisson** — Poisson regression

[XT] **xtpoisson** — Fixed-effects, random-effects, and population-averaged Poisson models

[U] **20 Estimation and postestimation commands**

Title

expoisson postestimation — Postestimation tools for expoisson

Description

The following postestimation command is of special interest after expoisson:

command	description
estat se	report odds ratio or coefficient asymptotic standard errors

For information about this command, see below.

The following standard postestimation command is also available:

command	description
estat summarize	estimation sample summary

Special-interest postestimation command

estat se reports regression coefficients or incidence-rate asymptotic standard errors. The estimates are stored in the matrix r(estimates).

Syntax for estat se

    estat se [, irr]

Option for estat se

irr reports estimated coefficients transformed to incidence-rate ratios, that is, $\exp(\beta)$ rather than β. Standard errors and confidence intervals are similarly transformed. The default is to report the regression coefficients and their asymptotic standard errors.

Remarks

▷ Example 1

To demonstrate estat se after expoisson, we use the British physicians smoking data.

```
. use http://www.stata-press.com/data/r10/smokes
(cigarette smoking and lung cancer among British physicians (45-49 years))

. expoisson cases smokes, exposure(peryrs) irr nolog
Exact Poisson regression
                                                 Number of obs =         7
```

cases	IRR	Suff.	2*Pr(Suff.)	[95% Conf. Interval]	
smokes	1.077718	797.4	0.0000	1.04552	1.111866
peryrs	(exposure)				

```
. estat se, irr
```

cases	IRR	Std. Err.
smokes	1.077718	.0168547

◁

Methods and Formulas

All postestimation commands listed above are implemented as ado-files.

Also See

[R] **expoisson** — Exact Poisson regression

[U] **20 Estimation and postestimation commands**

Title

fracpoly — Fractional polynomial regression

Syntax

Fractional polynomial regression

> fracpoly *regression_cmd yvar xvar*₁ [# [#...]] [*xvar*₂ [# [#...]]] [...]
>
> [*xvarlist*] [*if*] [*in*] [*weight*] [, *fracpoly_options*]

Display table showing the best fractional polynomial model for each degree

> fracpoly [, <u>compare</u>]

Create variables containing fractional polynomial powers

> fracgen *varname* # [# ...] [*if*] [*in*] [, *fracgen_options*]

where *regression_cmd* may be clogit, cnreg, glm, logistic, logit, mlogit, nbreg, ologit, oprobit, poisson, probit, qreg, regress, rreg, stcox, streg, or xtgee.

fracpoly_options	description
Model	
<u>degree</u>(#)	degree of fractional polynomial to fit; default is degree(2)
Model 2	
<u>noscaling</u>	suppress scaling of first independent variable
<u>noconstant</u>	suppress constant term
<u>powers</u>(*numlist*)	list of fractional polynomial powers from which models are chosen
<u>adj</u>ust(*adj_list*)	specification of adjustment for the independent variables
regression_cmd_options	options appropriate to the regression command in use
Reporting	
log	display iteration log
<u>compare</u>	compare models by degree
all	include out-of-sample observations in generated variables

All weight types supported by *regression_cmd* are allowed; see [U] **11.1.6 weight**.

See [U] **20 Estimation and postestimation commands** for more capabilities of estimation commands.

where *adj_list* is a comma-separated list with elements *varlist*: {mean|#|no}, except that the first element may optionally be of the form {mean|#|no} to specify the default for all variables.

fracgen_options	description
Main	
<u>ad</u>just(no|mean|#)	adjust *varname* as specified; default is adjust(no)
<u>nosc</u>aling	suppress scaling of *varname*
<u>restr</u>ict([*varname*] [*if*])	compute adjustments and scalings using specified subsample
<u>repl</u>ace	replace variables if they exist

Description

fracpoly fits fractional polynomials (FPs) in $xvar_1$ to *yvar*. After execution, fracpoly leaves variables in the dataset named I*xvar*__1, I*xvar*__2, . . ., where *xvar* represents the first four letters of the name of $xvar_1$. The new variables contain the best-fitting FP powers of $xvar_1$.

Covariates other than $xvar_1$, which are optional, are specified in $xvar_2$, . . . and *xvarlist*. They may be modeled linearly and with specified FP transformations. Fractional polynomial powers are specified by typing numbers after the variable's name. A variable name typed without numbers is entered linearly.

fracgen creates new variables named *varname*_1, *varname*_2, . . ., containing FP powers of *varname* by using the powers (# [#. . .]) specified.

See [R] **fracpoly postestimation** for information on fracplot and fracpred.

See [R] **mfp** for multivariable FP model fitting.

Options for fracpoly

⌐ Model └_____

degree(#) determines the degree of FP to be fitted. The default is degree(2), i.e., a model with two power terms.

⌐ Model 2 └_____

noscaling suppresses scaling of $xvar_1$ and its powers.

noconstant suppresses the regression constant if this is permitted by *regression_cmd*.

powers(*numlist*) is the set of FP powers from which models are to be chosen. The default is powers(-2,-1,-.5,0,.5,1,2,3) (0 means log).

adjust(*adj_list*) defines the adjustments for the covariates $xvar_1$, $xvar_2$, . . ., *xvarlist*. The default is adjust(mean). A typical item in *adj_list* is *varlist*:{mean\|#\|no}. Items are separated by commas. The first item is special because *varlist*: is optional, and if omitted, the default is (re)set to the specified value (mean or # or no). For example, adjust(no, age:mean) sets the default to no and sets the adjustment for age to mean.

regression_cmd_options are options appropriate to the regression command in use. For example, for stcox, *regression_cmd_options* may include efron or some alternate method for handling tied failures.

⌐ Reporting └_____

log displays deviances and (for regress) residual standard deviations for each FP model fitted.

compare reports a closed-test comparison between FP models.

all includes out-of-sample observations when generating the best-fitting FP powers of $xvar_1$, $xvar_2$, etc. By default, the generated FP variables contain missing values outside the estimation sample.

Options for fracgen

```
                ⌐ Main └
```

adjust(no | mean | #) specifies whether *varname* is to be adjusted; the default is adjust(no).

noscaling suppresses scaling of *varname*.

restrict([*varname*] [*if*]) specifies that adjustments and scaling be computed using the subsample identified by *varname* and *if*.

The subsample is defined by the observations for which $varname \neq 0$ that also meet the *if* conditions. Typically, $varname = 1$ defines the subsample and $varname = 0$ indicates observations not belonging to the subsample. For observations whose subsample status is uncertain, *varname* should be set to a missing value; such observations are dropped from the subsample.

By default fracgen computes the adjustments and scaling by using the sample of observations identified in the [*if*] [*in*] options. The restrict() option identifies a subset of this sample.

replace specifies that any variables named *varname_1*, *varname_2*, ..., may be replaced.

Remarks

Remarks are presented under the following headings:

> *Introduction*
> *fracpoly*
> *Adjustment*
> *Output with the compare option*
> *fracgen*
> *Models with several continuous covariates*
> *Examples*

Introduction

Regression models based on FP functions of a continuous covariate are described by Royston and Altman (1994a). Detailed examples using an earlier and rather more complex version of this set of commands are presented by Royston and Altman (1994b).

FPs increase the flexibility afforded by the family of conventional polynomial models. Although polynomials are popular in data analysis, linear and quadratic functions are severely limited in their range of curve shapes, whereas cubic and higher-order curves often produce undesirable artifacts, such as edge effects and waves.

A polynomial of degree m may be written as

$$\beta_0 + \beta_1 x + \beta_2 x^2 + \cdots + \beta_m x^m$$

whereas FP of degree m has m integer and/or fractional powers $p_1 < \cdots < p_m$,

$$\beta_0 + \beta_1 x^{(p_1)} + \beta_2 x^{(p_2)} + \cdots + \beta_m x^{(p_m)}$$

where for a power p

$$x^{(p)} = \begin{cases} x^p & \text{if } p \neq 0 \\ \log x & \text{if } p = 0 \end{cases}$$

x must be positive. An FP of first degree ($m = 1$) involves one power or log transformation of x.

This family of FP functions may be extended in a mathematically natural way to include repeated powers. An FP of degree m with exactly m repeated powers of p is defined as

$$\beta_0 + \beta_1 x^{(p)} + \beta_2 x^{(p)} \log x + \cdots + \beta_m x^{(p)} (\log x)^{m-1}$$

For example, an FP of second degree ($m = 2$) with repeated powers of 0.5 is

$$\beta_0 + \beta_1 x^{0.5} + \beta_2 x^{0.5} \log x$$

A general FP may include some unique and some repeated powers. For example, one with powers $(-1, 1, 3, 3)$ is

$$\beta_0 + \beta_1 x^{-1} + \beta_2 x + \beta_3 x^3 + \beta_4 x^3 \log x$$

The permitted powers are restricted to the set $\{-2, -1, -0.5, 0, 0.5, 1, 2, 3\}$ because our experience using FPs in data analysis indicates that including extra powers in the set is not often worthwhile.

Now we consider using FPs in regression modeling. If the values of the powers $p_1, \ldots, p_m$ were known, the FP would resemble a conventional multiple linear regression model with coefficients $\beta_0, \beta_1, \ldots, \beta_m$. However, the powers are not (usually) known and must be estimated, together with the coefficients, from the data. Estimation involves a systematic search for the best power or combination of powers from the permitted set. For each possible combination, a linear regression model as just described is fitted, and the corresponding deviance (defined as minus twice the log likelihood) is noted. The model with the lowest deviance is deemed to have the best fit, and the corresponding powers and regression coefficients constitute the final FP model.

fracpoly

fracpoly finds and reports a multiple regression model comprising the best-fitting powers of $xvar_1$ together with other covariates specified by $xvar_2, \ldots, xvarlist$. The model that is fitted depends on the type of *regression_cmd* used.

The regression output for the best-fitting model may be reproduced by typing *regression_cmd* without variables or options. predict, test, etc., may be used after fracpoly; the results will depend on *regression_cmd*.

The standard errors of the fitted values (as estimated after use of fracpoly by using predict or fracpred with the stdp option) are somewhat too low since no allowance has been made for the estimation of the powers.

If $xvar_1$ has any negative or zero values, fracpoly subtracts the minimum of *xvar* from *xvar* and then adds the rounding (or counting) interval. The interval is defined as the smallest positive difference between the ordered values of *xvar*. After this change of origin, the minimum value of $xvar_1$ is positive, so FPs (which require $xvar_1 > 0$) can be used. Unless the noscaling option is used, fracpoly scales the resulting variable by a power of 10 calculated from the data. The scaling is designed to improve numerical stability when fitting FP models.

After execution, fracpoly leaves in the dataset variables named I$xvar$__1, I$xvar$__2, ..., which are the best-fitting FP powers of $xvar_1$ (calculated, if necessary, after a change in origin and scale as just described, and if adjustment is specified, with a constant added to or subtracted from the values after FP transformation). Other variables, whose names follow the same convention, are left in the dataset if $xvar_2$ has been specified.

Adjustment

As discussed by Garrett (1995, 1998), covariate adjustment is a sensible, indeed often essential, step when reporting and interpreting the results of multiple regression models. For this and other reasons, adjustment has been introduced as the default option in `fracpoly`. As written, the familiar straight-line regression function $E[y|x] = \beta_0 + \beta_1 x$ is "adjusted" to 0 in that $\beta_0 = E[y|0]$. This is fine if $x = 0$ is a sensible base point. However, the sample values of x may not even encompass 0 (this is usually the case when FP models are contemplated). Then β_0 is a meaningless intercept, and the standard error of its estimate $\widehat{\beta}_0$ will be large. For an FP model $E[y|x] = \beta_0 + \beta_1 x^{(p)}$, the point $x^{(p)} = 0$ may even correspond to $x = \infty$ (consider $p < 0$). The scheme adopted by `fracpoly` is to adjust to the mean of x. For example, for the FP $E[y|x] = \beta_0 + \beta_1 x^p + \beta_1 x^q$, `fracpoly` actually fits the model

$$E[y|x] = \beta_0 + \beta_1 \left(x^p - \overline{x}^p\right) + \beta_2 \left(x^q - \overline{x}^q\right)$$

where $\overline{x}$ is the sample mean of the x values, and $E[y|\overline{x}] = \beta_0$, giving β_0 a respectable interpretation as the predicted value of y at the mean of x. This approach has the advantage that plots of the fitted values and 95% confidence intervals for $E[y|x]$ as a function of x, even within a multiple regression model, are always sensible (provided that the other predictors are suitably adjusted—otherwise, the confidence limits can be alarmingly wide).

Sometimes adjustment to the mean is not appropriate, an example being a binary covariate where often you will want to adjust to the lower value, usually 0 (i.e., not adjust). You should then use the `adjust()` option to override the default. An example is `adjust(x1:mean,x2-x5:no,x6:1)`.

Output with the compare option

If the `compare` option is used, `fracpoly` displays a table showing the best FP model for each degree $k < m$ (including the model without x and the model linear in x). Deviance differences between each FP model and the degree m model are also reported along with the corresponding p-values (Royston and Altman 1994a and Royston and Sauerbrei 2005).

The `compare` option implements a closed-test approach to selecting an FP model. It has the advantage of preserving the type 1 error probability at a nominal value. For example, suppose a nominal 5% significance level was chosen, and the test of FP2 versus the null model (i.e., omitting x) was not significant. No further tests among FP models would then be done, and x would be considered nonsignificant, regardless of the results of any further model comparisons.

fracgen

The basic syntax of `fracgen` is

> `fracgen` *varname* # $\big[$ # ... $\big]$

Each power (represented by # in the syntax diagram) should be separated by a space. `fracgen` creates new variables called *varname*_1, *varname*_2, etc. Each variable is labeled according to its power, preliminary linear transformation, and adjustment, if applied.

Positive or negative powers of *varname* are defined in the usual way. A power of zero is interpreted as log.

Models with several continuous covariates

fracpoly estimates powers for FP models in just one continuous covariate ($xvar_1$), though other covariates of any kind ($xvar_2, \ldots, xvarlist$) may be included as linear or predefined FP terms. An algorithm was suggested by Royston and Altman (1994a) for the joint estimation of FP models in several continuous covariates. It was later refined by Sauerbrei and Royston (1999) and is implemented in the Stata command mfp. See [R] **mfp** as well as Royston and Ambler (1998).

Examples

▷ Example 1

Consider the serum immunoglobulin G (IgG) dataset from Isaacs et al. (1983), which consists of 298 independent observations in young children. The dependent variable sqrtigg is the square root of the IgG concentration, and the independent variable age is the age of each child. (Preliminary Box–Cox analysis shows that a square root transformation removes the skewness in IgG.) The aim is to find a model that accurately predicts the mean of sqrtigg given age. We use fracpoly to find the best FP model of degree 2 (the default option) and graph the resulting fit and 95% confidence interval:

```
. use http://www.stata-press.com/data/r10/igg
(Immunoglobulin in children)

. fracpoly regress sqrtigg age
........
-> gen double Iage__1 = age^-2-.1299486216 if e(sample)
-> gen double Iage__2 = age^2-7.695349038 if e(sample)
```

Source	SS	df	MS		Number of obs = 298
					F(2, 295) = 64.49
Model	22.2846976	2	11.1423488		Prob > F = 0.0000
Residual	50.9676492	295	.172771692		R-squared = 0.3042
					Adj R-squared = 0.2995
Total	73.2523469	297	.246640898		Root MSE = .41566

sqrtigg	Coef.	Std. Err.	t	P>|t|	[95% Conf. Interval]
Iage__1	-.1562156	.027416	-5.70	0.000	-.2101713 -.10226
Iage__2	.0148405	.0027767	5.34	0.000	.0093757 .0203052
_cons	2.283145	.0305739	74.68	0.000	2.222974 2.343315

Deviance: 319.45. Best powers of age among 44 models fit: -2 2.

(*Continued on next page*)

```
. fracplot age, msize(small)
```

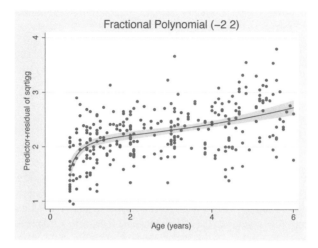

The new variables created by fracpoly contain the best-fitting FP powers of age, as centered by fracpoly. For example, Iage__1 is centered by subtracting the mean of age raised to the power -2. In general, the variables created by fracpoly are centered and possibly scaled, and this adjustment is reflected in the estimated regression coefficients and intercept. Adjustment does have its advantages (see the *Adjustment* section earlier in this entry); however, sometimes you may want estimation for unadjusted variables. To obtain regression results for uncentered and unscaled FP variables, specify options adjust(no) and noscaling to fracpoly.

The fitted curve has an asymmetric S shape. This model has powers $(-2, 2)$ and deviance 319.45. As many as 44 models have been quietly fitted in the search for the best powers. Now let us look at models of degree ≤ 4:

```
. fracpoly regress sqrtigg age, degree(4) compare
.............................................................................
> .............................................................................
> ........
-> gen double Iage__1 = ln(age)-1.020308063 if e(sample)
-> gen double Iage__2 = age^3-21.34727694 if e(sample)
-> gen double Iage__3 = age^3*ln(age)-21.78079878 if e(sample)
-> gen double Iage__4 = age^3*ln(age)^2-22.22312461 if e(sample)
```

Source	SS	df	MS
Model	22.5754541	4	5.64386353
Residual	50.6768927	293	.172958678
Total	73.2523469	297	.246640898

Number of obs = 298
F(4, 293) = 32.63
Prob > F = 0.0000
R-squared = 0.3082
Adj R-squared = 0.2987
Root MSE = .41588

sqrtigg	Coef.	Std. Err.	t	P>|t|	[95% Conf. Interval]
Iage__1	.8761824	.1898721	4.61	0.000	.5024962 1.249869
Iage__2	-.1922029	.0684934	-2.81	0.005	-.3270044 -.0574015
Iage__3	.2043794	.074947	2.73	0.007	.0568767 .3518821
Iage__4	-.0560067	.0212969	-2.63	0.009	-.097921 -.0140924
_cons	2.238735	.0482705	46.38	0.000	2.143734 2.333736

Deviance: 317.74. Best powers of age among 494 models fit: 0 3 3 3.

Fractional polynomial model comparisons:

age	df	Deviance	Res. SD	Dev. dif.	P [*]	Powers
Not in model	0	427.539	.49663	109.795	0.000	
Linear	1	337.561	.42776	19.818	0.007	1
m = 1	2	327.436	.420554	9.692	0.149	0
m = 2	4	319.448	.415658	1.705	0.798	-2 2
m = 3	6	319.275	.416243	1.532	0.476	-2 1 1
m = 4	8	317.744	.415883	—	—	0 3 3 3

[*] P-value from deviance difference comparing reported model with m = 4 model

It appears that the degree 4 FP model is not significantly different from the other FP models (at the 5% level).

Let us compare the curve shape from the $m = 2$ model with that from a conventional quartic polynomial, whose fit turns out to be significantly better than a cubic (not shown). We use the ability of `fracpoly` both to generate the required powers of `age`, namely, $(1, 2, 3, 4)$ for the quartic and $(-2, 2)$ for the second-degree FP, and to fit the model. We fit both models and graph the resulting curves:

```
. fracpoly regress sqrtigg age 1 2 3 4
-> gen double Iage__1 = age-2.774049213 if e(sample)
-> gen double Iage__2 = age^2-7.695349038 if e(sample)
-> gen double Iage__3 = age^3-21.34727694 if e(sample)
-> gen double Iage__4 = age^4-59.21839681 if e(sample)
```

Source	SS	df	MS	
Model	22.5835458	4	5.64588646	
Residual	50.668801	293	.172931061	
Total	73.2523469	297	.246640898	

```
Number of obs =     298
F(  4,   293) =   32.65
Prob > F      =  0.0000
R-squared     =  0.3083
Adj R-squared =  0.2989
Root MSE      =  .41585
```

sqrtigg	Coef.	Std. Err.	t	P>|t|	[95% Conf. Interval]	
Iage__1	2.047831	.4595962	4.46	0.000	1.143302	2.952359
Iage__2	-1.058902	.2822803	-3.75	0.000	-1.614456	-.5033479
Iage__3	.2284917	.0667591	3.42	0.001	.0971037	.3598798
Iage__4	-.0168534	.0053321	-3.16	0.002	-.0273475	-.0063594
_cons	2.240012	.0480157	46.65	0.000	2.145512	2.334511

Deviance: 317.70.

```
. predict fit1
(option xb assumed; fitted values)

. fracpoly regress sqrtigg age -2 2
-> gen double Iage__1 = age^-2-.1299486216 if e(sample)
-> gen double Iage__2 = age^2-7.695349038 if e(sample)
```

Source	SS	df	MS	
Model	22.2846976	2	11.1423488	
Residual	50.9676492	295	.172771692	
Total	73.2523469	297	.246640898	

```
Number of obs =     298
F(  2,   295) =   64.49
Prob > F      =  0.0000
R-squared     =  0.3042
Adj R-squared =  0.2995
Root MSE      =  .41566
```

sqrtigg	Coef.	Std. Err.	t	P>|t|	[95% Conf. Interval]	
Iage__1	-.1562156	.027416	-5.70	0.000	-.2101713	-.10226
Iage__2	.0148405	.0027767	5.34	0.000	.0093757	.0203052
_cons	2.283145	.0305739	74.68	0.000	2.222974	2.343315

Deviance: 319.45.

```
. predict fit2
(option xb assumed; fitted values)

. scatter sqrtigg fit1 fit2 age, c(. l l) m(o i i) msize(small)
> clpattern(. - _.) ytitle("Square root of IgG") xtitle("Age, years")
```

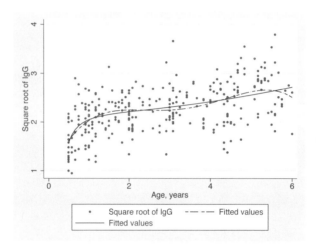

The quartic curve has an unsatisfactory wavy appearance that is implausible for the known behavior of IgG, the serum level of which increases throughout early life. The FP curve increases monotonically and is therefore biologically the more plausible curve. The two models have approximately the same deviance.

◁

▷ Example 2

Data from Smith et al. (1992) contain times to complete healing of leg ulcers in a randomized controlled clinical trial of two treatments in 192 elderly patients. Several covariates were available, of which an important one is mthson, the number of months since the recorded onset of the ulcer. Since the response variable is time to an event of interest and some (in fact, about one-half) of the times are censored, using Cox regression to analyze the data is appropriate. We consider FPs in mthson, adjusting for four other covariates: age; ulcarea, the area of tissue initially affected by the ulcer; deepppg, a binary variable indicating the presence or absence of deep vein involvement; and treat, a binary variable indicating treatment type. We fit FPs of degree 1 and 2:

```
. use http://www.stata-press.com/data/r10/legulcer, clear
(Leg ulcer clinical trial)

. stset ttevent, fail(cens)
 (output omitted )
```

```
. fracpoly stcox mthson age ulcarea deepppg treat, compare nohr
-> gen double Iage__1 = age-73.453125 if e(sample)
-> gen double Iulca__1 = ulcarea-1326.203125 if e(sample)
-> gen double Itrea__1 = treat-1 if e(sample)
........
-> gen double Imths__1 = X^.5-.4930242557 if e(sample)
-> gen double Imths__2 = X^.5*ln(X)+.6973304564 if e(sample)
   (where: X = (mthson+1)/100)
           failure _d:  censored
     analysis time _t:  ttevent
Iteration 0:   log likelihood = -422.65089
Iteration 1:   log likelihood = -390.49313
Iteration 2:   log likelihood = -383.44258
Iteration 3:   log likelihood = -374.28707
Iteration 4:   log likelihood = -369.31417
Iteration 5:   log likelihood = -368.38104
Iteration 6:   log likelihood = -368.35448
Iteration 7:   log likelihood = -368.35446
Refining estimates:
Iteration 0:   log likelihood = -368.35446

Cox regression -- Breslow method for ties
```

No. of subjects =	192	Number of obs =	192
No. of failures =	92		
Time at risk =	13825		
		LR chi2(6) =	108.59
Log likelihood =	-368.35446	Prob > chi2 =	0.0000

_t	Coef.	Std. Err.	z	P>|z|	[95% Conf. Interval]	
Imths__1	-2.81425	.6996385	-4.02	0.000	-4.185516	-1.442984
Imths__2	1.541451	.4703143	3.28	0.001	.6196521	2.46325
Iage__1	-.0261111	.0087983	-2.97	0.003	-.0433556	-.0088667
Iulca__1	-.0017491	.000359	-4.87	0.000	-.0024527	-.0010455
deepppg	-.5850499	.2163173	-2.70	0.007	-1.009024	-.1610758
Itrea__1	-.1624663	.2171048	-0.75	0.454	-.5879838	.2630513

Deviance: 736.71. Best powers of mthson among 44 models fit: .5 .5.

Fractional polynomial model comparisons:

mthson	df	Deviance	Dev. dif.	P [*]	Powers
Not in model	0	754.345	17.636	0.001	
Linear	1	751.680	14.971	0.002	1
m = 1	2	738.969	2.260	0.323	-.5
m = 2	4	736.709	—	—	.5 .5

[*] P-value from deviance difference comparing reported model with m = 2 model

The best-fit FP of degree 2 has powers $(0.5, 0.5)$ and deviance 736.71. However, this model does not fit significantly better than the FP of degree 1 (at the 5% level), which has power -0.5 and deviance 738.97. We prefer the model with $m = 1$, for which the partial linear predictor is shown on the next page.

(*Continued on next page*)

```
. quietly fracpoly stcox mthson age ulcarea deepppg treat, degree(1) nohr
. fracplot, ytitle(Partial linear predictor) m(i) ciopts(bcolor(white))
```

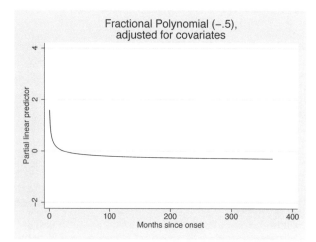

The hazard for healing is much higher for patients whose ulcer is of recent onset than for those who have had an ulcer for many months.

`fracpoly` has automatically adjusted the predictors to their mean values, but since in Cox regression there is no constant term, we cannot see the effects of adjustment in the table of regression estimates. The effects would be present if we were to graph the baseline hazard or survival function since these functions are defined with all predictors set equal to 0.

◁

Saved Results

In addition to what *regression_cmd* saves, `fracpoly` saves the following in `e()`:

Scalars

e(fp_N)	number of nonmissing observations
e(fp_dev)	deviance for FP model of degree m
e(fp_df)	FP model degrees of freedom
e(fp_d0)	deviance for model without *xvar*$_1$
e(fp_s0)	residual SD for model without *xvar*$_1$
e(fp_dlin)	deviance for model linear in *xvar*$_1$
e(fp_slin)	residual SD for model linear in *xvar*$_1$
e(fp_d1), e(fp_d2), ...	deviances for FP models of degree 1,2,...,m
e(fp_s1), e(fp_s2), ...	residual SDs for FP models of degree 1,2,...,m

Macros

e(fp_cmd)	fracpoly
e(cmdline)	command as typed
e(fp_depv)	*yvar*
e(fp_rhs)	*xvar*$_1$
e(fp_base)	variables in *xvar*$_2$, ..., *xvarlist* after adjustment and FP transformation
e(fp_xp)	I*xvar*__1, I*xvar*__2, etc.
e(fp_fvl)	variables in model finally estimated
e(fp_wgt)	weight type or ""
e(fp_wexp)	weight expression if 'e(fp_wgt)' != ""
e(fp_pwrs)	powers for FP model of degree m
e(fp_x1), e(fp_x2), ...	*xvar*$_1$ and variables in model
e(fp_k1), e(fp_k2), ...	powers for FP models of degree 1,2,...,m

Residual SDs are stored only when *regression_cmd* is `regress`.

Methods and Formulas

`fracpoly` and `fracgen` are implemented as ado-files.

The general definition of an FP, accommodating possible repeated powers, may be written for functions $H_1(x), \ldots, H_m(x)$ as

$$\beta_0 + \sum_{j=1}^{m} \beta_j H_j(x)$$

where $H_1(x) = x^{(p_1)}$ and for $j = 2, \ldots, m$,

$$H_j(x) = \begin{cases} x^{(p_j)} & \text{if } p_j \neq p_{j-1} \\ H_{j-1}(x) \log x & \text{if } p_j = p_{j-1} \end{cases}$$

For example, an FP of degree 3 with powers $(1, 3, 3)$ has $H_1(x) = x$, $H_2(x) = x^3$, and $H_3(x) = x^3 \log x$ and equals $\beta_0 + \beta_1 x + \beta_2 x^3 + \beta_3 x^3 \log x$.

An FP model of degree m is taken to have $2m + 1$ degrees of freedom (df): one for β_0 and one for each β_j and its associated power. Since the powers in an FP are chosen from a finite set rather than from the entire real line, the df defined in this way are approximate.

The deviance D of a model is defined as -2 times its maximized log likelihood. For normal-errors models, we use the formula

$$D = n \left(1 - \bar{l} + \log \frac{2\pi \text{RSS}}{n} \right)$$

where n is the sample size, $\bar{l}$ is the mean of the log-normalized weights ($\bar{l} = 0$ if the weights are all equal), and RSS is the residual sum of squares as fitted by `regress`.

The `compare` option causes `fracpoly` to report a table comparing FP models of degree $k < m$ to the degree m FP model. Suppose that we are comparing FP regression models with degrees k and m. The p-values reported by `compare` are calculated differently for normal and nonnormal regressions. Let D_k and D_m be the deviances of the models with degrees k and m, respectively. For normal-errors models such as `regress`, a variance ratio F is calculated as

$$F = \frac{n_2}{n_1} \left\{ \exp \left(\frac{D_k - D_m}{n} \right) - 1 \right\}$$

where n_1 is the numerator df (here, $2m - 2k$) and n_2 is the denominator df (equal to $rdf - 2m$, where rdf is the residual df for the regression model involving only the covariates in $xvar_2$, if any, but not x). The p-value is obtained by referring F to an F distribution on $(2, rdf)$ df.

For nonnormal models (`clogit`, `glm`, `logistic`, ... `stcox`, or `streg`; not `regress`), the p-value is obtained by referring $D_k - D_m$ to a χ^2 distribution on $2m - 2k$ df. These p-values for comparing models are approximate and are typically somewhat conservative (Royston and Altman 1994a).

The component-plus-residual values graphed by `fracplot` are calculated as follows: let the data consist of triplets $(y_i, x_i, \mathbf{z}_i)$, $i = 1, \ldots, n$, where $\mathbf{z}_i$ is the vector of covariates for the ith observation, after applying possible FP transformation and adjustment as described earlier. Let $\widehat{\eta}_i = \widehat{\beta}_0 + \{\mathbf{H}(x_i) - \mathbf{H}(x_0)\}' \widehat{\beta} + \mathbf{z}_i' \widehat{\gamma}$ be the linear predictor from the FP model, as given by the `fracpred` command, or equivalently, by the `predict` command with the `xb` option, after the use of `fracpoly`. Here $\mathbf{H}(x_i) = \{H_1(x_i), \ldots, H_m(x_i)\}'$ is the vector of FP functions described above, $\mathbf{H}(x_0) = \{H_1(x_0), \ldots, H_m(x_0)\}'$ is the vector of adjustments to x_0 (x_0 is often chosen to be the mean of the x_i), $\widehat{\beta}$ is the estimated parameter vector, and $\widehat{\gamma}$ is the estimated parameter vector for the covariates. The values $\widehat{\eta}_i^* = \widehat{\beta}_0 + \{\mathbf{H}(x_i) - \mathbf{H}(x_0)\}' \widehat{\beta}$ represent the behavior of the FP model for x at fixed values $\mathbf{z} = \mathbf{0}$ of the (adjusted) covariates. The ith component-plus-residual is defined as $\widehat{\eta}_i^* + d_i$, where d_i is the deviance residual for the ith observation. For normal-errors models, $d_i = \sqrt{w_i}(y_i - \widehat{\eta}_i)$, where w_i is the case weight (or 1, if *weight* is not specified). For logistic, Cox, and generalized linear regression models, see [R] **logistic**, [R] **probit**, [ST] **stcox**, and [R] **glm**, respectively, for the formula for d_i. The formula for `poisson` models is the same as that for `glm` with `family(poisson)`. For `stcox`, d_i is the partial martingale residual (see [ST] **stcox**).

Acknowledgment

`fracpoly` and `fracgen` were written by Patrick Royston of the MRC Clinical Trials Unit, London.

References

Becketti, S. 1995. sg26.2: Calculating and graphing fractional polynomials. *Stata Technical Bulletin* 24: 14–16. Reprinted in *Stata Technical Bulletin Reprints*, vol. 4, pp. 129–132.

Garrett, J. M. 1995. sg33: Calculation of adjusted means and adjusted proportions. *Stata Technical Bulletin* 24: 22–25. Reprinted in *Stata Technical Bulletin Reprints*, vol. 4, pp. 161–165.

——.1998. sg33.1: Enhancements for calculation of adjusted means and adjusted proportions. *Stata Technical Bulletin* 43: 16–24. Reprinted in *Stata Technical Bulletin Reprints*, vol. 8, pp. 111–123.

Isaacs, D., D. G. Altman, C. E. Tidmarsh, H. B. Valman, and A. D. B. Webster. 1983. Serum immunoglobulin concentrations in preschool children measured by laser nephelometry: Reference ranges for IgG, IgA, IgM. *Journal of Clinical Pathology* 36: 1193–1196.

Royston, J. P. 1995. sg26.3: Fractional polynomial utilities. *Stata Technical Bulletin* 25: 9–13. Reprinted in *Stata Technical Bulletin Reprints*, vol. 5, pp. 82–87.

Royston, P., and D. G. Altman. 1994a. Regression using fractional polynomials of continuous covariates: Parsimonious parametric modelling (with discussion). *Applied Statistics* 43: 429–467.

——. 1994b. sg26: Using fractional polynomials to model curved regression relationships. *Stata Technical Bulletin* 21: 11–23. Reprinted in *Stata Technical Bulletin Reprints*, vol. 4, pp. 110–128.

Royston, P., and G. Ambler. 1998. sg81: Multivariable fractional polynomials. *Stata Technical Bulletin* 43: 24–32. Reprinted in *Stata Technical Bulletin Reprints*, vol. 8, pp. 123–132.

——. 1999a. sg81.1: Multivariable fractional polynomials: Update. *Stata Technical Bulletin* 49: 17–23. Reprinted in *Stata Technical Bulletin Reprints*, vol. 9, pp. 161–168.

——. 1999b. sg81.2: Multivariable fractional polynomials: Update. *Stata Technical Bulletin* 50: 25. Reprinted in *Stata Technical Bulletin Reprints*, vol. 9, p. 168.

——. 1999c. sg112: Nonlinear regression models involving power or exponential functions of covariates. *Stata Technical Bulletin* 49: 25–30. Reprinted in *Stata Technical Bulletin Reprints*, vol. 9, pp. 173–179.

——. 1999d. sg112.1: Nonlinear regression models involving power or exponential functions of covariates: Update. *Stata Technical Bulletin* 50: 26. Reprinted in *Stata Technical Bulletin Reprints*, vol. 9, p. 180.

Royston, P., and W. Sauerbrei. 2005. Building multivariable regression models with continuous covariates in clinical epidemiology—with an emphasis on fractional polynomials. *Methods of Information in Medicine* 44: 561–571.

Sauerbrei, W., and P. Royston. 1999. Building multivariable prognostic and diagnostic models: Transformation of the predictors by using fractional polynomials. *Journal of the Royal Statistical Society, Series A* 162: 71–94.

Smith, J. M., C. J. Dore, A. Charlett, and J. D. Lewis. 1992. A randomized trial of Biofilm dressing for venous leg ulcers. *Phlebology* 7: 108–113.

Also See

[R] **fracpoly postestimation** — Postestimation tools for fracpoly

[R] **mfp** — Multivariable fractional polynomial models

[U] **20 Estimation and postestimation commands**

Title

---

fracpoly postestimation — Postestimation tools for fracpoly

Description

The following postestimation commands are of special interest after `fracpoly`:

command	description
fracplot	plot data and fit from most recently fitted fractional polynomial model
fracpred	create variable containing prediction, deviance residuals, or SEs of fitted values

For information about these commands, see below.

The following standard postestimation commands are also available if available after *regression_cmd*:

command	description
adjust	adjusted predictions of $\mathbf{x}\beta$, probabilities, or $\exp(\mathbf{x}\beta)$
estat	AIC, BIC, VCE, and estimation sample summary
estimates	cataloging estimation results
lincom	point estimates, standard errors, testing, and inference for linear combinations of coefficients
linktest	link test for model specification
lrtest	likelihood-ratio test
mfx	marginal effects or elasticities
nlcom	point estimates, standard errors, testing, and inference for nonlinear combinations of coefficients
predict	predictions, residuals, influence statistics, and other diagnostic measures
predictnl	point estimates, standard errors, testing, and inference for generalized predictions
test	Wald tests for simple and composite linear hypotheses
testnl	Wald tests of nonlinear hypotheses

See the corresponding entries in the *Stata Base Reference Manual* for details.

Special-interest postestimation commands

`fracplot` plots the data and fit, with 95% confidence limits, from the most recently fitted fractional polynomial model. The data and fit are plotted against *varname*, which may be $xvar_1$ or another of the covariates ($xvar_2, \ldots$, or a variable from *xvarlist*). If *varname* is not specified, $xvar_1$ is assumed.

`fracpred` creates *newvar* containing the fitted index or deviance residuals for the whole model, or the fitted index or its standard error for *varname*, which may be $xvar_1$ or another covariate.

Syntax for predict

The behavior of `predict` following `fracpoly` is determined by *regression_cmd*. See the corresponding *regression_cmd* **postestimation** entry for available `predict` options.

Also see information on `fracpred` below.

Syntax for fracplot and fracpred

Plot data and fit from most recently fitted fractional polynomial model

> `fracplot` [*varname*] [*if*] [*in*] [, *fracplot_options*]

Create variable containing the prediction, deviance residuals, or SEs of fitted values

> `fracpred` *newvar* [, *fracpred_options*]

fracplot_options	description
Plot	
marker_options	change look of markers (color, size, etc.)
marker_label_options	add marker labels; change look or position
Fitted line	
`lineopts`(*cline_options*)	affect rendition of the fitted line
CI plot	
`ciopts`(*area_options*)	affect rendition of the confidence bands
Add plots	
`addplot`(*plot*)	add other plots to the generated graph
Y axis, X axis, Titles, Legend, Overall	
twoway_options	any options other than by() documented in [G] ***twoway_options***

fracpred_options	description
<u>f</u>or(*varname*)	compute prediction for *varname*
<u>dr</u>esid	compute deviance residuals
<u>st</u>dp	compute standard errors of the fitted values *varname*

`fracplot` and `fracpred, dresid` are not allowed after `fracpoly` with `clogit` or `mlogit`.

(Continued on next page)

Options for fracplot

┌─ Plot ┐

marker_options affect the rendition of markers drawn at the plotted points, including their shape, size, color, and outline; see [G] ***marker_options***.

marker_label_options specify if and how the markers are to be labeled; see [G] ***marker_label_options***.

┌─ Fitted line ┐

lineopts(*cline_options*) affect the rendition of the fitted line; see [G] ***cline_options***.

┌─ CI plot ┐

ciopts(*area_options*) affect the rendition of the confidence bands; see [G] ***area_options***.

┌─ Add plots ┐

addplot(*plot*) provides a way to add other plots to the generated graph. See [G] ***addplot_option***.

┌─ Y axis, X axis, Titles, Legend, Overall ┐

twoway_options are any of the options documented in [G] ***twoway_options***, excluding by(). These include options for titling the graph (see [G] ***title_options***) and for saving the graph to disk (see [G] ***saving_option***).

Options for fracpred

for(*varname*) specifies (partial) prediction for variable *varname*. The fitted values are adjusted to the value specified by the adjust() option in fracpoly.

dresid specifies that deviance residuals be calculated.

stdp specifies calculation of the standard errors of the fitted values *varname*, adjusted for all the other predictors at the values specified by adjust().

Remarks

fracplot actually produces a component-plus-residual plot. For normal-error models with constant weights and one covariate, this amounts to a plot of the observations with the fitted line inscribed. For other normal-error models, weighted residuals are calculated and added to the fitted values.

For models with additional covariates, the line is the partial linear predictor for the variable in question (*xvar*$_1$ or a covariate) and includes the intercept β_0.

For generalized linear and Cox models, the fitted values are plotted on the scale of the "index" (linear predictor). Deviance residuals are added to the (partial) linear predictor to give component-plus-residual values. These values are plotted as small circles.

See [R] **fracpoly** for examples using fracplot.

Methods and Formulas

All postestimation commands listed above and `fracplot` and `fracpred` are implemented as ado-files.

See the *Methods and Formulas* of [R] **fracpoly** for notation.

The component-plus-residual values graphed by `fracplot` are calculated as follows: let the data consist of triplets $(y_i, x_i, \mathbf{z}_i)$, $i = 1, \ldots, n$, where $\mathbf{z}_i$ is the vector of covariates for the ith observation, after applying possible fractional polynomial transformation and adjustment as described earlier. Let $\widehat{\eta}_i = \widehat{\beta}_0 + \{\mathbf{H}(x_i) - \mathbf{H}(x_0)\}'\widehat{\beta} + \mathbf{z}_i'\widehat{\gamma}$ be the linear predictor from the FP model, as given by the `fracpred` command or, equivalently, by the `predict` command with the xb option, following `fracpoly`. Here $\mathbf{H}(x_i) = \{H_1(x_i), \ldots, H_m(x_i)\}'$ is the vector of FP functions described above, $\mathbf{H}(x_0) = \{H_1(x_0), \ldots, H_m(x_0)\}'$ is the vector of adjustments to x_0 (often, x_0 is chosen to be the mean of the x_i), $\widehat{\beta}$ is the estimated parameter vector, and $\widehat{\gamma}$ is the estimated parameter vector for the covariates. The values $\widehat{\eta}_i^* = \widehat{\beta}_0 + \{\mathbf{H}(x_i) - \mathbf{H}(x_0)\}'\widehat{\beta}$ represent the behavior of the FP model for x at fixed values $\mathbf{z} = \mathbf{0}$ of the (adjusted) covariates. The ith component-plus-residual is defined as $\widehat{\eta}_i^* + d_i$, where d_i is the deviance residual for the ith observation. For normal-errors models, $d_i = \sqrt{w_i}(y_i - \widehat{\eta}_i)$, where w_i is the case weight (or 1, if *weight* is not specified). For logistic, Cox, and generalized linear regression models, see [R] **logistic**, [R] **probit**, [ST] **stcox**, and [R] **glm**, respectively, for the formula for d_i. The formula for `poisson` models is the same as that for `glm` with `family(poisson)`. For `stcox`, d_i is the partial martingale residual (see [ST] **stcox**).

`fracplot` plots the values of d_i and the curve represented by $\widehat{\eta}_i^*$ against x_i. The confidence interval for $\widehat{\eta}_i^*$ is obtained from the variance–covariance matrix of the entire model and takes into account the uncertainty in estimating β_0, β, and γ (but not in estimating the FP powers for x).

`fracpred` with the for(*varname*) option calculates the predicted index at $x_i = x_0$ and $\mathbf{z}_i = \mathbf{0}$; that is $\widehat{\eta}_i = \widehat{\beta}_0 + \{\mathbf{H}(x_i) - \mathbf{H}(x_0)\}'\widehat{\beta}$. The standard error is calculated from the variance–covariance matrix of $(\widehat{\beta}_0, \widehat{\beta})$, again ignoring estimation of the powers.

Acknowledgment

`fracplot` and `fracpred` were written by Patrick Royston of the MRC Clinical Trials Unit, London.

Also See

[R] **fracpoly** — Fractional polynomial regression

[U] **20 Estimation and postestimation commands**

Title

frontier — Stochastic frontier models

Syntax

> frontier *depvar* [*indepvars*] [*if*] [*in*] [*weight*] [, *options*]

options	description
Model	
<u>nocons</u>tant	suppress constant term
<u>d</u>istribution(<u>h</u>normal)	half-normal distribution for the inefficiency term
<u>d</u>istribution(<u>e</u>xponential)	exponential distribution for the inefficiency term
<u>d</u>istribution(<u>t</u>normal)	truncated-normal distribution for the inefficiency term
<u>uf</u>rom(*matrix*)	specify untransformed log-likelihood; only with d(tnormal)
cm(*varlist* [, <u>nocons</u>tant])	fit conditional mean model; only with d(tnormal); use noconstant to suppress constant term
Model 2	
<u>c</u>onstraints(*constraints*)	apply specified linear constraints
<u>col</u>linear	keep collinear variables
<u>u</u>het(*varlist* [, <u>nocons</u>tant])	explanatory variables for technical inefficiency variance function; use noconstant to suppress constant term
<u>v</u>het(*varlist* [, <u>nocons</u>tant])	explanatory variables for idiosyncratic error variance function; use noconstant to suppress constant term
cost	fit cost frontier model; default is production frontier model
SE	
vce(*vcetype*)	*vcetype* may be oim, opg, <u>boot</u>strap, or <u>jack</u>knife
Reporting	
<u>l</u>evel(#)	set confidence level; default is level(95)
Max options	
maximize_options	control the maximization process; seldom used

bootstrap, by, jackknife, rolling, statsby, and xi are allowed; see [U] **11.1.10 Prefix commands**.
Weights are not allowed with the bootstrap prefix.
fweights, iweights, and pweights are allowed; see [U] **11.1.6 weight**.
See [U] **20 Estimation and postestimation commands** for more capabilities of estimation commands.

Description

frontier fits stochastic production or cost frontier models; the default is a production frontier model. It provides estimators for the parameters of a linear model with a disturbance that is assumed to be a mixture of two components, which have a strictly nonnegative and symmetric distribution, respectively. frontier can fit models in which the nonnegative distribution component (a measurement of inefficiency) is assumed to be from a half-normal, exponential, or truncated-normal distribution. See Kumbhakar and Lovell (2000) for a detailed introduction to frontier analysis.

Options

<u>Model</u>

noconstant; see [R] **estimation options**.

distribution(*distname*) specifies the distribution for the inefficiency term as half-normal (hnormal), exponential, or truncated-normal (tnormal). The default is hnormal.

ufrom(*matrix*) specifies a $1 \times K$ matrix of untransformed starting values when the distribution is truncated-normal (tnormal). frontier can estimate the parameters of the model by maximizing either the log likelihood or a transformed log-likelihood (see *Methods and Formulas*). frontier automatically transforms the starting values before passing them on to the transformed log-likelihood. The matrix must have the same number of columns as there are parameters to estimate.

cm(*varlist* [, noconstant]) may be used only with distribution(tnormal). Here frontier will fit a conditional mean model in which the mean of the truncated-normal distribution is modeled as a linear function of the set of covariates specified in *varlist*. Specifying noconstant suppresses the constant in the mean function.

<u>Model 2</u>

constraints(*constraints*), collinear; see [R] **estimation options**.

By default, when fitting the truncated-normal model or the conditional mean model, frontier maximizes a transformed log-likelihood. When constraints are applied, frontier will maximize the untransformed log-likelihood with constraints defined in the untransformed metric.

uhet(*varlist* [, noconstant]) specifies that the technical inefficiency component is heteroskedastic, with the variance function depending on a linear combination of $varlist_u$. Specifying noconstant suppresses the constant term from the variance function. This option may not be specified with distribution(tnormal).

vhet(*varlist* [, noconstant]) specifies that the idiosyncratic error component is heteroskedastic, with the variance function depending on a linear combination of $varlist_v$. Specifying noconstant suppresses the constant term from the variance function. This option may not be specified with distribution(tnormal).

cost specifies that frontier fit a cost frontier model.

<u>SE</u>

vce(*vcetype*) specifies the type of standard error reported, which includes types that are derived from asymptotic theory and that use bootstrap or jackknife methods; see [R] **vce_option**.

<u>Reporting</u>

level(*#*); see [R] **estimation options**.

<u>Max options</u>

maximize_options: <u>difficult</u>, <u>tech</u>nique(*algorithm_spec*), <u>iter</u>ate(*#*), [<u>no</u>]<u>log</u>, <u>trace</u>, <u>grad</u>ient, <u>showstep</u>, <u>hess</u>ian, <u>shownr</u>tolerance, <u>tol</u>erance(*#*), <u>ltol</u>erance(*#*), <u>gtol</u>erance(*#*), <u>nrtol</u>erance(*#*), <u>nonrtol</u>erance, from(*init_specs*); see [R] **maximize**. These options are seldom used.

Setting the optimization type to technique(bhhh) resets the default *vcetype* to vce(opg).

Remarks

Stochastic production frontier models were introduced by Aigner, Lovell, and Schmidt (1977) and Meeusen and van den Broeck (1977). Since then, stochastic frontier models have become a popular subfield in econometrics. Kumbhakar and Lovell (2000) provide a good introduction.

frontier fits three stochastic frontier models with distinct parameterizations of the inefficiency term and can fit stochastic production or cost frontier models.

Let's review the nature of the stochastic frontier problem. Suppose that a producer has a production function $f(\mathbf{z}_i, \beta)$. In a world without error or inefficiency, the ith firm would produce

$$q_i = f(\mathbf{z}_i, \beta)$$

Stochastic frontier analysis assumes that each firm potentially produces less than it might due to a degree of inefficiency. Specifically,

$$q_i = f(\mathbf{z}_i, \beta)\xi_i$$

where ξ_i is the level of efficiency for firm i; ξ_i must be in the interval $(0, 1]$. If $\xi_i = 1$, the firm is achieving the optimal output with the technology embodied in the production function $f(\mathbf{z}_i, \beta)$. When $\xi_i < 1$, the firm is not making the most of the inputs $\mathbf{z}_i$ given the technology embodied in the production function $f(\mathbf{z}_i, \beta)$. Since the output is assumed to be strictly positive (i.e., $q_i > 0$), the degree of technical efficiency is assumed to be strictly positive (i.e., $\xi_i > 0$).

Output is also assumed to be subject to random shocks, implying that

$$q_i = f(\mathbf{z}_i, \beta)\xi_i \exp(v_i)$$

Taking the natural log of both sides yields

$$\ln(q_i) = \ln\left\{f(\mathbf{z}_i, \beta)\right\} + \ln(\xi_i) + v_i$$

Assuming that there are k inputs and that the production function is linear in logs, defining $u_i = -\ln(\xi_i)$ yields

$$\ln(q_i) = \beta_0 + \sum_{j=1}^{k} \beta_j \ln(z_{ji}) + v_i - u_i \tag{1}$$

Since u_i is subtracted from $\ln(q_i)$, restricting $u_i \geq 0$ implies that $0 < \xi_i \leq 1$, as specified above.

Kumbhakar and Lovell (2000) provide a detailed version of the above derivation, and they show that performing an analogous derivation in the dual cost function problem allows us to specify the problem as

$$\ln(c_i) = \beta_0 + \beta_q \ln(q_i) + \sum_{j=1}^{k} \beta_j \ln(p_{ji}) + v_i + u_i \tag{2}$$

where q_i is output, z_{ji} are input quantities, c_i is cost, and the p_{ji} are input prices.

Intuitively, the inefficiency effect is required to lower output or raise expenditure, depending on the specification.

❏ Technical Note

The model that `frontier` actually fits is of the form

$$y_i = \beta_0 + \sum_{j=1}^{k} \beta_j x_{ji} + v_i - s u_i$$

where

$$s = \begin{cases} 1, & \text{for production functions} \\ -1, & \text{for cost functions} \end{cases}$$

so, in the context of the discussion above, $y_i = \ln(q_i)$, and $x_{ji} = \ln(z_{ji})$ for a production function; and for a cost function, $y_i = \ln(c_i)$, and the x_{ji} are the $\ln(p_{ji})$ and $\ln(q_i)$. You must take the natural logarithm of the data before fitting a stochastic frontier production or cost model. `frontier` performs no transformations on the data.

❏

Different specifications of the u_i and the v_i terms give rise to distinct models. `frontier` provides estimators for the parameters of three basic models in which the idiosyncratic component, v_i, is assumed to be independently $N(0, \sigma_v)$ distributed over the observations. The basic models differ in their specification of the inefficiency term, u_i, as follows:

`exponential`: the u_i are independently exponentially distributed with variance σ_u^2;

`hnormal`: the u_i are independently half-normally $N^+(0, \sigma_u^2)$ distributed;

`tnormal`: the u_i are independently $N^+(\mu, \sigma_u^2)$ distributed with truncation point at 0.

For half-normal or exponential distributions, `frontier` can fit models with heteroskedastic error components, conditional on a set of covariates. For a truncated-normal distribution, `frontier` can also fit a conditional mean model, in which the mean is modeled as a linear function of a set of covariates.

▷ Example 1

For our first example, we demonstrate the half-normal and exponential models by reproducing a study found in Greene (2003), which uses data originally published in Zellner and Revankar (1970). In this study of the transportation-equipment manufacturing industry, observations on value added, capital, and labor are used to estimate a Cobb–Douglas production function. The variable `lnv` is the log-transformed value added, `lnk` is the log-transformed capital, and `lnl` is the log-transformed labor. OLS estimates are compared to those from stochastic frontier models using both the half-normal and exponential distribution for the inefficiency term.

(Continued on next page)

```
. use http://www.stata-press.com/data/r10/greene9

. regress lnv lnk lnl
```

Source	SS	df	MS		Number of obs =	25
					F(2, 22) =	397.54
Model	44.1727741	2	22.086387		Prob > F =	0.0000
Residual	1.22225984	22	.055557265		R-squared =	0.9731
					Adj R-squared =	0.9706
Total	45.3950339	24	1.89145975		Root MSE =	.23571

lnv	Coef.	Std. Err.	t	P>|t|	[95% Conf. Interval]	
lnk	.2454281	.1068574	2.30	0.032	.0238193	.4670368
lnl	.805183	.1263336	6.37	0.000	.5431831	1.067183
_cons	1.844416	.2335928	7.90	0.000	1.359974	2.328858

```
. frontier lnv lnk lnl

Iteration 0:   log likelihood = 2.3357572
Iteration 1:   log likelihood = 2.4673009
Iteration 2:   log likelihood = 2.4695125
Iteration 3:   log likelihood = 2.4695222
Iteration 4:   log likelihood = 2.4695222
```

Stoc. frontier normal/half-normal model			Number of obs =	25
			Wald chi2(2) =	743.71
Log likelihood = 2.4695222			Prob > chi2 =	0.0000

lnv	Coef.	Std. Err.	z	P>|z|	[95% Conf. Interval]	
lnk	.2585478	.098764	2.62	0.009	.0649738	.4521218
lnl	.7802451	.1199399	6.51	0.000	.5451672	1.015323
_cons	2.081135	.281641	7.39	0.000	1.529128	2.633141
/lnsig2v	-3.48401	.6195353	-5.62	0.000	-4.698277	-2.269743
/lnsig2u	-3.014599	1.11694	-2.70	0.007	-5.203761	-.8254368
sigma_v	.1751688	.0542616			.0954514	.3214633
sigma_u	.2215073	.1237052			.074134	.6618486
sigma2	.0797496	.0426989			-.0039388	.163438
lambda	1.264536	.1678684			.9355204	1.593552

```
Likelihood-ratio test of sigma_u=0: chibar2(01) = 0.43   Prob>=chibar2 = 0.256

. predict double u_h, u
```

```
. frontier lnv lnk lnl, distribution(exponential)

Iteration 0:   log likelihood =  2.7270659
Iteration 1:   log likelihood =  2.8551532
Iteration 2:   log likelihood =  2.8604815
Iteration 3:   log likelihood =  2.8604897
Iteration 4:   log likelihood =  2.8604897

Stoc. frontier normal/exponential model          Number of obs   =         25
                                                 Wald chi2(2)    =     845.68
Log likelihood =  2.8604897                      Prob > chi2     =     0.0000
```

lnv	Coef.	Std. Err.	z	P>|z|	[95% Conf. Interval]	
lnk	.2624859	.0919988	2.85	0.004	.0821717	.4428002
lnl	.7703795	.1109569	6.94	0.000	.5529079	.9878511
_cons	2.069242	.2356159	8.78	0.000	1.607444	2.531041
/lnsig2v	-3.527598	.4486176	-7.86	0.000	-4.406873	-2.648324
/lnsig2u	-4.002457	.9274575	-4.32	0.000	-5.820241	-2.184674
sigma_v	.1713925	.0384448			.1104231	.2660258
sigma_u	.1351691	.0626818			.0544692	.3354317
sigma2	.0476461	.0157921			.016694	.0785981
lambda	.7886525	.087684			.616795	.9605101

```
Likelihood-ratio test of sigma_u=0: chibar2(01) = 1.21   Prob>=chibar2 = 0.135

. predict double u_e, u

. list state u_h u_e
```

	state	u_h	u_e
1.	Alabama	.2011338	.14592865
2.	California	.14480966	.0972165
3.	Connecticut	.1903485	.13478797
4.	Florida	.51753139	.5903303
5.	Georgia	.10397912	.07140994
6.	Illinois	.12126696	.0830415
7.	Indiana	.21128212	.15450664
8.	Iowa	.24933153	.20073081
9.	Kansas	.10099517	.06857629
10.	Kentucky	.05626919	.04152443
11.	Louisiana	.20332731	.15066405
12.	Maine	.22263164	.17245793
13.	Maryland	.13534062	.09245501
14.	Massachusetts	.15636999	.10932923
15.	Michigan	.15809566	.10756915
16.	Missouri	.10288047	.0704146
17.	NewJersey	.09584337	.06587986
18.	NewYork	.27787793	.22249416
19.	Ohio	.22914231	.16981857
20.	Pennsylvania	.1500667	.10302905
21.	Texas	.20297875	.14552218
22.	Virginia	.14000132	.09676078
23.	Washington	.11047581	.07533251
24.	WestVirginia	.15561392	.11236153
25.	Wisconsin	.14067066	.0970861

The parameter estimates and the estimates of the inefficiency terms closely match those published in Greene (2003), but the standard errors of the parameter estimates are estimated differently (see the *Technical Note* below).

The output from `frontier` includes estimates of the standard deviations of the two error components, σ_v and σ_u, which are labeled `sigma_v` and `sigma_u`, respectively. In the log likelihood, they are parameterized as $\ln\sigma_v^2$ and $\ln\sigma_u^2$, and these estimates are labeled `/lnsig2v` and `/lnsig2u` in the output. `frontier` also reports two other useful parameterizations. The estimate of the total error variance, $\sigma_S^2 = \sigma_v^2 + \sigma_u^2$, is labeled `sigma2`, and the estimate of the ratio of the standard deviation of the inefficiency component to the standard deviation of the idiosyncratic component, $\lambda = \sigma_u/\sigma_v$, is labeled `lambda`.

At the bottom of the output, `frontier` reports the results of a test that there is no technical inefficiency component in the model. This is a test of the null hypothesis $H_0 : \sigma_u^2 = 0$ against the alternative hypotheses $H_1 : \sigma_u^2 > 0$. If the null hypothesis is true, the stochastic frontier model reduces to an OLS model with normal errors. However, since the test lies on the boundary of the parameter space of σ_u^2, the standard likelihood-ratio test is not valid, and a one-sided generalized likelihood-ratio test must be constructed; see Gutierrez, Carter, and Drukker (2001). For this example, the output shows LR = .43 with a p-value 0.256 for the half-normal model and LR = 1.21 with a p-value 0.135 for the exponential model. There are several possible reasons for the failure to reject the null hypothesis, but the fact that the test is based on an asymptotic distribution and the sample size was 25 is certainly a leading candidate among those possibilities.

◁

❏ Technical Note

`frontier` maximizes the log-likelihood function of a stochastic frontier model by using the Newton–Raphson method, and the estimated variance–covariance matrix is calculated as the inverse of the negative Hessian (matrix of second partial derivatives); see [R] **ml**. When comparing the results with those published using other software, be aware of the difference in the optimization methods, which may result in different, yet asymptotically equivalent, variance estimates.

❏

▷ Example 2

Often the error terms may not have constant variance. `frontier` allows you to model heteroskedasticity in either error term as a linear function of a set of covariates. The variance of either the technical inefficiency or the idiosyncratic component may be modeled as

$$\sigma_i^2 = \exp(\mathbf{w}_i\boldsymbol{\delta})$$

The default constant included in $\mathbf{w}_i$ may be suppressed by appending a `noconstant` option to the list of covariates. Also, you can simultaneously specify covariates for both σ_{u_i} and σ_{v_i}.

In the example below, we use a sample of 756 observations of fictional firms producing a manufactured good by using capital and labor. The firms are hypothesized to use a constant returns-to-scale technology, but the sizes of the firms differ. Believing that this size variation will introduce heteroskedasticity into the idiosyncratic error term, we estimate the parameters of a Cobb–Douglas production function. To do this, we use a conditional heteroskedastic half-normal model, with the size of the firm as an explanatory variable in the variance function for the idiosyncratic error. We also perform a test of the hypothesis that the firms use a constant returns-to-scale technology.

```
. use http://www.stata-press.com/data/r10/frontier1, clear

. frontier lnoutput lnlabor lncapital, vhet(size)

Iteration 0:   log likelihood = -1508.3692
Iteration 1:   log likelihood =  -1501.583
Iteration 2:   log likelihood = -1500.3942
Iteration 3:   log likelihood = -1500.3794
Iteration 4:   log likelihood = -1500.3794
```

Stoc. frontier normal/half-normal model		Number of obs	=	756
		Wald chi2(2)	=	9.68
Log likelihood = -1500.3794		Prob > chi2	=	0.0079

lnoutput	Coef.	Std. Err.	z	P>|z|	[95% Conf. Interval]	
lnoutput						
lnlabor	.7090933	.2349374	3.02	0.003	.2486244	1.169562
lncapital	.3931345	.5422173	0.73	0.468	-.6695919	1.455861
_cons	1.252199	3.14656	0.40	0.691	-4.914946	7.419344
lnsig2v						
size	-.0016951	.0004748	-3.57	0.000	-.0026256	-.0007645
_cons	3.156091	.9265826	3.41	0.001	1.340023	4.97216
lnsig2u						
_cons	1.947487	.1017653	19.14	0.000	1.748031	2.146943
sigma_u	2.647838	.134729			2.396514	2.925518

```
. test _b[lnlabor] + _b[lncapital] = 1

 ( 1)  [lnoutput]lnlabor + [lnoutput]lncapital = 1

          chi2( 1) =    0.03
        Prob > chi2 =    0.8622
```

The output above indicates that the variance of the idiosyncratic error term is a function of firm size. Also, we failed to reject the hypothesis that the firms use a constant returns-to-scale technology.

◁

❑ Technical Note

In small samples, the conditional heteroskedastic estimators will lack precision for the variance parameters and may fail to converge altogether.

❑

▷ Example 3

Let's turn our attention to the truncated-normal model. Once again, we will use fictional data. For this example, we have 1,231 observations on the quantity of output, the total cost of production for each firm, the prices that each firm paid for labor and capital services, and a categorical variable measuring the quality of each firm's management. After taking the natural logarithm of the costs (lncost), prices (lnp_k and lnp_l), and output (lnout), we fit a stochastic cost frontier model and specify the distribution for the inefficiency term to be truncated normal.

```
. use http://www.stata-press.com/data/r10/frontier2

. frontier lncost lnp_k lnp_l lnout, distribution(tnormal) cost

Iteration 0:    log likelihood = -2386.9523
Iteration 1:    log likelihood = -2386.5146
Iteration 2:    log likelihood = -2386.2704
Iteration 3:    log likelihood = -2386.2504
Iteration 4:    log likelihood = -2386.2493
Iteration 5:    log likelihood = -2386.2493

Stoc. frontier normal/truncated-normal model          Number of obs    =      1231
                                                      Wald chi2(3)     =      8.82
Log likelihood = -2386.2493                           Prob > chi2      =    0.0318
```

lncost	Coef.	Std. Err.	z	P>|z|	[95% Conf. Interval]	
lnp_k	.3410717	.2363861	1.44	0.149	-.1222366	.80438
lnp_l	.6608628	.4951499	1.33	0.182	-.3096131	1.631339
lnout	.7528653	.3468968	2.17	0.030	.0729601	1.432771
_cons	2.602609	1.083004	2.40	0.016	.4799595	4.725259
/mu	1.095705	.881517	1.24	0.214	-.632037	2.823446
/lnsigma2	1.5534	.1873464	8.29	0.000	1.186208	1.920592
/ilgtgamma	1.257862	.2589522	4.86	0.000	.7503255	1.765399
sigma2	4.727518	.8856833			3.274641	6.825001
gamma	.7786579	.0446303			.6792496	.8538846
sigma_u2	3.681119	.7503408			2.210478	5.15176
sigma_v2	1.046399	.2660035			.5250413	1.567756

```
H0: No inefficiency component:          z =    5.595          Prob>=z = 0.000
```

In addition to the coefficients, the output reports estimates for several parameters. sigma_v2 is the estimate of σ_v^2. sigma_u2 is the estimate of σ_u^2. gamma is the estimate of $\gamma = \sigma_u^2/\sigma_S^2$. sigma2 is the estimate of $\sigma_S^2 = \sigma_v^2 + \sigma_u^2$. Since γ must be between 0 and 1, the optimization is parameterized in terms of the inverse logit of γ, and this estimate is reported as ilgtgamma. Since σ_S^2 must be positive, the optimization is parameterized in terms of $\ln(\sigma_S^2)$, whose estimate is reported as lnsigma2. Finally, mu is the estimate of μ, the mean of the truncated-normal distribution.

In the output above, the generalized log-likelihood test for the presence of the inefficiency term has been replaced with a test based on the third moment of the OLS residuals. When $\mu = 0$ and $\sigma_u = 0$, the truncated-normal model reduces to a linear regression model with normally distributed errors. However, the distribution of the test statistic under the null hypothesis is not well established, as it becomes impossible to evaluate the log likelihood as σ_u approaches zero, prohibiting the use of the likelihood-ratio test.

However, Coelli (1995) noted that the presence of an inefficiency term would negatively skew the residuals from an OLS regression. By identifying negative skewness in the residuals with the presence of an inefficiency term, Coelli derived a one-sided test for the presence of the inefficiency term. The results of this test are given at the bottom of the output. For this example, the null hypothesis of no inefficiency component is rejected.

In the example below, we fit a truncated model and detect a statistically significant inefficiency term in the model. We might question whether the inefficiency term is identically distributed over all firms or whether there might be heterogeneity across firms. frontier provides an extension to the truncated normal model by allowing the mean of the inefficiency term to be modeled as a linear function of a set of covariates. In our dataset, we have a categorical variable that measures the quality of a firm's management. We refit the model, including the cm() option, specifying a set of binary indicator variables representing the different categories of the quality-measurement variable as covariates.

```
. xi: frontier lncost lnp_k lnp_l lnout, distribution(tnormal) cm(i.quality) cost
i.quality        _Iquality_1-5       (naturally coded; _Iquality_1 omitted)

Iteration 0:   log likelihood = -2386.9523
Iteration 1:   log likelihood =  -2384.936
Iteration 2:   log likelihood = -2382.3942
Iteration 3:   log likelihood =  -2382.324
Iteration 4:   log likelihood = -2382.3233
Iteration 5:   log likelihood = -2382.3233

Stoc. frontier normal/truncated-normal model      Number of obs    =       1231
                                                  Wald chi2(3)     =       9.31
Log likelihood = -2382.3233                       Prob > chi2      =     0.0254
```

	Coef.	Std. Err.	z	P>|z|	[95% Conf.	Interval]
lncost						
lnp_k	.3611204	.2359749	1.53	0.126	-.1013819	.8236227
lnp_l	.680446	.4934935	1.38	0.168	-.2867835	1.647675
lnout	.7605533	.3466102	2.19	0.028	.0812098	1.439897
_cons	2.550769	1.078911	2.36	0.018	.4361417	4.665396
mu						
_Iquality_2	.5056067	.3382907	1.49	0.135	-.1574309	1.168644
_Iquality_3	.783223	.376807	2.08	0.038	.0446947	1.521751
_Iquality_4	.5577511	.3355061	1.66	0.096	-.0998288	1.215331
_Iquality_5	.6792882	.3428073	1.98	0.048	.0073981	1.351178
_cons	.6014025	.990167	0.61	0.544	-1.339289	2.542094
/lnsigma2	1.541784	.1790926	8.61	0.000	1.190769	1.892799
/ilgtgamma	1.242302	.2588968	4.80	0.000	.734874	1.749731
sigma2	4.67292	.8368852			3.289611	6.637923
gamma	.7759645	.0450075			.6758739	.8519189
sigma_u2	3.62602	.7139576			2.226689	5.025351
sigma_v2	1.0469	.2583469			.5405491	1.553251

The conditional mean model was developed in the context of panel-data estimators, and we can apply frontier's conditional mean model to panel data.

◁

(*Continued on next page*)

Saved Results

frontier saves the following in e():

Scalars

e(N)	number of observations	e(sigma_u)	standard deviation of technical inefficiency
e(df_m)	model degrees of freedom		
e(k)	number of parameters	e(sigma_v)	standard deviation of v_i
e(k_eq)	number of equations	e(p)	significance
e(k_eq_model)	number of equations in model Wald test	e(chi2_c)	LR test statistic
		e(p_z)	p-value for z
e(k_dv)	number of dependent variables	e(rank)	rank of e(V)
e(chi2)	χ^2	e(ic)	number of iterations
e(ll)	log likelihood	e(rc)	return code
e(ll_c)	log likelihood for H_0: $\sigma_u=0$	e(converged)	1 if converged, 0 otherwise
e(z)	test for negative skewness of OLS residuals		

Macros

e(cmd)	frontier	e(v_hetvar)	*varlist* in vhet()
e(cmdline)	command as typed	e(vce)	*vcetype* specified in vce()
e(depvar)	name of dependent variable	e(vcetype)	title used to label Std. Err.
e(function)	production or cost	e(opt)	type of optimization
e(wtype)	weight type	e(ml_method)	type of ml method
e(wexp)	weight expression	e(user)	name of likelihood-evaluator program
e(title)	title in estimation output	e(technique)	maximization technique
e(chi2type)	Wald; type of model χ^2 test	e(crittype)	optimization criterion
e(dist)	distribution assumption for u_i	e(properties)	b V
e(het)	heteroskedastic components	e(predict)	program used to implement predict
e(u_hetvar)	*varlist* in uhet()		

Matrices

e(b)	coefficient vector	e(V)	variance–covariance matrix of the estimators
e(ilog)	iteration log (up to 20 iterations)		
e(gradient)	gradient vector		

Functions

e(sample)	marks estimation sample

Methods and Formulas

frontier is implemented as an ado-file.

Consider an equation of the form

$$y_i = \mathbf{x}_i\boldsymbol{\beta} + v_i - su_i$$

where

y_i is the dependent variable,

$\mathbf{x}_i$ is a $1 \times k$ vector of observations on the independent variables included as covariates,

β is a $k \times 1$ vector of coefficients, and

$$s = \begin{cases} 1, & \text{for production functions} \\ -1, & \text{for cost functions} \end{cases}$$

The log-likelihood functions are as follows.

Normal-half normal model:

$$\ln L = \sum_{i=1}^{N} \left\{ \frac{1}{2} \ln \left(\frac{2}{\pi} \right) - \ln \sigma_S + \ln \Phi \left(-\frac{s \epsilon_i \lambda}{\sigma_S} \right) - \frac{\epsilon_i^2}{2 \sigma_S^2} \right\}$$

Normal-exponential model:

$$\ln L = \sum_{i=1}^{N} \left\{ -\ln \sigma_u + \frac{\sigma_v^2}{2 \sigma_u^2} + \ln \Phi \left(\frac{-s \epsilon_i - \frac{\sigma_v^2}{\sigma_u}}{\sigma_v} \right) + \frac{s \epsilon_i}{\sigma_u} \right\}$$

Normal-truncated normal model:

$$\ln L = \sum_{i=1}^{N} \left\{ -\frac{1}{2} \ln (2\pi) - \ln \sigma_S - \ln \Phi \left(\frac{\mu}{\sigma_S \sqrt{\gamma}} \right) \right.$$
$$\left. + \ln \Phi \left[\frac{(1-\gamma)\mu - s \gamma \epsilon_i}{\{ \sigma_S^2 \gamma (1-\gamma) \}^{1/2}} \right] - \frac{1}{2} \left(\frac{\epsilon_i + s\mu}{\sigma_S} \right)^2 \right\}$$

where $\sigma_S = (\sigma_u^2 + \sigma_v^2)^{1/2}$, $\lambda = \sigma_u / \sigma_v$, $\gamma = \sigma_u^2 / \sigma_S^2$, $\epsilon_i = y_i - \mathbf{x}_i \beta$, and $\Phi()$ is the cumulative distribution function of the standard normal distribution.

To obtain estimation for u_i, you can use either the mean or the mode of the conditional distribution $f(u|\epsilon)$.

$$E\left(u_i \mid \epsilon_i \right) = \mu_{*i} + \sigma_* \left\{ \frac{\phi(-\mu_{*i}/\sigma_*)}{\Phi(\mu_{*i}/\sigma_*)} \right\}$$

$$M\left(u_i \mid \epsilon_i \right) = \begin{cases} \mu_{*i}, & \text{if } \mu_{*i} >= 0 \\ 0, & \text{otherwise} \end{cases}$$

Then the technical efficiency $(s = 1)$ or cost efficiency $(s = -1)$ will be estimated by

$$E_i = E\left\{ \exp(-s u_i) \mid \epsilon_i \right\}$$
$$= \left\{ \frac{1 - \Phi\left(s\sigma_* - \mu_{*i}/\sigma_* \right)}{1 - \Phi\left(-\mu_{*i}/\sigma_* \right)} \right\} \exp \left(-s \mu_{*i} + \frac{1}{2} \sigma_*^2 \right)$$

where μ_{*i} and σ_* are defined for the normal/half-normal model as

$$\mu_{*i} = -s\epsilon_i\sigma_u^2/\sigma_S^2$$

$$\sigma_* = \sigma_u\sigma_v/\sigma_S$$

for the normal/exponential model as

$$\mu_{*i} = -s\epsilon_i - \sigma_v^2/\sigma_u$$

$$\sigma_* = \sigma_v$$

and for the normal/truncated-normal model as

$$\mu_{*i} = \frac{-s\epsilon_i\sigma_u^2 + \mu\sigma_v^2}{\sigma_S^2}$$

$$\sigma_* = \sigma_u\sigma_v/\sigma_S$$

In the half-normal and exponential models, when heteroskedasticity is assumed, the standard deviations, σ_u or σ_v, will be replaced in the above equations by

$$\sigma_i^2 = \exp(\mathbf{w}_i\boldsymbol{\delta})$$

where $\mathbf{w}$ is the vector of explanatory variables in the variance function.

In the conditional mean model, the mean parameter of the truncated normal distribution, μ, is modeled as a linear combination of the set of covariates $\mathbf{w}$.

$$\mu = \mathbf{w}_i\boldsymbol{\delta}$$

Therefore, the log-likelihood function can be rewritten as

$$\ln L = \sum_{i=1}^{N}\left\{-\frac{1}{2}\ln(2\pi) - \ln\sigma_S - \ln\Phi\left(\frac{\mathbf{w}_i\boldsymbol{\delta}}{\sqrt{\sigma_S^2\gamma}}\right)\right.$$
$$\left. + \ln\Phi\left(\frac{(1-\gamma)\mathbf{w}_i\boldsymbol{\delta} - s\gamma\epsilon_i}{\sqrt{\sigma_S^2\gamma(1-\gamma)}}\right) - \frac{1}{2}\left(\frac{\epsilon_i + s\mathbf{w}_i\boldsymbol{\delta}}{\sigma_S}\right)^2\right\}$$

The z test reported in the output of the truncated-normal model is a third-moment test developed by Coelli (1995) as an extension of a test previously developed by Pagan and Hall (1983). Coelli shows that under the null of normally distributed errors, the statistic

$$z = \frac{m_3}{\left(\frac{6m_2^3}{N}\right)^{1/2}}$$

has a standard normal distribution, where m_3 is the third moment from the OLS regression. Since the residuals are either negatively skewed (production function) or positively skewed (cost function), a one-sided p-value is used.

References

Aigner, D. J., C. A. K. Lovell, and P. Schmidt. 1977. Formulation and estimation of stochastic frontier production function models. *Journal of Econometrics* 6: 21–37.

Caudill, S. B., J. M. Ford, and D. M. Gropper. 1995. Frontier estimation and firm-specific inefficiency measures in the presence of heteroskedasticity. *Journal of Business and Economic Statistics* 13: 105–111.

Coelli, T. J. 1995. Estimators and hypothesis tests for a stochastic frontier function: A Monte Carlo analysis. *Journal of Productivity Analysis* 6: 247–268.

Gould, W. W., J. S. Pitblado, and W. M. Sribney. 2006. *Maximum Likelihood Estimation with Stata*. 3rd ed. College Station, TX: Stata Press.

Greene, W. H. 2003. *Econometric Analysis*. 5th ed. Upper Saddle River, NJ: Prentice Hall.

Gutierrez, R. G., S. L. Carter, and D. M. Drukker. 2001. sg160: On boundary-value likelihood-ratio tests. *Stata Technical Bulletin* 60: 15–18. Reprinted in *Stata Technical Bulletin Reprints*, vol. 10, pp. 269–273.

Kumbhakar, S. C., and C. A. K. Lovell. 2000. *Stochastic Frontier Analysis*. Cambridge: Cambridge University Press.

Meeusen, W., and J. van den Broeck. 1977. Efficiency estimation from Cobb–Douglas production functions with composed error. *International Economic Review* 18: 435–444.

Pagan, A. R., and A. D. Hall. 1983. Diagnostic tests as residual analysis. *Econometric Reviews* 2: 159–218.

Petrin, A., B. P. Poi, and J. Levinsohn. 2004. Production function estimation in Stata using inputs to control for unobservables. *Stata Journal* 4: 113–123.

Stevenson, R. E. 1980. Likelihood functions for generalized stochastic frontier estimation. *Journal of Econometrics* 13: 57–66.

Zellner, A., and N. Revankar. 1970. Generalized production functions. *Review of Economic Studies* 37: 241–250.

Also See

[R] **frontier postestimation** — Postestimation tools for frontier

[R] **regress** — Linear regression

[XT] **xtfrontier** — Stochastic frontier models for panel data

[U] **20 Estimation and postestimation commands**

Title

> **frontier postestimation** — Postestimation tools for frontier

Description

The following postestimation commands are available for `frontier`:

command	description
adjust	adjusted predictions of $\mathbf{x}\beta$
estat	AIC, BIC, VCE, and estimation sample summary
estimates	cataloging estimation results
lincom	point estimates, standard errors, testing, and inference for linear combinations of coefficients
linktest	link test for model specification
lrtest	likelihood-ratio test
mfx	marginal effects or elasticities
nlcom	point estimates, standard errors, testing, and inference for nonlinear combinations of coefficients
predict	predictions, residuals, influence statistics, and other diagnostic measures
predictnl	point estimates, standard errors, testing, and inference for generalized predictions
test	Wald tests for simple and composite linear hypotheses
testnl	Wald tests of nonlinear hypotheses

See the corresponding entries in the *Stata Base Reference Manual* for details.

Syntax for predict

predict [*type*] *newvar* [*if*] [*in*] [, *statistic*]

predict [*type*] { *stub** | *newvar*$_{\text{xb}}$ *newvar*$_v$ *newvar*$_u$ } [*if*] [*in*] , <u>sc</u>ores

statistic	description
Main	
xb	linear prediction; the default
stdp	standard error of the prediction
u	estimates of minus the natural log of the technical efficiency via $E\left(u_i \mid \epsilon_i\right)$
m	estimates of minus the natural log of the technical efficiency via $M\left(u_i \mid \epsilon_i\right)$
te	estimates of the technical efficiency via $E\left\{\exp(-su_i) \mid \epsilon_i\right\}$ $$s = \begin{cases} 1, & \text{for production functions} \\ -1, & \text{for cost functions} \end{cases}$$

These statistics are available both in and out of sample; type `predict ... if e(sample) ...` if wanted only for the estimation sample.

Options for predict

xb, the default, calculates the linear prediction.

stdp calculates the standard error of the linear prediction.

u produces estimates of minus the natural log of the technical efficiency via $E\left(u_i \mid \epsilon_i\right)$.

m produces estimates of minus the natural log of the technical efficiency via $M\left(u_i \mid \epsilon_i\right)$.

te produces estimates of the technical efficiency via $E\left\{\exp(-su_i) \mid \epsilon_i\right\}$.

scores calculates equation-level score variables.

   The first new variable will contain $\partial \ln L/\partial(\mathbf{x}_i\boldsymbol{\beta})$.

   The second new variable will contain $\partial \ln L/\partial(\texttt{lnsig2v})$.

   The third new variable will contain $\partial \ln L/\partial(\texttt{lnsig2u})$.

Methods and Formulas

All postestimation commands listed above are implemented as ado-files.

Also See

[R] **frontier** — Stochastic frontier models

[U] **20 Estimation and postestimation commands**

Title

---

gllamm — Generalized linear and latent mixed models

Description

GLLAMM stands for generalized linear latent and mixed models, and `gllamm` is a Stata command for fitting such models written by Sophia Rabe-Hesketh (University of California–Berkeley) as part of joint work with Anders Skrondal (London School of Economics and Political Science) and Andrew Pickles (University of Manchester).

Remarks

Generalized linear latent and mixed models are a class of multilevel latent variable models, where a latent variable is a factor or a random effect (intercept or coefficient), or a disturbance (residual). The `gllamm` command for fitting such models is not an official command of Stata; it has been independently developed by highly regarded authors and is itself highly regarded. You can learn more about `gllamm` by visiting http://www.gllamm.org.

`gllamm` is available from the Statistical Software Components (SSC) archive. To install, type

> . ssc describe gllamm

> . ssc install gllamm

If you later wish to uninstall `gllamm`, type `ado uninstall gllamm`.

References

The references below are restricted to works by the primary authors of `gllamm`. There are many other books and articles that use or discuss `gllamm`; see *http://www.gllamm.org/pub.html* for a list.

Rabe-Hesketh, S., and B. S. Everitt. 2004. *Handbook of Statistical Analysis Using Stata.* 3rd ed. Boca Raton, FL: Chapman & Hall/CRC.

Rabe-Hesketh, S., A. Pickles, and C. Taylor. 2000. sg129: Generalized linear latent and mixed models. *Stata Technical Bulletin* 53: 47–57. Reprinted in *Stata Technical Bulletin Reprints*, vol. 9, pp. 293–307.

Rabe-Hesketh, S., and S. Skrondal. 2006. *Multilevel and Longitudinal Modeling Using Stata.* College Station, TX: Stata Press.

Rabe-Hesketh, S., S. Skrondal, and A. Pickles. 2002. Reliable estimation of generalized linear mixed models using adaptive quadrature. *Stata Journal* 2: 1–21.

——. 2003. Maximum likelihood estimation of generalized linear models with covariate measurement error. *Stata Journal* 3: 386–411.

Skrondal, A., and S. Rabe-Hesketh. 2004. *Generalized Latent Variable Modeling: Multilevel, Longitudinal, and Structural Equation Models.* Boca Raton, FL: Chapman & Hall/CRC.

Also See

[XT] **xtmelogit** — Multilevel mixed-effects logistic regression

[XT] **xtmepoisson** — Multilevel mixed-effects Poisson regression

[XT] **xtmixed** — Multilevel mixed-effects linear regression

Title

> **glm** — Generalized linear models

Syntax

glm *depvar* [*indepvars*] [*if*] [*in*] [*weight*] [, *options*]

options	description
Model	
<u>f</u>amily(*familyname*)	distribution of *depvar*; default is family(gaussian)
<u>l</u>ink(*linkname*)	link function; default is canonical link for family() specified
Model 2	
<u>noconst</u>ant	suppress constant term
<u>exp</u>osure(*varname*)	include ln(*varname*) in model with coefficient constrained to 1
<u>off</u>set(*varname*)	include *varname* in model with coefficient constrained to 1
mu(*varname*)	use *varname* as the initial estimate for the mean of *depvar*
<u>init</u>(*varname*)	synonym for mu(*varname*)
SE/Robust	
vce(*vcetype*)	*vcetype* may be oim, <u>r</u>obust, <u>cl</u>uster *clustvar*, eim, opg, <u>boot</u>strap, jackknife, hac *kernel*, jackknife1, or <u>unbiased</u>
<u>vf</u>actor(#)	multiply variance matrix by scalar #
disp(#)	quasilikelihood multiplier
<u>scale</u>(x2 | dev | #)	set the scale parameter
Reporting	
<u>level</u>(#)	set confidence level; default is level(95)
<u>ef</u>orm	report exponentiated coefficients
Max options	
ml	use maximum likelihood optimization; the default
irls	use iterated, reweighted least-squares optimization of the deviance
maximize_options	control the maximization process; seldom used
fisher(#)	use the Fisher scoring Hessian or expected information matrix (EIM)
search	search for good starting values
[†] <u>nohead</u>er	suppress header table from above coefficient table
[†] notable	suppress coefficient table
[†] nodisplay	suppress the output; iteration log is still displayed

familyname	description
<u>gau</u>ssian	Gaussian (normal)
<u>i</u>gaussian	inverse Gaussian
<u>b</u>inomial $\left[\, varname_N \mid \#_N \,\right]$	Bernoulli/binomial
<u>po</u>isson	Poisson
<u>nb</u>inomial $\left[\,\#_k\,\right]$	negative binomial
<u>ga</u>mma	gamma

linkname	description
<u>i</u>dentity	identity
<u>l</u>og	log
<u>l</u>ogit	logit
<u>p</u>robit	probit
<u>c</u>loglog	cloglog
<u>p</u>ower #	power
<u>o</u>power #	odds power
<u>nb</u>inomial	negative binomial
<u>l</u>oglog	log-log
<u>l</u>ogc	log-complement

depvar and *indepvars* may contain time-series operators; see [U] **11.4.3 Time-series varlists**.

bootstrap, by, jackknife, nestreg, rolling, statsby, stepwise, svy, and xi are allowed; see
[U] **11.1.10 Prefix commands**.

Weights are not allowed with the bootstrap prefix.

aweights are not allowed with the jackknife prefix.

vce(), vfactor(), disp(), scale(), irls, fisher(), noheader, notable, nodisplay, and weights
are not allowed with the svy prefix.

†noheader, notable, and nodisplay do not appear in the dialog box.

fweights, aweights, iweights, and pweights are allowed; see [U] **11.1.6 weight**.

See [U] **20 Estimation and postestimation commands** for more capabilities of estimation commands.

Description

glm fits generalized linear models. It can fit models by using either IRLS (maximum quasilikelihood)
or Newton–Raphson (maximum likelihood) optimization, which is the default. Previous versions of
glm used only IRLS.

See [R] **logistic** and [R] **regress** for lists of related estimation commands.

Options

┌─ Model ──

family(*familyname*) specifies the distribution of *depvar*; family(gaussian) is the default.

link(*linkname*) specifies the link function; the default is the canonical link for the family()
specified.

⌐‾‾ Model 2 ⌐_____

noconstant, exposure(*varname*), offset(*varname*); see [R] **estimation options**.

mu(*varname*) specifies *varname* as the initial estimate for the mean of *depvar*. This option can be useful with models that experience convergence difficulties, such as family(binomial) models with power or odds-power links. init(*varname*) is a synonym.

⌐‾‾ SE/Robust ⌐_____

vce(*vcetype*) specifies the type of standard error reported, which includes types that are derived from asymptotic theory, that are robust to some kinds of misspecification, that allow for intragroup correlation, and that use bootstrap or jackknife methods; see [R] *vce_option*.

In addition to the standard *vcetype*s, glm allows the following alternatives:

vce(eim) specifies that the expected information matrix estimate of variance be used.

vce(jackknife1) specifies that the one-step jackknife estimate of variance be used.

vce(hac *kernel* $\left[\#\right]$) specifies that a heteroskedasticity- and autocorrelation-consistent (HAC) variance estimate be used. HAC refers to the general form for combining weighted matrices to form the variance estimate. There are three kernels built into glm. *kernel* is a user-written program or one of

n̲west │ g̲allant │ a̲nderson

specifies the number of lags. If # is not specified, $N - 2$ is assumed. If you wish to specify vce(hac ...), you must tsset your data before calling glm.

vce(unbiased) specifies that the unbiased sandwich estimate of variance be used.

vfactor(#) specifies a scalar by which to multiply the resulting variance matrix. This option allows you to match output with other packages, which may apply degrees of freedom or other small-sample corrections to estimates of variance.

disp(#) multiplies the variance of *depvar* by # and divides the deviance by #. The resulting distributions are members of the quasilikelihood family.

scale(x2│dev│#) overrides the default scale parameter. This option is allowed only with Hessian (information matrix) variance estimates.

By default, scale(1) is assumed for the discrete distributions (binomial, Poisson, and negative binomial), and scale(x2) is assumed for the continuous distributions (Gaussian, gamma, and inverse Gaussian).

scale(x2) specifies that the scale parameter be set to the Pearson chi-squared (or generalized chi-squared) statistic divided by the residual degrees of freedom, which is recommended by McCullagh and Nelder (1989) as a good general choice for continuous distributions.

scale(dev) sets the scale parameter to the deviance divided by the residual degrees of freedom. This option provides an alternative to scale(x2) for continuous distributions and overdispersed or underdispersed discrete distributions.

scale(#) sets the scale parameter to #. For example, using scale(1) in family(gamma) models results in exponential-errors regression. Additional use of link(log) rather than the default link(power -1) for family(gamma) essentially reproduces Stata's streg, dist(exp) nohr command (see [ST] **streg**) if all the observations are uncensored.

⌐ Reporting ⌐
level(*#*); see [R] **estimation options**.

eform displays the exponentiated coefficients and corresponding standard errors and confidence intervals. For family(binomial) link(logit) (i.e., logistic regression), exponentiation results in odds ratios; for family(poisson) link(log) (i.e., Poisson regression), exponentiated coefficients are rate ratios.

⌐ Max options ⌐

ml requests that optimization be carried out using Stata's ml commands and is the default.

irls requests iterated, reweighted least-squares (IRLS) optimization of the deviance instead of Newton–Raphson optimization of the log likelihood. If the irls option is not specified, the optimization is carried out using Stata's ml commands, in which case all options of ml maximize are also available.

maximize_options: <u>dif</u>ficult, <u>tech</u>nique(*algorithm_spec*), <u>iter</u>ate(*#*), [<u>no</u>]<u>log</u>, <u>trace</u>, <u>grad</u>ient, showstep, <u>hess</u>ian, <u>shownr</u>tolerance, <u>tol</u>erance(*#*), <u>ltol</u>erance(*#*), <u>gtol</u>erance(*#*), <u>nrtol</u>erance(*#*), <u>nonrtol</u>erance, from(*init_specs*); see [R] **maximize**. These options are seldom used.

Setting the optimization type to technique(bhhh) resets the default *vcetype* to vce(opg).

fisher(*#*) specifies the number of Newton–Raphson steps that should use the Fisher scoring Hessian or expected information matrix (EIM) before switching to the observed information matrix (OIM). This option is useful only for Newton–Raphson optimization (and not when using irls).

search specifies that the command should search for good starting values. This option is useful only for Newton–Raphson optimization (and not when using irls).

The following options are available with glm but are not shown in the dialog box:

noheader suppresses the header information from the output. The coefficient table is still displayed.

notable suppresses the table of coefficients from the output. The header information is still displayed.

nodisplay suppresses the output. The iteration log is still displayed.

Remarks

Remarks are presented under the following headings:

> *General use*
> *Variance estimators*
> *User-defined functions*

General use

glm fits generalized linear models of y with covariates $\mathbf{x}$:

$$g\{E(y)\} = \mathbf{x}\boldsymbol{\beta}, \qquad y \sim F$$

$g(\ )$ is called the link function, and F is the distributional family. Substituting various definitions for $g(\ )$ and F results in a surprising array of models. For instance, if y is distributed as Gaussian (normal) and $g(\ )$ is the identity function, we have

$$E(y) = \mathbf{x}\boldsymbol{\beta}, \qquad y \sim \text{Normal}$$

or linear regression. If $g(\ )$ is the logit function and y is distributed as Bernoulli, we have

$$\text{logit}\{E(y)\} = \mathbf{x}\boldsymbol{\beta}, \qquad y \sim \text{Bernoulli}$$

or logistic regression. If $g(\ )$ is the natural log function and y is distributed as Poisson, we have

$$\ln\{E(y)\} = \mathbf{x}\boldsymbol{\beta}, \qquad y \sim \text{Poisson}$$

or Poisson regression, also known as the log-linear model. Other combinations are possible.

Although `glm` can be used to perform linear regression (and, in fact, does so by default), this regression should be viewed as an instructional feature; `regress` produces such estimates more quickly, and many postestimation commands are available to explore the adequacy of the fit; see [R] **regress** and [R] **regress postestimation**.

In any case, you specify the link function by using the `link()` option and the distributional family by using `family()`. The available link functions are

Link function	glm option
identity	link(identity)
log	link(log)
logit	link(logit)
probit	link(probit)
complementary log-log	link(cloglog)
odds power	link(opower #)
power	link(power #)
negative binomial	link(nbinomial)
log-log	link(loglog)
log-complement	link(logc)

Define $\mu = E(y)$ and $\eta = g(\mu)$, meaning that $g(\cdot)$ maps $E(y)$ to $\eta = \mathbf{x}\boldsymbol{\beta} + \text{offset}$.

Link functions are defined as follows:

    `identity` is defined as $\eta = g(\mu) = \mu$.

    `log` is defined as $\eta = \ln(\mu)$.

    `logit` is defined as $\eta = \ln\{\mu/(1-\mu)\}$, the natural log of the odds.

    `probit` is defined as $\eta = \Phi^{-1}(\mu)$, where $\Phi^{-1}(\ )$ is the inverse Gaussian cumulative.

    `cloglog` is defined as $\eta = \ln\{-\ln(1-\mu)\}$.

    `opower` is defined as $\eta = \left[\{\mu/(1-\mu)\}^n - 1\right]/n$, the power of the odds. The function is generalized so that `link(opower 0)` is equivalent to `link(logit)`, the natural log of the odds.

    `power` is defined as $\eta = \mu^n$. Specifying `link(power 1)` is equivalent to specifying `link(identity)`. The power function is generalized so that $\mu^0 \equiv \ln(\mu)$. Thus `link(power 0)` is equivalent to `link(log)`. Negative powers are, of course, allowed.

    `nbinomial` is defined as $\eta = \ln\{\mu/(\mu+k)\}$, where $k=1$ if `family(nbinomial)` is specified and $k=\#_k$ if `family(nbinomial #_k)` is specified.

    `loglog` is defined as $\eta = -\ln\{-\ln(\mu)\}$.

    `logc` is defined as $\eta = \ln(1-\mu)$.

The available distributional families are

Family	glm option
Gaussian (normal)	family(gaussian)
inverse Gaussian	family(igaussian)
Bernoulli/binomial	family(binomial)
Poisson	family(poisson)
negative binomial	family(nbinomial)
gamma	family(gamma)

family(normal) is a synonym for family(gaussian).

The binomial distribution can be specified as (1) family(binomial), (2) family(binomial $\#_N$), or (3) family(binomial $varname_N$). In case 2, $\#_N$ is the value of the binomial denominator N, the number of trials. Specifying family(binomial 1) is the same as specifying family(binomial). In case 3, $varname_N$ is the variable containing the binomial denominator, allowing the number of trials to vary across observations.

The negative binomial distribution can be specified as either (1) family(nbinomial) or (2) family(nbinomial $\#_k$). Omitting $\#_k$ is equivalent to specifying family(nbinomial 1). The value $\#_k$ enters the variance and deviance functions. Typical values range between .01 and 2; see the technical note below.

You do not have to specify both family() and link(); the default link() is the canonical link for the specified family() (except for nbinomial):

Family	Default link
family(gaussian)	link(identity)
family(igaussian)	link(power -2)
family(binomial)	link(logit)
family(poisson)	link(log)
family(nbinomial)	link(log)
family(gamma)	link(power -1)

If you specify both family() and link(), not all combinations make sense. You may choose from the following combinations:

	identity	log	logit	probit	cloglog	power	opower	nbinomial	loglog	logc
Gaussian	x	x				x				
inverse Gaussian	x	x				x				
binomial	x	x	x	x	x	x	x		x	x
Poisson	x	x				x				
negative binomial	x	x				x		x		
gamma	x	x				x				

❑ Technical Note

Some family() and link() combinations result in models already fitted by Stata. These are

family()	link()	Options	Equivalent Stata command
gaussian	identity	*nothing* | irls | irls vce(oim)	regress
gaussian	identity	t(*var*) vce(hac nwest #) vfactor(#$_v$)	newey, t(*var*) lag(#) (see note 1)
binomial	cloglog	*nothing* | irls vce(oim)	cloglog (see note 2)
binomial	probit	*nothing* | irls vce(oim)	probit (see note 2)
binomial	logit	*nothing* | irls | irls vce(oim)	logit or logistic (see note 3)
poisson	log	*nothing* | irls | irls vce(oim)	poisson (see note 3)
nbinomial	log	*nothing* | irls vce(oim)	nbreg (see note 4)
gamma	log	scale(1)	streg, dist(exp) nohr (see note 5)

Notes:

1. The variance factor $\#_v$ should be set to $n/(n - k)$, where n is the number of observations and k the number of regressors. If the number of regressors is not specified, the estimated standard errors will, as a result, differ by this factor.

2. Since the link is not the canonical link for the binomial family, you must specify the vce(oim) option if using irls to get equivalent standard errors. If irls is used without vce(oim), the regression coefficients will be the same but the standard errors will be only asymptotically equivalent. If no options are specified (*nothing*), glm will optimize using Newton–Raphson, making it equivalent to the other Stata command.

   See [R] **cloglog** and [R] **probit** for more details about these commands.

3. Since the canonical link is being used, the standard errors will be equivalent whether the EIM or the OIM estimator of variance is used.

4. Family negative binomial, log-link models—also known as negative binomial regression models—are used for data with an overdispersed Poisson distribution. Although glm can be used to fit such models, using Stata's maximum likelihood nbreg command is probably better. In the GLM approach, you specify family(nbinomial $\#_k$) and then search for a $\#_k$ that results in the deviance-based dispersion being 1. nbreg, on the other hand, finds the maximum likelihood estimate of $\#_k$ and reports a confidence interval for it; see [R] **nbreg** and Rogers (1993). Of course, glm allows links other than log, and for those links, including the canonical nbinomial link, you will need to use glm. Since the default link for family(nbinomial) is a noncanonical link, standard errors will be only asymptotically equivalent if you use glm, irls without the vce(oim) option.

5. glm can be used to estimate parameters from exponential regressions, but this method requires specifying scale(1). However, censoring is not available. Censored exponential regression may be modeled using glm with family(poisson). The log of the original response is entered into a Poisson model as an offset, whereas the new response is the censor variable. The result of such modeling is identical to the log relative hazard parameterization of streg, dist(exp) nohr. See [ST] **streg** for details about the streg command.

In general, where there is overlap between a capability of glm and that of some other Stata command, we recommend using the other Stata command. Our recommendation is not due to some inferiority of the GLM approach. Rather, those other commands, by being specialized, provide options and ancillary commands that are missing in the broader `glm` framework. Nevertheless, `glm` does produce the same answers where it should.

Special note. When equivalence is expected, for some datasets, you may still see very slight differences in the results, most often only in the later digits of the standard errors. When you compare `glm` output to an equivalent Stata command, these tiny discrepancies arise for many reasons:

a. `glm` uses a general methodology for starting values, whereas the equivalent Stata command may be more specialized in its treatment of starting values.

b. When using a canonical link, `glm, irls` should be equivalent to the maximum likelihood method of the equivalent Stata command, yet the convergence criterion is different (one is for deviance, the other for log likelihood). These discrepancies are easily resolved by adjusting one convergence criterion to correspond to the other.

c. When both `glm` and the equivalent Stata command use Newton–Raphson, small differences may still occur if the Stata command has a different default convergence criterion from that of `glm`. Adjusting the convergence criterion will resolve the difference. See [R] **ml** and [R] **maximize** for more details.

❏

▷ Example 1

In [R] **logistic**, we fit a model based on data from a study of risk factors associated with low birthweight (Hosmer and Lemeshow 2000, 25). We can replicate the estimation by using `glm`:

```
. use http://www.stata-press.com/data/r10/lbw
(Hosmer & Lemeshow data)

. xi: glm low age lwt i.race smoke ptl ht ui, family(binomial) link(logit)
i.race          _Irace_1-3          (naturally coded; _Irace_1 omitted)

Iteration 0:   log likelihood =   -101.0213
Iteration 1:   log likelihood = -100.72519
Iteration 2:   log likelihood =    -100.724
Iteration 3:   log likelihood =    -100.724

Generalized linear models                      No. of obs      =        189
Optimization     : ML                          Residual df     =        180
                                               Scale parameter =          1
Deviance         =  201.4479911                (1/df) Deviance =   1.119156
Pearson          =  182.0233425                (1/df) Pearson  =   1.011241

Variance function: V(u) = u*(1-u)              [Bernoulli]
Link function    : g(u) = ln(u/(1-u))          [Logit]
                                               AIC             =     1.1611
Log likelihood   = -100.7239956                BIC             =  -742.0665
```

low	Coef.	OIM Std. Err.	z	P>|z|	[95% Conf. Interval]	
age	-.0271003	.0364504	-0.74	0.457	-.0985418	.0443412
lwt	-.0151508	.0069259	-2.19	0.029	-.0287253	-.0015763
_Irace_2	1.262647	.5264101	2.40	0.016	.2309024	2.294392
_Irace_3	.8620792	.4391532	1.96	0.050	.0013548	1.722804
smoke	.9233448	.4008266	2.30	0.021	.137739	1.708951
ptl	.5418366	.346249	1.56	0.118	-.136799	1.220472
ht	1.832518	.6916292	2.65	0.008	.4769494	3.188086
ui	.7585135	.4593768	1.65	0.099	-.1418484	1.658875
_cons	.4612239	1.20459	0.38	0.702	-1.899729	2.822176

glm, by default, presents coefficient estimates, whereas logistic presents the exponentiated coefficients—the odds ratios. glm's eform option reports exponentiated coefficients, and glm, like Stata's other estimation commands, replays results.

```
. glm, eform
Generalized linear models                        No. of obs      =        189
Optimization      : ML                           Residual df     =        180
                                                 Scale parameter =          1
Deviance       =  201.4479911                    (1/df) Deviance =   1.119156
Pearson        =  182.0233425                    (1/df) Pearson  =   1.011241
Variance function: V(u) = u*(1-u)                [Bernoulli]
Link function     : g(u) = ln(u/(1-u))           [Logit]
                                                 AIC             =     1.1611
Log likelihood    = -100.7239956                 BIC             =  -742.0665
```

		OIM					
low	Odds Ratio	Std. Err.	z	P>|z|	[95% Conf.	Interval]
age	.9732636	.0354759	-0.74	0.457	.9061578	1.045339
lwt	.9849634	.0068217	-2.19	0.029	.9716834	.9984249
_Irace_2	3.534767	1.860737	2.40	0.016	1.259736	9.918406
_Irace_3	2.368079	1.039949	1.96	0.050	1.001356	5.600207
smoke	2.517698	1.00916	2.30	0.021	1.147676	5.523162
ptl	1.719161	.5952579	1.56	0.118	.8721455	3.388787
ht	6.249602	4.322408	2.65	0.008	1.611152	24.24199
ui	2.1351	.9808153	1.65	0.099	.8677528	5.2534

These results are the same as reported in [R] **logistic**.

Included in the output header are values for the Akaike (1973) information criterion (AIC) and the Bayesian information criterion (BIC) (Raftery 1995). Both are measures of model fit adjusted for the number of parameters that can be compared across models. In both cases, a smaller value generally indicates a better model fit. AIC is based on the log likelihood and thus is available only when Newton–Raphson optimization is used. BIC is based on the deviance and thus is always available. ◁

❑ Technical Note

The values for AIC and BIC reported in the output after glm are different from those reported by estat ic:

```
. estat ic
```

Model	Obs	ll(null)	ll(model)	df	AIC	BIC
.	189	.	-100.724	9	219.448	248.6237

Note: N=Obs used in calculating BIC; see **[R] BIC note**

There are various definitions of these information criteria (IC) in the literature; glm and estat ic use different definitions. glm bases its computation of the BIC on deviance, whereas estat ic uses the likelihood. Both glm and estat ic use the likelihood to compute the AIC; however, the AIC from estat ic is equal to N, the number of observations, times the AIC from glm. Refer to *Methods and Formulas* and [R] **estat** for the references and formulas used by glm and estat, respectively, to compute AIC and BIC. Inferences based on comparison of IC values reported by glm for different GLM models will be equivalent to those based on comparison of IC values reported by estat ic after glm.

❑

▷ Example 2

We use data from an early insecticide experiment, given in Pregibon (1980). The variables are ldose, the log dose of insecticide; n, the number of flour beetles subjected to each dose; and r, the number killed.

```
. use http://www.stata-press.com/data/r10/ldose
. list, sep(4)
```

	ldose	n	r
1.	1.6907	59	6
2.	1.7242	60	13
3.	1.7552	62	18
4.	1.7842	56	28
5.	1.8113	63	52
6.	1.8369	59	53
7.	1.861	62	61
8.	1.8839	60	60

The aim of the analysis is to estimate a dose–response relationship between p, the proportion killed, and X, the log dose.

As a first attempt, we will formulate the model as a linear logistic regression of p on ldose; that is, we will take the logit of p and represent the dose–response curve as a straight line in X:

$$\ln\{p/(1-p)\} = \beta_0 + \beta_1 X$$

Since the data are grouped, we cannot use Stata's logistic command to fit the model. Stata does, however, already have a command for performing logistic regression on data organized in this way, so we could type

```
. blogit r n ldose
  (output omitted)
```

Instead, we will fit the model by using glm:

```
. glm r ldose, family(binomial n) link(logit)
Iteration 0:    log likelihood = -18.824848
Iteration 1:    log likelihood = -18.715271
Iteration 2:    log likelihood = -18.715123
Iteration 3:    log likelihood = -18.715123
Generalized linear models                     No. of obs      =           8
Optimization     : ML                         Residual df     =           6
                                              Scale parameter =           1
Deviance         =   11.23220702              (1/df) Deviance =    1.872035
Pearson          =    10.0267936              (1/df) Pearson  =    1.671132
Variance function: V(u) = u*(1-u/n)           [Binomial]
Link function    : g(u) = ln(u/(n-u))         [Logit]
                                              AIC             =    5.178781
Log likelihood   = -18.71512262               BIC             =   -1.244442
```

	Coef.	OIM Std. Err.	z	P>|z|	[95% Conf. Interval]	
ldose	34.27034	2.912141	11.77	0.000	28.56265	39.97803
_cons	-60.71747	5.180713	-11.72	0.000	-70.87149	-50.56346

The only difference between `blogit` and `glm` here is how they obtain the answer. `blogit` expands the data to contain 481 observations (the sum of n) so that it can run Stata's standard, individual-level logistic command. `glm`, on the other hand, uses the information on the binomial denominator directly. We specified `family(binomial n)`, meaning that variable n contains the denominator. Parameter estimates and standard errors from the two approaches do not differ.

An alternative model, which gives asymmetric sigmoid curves for p, involves the complementary log-log or cloglog function:

$$\ln\{-\ln(1-p)\} = \beta_0 + \beta_1 X$$

We fit this model by using `glm`:

```
. glm r ldose, family(binomial n) link(cloglog)
Iteration 0:    log likelihood = -14.883594
Iteration 1:    log likelihood = -14.822264
Iteration 2:    log likelihood = -14.822228
Iteration 3:    log likelihood = -14.822228
```

```
Generalized linear models                     No. of obs      =          8
Optimization     : ML                         Residual df     =          6
                                              Scale parameter =          1
Deviance         =   3.446418004              (1/df) Deviance =    .574403
Pearson          =   3.294675153              (1/df) Pearson  =   .5491125

Variance function: V(u) = u*(1-u/n)           [Binomial]
Link function    : g(u) = ln(-ln(1-u/n))      [Complementary log-log]

                                              AIC             =   4.205557
Log likelihood   = -14.82222811               BIC             =  -9.030231
```

r	Coef.	OIM Std. Err.	z	P>|z|	[95% Conf. Interval]	
ldose	22.04118	1.793089	12.29	0.000	18.52679	25.55557
_cons	-39.57232	3.229047	-12.26	0.000	-45.90114	-33.24351

The complementary log-log model is preferred; the deviance for the logistic model, 11.23, is much higher than the deviance for the cloglog model, 3.45. This change also is evident by comparing log likelihoods, or equivalently, AIC values.

This example also shows the advantage of the `glm` command—we can vary assumptions easily. Note the minor difference in what we typed to obtain the logistic and cloglog models:

```
. glm r ldose, family(binomial n) link(logit)
. glm r ldose, family(binomial n) link(cloglog)
```

If we were performing this work for ourselves, we would have typed the commands in a more abbreviated form:

```
. glm r ldose, f(b n) l(l)
. glm r ldose, f(b n) l(cl)
```

◁

(*Continued on next page*)

❏ Technical Note

Stata's `xi` command prefix may be used with `glm`. Say that, in the example above, we had `ldose`, the log dose of insecticide; `n`, the number of flour beetles subjected to each dose; and `r`, the number killed—all as before—except that now we have results for three different kinds of beetles. Our hypothetical data include `beetle`, which contains the values 1, 2, and 3.

```
. use http://www.stata-press.com/data/r10/beetle
. list, sep(0)
```

	beetle	ldose	n	r
1.	1	1.6907	59	6
2.	1	1.7242	60	13
3.	1	1.7552	62	18
4.	1	1.7842	56	28
5.	1	1.8113	63	52
	(output omitted)			
23.	3	1.861	64	23
24.	3	1.8839	58	22

Let us assume that, at first, we wish merely to add a shift factor for the type of beetle. We could type

```
. xi: glm r i.beetle ldose, f(bin n) l(cloglog)
i.beetle        _Ibeetle_1-3        (naturally coded; _Ibeetle_1 omitted)

Iteration 0:   log likelihood = -79.012269
Iteration 1:   log likelihood =  -76.94951
Iteration 2:   log likelihood = -76.945645
Iteration 3:   log likelihood = -76.945645
```

```
Generalized linear models                   No. of obs        =        24
Optimization      : ML                      Residual df       =        20
                                            Scale parameter   =         1
Deviance          =  73.76505595            (1/df) Deviance   = 3.688253
Pearson           =   71.8901173            (1/df) Pearson    = 3.594506

Variance function: V(u) = u*(1-u/n)         [Binomial]
Link function     : g(u) = ln(-ln(1-u/n))   [Complementary log-log]
                                            AIC               =   6.74547
Log likelihood    = -76.94564525            BIC               =  10.20398
```

	Coef.	OIM Std. Err.	z	P>|z|	[95% Conf. Interval]
_Ibeetle_2	-.0910396	.1076132	-0.85	0.398	-.3019576 .1198783
_Ibeetle_3	-1.836058	.1307125	-14.05	0.000	-2.09225 -1.579867
ldose	19.41558	.9954265	19.50	0.000	17.46458 21.36658
_cons	-34.84602	1.79333	-19.43	0.000	-38.36089 -31.33116

We find strong evidence that the insecticide works differently on the third kind of beetle. We now check whether the curve is merely shifted or also differently sloped:

```
. xi: glm r i.beetle*ldose, f(bin n) l(cloglog)
i.beetle          _Ibeetle_1-3        (naturally coded; _Ibeetle_1 omitted)
i.beetle*ldose    _IbeeXldose_#       (coded as above)

Iteration 0:    log likelihood = -67.270188
Iteration 1:    log likelihood = -65.149316
Iteration 2:    log likelihood = -65.147978
Iteration 3:    log likelihood = -65.147978

Generalized linear models                  No. of obs      =        24
Optimization      : ML                     Residual df     =        18
                                           Scale parameter =         1
Deviance        =    50.16972096           (1/df) Deviance =  2.787207
Pearson         =    49.28422567           (1/df) Pearson  =  2.738013

Variance function: V(u) = u*(1-u/n)        [Binomial]
Link function    : g(u) = ln(-ln(1-u/n))   [Complementary log-log]

                                           AIC             =  5.928998
Log likelihood  = -65.14797776             BIC             = -7.035248
```

r	Coef.	OIM Std. Err.	z	P>|z|	[95% Conf. Interval]	
_Ibeetle_2	-.79933	4.470882	-0.18	0.858	-9.562098	7.963438
_Ibeetle_3	17.78741	4.586429	3.88	0.000	8.798172	26.77664
ldose	22.04118	1.793089	12.29	0.000	18.52679	25.55557
_IbeeXldos~2	.3838708	2.478477	0.15	0.877	-4.473855	5.241596
_IbeeXldos~3	-10.726	2.526412	-4.25	0.000	-15.67768	-5.774321
_cons	-39.57232	3.229047	-12.26	0.000	-45.90114	-33.24351

We find that the (complementary log-log) dose–response curve for the third kind of beetle has roughly half the slope of that for the first kind.

We see that xi can be combined with glm just as it can with any other Stata estimation command; see [R] **xi**. Also see [U] **25 Dealing with categorical variables**; what is said there concerning linear regression is applicable to any GLM model.

❏

Variance estimators

glm offers many variance options and gives different types of standard errors when used in various combinations. We highlight some of them here, but for a full explanation, see Hardin and Hilbe (2007).

▷ Example 3

Continuing with our flour beetle data, we rerun the most recently displayed model, this time requesting estimation via iterated reweighted least squares (IRLS).

(Continued on next page)

```
. xi: glm r i.beetle*ldose, f(bin n) l(cloglog) ltol(1e-15) irls
i.beetle           _Ibeetle_1-3        (naturally coded; _Ibeetle_1 omitted)
i.beetle*ldose     _IbeeXldose_#       (coded as above)

Iteration 1:    deviance =  54.41414
Iteration 2:    deviance =  50.19424
Iteration 3:    deviance =  50.16973
  (output omitted )
Iteration 10:   deviance =  50.16972
Iteration 11:   deviance =  50.16972
Iteration 12:   deviance =  50.16972
```

```
Generalized linear models                    No. of obs        =        24
Optimization      : MQL Fisher scoring       Residual df       =        18
                    (IRLS EIM)               Scale parameter =           1
Deviance          =  50.16972096            (1/df) Deviance  =  2.787207
Pearson           =  49.28422567            (1/df) Pearson   =  2.738013

Variance function: V(u) = u*(1-u/n)          [Binomial]
Link function    : g(u) = ln(-ln(1-u/n))     [Complementary log-log]
                                             BIC              = -7.035248
```

r	Coef.	EIM Std. Err.	z	P>|z|	[95% Conf. Interval]	
_Ibeetle_2	-.79933	4.586649	-0.17	0.862	-9.788997	8.190337
_Ibeetle_3	17.78741	4.624834	3.85	0.000	8.7229	26.85192
ldose	22.04118	1.799356	12.25	0.000	18.5145	25.56785
_IbeeXldos~2	.3838708	2.544068	0.15	0.880	-4.602411	5.370152
_IbeeXldos~3	-10.726	2.548176	-4.21	0.000	-15.72033	-5.731665
_cons	-39.57232	3.240274	-12.21	0.000	-45.92314	-33.2215

Note our use of the ltol() option, which, although unrelated to our discussion on variance estimation, was used so that the regression coefficients would match those of the previous Newton–Raphson (NR) fit.

Since IRLS uses the expected information matrix (EIM) for optimization, the variance estimate is also based on EIM. If we want optimization via IRLS, but the variance estimate based on OIM, we specify glm, irls vce(oim):

```
. xi: glm r i.beetle*ldose, f(b n) l(cl) ltol(1e-15) irls vce(oim) noheader nolog
i.beetle           _Ibeetle_1-3        (naturally coded; _Ibeetle_1 omitted)
i.beetle*ldose     _IbeeXldose_#       (coded as above)
```

r	Coef.	OIM Std. Err.	z	P>|z|	[95% Conf. Interval]	
_Ibeetle_2	-.79933	4.470882	-0.18	0.858	-9.562098	7.963438
_Ibeetle_3	17.78741	4.586429	3.88	0.000	8.798172	26.77664
ldose	22.04118	1.793089	12.29	0.000	18.52679	25.55557
_IbeeXldos~2	.3838708	2.478477	0.15	0.877	-4.473855	5.241596
_IbeeXldos~3	-10.726	2.526412	-4.25	0.000	-15.67768	-5.774321
_cons	-39.57232	3.229047	-12.26	0.000	-45.90114	-33.24351

This approach is identical to NR except for the convergence path. Since the cloglog link is not the canonical link for the binomial family, EIM and OIM produce different results. Both estimators, however, are asymptotically equivalent.

Going back to NR, we can also specify vce(robust) to get the Huber/White/sandwich estimator of variance:

```
. xi: glm r i.beetle*ldose, f(b n) l(cl) vce(robust) noheader nolog
i.beetle              _Ibeetle_1-3        (naturally coded; _Ibeetle_1 omitted)
i.beetle*ldose        _IbeeXldose_#       (coded as above)
```

r	Coef.	Robust Std. Err.	z	P>|z|	[95% Conf. Interval]	
_Ibeetle_2	-.79933	5.733049	-0.14	0.889	-12.0359	10.43724
_Ibeetle_3	17.78741	5.158477	3.45	0.001	7.676977	27.89784
ldose	22.04118	.8998551	24.49	0.000	20.27749	23.80486
_IbeeXldos~2	.3838708	3.174427	0.12	0.904	-5.837892	6.605633
_IbeeXldos~3	-10.726	2.800606	-3.83	0.000	-16.21508	-5.236912
_cons	-39.57232	1.621306	-24.41	0.000	-42.75003	-36.39462

The sandwich estimator gets its name from the form of the calculation—it is the multiplication of three matrices, with the outer two matrices (the "bread") set to the OIM variance matrix. When irls is used along with vce(robust), the EIM variance matrix is instead used as the bread, and the resulting variance is labeled "Semi-Robust".

```
. xi: glm r i.beetle*ldose, f(b n) l(cl) irls ltol(1e-15) vce(robust) noheader nolog
i.beetle              _Ibeetle_1-3        (naturally coded; _Ibeetle_1 omitted)
i.beetle*ldose        _IbeeXldose_#       (coded as above)
```

r	Coef.	Semi-Robust Std. Err.	z	P>|z|	[95% Conf. Interval]	
_Ibeetle_2	-.79933	6.288963	-0.13	0.899	-13.12547	11.52681
_Ibeetle_3	17.78741	5.255307	3.38	0.001	7.487194	28.08762
ldose	22.04118	.9061566	24.32	0.000	20.26514	23.81721
_IbeeXldos~2	.3838708	3.489723	0.11	0.912	-6.455861	7.223603
_IbeeXldos~3	-10.726	2.855897	-3.76	0.000	-16.32345	-5.128542
_cons	-39.57232	1.632544	-24.24	0.000	-42.77205	-36.3726

The outer product of the gradient (OPG) estimate of variance is one that avoids the calculation of second derivatives. It is equivalent to the "middle" part of the sandwich estimate of variance and can be specified by using glm, vce(opg), regardless of whether NR or IRLS optimization is used.

```
. xi: glm r i.beetle*ldose, f(b n) l(cl) vce(opg) noheader nolog
i.beetle              _Ibeetle_1-3        (naturally coded; _Ibeetle_1 omitted)
i.beetle*ldose        _IbeeXldose_#       (coded as above)
```

r	Coef.	OPG Std. Err.	z	P>|z|	[95% Conf. Interval]	
_Ibeetle_2	-.79933	6.664045	-0.12	0.905	-13.86062	12.26196
_Ibeetle_3	17.78741	6.838505	2.60	0.009	4.384183	31.19063
ldose	22.04118	3.572983	6.17	0.000	15.03826	29.0441
_IbeeXldos~2	.3838708	3.700192	0.10	0.917	-6.868372	7.636114
_IbeeXldos~3	-10.726	3.796448	-2.83	0.005	-18.1669	-3.285097
_cons	-39.57232	6.433101	-6.15	0.000	-52.18097	-26.96368

The OPG estimate of variance is a component of the BHHH (Berndt et al. 1974) optimization technique. This method of optimization is also available with glm with the technique() option; however, the technique() option is not allowed with the irls option.

◁

▷ Example 4

The Newey–West (1987) estimator of variance is a sandwich estimator with the "middle" of the sandwich modified to take into account possible autocorrelation between the observations. These estimators are a generalization of those given by the Stata command newey in the case of linear regression. See [TS] **newey** for more details.

For example, consider the dataset given in [TS] **newey**, which has time-series measurements on usr and idle. We want to perform a linear regression with Newey–West standard errors.

```
. use http://www.stata-press.com/data/r10/idle2
. list usr idle time
```

	usr	idle	time
1.	0	100	1
2.	0	100	2
3.	0	97	3
4.	1	98	4
5.	2	94	5
	(output omitted)		
29.	1	98	29
30.	1	98	30

Examining *Methods and Formulas* of [TS] **newey**, we see that the variance estimate is multiplied by a correction factor of $n/(n-k)$, where k is the number of regressors. glm, vce(hac ...) does not make this correction, so to get the same standard errors, we must use the vfactor() option within glm to make the correction manually.

```
. display 30/28
1.0714286
. tsset time
        time variable:  time, 1 to 30
                delta:  1 unit
. glm usr idle, vce(hac nwest 3) vfactor(1.0714286)
Iteration 0:   log likelihood = -71.743396
```

Generalized linear models		No. of obs	=	30
Optimization : ML		Residual df	=	28
		Scale parameter	=	7.493297
Deviance = 209.8123165		(1/df) Deviance	=	7.493297
Pearson = 209.8123165		(1/df) Pearson	=	7.493297
Variance function: V(u) = 1		[Gaussian]		
Link function : g(u) = u		[Identity]		
HAC kernel (lags): Newey-West (3)				
		AIC	=	4.916226
Log likelihood = -71.74339627		BIC	=	114.5788

usr	Coef.	HAC Std. Err.	z	P>|z|	[95% Conf. Interval]
idle	-.2281501	.0690928	-3.30	0.001	-.3635694 -.0927307
_cons	23.13483	6.327033	3.66	0.000	10.73407 35.53558

The glm command above reproduces the results given in [TS] **newey**. We may now generalize this output to models other than simple linear regression and to different kernel weights.

```
. glm usr idle, fam(gamma) link(log) vce(hac gallant 3)
Iteration 0:   log likelihood =  -61.76593
Iteration 1:   log likelihood = -60.963233
Iteration 2:   log likelihood =  -60.95097
Iteration 3:   log likelihood = -60.950965
```

Generalized linear models		No. of obs	=	30
Optimization : ML		Residual df	=	28
		Scale parameter	=	.431296
Deviance = 9.908506707		(1/df) Deviance	=	.3538752
Pearson = 12.07628677		(1/df) Pearson	=	.431296

Variance function: V(u) = u^2   [Gamma]
Link function   : g(u) = ln(u)   [Log]

HAC kernel (lags): Gallant (3)

		AIC	=	4.196731
Log likelihood = -60.95096484		BIC	=	-85.32502

usr	Coef.	HAC Std. Err.	z	P>|z|	[95% Conf. Interval]	
idle	-.0796609	.0184647	-4.31	0.000	-.115851	-.0434708
_cons	7.771011	1.510198	5.15	0.000	4.811078	10.73094

glm also offers variance estimators based on the bootstrap (resampling your data with replacement) and the jackknife (refitting the model with each observation left out in succession). Also included is the one-step jackknife estimate, which, instead of performing full reestimation when each observation is omitted, calculates a one-step NR estimate, with the full data regression coefficients as starting values.

```
. set seed 1
. glm usr idle, fam(gamma) link(log) vce(bootstrap, reps(100) nodots)
```

Generalized linear models		No. of obs	=	30
Optimization : ML		Residual df	=	28
		Scale parameter	=	.431296
Deviance = 9.908506707		(1/df) Deviance	=	.3538752
Pearson = 12.07628677		(1/df) Pearson	=	.431296

Variance function: V(u) = u^2   [Gamma]
Link function   : g(u) = ln(u)   [Log]

		AIC	=	4.196731
Log likelihood = -60.95096484		BIC	=	-85.32502

usr	Observed Coef.	Bootstrap Std. Err.	z	P>|z|	Normal-based [95% Conf. Interval]	
idle	-.0796609	.0216591	-3.68	0.000	-.1221119	-.0372099
_cons	7.771011	1.80278	4.31	0.000	4.237627	11.3044

◁

See Hardin and Hilbe (2007) for a full discussion of the variance options that go with glm and, in particular, of how the different variance estimators are modified when vce(cluster *clustvar*) is specified. Finally, not all variance options are supported with all types of weights. See help glm for a current table of the variance options that are supported with the different weights.

User-defined functions

glm may be called with a user-written link function, variance (family) function, Newey–West kernel-weight function, or any combination of the three.

Syntax of link functions

```
program progname
        version 10
        args todo eta mu return
        if 'todo' == -1 {
                /*  Set global macros for output */
                global SGLM_lt " title for link function "
                global SGLM_lf " subtitle showing link definition "
                exit
        }
        if 'todo' == 0 {
                /*  set  η=g(μ) */
                /*  Intermediate calculations go here */
                generate double 'eta' = ...
                exit
        }
        if 'todo' == 1 {
                /*  set  μ=g⁻¹(η) */
                /*  Intermediate calculations go here */
                generate double 'mu' = ...
                exit
        }
        if 'todo' == 2 {
                /*  set return = ∂μ/∂η */
                /*  Intermediate calculations go here */
                generate double 'return' = ...
                exit
        }
        if 'todo' == 3 {
                /*  set return = ∂²μ/∂η² */
                /*  Intermediate calculations go here */
                generate double 'return' = ...
                exit
        }
        display as error "Unknown call to glm link function"
        exit 198
end
```

Syntax of variance functions

```
program progname
        version 10
        args todo eta mu return
        if 'todo' == -1 {
                /*  Set global macros for output */
                /*  Also check that depvar is in proper range */
                /*  Note: For this call, eta contains indicator for whether each obs. is in est. sample */
                global SGLM_vt " title for variance function "
                global SGLM_vf " subtitle showing function definition "
                global SGLM_mu " program to call to enforce boundary conditions on μ "
                exit
        }
```

```
if 'todo' == 0 {
        /*  set η to initial value. */
        /*  Intermediate calculations go here */
        generate double 'eta' = ...
        exit
}
if 'todo' == 1 {
        /*  set return = V(μ) */
        /*  Intermediate calculations go here */
        generate double 'return' = ...
        exit
}
if 'todo' == 2 {
        /*  set return = ∂V(μ)/∂μ */
        /*  Intermediate calculations go here */
        generate double 'return' = ...
        exit
}
if 'todo' == 3 {
        /*  set return = squared deviance (per observation) */
        /*  Intermediate calculations go here */
        generate double 'return' = ...
        exit
}
if 'todo' == 4 {
        /*  set return = Anscombe residual */
        /*  Intermediate calculations go here */
        generate double 'return' = ...
        exit
}
if 'todo' == 5 {
        /*  set return = log likelihood */
        /*  Intermediate calculations go here */
        generate double 'return' = ...
        exit
}
if 'todo' == 6 {
        /*  set return = adjustment for deviance residuals */
        /*  Intermediate calculations go here */
        generate double 'return' = ...
        exit
}
display as error "Unknown call to glm variance function"
exit 198
end
```

Syntax of Newey–West kernel-weight functions

```
program progname, rclass
        version 10
        args G j
        /* G  is the maximum lag */
        /* j  is the current lag */
        /* Intermediate calculations go here */
        return scalar wt = computed weight
        return local setype "Newey-West"
        return local sewtype " name of kernel "
end
```

Global macros available for user-written programs

Global macro	Description
SGLM_V	program name of variance (family) evaluator
SGLM_L	program name of link evaluator
SGLM_y	dependent variable name
SGLM_m	binomial denominator
SGLM_a	negative binomial k
SGLM_p	power if power() or opower() is used, or an argument from a user-specified link function
SGLM_s1	indicator; set to one if scale is equal to one
SGLM_ph	value of scale parameter

▷ Example 5

Suppose that we wish to perform Poisson regression with a log-link function. Although this regression is already possible with standard glm, we will write our own version for illustrative purposes.

Since we want a log link, $\eta = g(\mu) = \ln(\mu)$, and for a Poisson family the variance function is $V(\mu) = \mu$.

The Poisson density is given by

$$f(y_i) = \frac{e^{-\exp(\mu_i)} e^{\mu_i y_i}}{y_i!}$$

resulting in a log likelihood of

$$L = \sum_{i=1}^{n} \{-e^{\mu_i} + \mu_i y_i - \ln(y_i!)\}$$

The squared deviance of the ith observation for the Poisson family is given by

$$d_i^2 = \begin{cases} 2\widehat{\mu}_i & \text{if } y_i = 0 \\ 2\{y_i \ln(y_i/\widehat{\mu}_i) - (y_i - \widehat{\mu}_i)\} & \text{otherwise} \end{cases}$$

We now have enough information to write our own Poisson-log glm module. We create the file mylog.ado, which contains

```
program mylog
        version 10
        args todo eta mu return
        if 'todo' == -1 {
                global SGLM_lt "My Log"          //  Titles for output
                global SGLM_lf "ln(u)"
                exit
        }
        if 'todo' == 0  {
                gen double 'eta' = ln('mu')      //   η = ln(μ)
                exit
        }
```

```
        if 'todo' == 1  {
                gen double 'mu' = exp('eta')      //   μ = exp(η)
                exit
        }
        if 'todo' == 2 {
                gen double 'return' = 'mu'        //   ∂μ/∂η = exp(η) = μ
                exit
        }
        if 'todo' == 3 {
                gen double 'return' = 'mu'        //   ∂²μ/∂η² = exp(η) = μ
                exit
        }
        di as error "Unknown call to glm link function"
        exit 198
end
```

and we create the file `mypois.ado`, which contains

```
program mypois
        version 10
        args todo eta mu return
        if 'todo' == -1 {
                local y      "$SGLM_y"
                local touse "''eta''"                    // 'eta'  marks estimation sample here
                capture assert 'y'>=0 if 'touse'    // check range of y
                if _rc {
                        di as error '"dependent variable 'y' has negative values"'
                        exit 499
                }
                global SGLM_vt "My Poisson"         // Titles for output
                global SGLM_vf "u"
                global SGLM_mu "glim_mu 0 ."        // see note 1
                exit
        }
        if 'todo' == 0 {                            // Initialization of η; see note 2
                gen double 'eta' = ln('mu')
                exit
        }
        if 'todo' == 1 {
                gen double 'return' =  'mu'         //   V(μ) = μ
                exit
        }
        if 'todo' == 2 {                            //   ∂ V(μ)/∂μ
                gen byte 'return' = 1
                exit
        }
        if 'todo' == 3 {                            // squared deviance, defined above
                local y "$SGLM_y"
                if "'y'" == "" {
                        local y "'e(depvar)'"
                }
                gen double 'return' = cond('y'==0, 2*'mu', /*
                        */ 2*('y'*ln('y'/'mu')-('y'-'mu')))
                exit
        }
```

```
if 'todo' == 4 {                              //  Anscombe residual; see note 3
        local y "$SGLM_y"
        if "'y'" == "" {
                local y "'e(depvar)'"
        }
        gen double 'return' = 1.5*('y'^(2/3)-'mu'^(2/3)) / 'mu'^(1/6)
        exit
}
if 'todo' == 5 {                              //  log likelihood; see note 4
        local y "$SGLM_y"
        if "'y'" == "" {
                local y "'e(depvar)'"
        }
        gen double 'return' = -'mu'+'y'*ln('mu')-lngamma('y'+1)
        exit
}
if 'todo' == 6 {                              //  adjustment to residual; see note 5
        gen double 'return' = 1/(6*sqrt('mu'))
        exit
}
di as error "Unknown call to glm variance function"
error 198
end
```

Notes:

1. `glim_mu` is a Stata program that will, at each iteration, bring $\hat{\mu}$ back into its plausible range, should it stray out of it. Here `glim_mu` is called with the arguments zero and missing, meaning that zero is the lower bound of $\hat{\mu}$ and there exists no upper bound—such is the case for Poisson models.

2. Here the initial value of η is easy since we intend to fit this model with our user-defined log link. In general, however, the initialization may need to vary according to the link to obtain convergence. If so, the global macro SGLM_L is used to determine which link is being utilized.

3. The Anscombe formula is given here because we know it. If we were not interested in Anscombe residuals, we could merely set 'return' to missing. Also the local macro y is set either to SGLM_y if it is in current estimation or to e(depvar) if this function is being accessed by `predict`.

4. If we were not interested in ML estimation, we could omit this code entirely and just leave an `exit` statement in its place. Similarly, if we were not interested in deviance or IRLS optimization, we could set 'return' in the deviance portion of the code ('todo'==3) to missing.

5. This code defines the term to be added to the predicted residuals if the option `adjusted` is specified. Again, if we were not interested, we could set 'return' to missing.

   We can now test our Poisson-log module by running it on the airline data presented in [R] **poisson**.

```
. use http://www.stata-press.com/data/r10/airline

. list airline injuries n XYZowned
```

	airline	injuries	n	XYZowned
1.	1	11	0.0950	1
2.	2	7	0.1920	0
3.	3	7	0.0750	0
4.	4	19	0.2078	0
5.	5	9	0.1382	0
6.	6	4	0.0540	1
7.	7.	3	0.1292	0
8.	8	1	0.0503	0
9.	9	3	0.0629	1

```
. gen lnN=ln(n)

. glm injuries XYZowned lnN, fam(mypois) link(mylog) scale(1)

Iteration 0:    log likelihood = -22.557572
Iteration 1:    log likelihood = -22.332861
Iteration 2:    log likelihood = -22.332276
Iteration 3:    log likelihood = -22.332276
```

```
Generalized linear models                    No. of obs        =          9
Optimization     : ML                        Residual df       =          6
                                             Scale parameter   =          1
Deviance         =   12.70432823             (1/df) Deviance   =   2.117388
Pearson          =   12.7695081              (1/df) Pearson    =   2.128251

Variance function: V(u) = u                  [My Poisson]
Link function    : g(u) = ln(u)              [My Log]

                                             AIC               =   5.629395
Log likelihood   = -22.33227605              BIC               = -.4790192
```

		OIM				
injuries	Coef.	Std. Err.	z	P>|z|	[95% Conf.	Interval]
XYZowned	.6840668	.3895877	1.76	0.079	-.0795111	1.447645
lnN	1.424169	.3725155	3.82	0.000	.6940517	2.154286
_cons	4.863891	.7090501	6.86	0.000	3.474178	6.253603

```
(Standard errors scaled using dispersion equal to square root of 1)
```

These are precisely the results given in [R] **poisson** and are those that would have been given had we run glm, family(poisson) link(log). The only minor adjustment we needed to make was to specify the option scale(1). If scale() is left unspecified, glm assumes scale(1) for discrete distributions and scale(x2) for continuous ones. By default, glm assumes that any user-defined family is continuous since it has no way of checking. Thus we needed to specify scale(1) since our model is discrete.

Since we were careful in defining the squared deviance, we could have fit this model with IRLS. Since log is the canonical link for the Poisson family, we would not only get the same regression coefficients but also the same standard errors.

◁

(Continued on next page)

▷ Example 6

Suppose now that we wish to use our log link (mylog.ado) with glm's binomial family. This task requires some modification since our current function is not equipped to deal with the binomial denominator, which we are allowed to specify. This denominator is accessible to our link function through the global macro SGLM_m. We now make the modifications and store them in mylog2.ado.

```
program mylog2                                        // <-- changed
        version 10
        args todo eta mu return
        if `todo' == -1 {
                global SGLM_lt "My Log, Version 2"    // <-- changed
                if "$SGLM_m" == "1" {                 // <-- changed
                        global SGLM_lf "ln(u)"        // <-- changed
                }                                     // <-- changed
                else    global SGLM_lf "ln(u/$SGLM_m)" // <-- changed
                exit
        }
        if `todo' == 0  {
                gen double `eta' = ln(`mu'/$SGLM_m)    // <-- changed
                exit
        }
        if `todo' == 1  {
                gen double `mu' = $SGLM_m*exp(`eta')   // <-- changed
                exit
        }
        if `todo' == 2 {
                gen double `return' = `mu'
                exit
        }
        if `todo' == 3 {
                gen double `return' = `mu'
                exit
        }
        di as error "Unknown call to glm link function"
        exit 198
end
```

We can now run our new log link with glm's binomial family. Using a subset of the flour beetle data from earlier, we have

```
. use http://www.stata-press.com/data/r10/beetle2

. glm r ldose, fam(bin n) link(mylog2) irls

Iteration 1:    deviance =  85.11105
Iteration 2:    deviance =  52.14709
Iteration 3:    deviance =  51.01859
Iteration 4:    deviance =  51.00723
Iteration 5:    deviance =  51.00716
Iteration 6:    deviance =  51.00716

Generalized linear models                  No. of obs      =          8
Optimization      : MQL Fisher scoring      Residual df     =          6
                    (IRLS EIM)               Scale parameter =          1
Deviance          =   51.00716029           (1/df) Deviance =   8.501193
Pearson           =   42.07549795           (1/df) Pearson  =   7.012583

Variance function: V(u) = u*(1-u/n)         [Binomial]
Link function     : g(u) = ln(u/n)          [My Log, Version 2]

                                            BIC             =   38.53051
```

r	Coef.	EIM Std. Err.	z	P>|z|	[95% Conf. Interval]
ldose	7.175578	.50964	14.08	0.000	6.176702 8.174454
_cons	-13.4936	.9601108	-14.05	0.000	-15.37538 -11.61182

◁

For a more detailed discussion on user-defined functions, and for an example of a user-defined Newey–West kernel weight, see Hardin and Hilbe (2007).

John Ashworth Nelder was born in Somerset, England, in 1924. He studied mathematics and statistics at Cambridge and worked as a statistician at the National Vegetable Research Station and then Rothamsted Experimental Station. He is especially well known for his contributions to the theory of linear models and statistical computing, as the principal architect of both Genstat and Glim.

Robert William Maclagan Wedderburn (1947–1975) was born in Edinburgh and studied mathematics and statistics at Cambridge. At Rothamsted Experimental Station, he developed the theory of generalized linear models with Nelder and originated the concept of quasilikelihood. He died of anaphylactic shock from an insect bite on a canal holiday.

(Continued on next page)

Saved Results

glm saves the following in e():

Scalars

e(N)	number of observations	e(rank)	rank of e(V)
e(k)	number of parameters	e(deviance)	deviance
e(k_eq)	number of equations in e(b)	e(deviance_p)	Pearson deviance
e(k_eq_model)	number of equations in model Wald test	e(deviance_ps)	scaled Pearson deviance
		e(deviance_s)	scaled deviance
e(k_dv)	number of dependent variables	e(dispers)	dispersion
e(df)	residual degrees of freedom	e(dispers_p)	Pearson dispersion
e(phi)	scale parameter	e(dispers_ps)	scaled Pearson dispersion
e(disp)	dispersion parameter	e(dispers_s)	scaled dispersion
e(df_m)	model degrees of freedom	e(vf)	factor set by vfactor(),
e(aic)	model AIC		1 if not set
e(bic)	model BIC	e(power)	power set by power(), opower()
e(ll)	log likelihood, if NR	e(ic)	number of iterations
e(N_clust)	number of clusters	e(rc)	return code
e(chi2)	χ^2	e(converged)	1 if converged, 0 otherwise
e(p)	significance		

Macros

e(cmd)	glm	e(chi2type)	Wald or LR; type of model χ^2 test
e(cmdline)	command as typed	e(cons)	set if noconstant specified
e(depvar)	name of dependent variable	e(linkt)	link title
e(varfunc)	name of variance function used	e(linkf)	link form
e(varfunct)	Gaussian, Inverse Gaussian, Binomial, Poisson, Neg. Binomial, Bernoulli, Power, or Gamma	e(hac_kernel)	HAC kernel
		e(hac_lag)	HAC lag
		e(vce)	*vcetype* specified in vce()
		e(vcetype)	title used to label Std. Err.
e(varfuncf)	variance function	e(opt1)	optimization title, line 1
e(link)	name of link function used	e(opt2)	optimization title, line 2
e(m)	number of binomial trials	e(opt)	ml or irls
e(a)	neg. binomial k	e(ml_method)	type of ml method
e(wtype)	weight type	e(user)	name of likelihood-evaluator program
e(wexp)	weight expression	e(technique)	maximization technique
e(title)	title in estimation output	e(crittype)	optimization criterion
e(clustvar)	name of cluster variable	e(properties)	b V
e(offset)	offset	e(predict)	program used to implement predict

Matrices

e(b)	coefficient vector	e(V)	variance–covariance matrix of the estimators
e(ilog)	iteration log (up to 20 iterations)		
e(gradient)	gradient vector		

Functions

e(sample)	marks estimation sample

Methods and Formulas

`glm` is implemented as an ado-file.

The canonical reference on GLM is McCullagh and Nelder (1989). The term "generalized linear model" is from Nelder and Wedderburn (1972). Many people use the acronym GLIM for GLM models because of the classic GLM software tool GLIM by Baker and Nelder (1985). See Dobson (2002) for a concise introduction and overview. See Rabe-Hesketh and Everitt (2004) for more examples of GLM using Stata. Hoffmann (2004) focuses on applying generalized linear models, using real-world datasets, along with interpreting computer output, which for the most part is obtained using Stata.

This discussion highlights the details of parameter estimation and predicted statistics. For a more detailed treatment, and for information on variance estimation, see Hardin and Hilbe (2007).

`glm` obtains results by iteratively reweighted least squares (IRLS), as described in McCullagh and Nelder (1989), or by maximum likelihood using Newton–Raphson. The implementation here, however, allows user-specified weights, which we denote as v_j for the jth observation. Let M be the number of "observations" ignoring weights. Define

$$
w_j = \begin{cases} 1 & \text{if no weights are specified} \\ v_j & \text{if \texttt{fweights} or \texttt{iweights} are specified} \\ Mv_j/(\sum_k v_k) & \text{if \texttt{aweights} or \texttt{pweights} are specified} \end{cases}
$$

The number of observations is then $N = \sum_j w_j$ if `fweights` are specified and $N = M$ otherwise. Each IRLS step is performed by `regress` using w_j as the weights.

Let d_j^2 denote the squared deviance residual for the jth observation:

For the Gaussian family, $d_j^2 = (y_j - \widehat{\mu}_j)^2$.

For the Bernoulli family (binomial with denominator 1),

$$
d_j^2 = \begin{cases} -2\ln(1 - \widehat{\mu}_j) & \text{if } y_j = 0 \\ -2\ln(\widehat{\mu}_j) & \text{otherwise} \end{cases}
$$

For the binomial family with denominator m_j,

$$
d_j^2 = \begin{cases} 2y_j\ln(y_j/\widehat{\mu}_j) + 2(m_j - y_j)\ln\{(m_j - y_j)/(m_j - \widehat{\mu}_j)\} & \text{if } 0 < y_j < m_j \\ 2m_j\ln\{m_j/(m_j - \widehat{\mu}_j)\} & \text{if } y_j = 0 \\ 2y_j\ln(y_j/\widehat{\mu}_j) & \text{if } y_j = m_j \end{cases}
$$

For the Poisson family,

$$
d_j^2 = \begin{cases} 2\widehat{\mu}_j & \text{if } y_j = 0 \\ 2\{y_j\ln(y_j/\widehat{\mu}_j) - (y_j - \widehat{\mu}_j)\} & \text{otherwise} \end{cases}
$$

For the gamma family, $d_j^2 = -2\{\ln(y_j/\widehat{\mu}_j) - (y_j - \widehat{\mu}_j)/\widehat{\mu}_j\}$.

For the inverse Gaussian, $d_j^2 = (y_j - \widehat{\mu}_j)^2/(\widehat{\mu}_j^2 y_j)$.

For the negative binomial,

$$
d_j^2 = \begin{cases} 2\ln(1 + k\widehat{\mu}_j)/k & \text{if } y_j = 0 \\ 2y_j\ln(y_j/\widehat{\mu}_j) - 2\{(1 + ky_j)/k\}\ln\{(1 + ky_j)/(1 + k\widehat{\mu}_j)\} & \text{otherwise} \end{cases}
$$

Let $\phi = 1$ if the scale parameter is set to one; otherwise, define $\phi = \widehat{\phi}_0(n - k)/n$, where $\widehat{\phi}_0$ is the estimated scale parameter and k is the number of covariates in the model (including intercept).

Let $\ln L_j$ denote the log likelihood for the jth observation:

For the Gaussian family,

$$
\ln L_j = -\frac{1}{2}\left[\left\{\frac{(y_j - \widehat{\mu}_j)^2}{\phi}\right\} + \ln(2\pi\phi)\right]
$$

For the binomial family with denominator m_j (Bernoulli if all $m_j = 1$),

$$
\ln L_j = \phi \times \begin{cases} \ln\{\Gamma(m_j + 1)\} - \ln\{\Gamma(y_j + 1)\} + \ln\{\Gamma(m_j - y_j + 1)\} & \text{if } 0 < y_j < m_j \\ \quad +(m_j - y_j)\ln(1 - \widehat{\mu}_j/m_j) + y_j\ln(\widehat{\mu}_j/m_j) & \\ m_j\ln(1 - \widehat{\mu}_j/m_j) & \text{if } y_j = 0 \\ m_j\ln(\widehat{\mu}_j/m_j) & \text{if } y_j = m_j \end{cases}
$$

For the Poisson family,

$$
\ln L_j = \phi\left[y_j\ln(\widehat{\mu}_j) - \widehat{\mu}_j - \ln\{\Gamma(y_j + 1)\}\right]
$$

For the gamma family, $\ln L_j = -y_j/\widehat{\mu}_j + \ln(1/\widehat{\mu}_j)$.

For the inverse Gaussian,

$$
\ln L_j = -\frac{1}{2}\left\{\frac{(y_j - \widehat{\mu}_j)^2}{y_j\widehat{\mu}_j^2} + 3\ln(y_j) + \ln(2\pi)\right\}
$$

For the negative binomial (let $m = 1/k$),

$$
\ln L_j = \phi\left[\ln\{\Gamma(m + y_j)\} - \ln\{\Gamma(y_j + 1)\} - \ln\{\Gamma(m)\} \\ -m\ln(1 + \widehat{\mu}_j/m) + y_j\ln\{\widehat{\mu}_j/(\widehat{\mu}_j + m)\}\right]
$$

The overall deviance reported by glm is $D^2 = \sum_j w_j d_j^2$. The dispersion of the deviance is D^2 divided by the residual degrees of freedom.

The Akaike information criterion (AIC) and Bayesian information criterion (BIC) are given by

$$
\text{AIC} = \frac{-2\ln L + 2k}{N}
$$
$$
\text{BIC} = D^2 - (N - k)\ln(N)
$$

where $\ln L = \sum_j w_j \ln L_j$ is the overall log likelihood.

The Pearson deviance reported by `glm` is $\sum_j w_j r_j^2$. The corresponding Pearson dispersion is the Pearson deviance divided by the residual degrees of freedom. `glm` also calculates the scaled versions of all these quantities by dividing by the estimated scale parameter.

Acknowledgments

`glm` was written by James Hardin, University of South Carolina, and Joseph Hilbe, Arizona State University. The previous version of this routine was written by Patrick Royston of the MRC Clinical Trials Unit, London. The original version of this routine was published in Royston (1994). Royston's work, in turn, was based on a prior implementation by Joseph Hilbe, first published in Hilbe (1993). Roger Newson wrote an early implementation (Newson 1999) of robust variance estimates for GLM. Parts of this entry are excerpts from Hardin and Hilbe (2007).

References

Akaike, H. 1973. Information theory and an extension of the maximum likelihood principle. In *Second International Symposium on Information Theory*, ed. B. N. Petrov and F. Csaki, 267–281. Budapest: Akademiai Kiado.

Anscombe, F. J. 1953. Contribution to the discussion of H. Hotelling's paper. *Journal of the Royal Statistical Society, Series B* 15: 229–230.

Baker, R. J., and J. A. Nelder. 1985. *The Generalized Linear Interactive Modelling System, release 3.77*. Oxford: Numerical Algorithms Group.

Basu, A. 2005. Extended generalized linear models: Simultaneous estimation of flexible link and variance functions. *Stata Journal* 5: 501–516.

Berndt, E. K., B. H. Hall, R. E. Hall, and J. A. Hausman. 1974. Estimation and inference in nonlinear structural models. *Annals of Economic and Social Measurement* 3/4: 653–665.

Dobson, A. J. 2002. *An Introduction to Generalized Linear Models*. 2nd ed. Boca Raton, FL: Chapman & Hall/CRC.

Hardin, J. W. and J. Hilbe. 2007. *Generalized Linear Models and Extensions*. 2nd ed. College Station, TX: Stata Press.

Hilbe, J. 1993. sg16: Generalized linear models. *Stata Technical Bulletin* 11: 20–28. Reprinted in *Stata Technical Bulletin Reprints*, vol. 2, pp. 149–159.

———. 2000. sg126: Two-parameter log-gamma and log-inverse Gaussian models. *Stata Technical Bulletin* 53: 31–32. Reprinted in *Stata Technical Bulletin Reprints*, vol. 9, pp. 273–275.

Hoffmann, J. P. 2004. *Generalized Linear Models: An Applied Approach*. Boston: Pearson.

Hosmer, D. W., Jr., and S. Lemeshow. 2000. *Applied Logistic Regression*. 2nd ed. New York: Wiley.

McCullagh, P., and J. A. Nelder. 1989. *Generalized Linear Models*. 2nd ed. London: Chapman & Hall.

Nelder, J. A. 1975. Robert William Maclagan Wedderburn. *Journal of the Royal Statistical Society, Series A* 138: 587.

Nelder, J. A., and R. W. M. Wedderburn. 1972. Generalized linear models. *Journal of the Royal Statistical Society, Series A* 135: 370–384.

Newey, W. K., and K. D. West. 1987. A simple, positive semi-definite, heteroskedasticity and autocorrelation consistent covariance matrix. *Econometrica* 55: 703–708.

Newson, R. 1999. sg114: rglm—Robust variance estimates for generalized linear models. *Stata Technical Bulletin* 50: 27–33. Reprinted in *Stata Technical Bulletin Reprints*, vol. 9, pp. 181–190.

———. 2004. Generalized power calculations for generalized linear models and more. *Stata Journal* 4: 379–401.

Pregibon, D. 1980. Goodness of link tests for generalized linear models. *Applied Statistics* 29: 15–24.

Rabe-Hesketh, S., and B. S. Everitt. 2004. *A Handbook of Statistical Analysis Using Stata*. 3rd ed. Boca Raton, FL: Chapman & Hall/CRC.

Rabe-Hesketh, S., A. Pickles, and C. Taylor. 2000. sg129: Generalized linear latent and mixed models. *Stata Technical Bulletin* 53: 47–57. Reprinted in *Stata Technical Bulletin Reprints*, vol. 9, pp. 293–307.

Rabe-Hesketh, S., A. Skrondal, and A. Pickles. 2002. st0005: Reliable estimation of generalized linear mixed models using adaptive quadrature. *Stata Journal* 2: 1–21.

Raftery, A. 1995. Bayesian model selection in social research. In *Sociological Methodology*, vol. 25, ed. P. V. Marsden, 111–163. Oxford: Blackwell.

Rogers, W. H. 1993. sg16.4: Comparison of nbreg and glm for negative binomial. *Stata Technical Bulletin* 16: 7. Reprinted in *Stata Technical Bulletin Reprints*, vol. 3, pp. 82–84.

Royston, J. P. 1994. sg22: Generalized linear models: Revision of glm. *Stata Technical Bulletin* 18: 6–11. Reprinted in *Stata Technical Bulletin Reprints*, vol. 3, pp. 112–121.

Schonlau, M. 2005. Boosted regression (boosting): An introductory tutorial and a Stata plugin. *Stata Journal* 5: 330–354.

Senn, S. 2003. A conversation with John Nelder. *Statistical Science* 18: 118–131.

Also See

[R] **glm postestimation** — Postestimation tools for glm

[R] **cloglog** — Complementary log-log regression

[R] **logistic** — Logistic regression, reporting odds ratios

[R] **nbreg** — Negative binomial regression

[R] **poisson** — Poisson regression

[R] **regress** — Linear regression

[SVY] **svy estimation** — Estimation commands for survey data

[XT] **xtgee** — Fit population-averaged panel-data models by using GEE

[U] **20 Estimation and postestimation commands**

Title

> **glm postestimation** — Postestimation tools for glm

Description

The following postestimation commands are available for `glm`:

command	description
adjust[1]	adjusted predictions of $\mathbf{x}\beta$ or $\exp(\mathbf{x}\beta)$
estat	AIC, BIC, VCE, and estimation sample summary
estat (svy)	postestimation statistics for survey data
estimates	cataloging estimation results
lincom	point estimates, standard errors, testing, and inference for linear combinations of coefficients
linktest	link test for model specification
lrtest[2]	likelihood-ratio test
mfx	marginal effects or elasticities
nlcom	point estimates, standard errors, testing, and inference for nonlinear combinations of coefficients
predict	predictions, residuals, influence statistics, and other diagnostic measures
predictnl	point estimates, standard errors, testing, and inference for generalized predictions
suest	seemingly unrelated estimation
test	Wald tests for simple and composite linear hypotheses
testnl	Wald tests of nonlinear hypotheses

[1] `adjust` is not appropriate with time-series operators.

[2] `lrtest` is not appropriate with `svy` estimation results.

See the corresponding entries in the *Stata Base Reference Manual* for details, but see
[SVY] **estat** for details about `estat` (svy).

(Continued on next page)

Syntax for predict

predict [*type*] *newvar* [*if*] [*in*] [, *statistic options*]

statistic	description
Main	
<u>mu</u>	$g^{-1}(\mathbf{x}\widehat{\beta})$; the default
xb	the linear prediction $\eta = \mathbf{x}\widehat{\beta}$
<u>eta</u>	synonym of xb
stdp	standard error of the linear prediction
<u>a</u>nscombe	Anscombe (1953) residuals
<u>c</u>ooksd	Cook's distance
<u>d</u>eviance	deviance residuals
<u>h</u>at	diagonals of the "hat" matrix
<u>l</u>ikelihood	a weighted average of standardized deviance and standardized Pearson residuals
<u>p</u>earson	Pearson residuals
<u>r</u>esponse	differences between the observed and fitted outcomes
<u>sc</u>ore	first derivative of the log likelihood with respect to $\mathbf{x}_j\beta$
<u>w</u>orking	working residuals

options	description
Options	
<u>noo</u>ffset	modify calculations to ignore offset variable
<u>adj</u>usted	adjust deviance residual to speed up convergence
<u>stand</u>ardized	multiply residual by the factor $(1-h)^{-1/2}$
<u>stud</u>entized	multiply residual by one over the square root of the estimated scale parameter
<u>mod</u>ified	modify denominator of residual to be a reasonable estimate of the variance of *depvar*

These statistics are available both in and out of sample; type predict ... if e(sample) ... if wanted only for the estimation sample.

mu, xb, stdp, and score are the only statistics allowed with svy estimation results.

Options for predict

   Main

mu, the default, specifies that predict calculate $g^{-1}(\mathbf{x}\widehat{\beta})$, the inverse link of the linear prediction.

xb calculates the linear prediction $\eta = \mathbf{x}\widehat{\beta}$.

eta is a synonym for xb.

stdp calculates the standard error of the linear prediction.

anscombe calculates the Anscombe (1953) residuals to produce residuals that closely follow a normal distribution.

cooksd calculates Cook's distance, which measures the aggregate change in the estimated coefficients when each observation is left out of the estimation.

deviance calculates the deviance residuals. Deviance residuals are recommended by McCullagh and Nelder (1989) and by others as having the best properties for examining the goodness of fit of a GLM. They are approximately normally distributed if the model is correct. They may be plotted against the fitted values or against a covariate to inspect the model's fit. Also see the pearson option below.

hat calculates the diagonals of the "hat" matrix as an analog to simple linear regression.

likelihood calculates a weighted average of standardized deviance and standardized Pearson residuals.

pearson calculates the Pearson residuals. Pearson residuals often have markedly skewed distributions for nonnormal family distributions. Also see the deviance option above.

response calculates the differences between the observed and fitted outcomes.

score calculates the equation-level score, $\partial \ln L / \partial(\mathbf{x}_j \boldsymbol{\beta})$.

working calculates the working residuals, which are response residuals weighted according to the derivative of the link function.

⌐‾Options‾⌐

nooffset is relevant only if you specified offset(*varname*) for glm. It modifies the calculations made by predict so that they ignore the offset variable; the linear prediction is treated as $\mathbf{x}_j \mathbf{b}$ rather than as $\mathbf{x}_j \mathbf{b} + \text{offset}_j$.

adjusted adjusts the deviance residual to speed up the convergence to the limiting normal distribution. The adjustment deals with adding to the deviance residual a higher-order term that depends on the variance function family. This option is allowed only when deviance is specified.

standardized requests that the residual be multiplied by the factor $(1-h)^{-1/2}$, where h is the diagonal of the hat matrix. This operation is done to account for the correlation between *depvar* and its predicted value.

studentized requests that the residual be multiplied by one over the square root of the estimated scale parameter.

modified requests that the denominator of the residual be modified to be a reasonable estimate of the variance of *depvar*. The base residual is multiplied by the factor $(k/w)^{-1/2}$, where k is either one or the user-specified dispersion parameter and w is the specified weight (or one if left unspecified).

Remarks

Remarks are presented under the following headings:

> Predictions
> Other postestimation commands

Predictions

▷ Example 1

After glm estimation, predict may be used to obtain various predictions based on the model. In example 2 in [R] **glm**, we mentioned that the complementary log-log link seemed to fit the data better than the logit link. Now we go back and obtain the fitted values and deviance residuals:

```
. use http://www.stata-press.com/data/r10/ldose

. glm r ldose, f(binomial n) l(logit)
  (output omitted )

. predict mu_logit
(option mu assumed; predicted mean r)

. predict dr_logit, deviance

. quietly glm r ldose, f(binomial n) l(cloglog)

. predict mu_cl
(option mu assumed; predicted mean r)

. predict dr_cl, d

. format mu_logit dr_logit mu_cl dr_cl %9.5f

. list r mu_logit dr_logit mu_cl dr_cl, sep(4)
```

	r	mu_logit	dr_logit	mu_cl	dr_cl
1.	6	3.45746	1.28368	5.58945	0.18057
2.	13	9.84167	1.05969	11.28067	0.55773
3.	18	22.45139	-1.19611	20.95422	-0.80330
4.	28	33.89761	-1.59412	30.36942	-0.63439
5.	52	50.09584	0.60614	47.77644	1.28883
6.	53	53.29092	-0.12716	54.14273	-0.52366
7.	61	59.22216	1.25107	61.11331	-0.11878
8.	60	58.74297	1.59398	59.94723	0.32495

In six of the eight cases, $|\text{dr\_logit}| > |\text{dr\_cl}|$. The above represents only one of the many available options for predict. See Hardin and Hilbe (2007) for a more in-depth examination.

◁

Other postestimation commands

❏ Technical Note

After glm estimation, you may perform any of the postestimation commands that you would perform after any other kind of estimation in Stata; see [U] **20 Estimation and postestimation commands**. Below we test the joint significance of all the interaction terms.

```
. use http://www.stata-press.com/data/r10/beetle, clear

. xi: glm r i.beetle*ldose, f(bin n) l(cloglog)
  (output omitted )

. testparm _I*
 ( 1)   [r]_Ibeetle_2 = 0
 ( 2)   [r]_Ibeetle_3 = 0
 ( 3)   [r]_IbeeXldose_2 = 0
 ( 4)   [r]_IbeeXldose_3 = 0

        chi2(  4) =   249.69
      Prob > chi2 =    0.0000
```

If you wanted to print the variance–covariance matrix of the estimators, you would type correlate, _coef (to see it as a correlation matrix) or correlate, _coef cov (to see the covariances themselves).

If you use the linktest postestimation command, you must also specify the family() and link() options; see [R] **linktest**.

❏

Methods and Formulas

All postestimation commands listed above are implemented as ado-files.

We follow the terminology used in *Methods and Formulas* in [R] **glm**.

The deviance residual calculated by `predict` following `glm` is $d_j = \text{sign}(y_j - \widehat{\mu}_j)\sqrt{d_j^2}$.

The Pearson residual calculated by `predict` following `glm` is

$$r_j = \frac{y_j - \widehat{\mu}_j}{\sqrt{V(\widehat{\mu}_j)}}$$

where $V(\widehat{\mu}_j)$ is the family-specific variance function.

$$V(\widehat{\mu}_j) = \begin{cases} \widehat{\mu}_j(1 - \widehat{\mu}_j/m_j) & \text{if binomial or Bernoulli } (m_j = 1) \\ \widehat{\mu}_j^2 & \text{if gamma} \\ 1 & \text{if Gaussian} \\ \widehat{\mu}_j^3 & \text{if inverse Gaussian} \\ \widehat{\mu}_j + k\widehat{\mu}_j^2 & \text{if negative binomial} \\ \widehat{\mu}_j & \text{if Poisson} \end{cases}$$

The response residuals are given by $r_i^R = y_i - \mu_i$. The working residuals are

$$r_i^W = (y_i - \widehat{\mu}_i)\left(\frac{\partial \eta}{\partial \mu}\right)_i$$

and the score residuals are

$$r_i^S = \frac{y_i - \widehat{\mu}_i}{V(\widehat{\mu}_i)}\left(\frac{\partial \eta}{\partial \mu}\right)_i^{-1}$$

Define $\widehat{W} = V(\widehat{\mu})$ and X to be the covariate matrix. h_i, then, is the ith diagonal of the hat matrix given by

$$\widehat{H} = \widehat{W}^{1/2}X(X^T\widehat{W}X)^{-1}X^T\widehat{W}^{1/2}$$

As a result, the likelihood residuals are given by

$$r_i^L = \text{sign}(y_i - \widehat{\mu}_i)\left\{h_i(r_i')^2 + (1 - h_i)(d_i')^2\right\}^{1/2}$$

where r_i' and d_i' are the standardized Pearson and standardized deviance residuals, respectively. By *standardized*, we mean that the residual is divided by $\{1 - h_i\}^{1/2}$.

Cook's distance is an overall measure of the change in the regression coefficients caused by omitting the ith observation from the analysis. Computationally, Cook's distance is obtained as

$$C_i = \frac{(r_i')^2 h_i}{k(1 - h_i)}$$

where k is the number of regressors, including the constant.

Anscombe residuals are given by

$$r_i^A = \frac{A(y_i) - A(\widehat{\mu}_i)}{A'(\widehat{\mu}_i)\{V(\widehat{\mu}_i)\}^{1/2}}$$

where

$$A(\cdot) = \int \frac{d\mu}{V^{1/3}(\mu)}$$

Deviance residuals may be adjusted (`predict, adjusted`) to make the following correction,

$$d_i^a = d_i + \frac{1}{6}\rho_3(\theta)$$

where $\rho_3(\theta)$ is a family-specific correction. See Hardin and Hilbe (2007) for the exact forms of $\rho_3(\theta)$ for each family.

References

Anscombe, F. J. 1953. Contribution of discussion paper by H. Hotelling 'New light on the correlation coefficient and its transform'. *Journal of the Royal Statistical Society, Series B* 15: 229–230.

Hardin, J. W. and J. Hilbe. 2007. *Generalized Linear Models and Extensions.* 2nd ed. College Station, TX: Stata Press.

McCullagh, P., and J. A. Nelder. 1989. *Generalized Linear Models.* 2nd ed. London: Chapman & Hall/CRC.

Also See

[R] **glm** — Generalized linear models

[R] **regress postestimation** — Postestimation tools for regress

[U] **20 Estimation and postestimation commands**

Title

> **glogit** — Logit and probit regression for grouped data

Syntax

Logistic regression for grouped data

> blogit *pos_var pop_var* [*rhsvars*] [*if*] [*in*] [, *blogit_options*]

Probit regression for grouped data

> bprobit *pos_var pop_var* [*rhsvars*] [*if*] [*in*] [, *bprobit_options*]

Weighted least-squares logistic regression for grouped data

> glogit *pos_var pop_var* [*rhsvars*] [*if*] [*in*] [, *glogit_options*]

Weighted least-squares probit regression for grouped data

> gprobit *pos_var pop_var* [*rhsvars*] [*if*] [*in*] [, *gprobit_options*]

blogit_options	description
Model	
<u>nocons</u>tant	suppress constant term
asis	retain perfect predictor variables
<u>off</u>set(*varname*)	include *varname* in model with coefficient constrained to 1
SE/Robust	
vce(*vcetype*)	*vcetype* may be oim, <u>r</u>obust, <u>cl</u>uster *clustvar*, <u>boot</u>strap, or <u>jack</u>knife
Reporting	
<u>l</u>evel(#)	set confidence level; default is level(95)
or	report odds ratios
Max options	
maximize_options	control the maximization process; seldom used
† <u>noco</u>ef	do not display coefficient table; seldom used

bprobit_options	description
Model	
<u>nocon</u>stant	suppress constant term
asis	retain perfect predictor variables
<u>off</u>set(*varname*)	include *varname* in model with coefficient constrained to 1
SE/Robust	
vce(*vcetype*)	*vcetype* may be oim, <u>r</u>obust, <u>c</u>luster *clustvar*, <u>boot</u>strap, or <u>jack</u>knife
Reporting	
<u>l</u>evel(*#*)	set confidence level; default is level(95)
Max options	
maximize_options	control the maximization process; seldom used
[†] <u>nocoef</u>	do not display coefficient table; seldom used

glogit_options	description
SE	
vce(*vcetype*)	*vcetype* may be ols, <u>boot</u>strap, or <u>jack</u>knife
Reporting	
<u>l</u>evel(*#*)	set confidence level; default is level(95)
or	report odds ratios

gprobit_options	description
SE	
vce(*vcetype*)	*vcetype* may be ols, <u>boot</u>strap, or <u>jack</u>knife
Reporting	
<u>l</u>evel(*#*)	set confidence level; default is level(95)

[†] nocoef does not appear in the dialog box.

bootstrap, by, jackknife, rolling, statsby, and xi are allowed; see [U] **11.1.10 Prefix commands**.

See [U] **20 Estimation and postestimation commands** for more capabilities of estimation commands.

Description

blogit and bprobit produce maximum-likelihood logit and probit estimates on grouped ("blocked") data; glogit and gprobit produce weighted least-squares estimates. In the syntax diagrams above, *pos_var* and *pop_var* refer to variables containing the total number of positive responses and the total population.

See [R] **logistic** for a list of related estimation commands.

Options for blogit and bprobit

     ⌐ Model ⌐

noconstant; see [R] **estimation options**.

asis forces retention of perfect predictor variables and their associated perfectly predicted observations and may produce instabilities in maximization; see [R] **probit**.

offset(*varname*); see [R] **estimation options**.

     ⌐ SE/Robust ⌐

vce(*vcetype*) specifies the type of standard error reported, which includes types that are derived from asymptotic theory, that are robust to some kinds of misspecification, that allow for intragroup correlation, and that use bootstrap or jackknife methods; see [R] *vce_option*.

     ⌐ Reporting ⌐

level(*#*); see [R] **estimation options**.

or (blogit only) reports the estimated coefficients transformed to odds ratios, i.e., e^b rather than b. Standard errors and confidence intervals are similarly transformed. This option affects how results are displayed, not how they are estimated. or may be specified at estimation or when replaying previously estimated results.

     ⌐ Max options ⌐

maximize_options: <u>iter</u>ate(*#*), [<u>no</u>]<u>log</u>, <u>trace</u>, <u>tol</u>erance(*#*), <u>ltol</u>erance(*#*); see [R] **maximize**. These options are seldom used.

The following option is available with blogit and bprobit but is not shown in the dialog box:

nocoef specifies that the coefficient table not be displayed. This option is sometimes used by program writers but is useless interactively.

Options for glogit and gprobit

     ⌐ SE ⌐

vce(*vcetype*) specifies the type of standard error reported, which includes types that are derived from asymptotic theory and that use bootstrap or jackknife methods; see [R] *vce_option*.

vce(ols), the default, uses the standard variance estimator for ordinary least-squares regression.

     ⌐ Reporting ⌐

level(*#*); see [R] **estimation options**.

or (glogit only) reports the estimated coefficients transformed to odds ratios, i.e., e^b rather than b. Standard errors and confidence intervals are similarly transformed. This option affects how results are displayed, not how they are estimated. or may be specified at estimation or when replaying previously estimated results.

Remarks

Remarks are presented under the following headings:

Maximum likelihood estimates
Weighted least-squares estimates

Maximum likelihood estimates

blogit produces the same results as logit and logistic, and bprobit produces the same results as probit, but the "blocked" commands accept data in slightly different "shape". Consider the following two datasets:

```
. use http://www.stata-press.com/data/r10/xmpl1
. list, sepby(agecat)
```

	agecat	exposed	died	pop
1.	0	0	0	115
2.	0	0	1	5
3.	0	1	0	98
4.	0	1	1	8
5.	1	0	0	69
6.	1	0	1	16
7.	1	1	0	76
8.	1	1	1	22

```
. use http://www.stata-press.com/data/r10/xmpl2
. list
```

	agecat	exposed	deaths	pop
1.	0	0	5	120
2.	0	1	8	106
3.	1	0	16	85
4.	1	1	22	98

These two datasets contain the same information; observations 1 and 2 of xmpl1 correspond to observation 1 of xmpl2, observations 3 and 4 of xmpl1 correspond to observation 2 of xmpl2, and so on.

The first observation of xmpl1 says that, for agecat==0 and exposed==0, 115 subjects did not die (died==0). The second observation says that, for the same agecat and exposed groups, five subjects did die (died==1). In xmpl2, the first observation says that there were five deaths of a population of 120 in agecat==0 and exposed==0. These are two different ways of saying the same thing. Both datasets are transcriptions from the following table, reprinted in Rothman and Greenland (1998, 255), for age-specific deaths from all causes for tolbutamide and placebo treatment groups (University Group Diabetes Program 1970):

	Age through 54		Age 55 and above	
	Tolbutamide	Placebo	Tolbutamide	Placebo
Dead	8	5	22	16
Surviving	98	115	76	79

The data in xmpl1 are said to be "fully relational", which is computer jargon meaning that each observation corresponds to one cell of the table. Stata typically prefers data in this format. The second form of storing these data in xmpl2 is said to be "folded", which is computer jargon for something less than fully relational.

blogit and bprobit deal with "folded" data and produce the same results that logit and probit would have if the data had been stored in the "fully relational" representation.

▷ Example 1

For the tolbutamide data, the fully relational representation is preferred. We could then use logistic, logit, and any of the epidemiological table commands; see [R] **logistic**, [R] **logit**, and [ST] **epitab**. Nevertheless, there are occasions when the folded representation seems more natural. With blogit and bprobit, we avoid the tedium of having to unfold the data:

```
. use http://www.stata-press.com/data/r10/xmpl2
. blogit deaths pop agecat exposed, or
```

Logistic regression for grouped data				Number of obs	=	409
				LR chi2(2)	=	22.47
				Prob > chi2	=	0.0000
Log likelihood = -142.6212				Pseudo R2	=	0.0730

_outcome	Odds Ratio	Std. Err.	z	P>|z|	[95% Conf. Interval]	
agecat	4.216299	1.431519	4.24	0.000	2.167361	8.202223
exposed	1.404674	.4374454	1.09	0.275	.7629451	2.586175

If we had not specified the or option, results would have been presented as coefficients instead of as odds ratios. The estimated odds ratio of death for tolbutamide exposure is 1.40, although the 95% confidence interval includes 1. (By comparison, these data, in fully relational form and analyzed using the cs command [see [ST] **epitab**], produce a Mantel–Haenszel weighted odds ratio of 1.40 with a 95% confidence interval of 0.76 to 2.59.)

We can see the underlying coefficients by replaying the estimation results and not specifying the or option:

```
. blogit
```

Logistic regression for grouped data				Number of obs	=	409
				LR chi2(2)	=	22.47
				Prob > chi2	=	0.0000
Log likelihood = -142.6212				Pseudo R2	=	0.0730

_outcome	Coef.	Std. Err.	z	P>|z|	[95% Conf. Interval]	
agecat	1.438958	.3395203	4.24	0.000	.7735101	2.104405
exposed	.3398053	.3114213	1.09	0.275	-.2705692	.9501798
_cons	-2.968471	.33234	-8.93	0.000	-3.619846	-2.317097

◁

(Continued on next page)

▷ Example 2

bprobit works like blogit, substituting the probit for the logit likelihood function.

```
. bprobit deaths pop agecat exposed
```

Probit regression for grouped data

				Number of obs	=	409
				LR chi2(2)	=	22.58
				Prob > chi2	=	0.0000
Log likelihood = -142.56478				Pseudo R2	=	0.0734

_outcome	Coef.	Std. Err.	z	P>|z|	[95% Conf. Interval]	
agecat	.7542049	.170969	4.41	0.000	.4191118	1.089298
exposed	.1906236	.1666057	1.14	0.253	-.1359176	.5171649
_cons	-1.673973	.1619592	-10.34	0.000	-1.991407	-1.356539

◁

Weighted least-squares estimates

▷ Example 3

We have state data for the United States on the number of marriages (marriage), the total population aged 18 years or more (pop18p), and the median age (medage). The dataset excludes Nevada, so it has 49 observations. We create a new variable called medage2 equal to age squared by typing generate medage2=medage^2 and now wish to estimate a logit equation for the marriage rate:

```
. use http://www.stata-press.com/data/r10/census7
(1980 Census data by state)
. glogit marriage pop18p medage medage2
```

Weighted LS logistic regression for grouped data

Source	SS	df	MS		Number of obs	=	49
					F(2, 46)	=	12.89
Model	.715984037	2	.357992019		Prob > F	=	0.0000
Residual	1.27772776	46	.02777669		R-squared	=	0.3591
					Adj R-squared	=	0.3313
Total	1.9937118	48	.041535662		Root MSE	=	.16666

marriage	Coef.	Std. Err.	t	P>|t|	[95% Conf. Interval]	
medage	-.6459366	.2828377	-2.28	0.027	-1.215259	-.0766139
medage2	.0095415	.0046608	2.05	0.046	.0001598	.0189231
_cons	6.503859	4.288972	1.52	0.136	-2.129393	15.13711

◁

▷ Example 4

We could just as easily have fitted a grouped-probit model by typing gprobit rather than glogit:

```
. gprobit marriage pop18p medage medage2
Weighted LS probit regression for grouped data
```

Source	SS	df	MS		
Model	.108223109	2	.054111555		
Residual	.192322336	46	.00418092		
Total	.300545446	48	.006261363		

Number of obs = 49
F(2, 46) = 12.94
Prob > F = 0.0000
R-squared = 0.3601
Adj R-squared = 0.3323
Root MSE = .06466

marriage	Coef.	Std. Err.	t	P>|t|	[95% Conf. Interval]
medage	-.2755013	.1121041	-2.46	0.018	-.5011551 -.0498476
medage2	.0041082	.0018422	2.23	0.031	.0004001 .0078163
_cons	2.357718	1.704444	1.38	0.173	-1.073149 5.788585

◁

Saved Results

blogit and bprobit save the following in e():

Scalars

e(N)	number of observations	e(r2_p)	pseudo-R-squared
e(N_cdf)	number of completely determined failures	e(N_clust)	number of clusters
		e(ll)	log likelihood
e(N_cds)	number of completely determined successes	e(ll_0)	log likelihood, constant-only model
		e(chi2)	χ^2
e(df_m)	model degrees of freedom		

Macros

e(cmd)	blogit or bprobit	e(chi2type)	Wald or LR; type of model χ^2 test
e(cmdline)	command as typed	e(vce)	*vcetype* specified in vce()
e(depvar)	name of dependent variable	e(vcetype)	title used to label Std. Err.
e(title)	title in estimation output	e(crittype)	optimization criterion
e(clustvar)	name of cluster variable	e(properties)	b V
e(offset)	offset	e(predict)	program used to implement predict

Matrices

e(b)	coefficient vector	e(V)	variance–covariance matrix of the estimators

Functions

e(sample)	marks estimation sample

(Continued on next page)

glogit and gprobit save the following in e():

Scalars
    e(N)   number of observations   e(r2) R-squared
    e(mss)   model sum of squares   e(r2_a)   adjusted R-squared
    e(df_m)   model degrees of freedom   e(F) F statistic
    e(rss)   residual sum of squares   e(rmse)   root mean squared error
    e(df_r)   residual degrees of freedom

Macros
    e(cmd)   glogit or gprobit   e(vce)   *vcetype* specified in vce()
    e(cmdline) command as typed   e(vcetype)   title used to label Std. Err.
    e(depvar)   name of dependent variable   e(predict)   program used to implement predict
    e(title)   title in estimation output   e(properties) b V
    e(model)   ols

Matrices
    e(b)   coefficient vector   e(V)   variance–covariance matrix of the
                                                             estimators

Functions
    e(sample)   marks estimation sample

Methods and Formulas

blogit, bprobit, glogit, and gprobit are implemented as ado-files.

Maximum likelihood estimates

The results reported by blogit and bprobit are obtained by maximizing a weighted logit or probit-likelihood function. Let $F(\ )$ denote the normal or logistic-likelihood function. The likelihood of observing each observation in the data is then

$$F(\beta x)^s \left\{ 1 - F(\beta x) \right\}^{t-s}$$

where s is the number of successes and t is the population. The term above is counted as contributing $s + (t - s) = t$ degrees of freedom. All of this follows directly from the definition of logit and probit.

Weighted least-squares estimates

The logit function is defined as the log of the odds ratio. If there is one explanatory variable, the model can be written as

$$\log\left(\frac{p_j}{1 - p_j}\right) = \beta_0 + \beta_1 x_j + \epsilon_j \tag{1}$$

where p_j represents successes divided by population for the jth observation. (If there is more than one explanatory variable, we simply interpret β_1 as a row vector and x_j as a column vector.) The large-sample expectation of ϵ_j is zero, and its variance is

$$\sigma_j^2 = \frac{1}{n_j p_j (1 - p_j)}$$

where n_j represents the population for observation j. We can thus apply weighted least-squares to the observations, with weights proportional to $n_j p_j (1 - p_j)$.

As in any feasible generalized least-squares problem, estimation proceeds in two steps. First we fit equation (1) by OLS and compute the predicted probabilities as

$$\widehat{p_j} = \frac{\exp(\widehat{\beta_0} + \widehat{\beta_1} x_j)}{1 + \exp(\widehat{\beta_0} + \widehat{\beta_1} x_j)}$$

In the second step we fit equation (1) by using analytic weights equal to $n_j \widehat{p_j} (1 - \widehat{p_j})$.

For `gprobit`, write $\Phi(\cdot)$ for the cumulative normal distribution, and define z_j implicitly by $\Phi(z_j) = p_j$, where p_j is the fraction of successes for observation j. The probit model for one explanatory variable can be written as

$$\Phi^{-1}(p_j) = \beta_0 + \beta_1 x_j + \epsilon_j$$

(If there is more than one explanatory variable, we simply interpret β_1 as a row vector and x_j as a column vector.)

The expectation of ϵ_j is zero, and its variance is given by

$$\sigma_j^2 = \frac{p_j(1 - p_j)}{n_j \phi^2 \{\Phi^{-1}(p_j)\}}$$

where $\phi(\cdot)$ represents the normal density (Amemiya 1981, 1498). We can thus apply weighted least squares to the observations with weights proportional to $1/\sigma_j^2$. As for grouped logit, we use a two-step estimator to obtain the weighted least-squares estimates.

References

Amemiya, T. 1981. Qualitative response models: A survey. *Journal of Economic Literature* 19: 1483–1536.

Hosmer, D. W., Jr., and S. Lemeshow. 2000. *Applied Logistic Regression.* 2nd ed. New York: Wiley.

Judge, G. G., W. E. Griffiths, R. C. Hill, H. Lütkepohl, and T.-C. Lee. 1985. *The Theory and Practice of Econometrics.* 2nd ed. New York: Wiley.

Rothman, K. J., and S. Greenland. 1998. *Modern Epidemiology.* 2nd ed. Philadelphia: Lippincott–Raven.

University Group Diabetes Program. 1970. A study of the effects of hypoglycemic agents on vascular complications in patients with adult onset diabetes. *Diabetes* 19, supplement 2: 747–830.

Also See

[R] **glogit postestimation** — Postestimation tools for glogit, gprobit, blogit, and bprobit

[R] **logistic** — Logistic regression, reporting odds ratios

[R] **logit** — Logistic regression, reporting coefficients

[R] **probit** — Probit regression

[R] **scobit** — Skewed logistic regression

[U] **20 Estimation and postestimation commands**

Title

> **glogit postestimation** — Postestimation tools for glogit, gprobit, blogit, and bprobit

Description

The following postestimation commands are available for `glogit`, `gprobit`, `blogit`, and `bprobit`:

command	description
adjust	adjusted predictions of $\mathbf{x}\beta$, probabilities, or $\exp(\mathbf{x}\beta)$
*estat	AIC, BIC, VCE, and estimation sample summary
estimates	cataloging estimation results
lincom	point estimates, standard errors, testing, and inference for linear combinations of coefficients
*lrtest	likelihood-ratio test
mfx	marginal effects or elasticities
nlcom	point estimates, standard errors, testing, and inference for nonlinear combinations of coefficients
predict	predictions, residuals, influence statistics, and other diagnostic measures
predictnl	point estimates, standard errors, testing, and inference for generalized predictions
test	Wald tests for simple and composite linear hypotheses
testnl	Wald tests of nonlinear hypotheses

* `estat ic` and `lrtest` are not appropriate after `glogit` and `gprobit`.

See the corresponding entries in the *Stata Base Reference Manual* for details.

Syntax for predict

> predict $[type]$ *newvar* $[if]$ $[in]$ $[$, *statistic* $]$

statistic	description
Main	
n	predicted count; the default
<u>pr</u>	probability of a positive outcome
xb	linear prediction
stdp	standard error of the linear prediction

These statistics are available both in and out of sample; type `predict ... if e(sample) ...` if wanted only for the estimation sample.

Options for predict

> **Main**

n, the default, calculates the expected count, that is, the estimated probability times *pop_var*, which is the total population.

pr calculates the predicted probability of a positive outcome.

xb calculates the linear prediction.

stdp calculates the standard error of the linear prediction.

Methods and Formulas

All postestimation commands listed above are implemented as ado-files.

Also See

[R] **glogit** — Logit and probit regression for grouped data

[U] **20 Estimation and postestimation commands**

Title

> **grmeanby** — Graph means and medians by categorical variables

Syntax

> grmeanby *varlist* [*if*] [*in*] [*weight*], <u>summarize</u>(*varname*) [*options*]

options	description
Main	
* <u>summarize</u>(*varname*)	graph mean (or median) of *varname*
<u>median</u>	graph medians; default is to graph means
Plot	
cline_options	change the look of the lines
marker_options	change look of markers (color, size, etc.)
marker_label_options	add marker labels; change look or position
Y axis, X axis, Titles, Legend, Overall	
twoway_options	any options other than by() documented in [G] *twoway_options*

* <u>summarize</u>(*varname*) is required.

aweights and fweights are allowed; see [U] **11.1.6 weight**.

Description

grmeanby graphs the (optionally weighted) means or medians of *varname* according to the values of the variables in *varlist*. The variables in *varlist* may be string or numerical and, if numerical, may be labeled.

Options

⌐ Main ⌐

summarize(*varname*) is required; it specifies the name of the variable whose mean or median is to be graphed.

median is optional; it specifies that the graph is to be of medians, not means.

⌐ Plot ⌐

cline_options affect the rendition of the lines through the markers, including their color, pattern, and width; see [G] *cline_options*.

marker_options affect the rendition of markers drawn at the plotted points, including their shape, size, color, and outline; see [G] *marker_options*.

marker_label_options specify if and how the markers are to be labeled; see [G] *marker_label_options*.

Y axis, X axis, Titles, Legend, Overall

twoway_options are any of the options documented in [G] *twoway_options*, excluding by(). These include options for titling the graph (see [G] *title_options*) and for saving the graph to disk (see [G] *saving_option*).

Remarks

The idea of graphing means of categorical variables was shown in Chambers and Hastie (1992, 3). Since this was shown in the context of an S function for making such graphs, it doubtless has roots going back further than that. grmeanby is, in any case, another implementation of what we will assume is their idea.

▷ Example 1

Using a variation of our auto dataset, we graph the mean of mpg by foreign, rep77, rep78, and make:

```
. use http://www.stata-press.com/data/r10/auto1
(Automobile Models)
. grmeanby foreign rep77 rep78 make, sum(mpg)
```

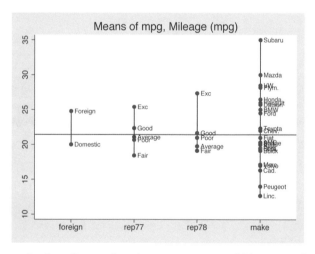

If we had wanted a graph of medians rather than means, we could have typed

```
. grmeanby foreign rep77 rep78 make, sum(mpg) median
```

◁

Methods and Formulas

grmeanby is implemented as an ado-file.

References

Chambers, J. M., and T. J. Hastie, eds. 1992. *Statistical Models in S*. Pacific Grove, CA: Wadsworth and Brooks/Cole.

Gould, W. W. 1993. gr12: Graphs of means and medians by categorical variables. *Stata Technical Bulletin* 12: 13. Reprinted in *Stata Technical Bulletin Reprints*, vol. 2, pp. 44–45.

Title

> **hausman** — Hausman specification test

Syntax

> hausman *name-consistent* [*name-efficient*] [, *options*]

options	description
Main	
<u>c</u>onstant	include estimated intercepts in comparison; default is to exclude
<u>a</u>lleqs	use all equations to perform test; default is first equation only
<u>sk</u>ipeqs(*eqlist*)	skip specified equations when performing test
<u>e</u>quations(*matchlist*)	associate/compare the specified (by number) pairs of equations
force	force performance of test, even though assumptions are not met
df(*#*)	use *#* degrees of freedom
sigmamore	base both (co)variance matrices on disturbance variance estimate from efficient estimator
sigmaless	base both (co)variance matrices on disturbance variance estimate from consistent estimator
Advanced	
<u>tc</u>onsistent(*string*)	consistent estimator column header
<u>te</u>fficient(*string*)	efficient estimator column header

where *name-consistent* and *name-efficient* are names under which estimation results were saved via estimates store.

A period (.) may be used to refer to the last estimation results, even if these were not already stored.

Not specifying *name-efficient* is equivalent to specifying the last estimation results as ".".

Description

hausman performs Hausman's (1978) specification test.

Options

> ⌐ Main ⌐

constant specifies that the estimated intercept(s) be included in the model comparison; by default, they are excluded. The default behavior is appropriate for models in which the constant does not have a common interpretation across the two models.

alleqs specifies that all the equations in the models be used to perform the Hausman test; by default, only the first equation is used.

skipeqs(*eqlist*) specifies in *eqlist* the names of equations to be excluded from the test. Equation numbers are not allowed in this context, as the equation names, along with the variable names, are used to identify common coefficients.

542

equations(*matchlist*) specifies, by number, the pairs of equations that are to be compared.

The *matchlist* in equations() should follow the syntax

$$\#_c : \#_e \; \left[, \#_c : \#_e \left[, \dots \right] \right]$$

where $\#_c$ ($\#_e$) is an equation number of the always-consistent (efficient under H_0) estimator. For instance, equations(1:1), equations(1:1, 2:2), or equations(1:2).

If equations() is not specified, then equations are matched on equation names.

equations() handles the situation in which one estimator uses equation names and the other does not. For instance, equations(1:2) means that equation 1 of the always-consistent estimator is to be tested against equation 2 of the efficient estimator. equations(1:1, 2:2) means that equation 1 is to be tested against equation 1 and that equation 2 is to be tested against equation 2. If equations() is specified, options alleqs and skipeqs are ignored.

force specifies that the Hausman test be performed, even though the assumptions of the Hausman test seem not to be met, for example, because the estimators were pweighted or the data were clustered.

df(#) specifies the degrees of freedom for the Hausman test. The default is the matrix rank of the variance of the difference between the coefficients of the two estimators.

sigmamore and sigmaless specify that the two covariance matrices used in the test be based on a common estimate of disturbance variance (σ^2).

sigmamore specifies that the covariance matrices be based on the estimated disturbance variance from the efficient estimator. This option provides a proper estimate of the contrast variance for so-called tests of exogeneity and overidentification in instrumental variables regression.

sigmaless specifies that the covariance matrices be based on the estimated disturbance variance from the consistent estimator.

These options can be specified only when both estimators save e(sigma) or e(rmse), or with command xtreg. e(sigma_e) is saved after the command xtreg with option fe or mle. e(rmse) is saved after the command xtreg with option re.

sigmamore or sigmaless are recommended when comparing fixed-effects and random-effects linear regression because they are much less likely to produce a non–positive definite differenced covariance matrix (although the tests are asymptotically equivalent whether or not one of the options is specified).

___ Advanced _____

tconsistent(*string*) and tefficient(*string*) are formatting options. They allow you to specify the headers of the columns of coefficients that default to the names of the models. These options will be of interest primarily to programmers.

Remarks

hausman is a general implementation of Hausman's (1978) specification test, which compares an estimator $\widehat{\theta}_1$ that is known to be consistent with an estimator $\widehat{\theta}_2$ that is efficient under the assumption being tested. The null hypothesis is that the estimator $\widehat{\theta}_2$ is indeed a (consistent and) efficient estimator of the true parameters. If this is the case, there should be no systematic difference between the two estimators. If there exists a systematic difference in the estimates, you have reason to doubt the assumptions on which the efficient estimator is based.

The assumption of efficiency is violated if the estimator is `pweighted` or the data are clustered, so `hausman` cannot be used. The test can be forced by specifying the `force` option with `hausman`. For an alternative to using `hausman` in these cases, see [R] **suest**.

To use `hausman`, you

```
. (compute the always-consistent estimator)
. estimates store name-consistent
. (compute the estimator that is efficient under H_0)
. hausman name-consistent .
```

Alternatively, you can turn this around:

```
. (compute the estimator that is efficient under H_0)
. estimates store name-efficient
. (fit the less-efficient model)
. (compute the always-consistent estimator)
. hausman . name-efficient
```

You can, of course, also compute and store both the always-consistent and efficient-under-H_0 estimators and perform the Hausman test with

```
. hausman name-consistent name-efficient
```

▷ Example 1

We are studying the factors that affect the wages of young women in the United States between 1970 and 1988 and have a panel-data sample of individual women over that time span.

```
. use http://www.stata-press.com/data/r10/nlswork4
(National Longitudinal Survey.  Young Women 14-26 years of age in 1968)
. describe
Contains data from http://www.stata-press.com/data/r10/nlswork4.dta
  obs:        28,534                        National Longitudinal Survey.
                                            Young Women 14-26 years of age
                                            in 1968
  vars:            6                        29 Jan 2007 16:35
  size:       485,078 (88.4% of memory free)
```

variable name	storage type	display format	value label	variable label
idcode	int	%8.0g		NLS ID
year	byte	%8.0g		interview year
age	byte	%8.0g		age in current year
msp	byte	%8.0g		1 if married, spouse present
ttl_exp	float	%9.0g		total work experience
ln_wage	float	%9.0g		ln(wage/GNP deflator)

```
Sorted by:  idcode  year
```

We believe that a random-effects specification is appropriate for individual-level effects in our model. We fit a fixed-effects model that will capture all temporally constant individual-level effects.

```
. xtreg ln_wage age msp ttl_exp, fe
Fixed-effects (within) regression              Number of obs      =      28494
Group variable: idcode                         Number of groups   =       4710

R-sq:  within  = 0.1373                         Obs per group: min =          1
       between = 0.2571                                        avg =        6.0
       overall = 0.1800                                        max =         15

                                                F(3,23781)         =    1262.01
corr(u_i, Xb)  = 0.1476                         Prob > F           =     0.0000
```

ln_wage	Coef.	Std. Err.	t	P>|t|	[95% Conf. Interval]	
age	-.005485	.000837	-6.55	0.000	-.0071256	-.0038443
msp	.0033427	.0054868	0.61	0.542	-.0074118	.0140971
ttl_exp	.0383604	.0012416	30.90	0.000	.0359268	.0407941
_cons	1.593953	.0177538	89.78	0.000	1.559154	1.628752

sigma_u	.37674223	
sigma_e	.29751014	
rho	.61591044	(fraction of variance due to u_i)

```
F test that all u_i=0:     F(4709, 23781) =     7.76        Prob > F = 0.0000
```

We assume that this model is consistent for the true parameters and save the results by using `estimates store` under a name, `fixed`:

```
. estimates store fixed
```

Now we fit a random-effects model as a fully efficient specification of the individual effects under the assumption that they are random and follow a normal distribution. We then compare these estimates with the previously saved results by using the `hausman` command.

```
. xtreg ln_wage age msp ttl_exp, re
Random-effects GLS regression                  Number of obs      =      28494
Group variable: idcode                         Number of groups   =       4710

R-sq:  within  = 0.1373                         Obs per group: min =          1
       between = 0.2552                                        avg =        6.0
       overall = 0.1797                                        max =         15

Random effects u_i ~ Gaussian                   Wald chi2(3)       =    5100.33
corr(u_i, X)       = 0 (assumed)                Prob > chi2        =     0.0000
```

ln_wage	Coef.	Std. Err.	z	P>|z|	[95% Conf. Interval]	
age	-.0069749	.0006882	-10.13	0.000	-.0083238	-.0056259
msp	.0046594	.0051012	0.91	0.361	-.0053387	.0146575
ttl_exp	.0429635	.0010169	42.25	0.000	.0409704	.0449567
_cons	1.609916	.0159176	101.14	0.000	1.578718	1.641114

sigma_u	.32648519	
sigma_e	.29751014	
rho	.54633481	(fraction of variance due to u_i)

```
. hausman fixed .
                    ────── Coefficients ──────
                     (b)          (B)            (b-B)      sqrt(diag(V_b-V_B))
                    fixed          .           Difference          S.E.

         age      -.005485     -.0069749        .0014899          .0004764
         msp      .0033427      .0046594       -.0013167          .0020206
     ttl_exp      .0383604      .0429635       -.0046031          .0007124
```

b = consistent under Ho and Ha; obtained from xtreg
B = inconsistent under Ha, efficient under Ho; obtained from xtreg

Test: Ho: difference in coefficients not systematic

$$\text{chi2(3)} = (b-B)'[(V\_b-V\_B)^{(-1)}](b-B)$$
$$= 275.44$$
$$\text{Prob>chi2} = 0.0000$$

Under the current specification, our initial hypothesis that the individual-level effects are adequately modeled by a random-effects model is resoundingly rejected. This result is based on the rest of our model specification, and random effects might be appropriate for some alternate model of wages.

◁

Jerry Allen Hausman was born in West Virginia in 1946. He studied economics at Brown and Oxford, has been at MIT since 1972, and has made many outstanding contributions to econometrics and applied microeconomics.

▷ Example 2

A stringent assumption of multinomial and conditional logit models is that outcome categories for the model have the property of independence of irrelevant alternatives (IIA). Stated simply, this assumption requires that the inclusion or exclusion of categories does not affect the relative risks associated with the regressors in the remaining categories.

One classic example of a situation in which this assumption would be violated involves the choice of transportation mode; see McFadden (1974). For simplicity, postulate a transportation model with the four possible outcomes: rides a train to work, takes a bus to work, drives the Ford to work, and drives the Chevrolet to work. Clearly, "drives the Ford" is a closer substitute to "drives the Chevrolet" than it is to "rides a train" (at least for most people). This means that excluding "drives the Ford" from the model could be expected to affect the relative risks of the remaining options and that the model would not obey the IIA assumption.

Using the data presented in [R] **mlogit**, we will use a simplified model to test for IIA. The choice of insurance type among indemnity, prepaid, and uninsured is modeled as a function of age and gender. The indemnity category is allowed to be the base category, and the model including all three outcomes is fitted. The results are then stored under the name allcats.

```
. use http://www.stata-press.com/data/r10/sysdsn3
(Health insurance data)

. mlogit insure age male

Iteration 0:   log likelihood = -555.85446
Iteration 1:   log likelihood = -551.32973
Iteration 2:   log likelihood = -551.32802
```

Multinomial logistic regression				Number of obs	=	615
				LR chi2(4)	=	9.05
				Prob > chi2	=	0.0598
Log likelihood = -551.32802				Pseudo R2	=	0.0081

insure	Coef.	Std. Err.	z	P>|z|	[95% Conf. Interval]	
Prepaid						
age	-.0100251	.0060181	-1.67	0.096	-.0218204	.0017702
male	.5095747	.1977893	2.58	0.010	.1219148	.8972345
_cons	.2633838	.2787574	0.94	0.345	-.2829708	.8097383
Uninsure						
age	-.0051925	.0113821	-0.46	0.648	-.027501	.017116
male	.4748547	.3618446	1.31	0.189	-.2343477	1.184057
_cons	-1.756843	.5309591	-3.31	0.001	-2.797504	-.7161824

```
(insure==Indemnity is the base outcome)

. estimates store allcats
```

Under the IIA assumption, we would expect no systematic change in the coefficients if we excluded one of the outcomes from the model. (For an extensive discussion, see Hausman and McFadden 1984.) We reestimate the parameters, excluding the uninsured outcome, and perform a Hausman test against the fully efficient full model.

```
. mlogit insure age male if insure != "Uninsure":insure

Iteration 0:   log likelihood =  -394.8693
Iteration 1:   log likelihood =  -390.4871
Iteration 2:   log likelihood = -390.48643
```

Multinomial logistic regression				Number of obs	=	570
				LR chi2(2)	=	8.77
				Prob > chi2	=	0.0125
Log likelihood = -390.48643				Pseudo R2	=	0.0111

insure	Coef.	Std. Err.	z	P>|z|	[95% Conf. Interval]	
Prepaid						
age	-.0101521	.0060049	-1.69	0.091	-.0219214	.0016173
male	.5144003	.1981735	2.60	0.009	.1259875	.9028132
_cons	.2678043	.2775562	0.96	0.335	-.2761959	.8118046

```
(insure==Indemnity is the base outcome)
```

(Continued on next page)

```
. hausman . allcats, alleqs constant
```

	Coefficients			
	(b) .	(B) allcats	(b-B) Difference	sqrt(diag(V_b-V_B)) S.E.
age	-.0101521	-.0100251	-.0001269	.
male	.5144003	.5095747	.0048256	.012334
_cons	.2678043	.2633838	.0044205	.

```
                           b = consistent under Ho and Ha; obtained from mlogit
              B = inconsistent under Ha, efficient under Ho; obtained from mlogit
    Test:  Ho:  difference in coefficients not systematic
                chi2(3) = (b-B)'[(V_b-V_B)^(-1)](b-B)
                        =        0.08
             Prob>chi2 =        0.9944
             (V_b-V_B is not positive definite)
```

The syntax of the `if` condition on the `mlogit` command simply identified the "Uninsured" category with the `insure` value label; see [U] **12.6.3 Value labels**. On examining the output from hausman, we see that there is no evidence that the IIA assumption has been violated.

Since the Hausman test is a standardized comparison of model coefficients, using it with `mlogit` requires that the base outcome be the same in both competing models. In particular, if the most-frequent category (the default base outcome) is being removed to test for IIA, you must use the `baseoutcome()` option in `mlogit` to manually set the base outcome to something else. Or you can use the `equation()` option of the `hausman` command to align the equations of the two models.

Having the missing values for the square root of the diagonal of the covariance matrix of the differences is not comforting, but it is also not surprising. This covariance matrix is guaranteed to be positive definite only asymptotically (it is a consequence of the assumption that one of the estimators is efficient), and assurances are not made about the diagonal elements. Negative values along the diagonal are possible, and the fourth column of the table is provided mainly for descriptive use.

We can also perform the Hausman IIA test against the remaining alternative in the model.

```
. mlogit insure age male if insure != "Prepaid":insure
Iteration 0:   log likelihood = -132.59913
Iteration 1:   log likelihood = -131.78009
Iteration 2:   log likelihood = -131.76808
Iteration 3:   log likelihood = -131.76807

Multinomial logistic regression              Number of obs    =        338
                                             LR chi2(2)       =       1.66
                                             Prob > chi2      =     0.4356
Log likelihood = -131.76807                  Pseudo R2        =     0.0063
```

insure	Coef.	Std. Err.	z	P>|z|	[95% Conf. Interval]	
Uninsure						
age	-.0041055	.0115807	-0.35	0.723	-.0268033	.0185923
male	.4591072	.3595663	1.28	0.202	-.2456298	1.163844
_cons	-1.801774	.5474476	-3.29	0.001	-2.874752	-.7287968

```
(insure==Indemnity is the base outcome)
```

```
. hausman . allcats, alleqs constant
```

	── Coefficients ── (b) .	(B) allcats	(b-B) Difference	sqrt(diag(V_b-V_B)) S.E.
age	-.0041055	-.0051925	.001087	.0021357
male	.4591072	.4748547	-.0157475	.
_cons	-1.801774	-1.756843	-.0449311	.1333464

```
                  b = consistent under Ho and Ha; obtained from mlogit
          B = inconsistent under Ha, efficient under Ho; obtained from mlogit
   Test:  Ho:  difference in coefficients not systematic

            chi2(3) = (b-B)'[(V_b-V_B)^(-1)](b-B)
                    =    -0.18    chi2<0 ==> model fitted on these
                                  data fails to meet the asymptotic
                                  assumptions of the Hausman test;
                                  see suest for a generalized test.
```

Here the χ^2 statistic is actually negative. We might interpret this result as strong evidence that we cannot reject the null hypothesis. Such a result is not an unusual outcome for the Hausman test, particularly when the sample is relatively small—there are only 45 uninsured individuals in this dataset.

Are we surprised by the results of the Hausman test in this example? Not really. Judging from the z-statistics on the original multinomial logit model, we were struggling to identify any structure in the data with the current specification. Even when we were willing to assume IIA and computed the efficient estimator under this assumption, few of the effects could be identified as statistically different from those on the base category. Trying to base a Hausman test on a contrast (difference) between two poor estimates is just asking too much of the existing data.

◁

In example 2, we encountered a case in which the Hausman was not well defined. Unfortunately, in our experience this happens fairly often. Stata provides an alternative to the Hausman test that overcomes this problem through an alternative estimator of the variance of the difference between the two estimators. This other estimator is guaranteed to be positive (semi)definite. This alternative estimator also allows a widening of the scope of problems to which Hausman-type tests can be applied by relaxing the assumption that one of the estimators is efficient. For instance, you can perform Hausman-type tests to clustered observations and survey estimators. See [R] **suest** for details.

Saved Results

hausman saves the following in r():

Scalars
      r(chi2) χ^2
      r(df)   degrees of freedom for the statistic
      r(p) p-value for the χ^2
      r(rank)   rank of (V_b-V_B)^(-1)

Methods and Formulas

hausman is implemented as an ado-file.

The Hausman statistic is distributed as χ^2 and is computed as

$$H = (\beta_c - \beta_e)'(V_c - V_e)^{-1}(\beta_c - \beta_e)$$

where

β_c is the coefficient vector from the consistent estimator
β_e is the coefficient vector from the efficient estimator
V_c is the covariance matrix of the consistent estimator
V_e is the covariance matrix of the efficient estimator

When the difference in the variance matrices is not positive definite, a Moore–Penrose generalized inverse is used. As noted in Gourieroux and Monfort (1989, 125–128), the choice of generalized inverse is not important asymptotically.

The number of degrees of freedom for the statistic is the rank of the difference in the variance matrices. When the difference is positive definite, this is the number of common coefficients in the models being compared.

Acknowledgment

Portions of hausman are based on an early implementation by Jeroen Weesie, Utrecht University, The Netherlands.

References

Baltagi, B. H. 2002. *Econometrics*. 3rd ed. New York: Springer.

Gourieroux, C., and A. Monfort. 1989. *Statistics and Econometric Models, Vol. 2*. New York: Cambridge University Press.

Hausman, J. A. 1978. Specification tests in econometrics. *Econometrica* 46: 1251–1271.

Hausman, J., and D. McFadden. 1984. Specification tests in econometrics. *Econometrica* 52: 1219–1240.

McFadden, D. 1974. Measurement of urban travel demand. *Journal of Public Economics* 3: 303–328.

Also See

[R] **lrtest** — Likelihood-ratio test after estimation

[R] **suest** — Seemingly unrelated estimation

[R] **test** — Test linear hypotheses after estimation

[XT] **xtreg** — Fixed-, between-, and random-effects, and population-averaged linear models

Title

> **heckman** — Heckman selection model

Syntax

Basic syntax

> heckman *depvar* $[$*indepvars*$]$, <u>sel</u>ect(*varlist$_s$*) $[$<u>two</u>step$]$
>
> or
>
> heckman *depvar* $[$*indepvars*$]$, <u>sel</u>ect(*depvar$_s$* = *varlist$_s$*) $[$<u>two</u>step$]$

Full syntax for maximum likelihood estimates only

> heckman *depvar* $[$*indepvars*$]$ $[$*if*$]$ $[$*in*$]$ $[$*weight*$]$,
>
>   <u>sel</u>ect($[$*depvar$_s$* =$]$ *varlist$_s$* $[$, <u>off</u>set(*varname*) <u>noc</u>onstant $]$)
>
> $[$*heckman_ml_options*$]$

Full syntax for Heckman's two-step consistent estimates only

> heckman *depvar* $[$*indepvars*$]$ $[$*if*$]$ $[$*in*$]$, <u>two</u>step
>
>   <u>sel</u>ect($[$*depvar$_s$* =$]$ *varlist$_s$* $[$, <u>noc</u>onstant $]$) $[$*heckman_ts_options*$]$

heckman_ml_options	description
Model	
* <u>sel</u>ect()	specify selection equation: dependent and independent variables; whether to have constant term and offset variable
<u>noc</u>onstant	suppress constant term
<u>off</u>set(*varname*)	include *varname* in model with coefficient constrained to 1
<u>c</u>onstraints(*constraints*)	apply specified linear constraints
<u>col</u>linear	keep collinear variables
SE/Robust	
vce(*vcetype*)	*vcetype* may be oim, <u>r</u>obust, <u>cl</u>uster *clustvar*, opg, <u>boot</u>strap, or <u>jack</u>knife
Reporting	
<u>l</u>evel(*#*)	set confidence level; default is level(95)
<u>fir</u>st	report first-step probit estimates
noskip	perform likelihood-ratio test
<u>nsh</u>azard(*newvar*)	generate nonselection hazard variable
<u>m</u>ills(*newvar*)	synonym for nshazard()

Max options

maximize_options	control the maximization process; seldom used

*select( ) is required.

The full specification is <u>sel</u>ect([*depvar_s* =] *varlist_s* [, <u>off</u>set(*varname*) <u>noc</u>onstant])

heckman_ts_options	description
Model	
*<u>sel</u>ect()	specify selection equation: dependent and independent variables; whether to have constant term
*<u>two</u>step	produce two-step consistent estimate
<u>noc</u>onstant	suppress constant term
<u>rhos</u>igma	truncate ρ to $[-1,1]$ with consistent σ
<u>rhot</u>runc	truncate ρ to $[-1,1]$
<u>rhol</u>imited	truncate ρ in limited cases
<u>rhof</u>orce	do not truncate ρ
SE	
vce(*vcetype*)	*vcetype* may be conventional, <u>boot</u>strap, or jackknife
Reporting	
<u>l</u>evel(*#*)	set confidence level; default is level(95)
<u>first</u>	report first-step probit estimates
<u>ns</u>hazard(*newvar*)	generate nonselection hazard variable
<u>mills</u>(*newvar*)	synonym for nshazard()

*select( ) and twostep are required.

The full specification is <u>sel</u>ect([*depvar_s* =] *varlist_s* [, <u>noc</u>onstant])

depvar, *indepvars*, *varlist_s*, and *depvar_s* may contain time-series operators; see [U] **11.4.3 Time-series varlists**.
bootstrap, by, jackknife, rolling, statsby, svy, and xi are allowed; see [U] **11.1.10 Prefix commands**.
Weights are not allowed with the bootstrap prefix.
aweights are not allowed with the jackknife prefix.
twostep, vce(), first, noskip, and weights are not allowed with the svy prefix.
pweights, aweights, fweights, and iweights are allowed with maximum likelihood estimation;
   see [U] **11.1.6 weight**. No weights are allowed if twostep is specified.
See [U] **20 Estimation and postestimation commands** for more capabilities of estimation commands.

Description

heckman fits regression models with selection by using either Heckman's two-step consistent estimator or full maximum likelihood.

Options for Heckman selection model (ML)

> **Model**

select(...) specifies the variables and options for the selection equation. It is an integral part of specifying a Heckman model and is required. The selection equation should contain at least one variable that is not in the outcome equation.

If *depvar_s* is specified, it should be coded as 0 or 1, 0 indicating an observation not selected and 1 indicating a selected observation. If *depvar_s* is not specified, observations for which *depvar* is not missing are assumed selected, and those for which *depvar* is missing are assumed not selected.

noconstant, offset(*varname*), constraints(*constraints*), collinear; see [R] **estimation options**.

> **SE/Robust**

vce(*vcetype*) specifies the type of standard error reported, which includes types that are derived from asymptotic theory, that are robust to some kinds of misspecification, that allow for intragroup correlation, and that use bootstrap or jackknife methods; see [R] **vce_option**.

> **Reporting**

level(*#*); see [R] **estimation options**.

first specifies that the first-step probit estimates of the selection equation be displayed before estimation.

noskip specifies that a full maximum-likelihood model with only a constant for the regression equation be fitted. This model is not displayed but is used as the base model to compute a likelihood-ratio test for the model test statistic displayed in the estimation header. By default, the overall model test statistic is an asymptotically equivalent Wald test that all the parameters in the regression equation are zero (except the constant). For many models, this option can substantially increase estimation time.

nshazard(*newvar*) and mills(*newvar*) are synonyms; either will create a new variable containing the nonselection hazard—what Heckman (1979) referred to as the inverse of the Mills' ratio—from the selection equation. The nonselection hazard is computed from the estimated parameters of the selection equation.

> **Max options**

maximize_options: difficult, technique(*algorithm_spec*), iterate(*#*), [no]log, trace, gradient, showstep, hessian, shownrtolerance, tolerance(*#*), ltolerance(*#*), gtolerance(*#*), nrtolerance(*#*), nonrtolerance, from(*init_specs*); see [R] **maximize**. These options are seldom used.

Setting the optimization type to technique(bhhh) resets the default *vcetype* to vce(opg).

Options for Heckman selection model (two-step)

> **Model**

select(...) specifies the variables and options for the selection equation. It is an integral part of specifying a Heckman model and is required. The selection equation should contain at least one variable that is not in the outcome equation.

If *depvar$_s$* is specified, it should be coded as 0 or 1, 0 indicating an observation not selected and 1 indicating a selected observation. If *depvar$_s$* is not specified, observations for which *depvar* is not missing are assumed selected, and those for which *depvar* is missing are assumed not selected.

twostep specifies that Heckman's (1979) two-step efficient estimates of the parameters, standard errors, and covariance matrix be produced.

noconstant; see [R] **estimation options**.

rhosigma, rhotrunc, rholimited, and rhoforce are rarely used options to specify how the two-step estimator (option twostep) handles unusual cases in which the two-step estimate of ρ is outside the admissible range for a correlation, $[-1, 1]$. When abs$(\rho) > 1$, the two-step estimate of the coefficient variance–covariance matrix may not be positive definite and thus may be unusable for testing. The default is rhosigma.

rhotrunc specifies that ρ be truncated to lie in the range $[-1, 1]$. If the two-step estimate is less than -1, ρ is set to -1; if the two-step estimate is greater than 1, ρ is set to 1. This truncated value of ρ is used in all computations to estimate the two-step covariance matrix.

rhosigma specifies that ρ be truncated, as with option rhotrunc, and that the estimate of σ be made consistent with $\widehat{\rho}$, the truncated estimate of ρ. So, $\widehat{\sigma} = \beta_m \widehat{\rho}$; see *Methods and Formulas* for the definition of β_m. Both the truncated ρ and the new estimate of $\widehat{\sigma}$ are used in all computations to estimate the two-step covariance matrix.

rholimited specifies that ρ be truncated only in computing the diagonal matrix **D** as it enters $\mathbf{V}_{\text{twostep}}$ and **Q**; see *Methods and Formulas*. In all other computations, the untruncated estimate of ρ is used.

rhoforce specifies that the two-step estimate of ρ be retained, even if it is outside the admissible range for a correlation. This option may, in rare cases, lead to a non–positive-definite covariance matrix.

These options have no effect when estimation is by maximum likelihood, the default. They also have no effect when the two-step estimate of ρ is in the range $[-1, 1]$.

___

    ⌐ SE ⌐

vce(*vcetype*) specifies the type of standard error reported, which includes types that are derived from asymptotic theory and that use bootstrap or jackknife methods; see [R] *vce_option*.

vce(conventional), the default, uses the two-step variance estimator derived by Heckman.

___

    ⌐ Reporting ⌐

level(*#*); see [R] **estimation options**.

first specifies that the first-step probit estimates of the selection equation be displayed before estimation.

nshazard(*newvar*) and mills(*newvar*) are synonyms; either will create a new variable containing the nonselection hazard—what Heckman (1979) referred to as the inverse of the Mills' ratio—from the selection equation. The nonselection hazard is computed from the estimated parameters of the selection equation.

Remarks

The Heckman selection model (Gronau 1974, Lewis 1974, Heckman 1976) assumes that there exists an underlying regression relationship,

$$y_j = \mathbf{x}_j \beta + u_{1j} \qquad \qquad regression\ equation$$

The dependent variable, however, is not always observed. Rather, the dependent variable for observation j is observed if

$$\mathbf{z}_j \gamma + u_{2j} > 0 \qquad \qquad selection\ equation$$

where

$$u_1 \sim N(0, \sigma)$$

$$u_2 \sim N(0, 1)$$

$$\mathrm{corr}(u_1, u_2) = \rho$$

When $\rho \neq 0$, standard regression techniques applied to the first equation yield biased results. heckman provides consistent, asymptotically efficient estimates for all the parameters in such models.

In one classic example, the first equation describes the wages of women. Women choose whether to work, and thus, from our point of view as researchers, whether we observe their wages in our data. If women made this decision randomly, we could ignore that not all wages are observed and use ordinary regression to fit a wage model. Such an assumption of random participation, however, is unlikely to be true; women who would have low wages may be unlikely to choose to work, and thus the sample of observed wages is biased upward. In the jargon of economics, women choose not to work when their personal reservation wage is greater than the wage offered by employers. Thus women who choose not to work might have even higher offer wages than those who do work—they may have high offer wages, but they have even higher reservation wages. We could tell a story that competency is related to wages, but competency is rewarded more at home than in the labor force.

In any case, in this problem—which is the paradigm for most such problems—a solution can be found if there are some variables that strongly affect the chances for observation (the reservation wage) but not the outcome under study (the offer wage). Such a variable might be the number of children in the home. (Theoretically, we do not need such identifying variables, but without them, we depend on functional form to identify the model. It would be difficult for anyone to take such results seriously since the functional-form assumptions have no firm basis in theory.)

▷ Example 1

In the syntax for heckman, *depvar* and *indepvars* are the dependent variable and regressors for the underlying regression model to be fitted ($\mathbf{y} = \mathbf{X}\beta$), and *varlist$_s$* are the variables ($\mathbf{Z}$) thought to determine whether *depvar* is observed or unobserved (selected or not selected). In our female wage example, the number of children at home would be included in the second list. By default, heckman assumes that missing values (see [U] **12.2.1 Missing values**) of *depvar* imply that the dependent variable is unobserved (not selected). With some datasets, it is more convenient to specify a binary variable (*depvar$_s$*) that identifies the observations for which the dependent is observed/selected (*depvar$_s \neq$ 0*) or not observed (*depvar$_s$* = 0); heckman will accommodate either type of data.

We have a (fictional) dataset on 2,000 women, 1,343 of whom work:

```
. use http://www.stata-press.com/data/r10/womenwk
. summarize age educ married children wage
```

Variable	Obs	Mean	Std. Dev.	Min	Max
age	2000	36.208	8.28656	20	59
education	2000	13.084	3.045912	10	20
married	2000	.6705	.4701492	0	1
children	2000	1.6445	1.398963	0	5
wage	1343	23.69217	6.305374	5.88497	45.80979

We will assume that the hourly wage is a function of education and age, whereas the likelihood of working (the likelihood of the wage being observed) is a function of marital status, the number of children at home, and (implicitly) the wage (via the inclusion of age and education, which we think determine the wage):

```
. heckman wage educ age, select(married children educ age)
Iteration 0:   log likelihood = -5178.7009
Iteration 1:   log likelihood = -5178.3049
Iteration 2:   log likelihood = -5178.3045
```

Heckman selection model				Number of obs	=	2000
(regression model with sample selection)				Censored obs	=	657
				Uncensored obs	=	1343
				Wald chi2(2)	=	508.44
Log likelihood = -5178.304				Prob > chi2	=	0.0000

	Coef.	Std. Err.	z	P>|z|	[95% Conf. Interval]	
wage						
education	.9899537	.0532565	18.59	0.000	.8855729	1.094334
age	.2131294	.0206031	10.34	0.000	.1727481	.2535108
_cons	.4857752	1.077037	0.45	0.652	-1.625179	2.59673
select						
married	.4451721	.0673954	6.61	0.000	.3130794	.5772647
children	.4387068	.0277828	15.79	0.000	.3842534	.4931601
education	.0557318	.0107349	5.19	0.000	.0346917	.0767718
age	.0365098	.0041533	8.79	0.000	.0283694	.0446502
_cons	-2.491015	.1893402	-13.16	0.000	-2.862115	-2.119915
/athrho	.8742086	.1014225	8.62	0.000	.6754241	1.072993
/lnsigma	1.792559	.027598	64.95	0.000	1.738468	1.84665
rho	.7035061	.0512264			.5885365	.7905862
sigma	6.004797	.1657202			5.68862	6.338548
lambda	4.224412	.3992265			3.441942	5.006881

LR test of indep. eqns. (rho = 0):   chi2(1) =   61.20   Prob > chi2 = 0.0000

heckman assumes that wage is the dependent variable and that the first variable list (educ and age) are the determinants of wage. The variables specified in the select() option (married, children, educ, and age) are assumed to determine whether the dependent variable is observed (the selection equation). Thus we fitted the model

$$\text{wage} = \beta_0 + \beta_1\text{educ} + \beta_2\text{age} + u_1$$

and we assumed that wage is observed if

$$\gamma_0 + \gamma_1 \text{married} + \gamma_2 \text{children} + \gamma_3 \text{educ} + \gamma_4 \text{age} + u_2 > 0$$

where u_1 and u_2 have correlation ρ.

The reported results for the wage equation are interpreted exactly as though we observed wage data for all women in the sample; the coefficients on age and education level represent the estimated marginal effects of the regressors in the underlying regression equation. The results for the two ancillary parameters require some explanation. heckman does not directly estimate ρ; to constrain ρ within its valid limits, and for numerical stability during optimization, it estimates the inverse hyperbolic tangent of ρ:

$$\text{atanh}\,\rho = \frac{1}{2} \ln \left(\frac{1+\rho}{1-\rho} \right)$$

This estimate is reported as /athrho. In the bottom panel of the output, heckman undoes this transformation for you: the estimated value of ρ is .7035061. The standard error for ρ is computed using the delta method, and its confidence intervals are the transformed intervals of /athrho.

Similarly, σ, the standard error of the residual in the wage equation, is not directly estimated; for numerical stability, heckman instead estimates $\ln \sigma$. The untransformed sigma is reported at the end of the output: 6.004797.

Finally, some researchers—especially economists—are used to the selectivity effect summarized not by ρ but by $\lambda = \rho\sigma$. heckman reports this, too, along with an estimate of the standard error and confidence interval.

◁

❏ Technical Note

If each of the equations in the model had contained many regressors, the heckman command could have become long. An alternate way of specifying our wage model would be to use Stata's global macros. The following lines are an equivalent way of specifying our model:

```
. global wageeq "wage educ age"
. global seleq "married children educ age"
. heckman $wageeq, select($seleq)
  (output omitted )
```

❏

❏ Technical Note

The reported model χ^2 test is a Wald test that all coefficients in the regression model (except the constant) are 0. heckman is an estimation command, so you can use test, testnl, or lrtest to perform tests against whatever nested alternate model you choose; see [R] test, [R] testnl, and [R] lrtest.

The estimation of ρ and σ in the forms atanh ρ and $\ln \sigma$ extends the range of these parameters to infinity in both directions, thus avoiding boundary problems during the maximization. Tests of ρ must be made in the transformed units. However, since atanh$(0) = 0$, the reported test for atanh $\rho = 0$ is equivalent to the test for $\rho = 0$.

The likelihood-ratio test reported at the bottom of the output is an equivalent test for $\rho = 0$ and is computationally the comparison of the joint likelihood of an independent probit model for the selection equation and a regression model on the observed wage data against the Heckman model likelihood. Since $\chi^2 = 61.20$, this clearly justifies the Heckman selection equation with these data.

❏

▷ Example 2

heckman supports the Huber/White/sandwich estimator of variance under the vce(robust) and vce(cluster *clustvar*) options or when pweights are used for population-weighted data; see [U] **20.15 Obtaining robust variance estimates**. We can obtain robust standard errors for our wage model by specifying clustering on county of residence (the county variable).

```
. heckman wage educ age, select(married children educ age) vce(cluster county)

Iteration 0:   log pseudolikelihood = -5178.7009
Iteration 1:   log pseudolikelihood = -5178.3049
Iteration 2:   log pseudolikelihood = -5178.3045

Heckman selection model                          Number of obs     =       2000
(regression model with sample selection)         Censored obs      =        657
                                                 Uncensored obs    =       1343

                                                 Wald chi2(1)      =          .
Log pseudolikelihood = -5178.304                 Prob > chi2       =          .
```

(Std. Err. adjusted for 10 clusters in county)

	Coef.	Robust Std. Err.	z	P>|z|	[95% Conf. Interval]	
wage						
education	.9899537	.0600061	16.50	0.000	.8723438	1.107564
age	.2131294	.020995	10.15	0.000	.17198	.2542789
_cons	.4857752	1.302103	0.37	0.709	-2.066299	3.03785
select						
married	.4451721	.0731472	6.09	0.000	.3018062	.5885379
children	.4387068	.0312386	14.04	0.000	.3774802	.4999333
education	.0557318	.0110039	5.06	0.000	.0341645	.0772991
age	.0365098	.004038	9.04	0.000	.0285954	.0444242
_cons	-2.491015	.1153305	-21.60	0.000	-2.717059	-2.264972
/athrho	.8742086	.1403337	6.23	0.000	.5991596	1.149258
/lnsigma	1.792559	.0258458	69.36	0.000	1.741902	1.843216
rho	.7035061	.0708796			.5364513	.817508
sigma	6.004797	.155199			5.708189	6.316818
lambda	4.224412	.5186709			3.207835	5.240988

Wald test of indep. eqns. (rho = 0): chi2(1) =   61.20   Prob > chi2 = 0.0000

The robust standard errors tend to be a bit larger, but we notice no systematic differences. This finding is not surprising since the data were not constructed to have any county-specific correlations or any other characteristics that would deviate from the assumptions of the Heckman model.

◁

▷ Example 3

Stata also produces Heckman's (1979) two-step efficient estimator of the model with the `twostep` option. Maximum likelihood estimation of the parameters can be time consuming with large datasets, and the two-step estimates may provide a good alternative in such cases. Continuing with the women's wage model, we can obtain the two-step estimates with Heckman's consistent covariance estimates by typing

```
. heckman wage educ age, select(married children educ age) twostep
```

Heckman selection model -- two-step estimates	Number of obs	=	2000
(regression model with sample selection)	Censored obs	=	657
	Uncensored obs	=	1343
	Wald chi2(4)	=	551.37
	Prob > chi2	=	0.0000

	Coef.	Std. Err.	z	P>|z|	[95% Conf. Interval]	
wage						
education	.9825259	.0538821	18.23	0.000	.8769189	1.088133
age	.2118695	.0220511	9.61	0.000	.1686502	.2550888
_cons	.7340391	1.248331	0.59	0.557	-1.712645	3.180723
select						
married	.4308575	.074208	5.81	0.000	.2854125	.5763025
children	.4473249	.0287417	15.56	0.000	.3909922	.5036576
education	.0583645	.0109742	5.32	0.000	.0368555	.0798735
age	.0347211	.0042293	8.21	0.000	.0264318	.0430105
_cons	-2.467365	.1925635	-12.81	0.000	-2.844782	-2.089948
mills						
lambda	4.001615	.6065388	6.60	0.000	2.812821	5.19041
rho	0.67284					
sigma	5.9473529					
lambda	4.0016155	.6065388				

◁

❑ Technical Note

The Heckman selection model depends strongly on the model being correct, much more so than ordinary regression. Running a separate probit or logit for sample inclusion followed by a regression, referred to in the literature as the two-part model (Manning, Duan, and Rogers 1987)—not to be confused with Heckman's two-step procedure—is an especially attractive alternative if the regression part of the model arose because of taking a logarithm of zero values. When the goal is to analyze an underlying regression model or to predict the value of the dependent variable that would be observed in the absence of selection, however, the Heckman model is more appropriate. When the goal is to predict an actual response, the two-part model is usually the better choice.

The Heckman selection model can be unstable when the model is not properly specified or if a specific dataset simply does not support the model's assumptions. For example, let's examine the solution to another simulated problem.

```
. use http://www.stata-press.com/data/r10/twopart

. heckman yt x1 x2 x3, select(z1 z2)

Iteration 0:    log likelihood = -111.94996
Iteration 1:    log likelihood = -110.82258
Iteration 2:    log likelihood = -110.17707
Iteration 3:    log likelihood = -107.70663  (not concave)
Iteration 4:    log likelihood = -107.07729  (not concave)
  (output omitted)
Iteration 42:   log likelihood = -104.08246
Iteration 43:   log likelihood = -104.08245

Heckman selection model                       Number of obs     =        150
(regression model with sample selection)      Censored obs      =         87
                                              Uncensored obs    =         63

                                              Wald chi2(3)      =    5.36e+10
Log likelihood = -104.0825                    Prob > chi2       =     0.0000
```

	Coef.	Std. Err.	z	P>|z|	[95% Conf.	Interval]
yt						
x1	.8974192	.0000267	33644.42	0.000	.8973669	.8974715
x2	-2.525303	.0000162	-1.6e+05	0.000	-2.525334	-2.525271
x3	2.855786	.0000391	72964.55	0.000	2.855709	2.855862
_cons	.697487	.0573018	12.17	0.000	.5851776	.8097964
select						
z1	-.6826644	.0618966	-11.03	0.000	-.8039796	-.5613493
z2	1.003701	.0910048	11.03	0.000	.8253354	1.182068
_cons	-.3605988	.1113424	-3.24	0.001	-.5788259	-.1423716
/athrho	17.93948	201.7724	0.09	0.929	-377.5271	413.4061
/lnsigma	-.5397114	.0906692	-5.95	0.000	-.7174197	-.3620031
rho	1	1.79e-13			-1	1
sigma	.5829165	.0528526			.4880098	.6962802
lambda	.5829165	.0528526			.4793274	.6865056

```
LR test of indep. eqns. (rho = 0):   chi2(1) =     25.67   Prob > chi2 = 0.0000
```

The model has converged to a value of ρ that is 1.0—within machine-rounding tolerances. Given the form of the likelihood for the Heckman selection model, this implies a division by zero, and it is surprising that the model solution turns out as well as it does. Reparameterizing ρ has allowed the estimation to converge, but we clearly have problems with the estimates. Moreover, if this had occurred in a large dataset, waiting 43 iterations for convergence might take considerable time.

This dataset was not intentionally developed to cause problems. It is actually generated by a "Heckman process" and when generated starting from different random values can be easily estimated. The luck of the draw here merely led to data that, despite the source, did not support the assumptions of the Heckman model.

The two-step model is generally more stable when the data are problematic. It even tolerates estimates of ρ less than -1 and greater than 1. For these reasons, the two-step model may be preferred when exploring a large dataset. Still, if the maximum likelihood estimates cannot converge, or converge to a value of ρ that is at the boundary of acceptable values, there is scant support for fitting a Heckman selection model on the data. Heckman (1979) discusses the implications of ρ being exactly 1 or 0, together with the implications of other possible covariance relationships among the model's determinants.

❏

Saved Results

heckman saves the following in e():

Scalars

e(N)	number of observations	e(selambda)	standard error of λ
e(k)	number of parameters	e(sigma)	sigma
e(k_eq)	number of equations	e(chi2)	χ^2
e(k_eq_model)	number of equations in model Wald test	e(chi2_c)	χ^2 for comparison test
		e(p_c)	p-value for comparison test
e(k_aux)	number of auxiliary parameters	e(p)	significance of comparison test
e(k_dv)	number of dependent variables	e(rho)	ρ
e(N_cens)	number of censored observations	e(rank)	rank of e(V)
e(df_m)	model degrees of freedom	e(rank0)	rank of e(V) for constant-only model
e(ll)	log likelihood		
e(ll_0)	log likelihood, constant-only model	e(ic)	number of iterations
e(N_clust)	number of clusters	e(rc)	return code
e(lambda)	λ	e(converged)	1 if converged, 0 otherwise

Macros

e(cmd)	heckman	e(chi2_ct)	Wald or LR; type of model χ^2 test corresponding to e(chi2_c)
e(cmdline)	command as typed		
e(depvar)	names of dependent variables	e(vce)	*vcetype* specified in vce()
e(wtype)	weight type	e(vcetype)	title used to label Std. Err.
e(wexp)	weight expression	e(diparm#)	display transformed parameter #
e(title)	title in estimation output	e(opt)	type of optimization
e(title2)	secondary title in estimation output	e(method)	ml or twostep
e(clustvar)	name of cluster variable	e(ml_method)	type of ml method
e(offset1)	offset for regression equation	e(user)	name of likelihood-evaluator program
e(offset2)	offset for selection equation	e(technique)	maximization technique
e(mills)	variable containing nonselection hazard (inverse of Mills')	e(crittype)	optimization criterion
		e(properties)	b V
e(chi2type)	Wald or LR; type of model χ^2 test	e(predict)	program used to implement predict

Matrices

e(b)	coefficient vector	e(V)	variance–covariance matrix of the estimators
e(ilog)	iteration log (up to 20 iterations)		
e(gradient)	gradient vector		

Functions

e(sample)	marks estimation sample

Methods and Formulas

heckman is implemented as an ado-file. Greene (2003, 782–787) and Johnston and DiNardo (1997, 446–450) provide good introductions to the Heckman selection model.

Regression estimates using the nonselection hazard (Heckman 1979) provide starting values for maximum likelihood estimation.

The regression equation is

$$y_j = \mathbf{x}_j \boldsymbol{\beta} + u_{1j}$$

The selection equation is

$$\mathbf{z}_j \boldsymbol{\gamma} + u_{2j} > 0$$

where

$$u_1 \sim N(0, \sigma)$$

$$u_2 \sim N(0, 1)$$

$$\text{corr}(u_1, u_2) = \rho$$

The log likelihood for observation j, $\ln L_j = l_j$, is

$$l_j = \begin{cases} w_j \ln\Phi\left\{ \dfrac{\mathbf{z}_j\boldsymbol{\gamma} + (y_j - \mathbf{x}_j\boldsymbol{\beta})\rho/\sigma}{\sqrt{1 - \rho^2}} \right\} - \dfrac{w_j}{2}\left(\dfrac{y_j - \mathbf{x}_j\boldsymbol{\beta}}{\sigma} \right)^2 - w_j \ln(\sqrt{2\pi}\sigma) & y_j \text{ observed} \\[2em] w_j \ln\Phi(-\mathbf{z}_j\boldsymbol{\gamma}) & y_j \text{ not observed} \end{cases}$$

where $\Phi()$ is the standard cumulative normal and w_j is an optional weight for observation j.

In the maximum likelihood estimation, σ and ρ are not directly estimated. Directly estimated are $\ln\sigma$ and atanh ρ:

$$\text{atanh } \rho = \frac{1}{2}\ln\left(\frac{1+\rho}{1-\rho}\right)$$

The standard error of $\lambda = \rho\sigma$ is approximated through the propagation of error (delta) method; that is,

$$\text{Var}(\lambda) \approx \mathbf{D}\,\text{Var}\left\{(\text{atanh } \rho \ \ln\sigma)\right\}\mathbf{D}'$$

where $\mathbf{D}$ is the Jacobian of λ with respect to atanh ρ and $\ln\sigma$.

The two-step estimates are computed using Heckman's (1979) procedure.

Probit estimates of the selection equation

$$\Pr(y_j \text{ observed} \mid \mathbf{z}_j) = \Phi(\mathbf{z}_j\boldsymbol{\gamma})$$

are obtained. From these estimates, the nonselection hazard—what Heckman (1979) referred to as the inverse of the Mills' ratio, m_j—for each observation j is computed as

$$m_j = \frac{\phi(\mathbf{z}_j\widehat{\boldsymbol{\gamma}})}{\Phi(\mathbf{z}_j\widehat{\boldsymbol{\gamma}})}$$

where ϕ is the normal density. We also define

$$\delta_j = m_j(m_j + \widehat{\boldsymbol{\gamma}}\mathbf{z}_j)$$

Following Heckman, the two-step parameter estimates of $\boldsymbol{\beta}$ are obtained by augmenting the regression equation with the nonselection hazard $\mathbf{m}$. Thus the regressors become $[\mathbf{X}\ \mathbf{m}]$, and we obtain the additional parameter estimate β_m on the variable containing the nonselection hazard.

A consistent estimate of the regression disturbance variance is obtained using the residuals from the augmented regression and the parameter estimate on the nonselection hazard,

$$\widehat{\sigma}^2 = \frac{\mathbf{e}'\mathbf{e} + \beta_m^2 \sum_{j=1}^{N} \delta_j}{N}$$

The two-step estimate of ρ is then

$$\widehat{\rho} = \frac{\beta_m}{\widehat{\sigma}}$$

Heckman derived consistent estimates of the coefficient covariance matrix on the basis of the augmented regression.

Let $\mathbf{W} = [\, \mathbf{X} \; \mathbf{m} \,]$ and $\mathbf{R}$ be a square diagonal matrix of dimension N, with $(1 - \widehat{\rho}^2 \delta_j)$ as the diagonal elements. The conventional VCE is

$$\mathbf{V}_{\text{twostep}} = \widehat{\sigma}^2 (\mathbf{W}'\mathbf{W})^{-1}(\mathbf{W}'\mathbf{R}\mathbf{W} + \mathbf{Q})(\mathbf{W}'\mathbf{W})^{-1}$$

where

$$\mathbf{Q} = \widehat{\rho}^2 (\mathbf{W}'\mathbf{D}\mathbf{Z})\mathbf{V_p}(\mathbf{Z}'\mathbf{D}\mathbf{W})$$

where $\mathbf{D}$ is the square diagonal matrix of dimension N with δ_j as the diagonal elements, $\mathbf{Z}$ is the data matrix of selection equation covariates, and $\mathbf{V}_p$ is the variance–covariance estimate from the probit estimation of the selection equation.

James Joseph Heckman was born in Chicago in 1944 and studied mathematics at Colorado College and economics at Princeton. He has taught economics at Columbia and (since 1973) at the University of Chicago. He has worked on developing a scientific basis for economic policy evaluation, with emphasis on models of individuals or disaggregated groups and the problems and possibilities created by heterogeneity, diversity, and unobserved counterfactual states. In 2000 he shared the Nobel Prize in Economics with Daniel L. McFadden.

References

Baum, C. F. 2006. *An Introduction to Modern Econometrics Using Stata*. College Station, TX: Stata Press.

Greene, W. H. 2003. *Econometric Analysis*. 5th ed. Upper Saddle River, NJ: Prentice Hall.

Gronau, R. 1974. Wage comparisons: A selectivity bias. *Journal of Political Economy* 82: 1119–1155.

Heckman, J. 1976. The common structure of statistical models of truncation, sample selection, and limited dependent variables and a simple estimator for such models. *Annals of Economic and Social Measurement* 5: 475–492.

———. 1979. Sample selection bias as a specification error. *Econometrica* 47: 153–161.

Johnston, J., and J. DiNardo. 1997. *Econometric Methods*. 4th ed. New York: McGraw–Hill.

Lewis, H. 1974. Comments on selectivity biases in wage comparisons. *Journal of Political Economy* 82: 1119–1155.

Manning, W. G., N. Duan, and W. H. Rogers. 1987. Monte Carlo evidence on the choice between sample selection and two-part models. *Journal of Econometrics* 35: 59–82.

Also See

[R] **heckman postestimation** — Postestimation tools for heckman

[R] **heckprob** — Probit model with sample selection

[R] **regress** — Linear regression

[R] **tobit** — Tobit regression

[SVY] **svy estimation** — Estimation commands for survey data

[U] **20 Estimation and postestimation commands**

Title

> **heckman postestimation** — Postestimation tools for heckman

Description

The following postestimation commands are available for `heckman`:

command	description
adjust[1]	adjusted predictions of $\mathbf{x}\beta$
estat[2]	AIC, BIC, VCE, and estimation sample summary
estat (svy)	postestimation statistics for survey data
estimates	cataloging estimation results
lincom	point estimates, standard errors, testing, and inference for linear combinations of coefficients
lrtest[3]	likelihood-ratio test; not available with two-step estimator
mfx	marginal effects or elasticities
nlcom	point estimates, standard errors, testing, and inference for nonlinear combinations of coefficients
predict	predictions, residuals, influence statistics, and other diagnostic measures
predictnl	point estimates, standard errors, testing, and inference for generalized predictions
suest[2]	seemingly unrelated estimation
test	Wald tests for simple and composite linear hypotheses
testnl	Wald tests of nonlinear hypotheses

[1] `adjust` is not appropriate with time-series operators.

[2] `estat ic` and `suest` are not appropriate after `heckman, twostep`.

[3] `lrtest` is not appropriate with `svy` estimation results.

See the corresponding entries in the *Stata Base Reference Manual* for details, but see
[SVY] **estat** for details about `estat` (svy).

Syntax for predict

After ML or twostep

> predict [*type*] *newvar* [*if*] [*in*] [, *statistic* <u>nooff</u>set]

After ML

> predict [*type*] { *stub** | *newvar*$_{\text{reg}}$ *newvar*$_{\text{sel}}$ *newvar*$_{\text{athrho}}$ *newvar*$_{\text{lnsigma}}$ }
>
> [*if*] [*in*] , <u>sc</u>ores

statistic	description
Main	
xb	linear prediction; the default
stdp	standard error of the prediction
stdf	standard error of the forecast
<u>xbs</u>el	linear prediction for selection equation
stdpsel	standard error of the linear prediction for selection equation
<u>pr</u>(*a*,*b*)	$\Pr(y_j \mid a < y_j < b)$
e(*a*,*b*)	$E(y_j \mid a < y_j < b)$
<u>ys</u>tar(*a*,*b*)	$E(y_j^*)$, $y_j^* = \max\{a, \min(y_j, b)\}$
<u>yc</u>ond	$E(y_j \mid y_j \text{ observed})$
<u>ye</u>xpected	$E(y_j^*)$, y_j taken to be 0 where unobserved
<u>ns</u>hazard or <u>mi</u>lls	nonselection hazard (also called the inverse of Mills' ratio)
psel	$\Pr(y_j \text{ observed})$

These statistics are available both in and out of sample; type predict ... if e(sample) ... if wanted only for the estimation sample.

stdf is not allowed with svy estimation results.

where *a* and *b* may be numbers or variables; *a* missing ($a \geq .$) means $-\infty$, and *b* missing ($b \geq .$) means $+\infty$; see [U] **12.2.1 Missing values**.

Options for predict

> Main

xb, the default, calculates the linear prediction $\mathbf{x}_j\mathbf{b}$.

stdp calculates the standard error of the prediction, which can be thought of as the standard error of the predicted expected value or mean for the observation's covariate pattern. The standard error of the prediction is also referred to as the standard error of the fitted value.

stdf calculates the standard error of the forecast, which is the standard error of the point prediction for 1 observation. It is commonly referred to as the standard error of the future or forecast value. By construction, the standard errors produced by stdf are always larger than those produced by stdp; see [R] **regress** *Methods and Formulas*.

xbsel calculates the linear prediction for the selection equation.

stdpsel calculates the standard error of the linear prediction for the selection equation.

pr(a,b) calculates $\Pr(a < \mathbf{x}_j\mathbf{b} + u_1 < b)$, the probability that $y_j|\mathbf{x}_j$ would be observed in the interval (a, b).

a and b may be specified as numbers or variable names; lb and ub are variable names; pr(20,30) calculates $\Pr(20 < \mathbf{x}_j\mathbf{b} + u_1 < 30)$; pr($lb$,$ub$) calculates $\Pr(lb < \mathbf{x}_j\mathbf{b} + u_1 < ub)$; and pr(20,$ub$) calculates $\Pr(20 < \mathbf{x}_j\mathbf{b} + u_1 < ub)$.

a missing ($a \geq .$) means $-\infty$; pr(.,30) calculates $\Pr(-\infty < \mathbf{x}_j\mathbf{b} + u_j < 30)$; pr($lb$,30) calculates $\Pr(-\infty < \mathbf{x}_j\mathbf{b} + u_j < 30)$ in observations for which $lb \geq .$ and calculates $\Pr(lb < \mathbf{x}_j\mathbf{b} + u_j < 30)$ elsewhere.

b missing ($b \geq .$) means $+\infty$; pr(20,.) calculates $\Pr(+\infty > \mathbf{x}_j\mathbf{b} + u_j > 20)$; pr(20,$ub$) calculates $\Pr(+\infty > \mathbf{x}_j\mathbf{b} + u_j > 20)$ in observations for which $ub \geq .$ and calculates $\Pr(20 < \mathbf{x}_j\mathbf{b} + u_j < ub)$ elsewhere.

e(a,b) calculates $E(\mathbf{x}_j\mathbf{b} + u_1 \mid a < \mathbf{x}_j\mathbf{b} + u_1 < b)$, the expected value of $y_j|\mathbf{x}_j$ conditional on $y_j|\mathbf{x}_j$ being in the interval (a, b), meaning that $y_j|\mathbf{x}_j$ is censored. a and b are specified as they are for pr().

ystar(a,b) calculates $E(y_j^*)$, where $y_j^* = a$ if $\mathbf{x}_j\mathbf{b} + u_j \leq a$, $y_j^* = b$ if $\mathbf{x}_j\mathbf{b} + u_j \geq b$, and $y_j^* = \mathbf{x}_j\mathbf{b} + u_j$ otherwise, meaning that y_j^* is truncated. a and b are specified as they are for pr().

ycond calculates the expected value of the dependent variable conditional on the dependent variable being observed, i.e., selected; $E(y_j \mid y_j \text{ observed})$.

yexpected calculates the expected value of the dependent variable (y_j^*), where that value is taken to be 0 when it is expected to be unobserved; $y_j^* = \Pr(y_j \text{ observed})E(y_j \mid y_j \text{ observed})$.

The assumption of 0 is valid for many cases where nonselection implies nonparticipation (e.g., unobserved wage levels, insurance claims from those who are uninsured) but may be inappropriate for some problems (e.g., unobserved disease incidence).

nshazard and mills are synonyms; both calculate the nonselection hazard—what Heckman (1979) referred to as the inverse of the Mills' ratio—from the selection equation.

psel calculates the probability of selection (or being observed):
$\Pr(y_j \text{ observed}) = \Pr(\mathbf{z}_j\boldsymbol{\gamma} + u_{2j} > 0)$.

nooffset is relevant when you specify offset(*varname*) for heckman. It modifies the calculations made by predict so that they ignore the offset variable; the linear prediction is treated as $\mathbf{x}_j\mathbf{b}$ rather than as $\mathbf{x}_j\mathbf{b} + \text{offset}_j$.

scores, not available with twostep, calculates equation-level score variables.

The first new variable will contain $\partial \ln L / \partial(\mathbf{x}_j\boldsymbol{\beta})$.

The second new variable will contain $\partial \ln L / \partial(\mathbf{z}_j\boldsymbol{\gamma})$.

The third new variable will contain $\partial \ln L / \partial(\text{atanh}\,\rho)$.

The fourth new variable will contain $\partial \ln L / \partial(\ln \sigma)$.

Remarks

▷ Example 1

The default statistic produced by `predict` after `heckman` is the expected value of the dependent variable from the underlying distribution of the regression model. In the wage model of [R] **heckman**, this is the expected wage rate among all women, regardless of whether they were observed to participate in the labor force.

```
. use http://www.stata-press.com/data/r10/womenwk
. heckman wage educ age, select(married children educ age) vce(cluster county)
  (output omitted)
. predict heckwage
(option xb assumed; fitted values)
```

It is instructive to compare these predicted wage values from the Heckman model with an ordinary regression model—a model without the selection adjustment:

```
. regress wage educ age
```

Source	SS	df	MS		Number of obs =	1343
					F(2, 1340) =	227.49
Model	13524.0337	2	6762.01687		Prob > F =	0.0000
Residual	39830.8609	1340	29.7245231		R-squared =	0.2535
					Adj R-squared =	0.2524
Total	53354.8946	1342	39.7577456		Root MSE =	5.452

wage	Coef.	Std. Err.	t	P>|t|	[95% Conf. Interval]	
education	.8965829	.0498061	18.00	0.000	.7988765	.9942893
age	.1465739	.0187135	7.83	0.000	.109863	.1832848
_cons	6.084875	.8896182	6.84	0.000	4.339679	7.830071

```
. predict regwage
(option xb assumed; fitted values)
. summarize heckwage regwage
```

Variable	Obs	Mean	Std. Dev.	Min	Max
heckwage	2000	21.15532	3.83965	14.6479	32.85949
regwage	2000	23.12291	3.241911	17.98218	32.66439

Since this dataset was concocted, we know the true coefficients of the wage regression equation to be 1, 0.2, and 1, respectively. We can compute the true mean wage for our sample.

```
. generate truewage = 1 + .2*age + 1*educ
. summarize truewage
```

Variable	Obs	Mean	Std. Dev.	Min	Max
truewage	2000	21.3256	3.797904	15	32.8

Whereas the mean of the predictions from `heckman` is within 18 cents of the true mean wage, ordinary regression yields predictions that are on average about $1.80 per hour too high because of the selection effect. The regression predictions also show somewhat less variation than the true wages.

The coefficients from `heckman` are so close to the true values that they are not worth testing. Conversely, the regression equation is significantly off but seems to give the right sense. Would we be led far astray if we relied on the OLS coefficients? The effect of age is off by more than 5 cents per year of age, and the coefficient on education level is off by about 10%. We can test the OLS coefficient on education level against the true value by using `test`.

```
. test educ = 1
( 1)   education = 1
       F(  1,  1340) =     4.31
             Prob > F =   0.0380
```

Not only is the OLS coefficient on education substantially lower than the true parameter, but the difference from the true parameter is also statistically significant beyond the 5% level. We can perform a similar test for the OLS age coefficient:

```
. test age = .2
( 1)   age = .2
       F(  1,  1340) =     8.15
             Prob > F =   0.0044
```

We find even stronger evidence that the OLS regression results are biased away from the true parameters.

◁

▷ Example 2

Several other interesting aspects of the Heckman model can be explored with `predict`. Continuing with our wage model, we can obtain the expected wages for women conditional on participating in the labor force with the `ycond` option. Let's get these predictions and compare them with actual wages for women participating in the labor force.

```
. use http://www.stata-press.com/data/r10/womenwk, clear
. heckman wage educ age, select(married children educ age)
  (output omitted)
. predict hcndwage, ycond
. summarize wage hcndwage if wage != .
```

Variable	Obs	Mean	Std. Dev.	Min	Max
wage	1343	23.69217	6.305374	5.88497	45.80979
hcndwage	1343	23.68239	3.335087	16.18337	33.7567

We see that the average predictions from `heckman` are close to the observed levels but do not have the same mean. These conditional wage predictions are available for all observations in the dataset but can be directly compared only with observed wages, where individuals are participating in the labor force.

What if we were interested in making predictions about mean wages for all women? Here the expected wage is 0 for those who are not expected to participate in the labor force, with expected participation determined by the selection equation. These values can be obtained with the `yexpected` option of `predict`. For comparison, a variable can be generated where the wage is set to 0 for nonparticipants.

```
. predict hexpwage, yexpected
. generate wage0 = wage
(657 missing values generated)
. replace wage0 = 0 if wage == .
(657 real changes made)
```

```
. summarize hexpwage wage0

    Variable |       Obs        Mean    Std. Dev.       Min        Max
-------------+--------------------------------------------------------
    hexpwage |      2000    15.92511    5.979336    2.492469   32.45858
       wage0 |      2000    15.90929    12.27081           0   45.80979
```

Again we note that the predictions from `heckman` are close to the observed mean hourly wage rate for all women. Why aren't the predictions using `ycond` and `yexpected` equal to their observed sample equivalents? For the Heckman model, unlike linear regression, the sample moments implied by the optimal solution to the model likelihood do not require that these predictions match observed data. Properly accounting for the additional variation from the selection equation requires that the model use more information than just the sample moments of the observed wages.

◁

Methods and Formulas

All postestimation commands listed above are implemented as ado-files.

Reference

Heckman, J. 1979. Sample selection bias as a specification error. *Econometrica* 47: 153–161.

Also See

[R] **heckman** — Heckman selection model

[U] **20 Estimation and postestimation commands**

Title

> **heckprob** — Probit model with sample selection

Syntax

>heckprob *depvar indepvars* [*if*] [*in*] [*weight*] ,
>
>   <u>sel</u>ect([*depvar_s* =] *varlist_s* [, <u>off</u>set(*varname*) <u>nocon</u>stant]) [*options*]

options	description
Model	
* <u>sel</u>ect()	specify selection equation: dependent and independent variables; whether to have constant term and offset variable
<u>nocon</u>stant	suppress constant term
<u>off</u>set(*varname*)	include *varname* in model with coefficient constrained to 1
<u>constraints</u>(*constraints*)	apply specified linear constraints
<u>coll</u>inear	keep collinear variables
SE/Robust	
vce(*vcetype*)	*vcetype* may be oim, <u>r</u>obust, <u>cl</u>uster *clustvar*, opg, <u>boot</u>strap, or <u>jack</u>knife
Reporting	
<u>level</u>(#)	set confidence level; default is level(95)
<u>first</u>	report first-step probit estimates
noskip	perform likelihood-ratio test
Max options	
maximize_options	control the maximization process; seldom used

* select( ) is required.

    The full specification is <u>sel</u>ect([*depvar_s* =] *varlist_s* [, <u>off</u>set(*varname*) <u>nocon</u>stant])

depvar, *indepvars*, *depvar_s*, and *varlist_s* may contain time-series operators; see [U] **11.4.3 Time-series varlists**.

bootstrap, by, jackknife, rolling, statsby, svy, and xi are allowed; see [U] **11.1.10 Prefix commands**.

Weights are not allowed with the bootstrap prefix.

vce(), first, noskip, and weights are not allowed with the svy prefix.

pweights, fweights, and iweights are allowed; see [U] **11.1.6 weight**.

See [U] **20 Estimation and postestimation commands** for more capabilities of estimation commands.

Description

heckprob fits maximum-likelihood probit models with sample selection.

Options

Model

select(...) specifies the variables and options for the selection equation. It is an integral part of specifying a selection model and is required. The selection equation should contain at least one variable that is not in the outcome equation.

If *depvar$_s$* is specified, it should be coded as 0 or 1, 0 indicating an observation not selected and 1 indicating a selected observation. If *depvar$_s$* is not specified, observations for which *depvar* is not missing are assumed selected, and those for which *depvar* is missing are assumed not selected.

noconstant, offset(*varname*), constraints(*constraints*), collinear; see [R] **estimation options**.

SE/Robust

vce(*vcetype*) specifies the type of standard error reported, which includes types that are derived from asymptotic theory, that are robust to some kinds of misspecification, that allow for intragroup correlation, and that use bootstrap or jackknife methods; see [R] **vce_option**.

Reporting

level(*#*); see [R] **estimation options**.

first specifies that the first-step probit estimates of the selection equation be displayed before estimation.

noskip specifies that a full maximum-likelihood model with only a constant for the regression equation be fitted. This model is not displayed but is used as the base model to compute a likelihood-ratio test for the model test statistic displayed in the estimation header. By default, the overall model test statistic is an asymptotically equivalent Wald test that all the parameters in the regression equation are zero (except the constant). For many models, this option can substantially increase estimation time.

Max options

maximize_options: difficult, technique(*algorithm_spec*), iterate(*#*), [no]log, trace, gradient, showstep, hessian, shownrtolerance, tolerance(*#*), ltolerance(*#*), gtolerance(*#*), nrtolerance(*#*), nonrtolerance, from(*init_specs*); see [R] **maximize**. These options are seldom used.

Setting the optimization type to technique(bhhh) resets the default *vcetype* to vce(opg).

Remarks

The probit model with sample selection (Van de Ven and Van Pragg 1981) assumes that there exists an underlying relationship

$$y_j^* = \mathbf{x}_j \boldsymbol{\beta} + u_{1j} \qquad \qquad \textit{latent equation}$$

such that we observe only the binary outcome

$$y_j^{\text{probit}} = (y_j^* > 0) \qquad \qquad \textit{probit equation}$$

The dependent variable, however, is not always observed. Rather, the dependent variable for observation j is observed if

$$y_j^{\text{select}} = (\mathbf{z}_j \boldsymbol{\gamma} + u_{2j} > 0) \qquad \qquad \textit{selection equation}$$

where

$$u_1 \sim N(0, 1)$$

$$u_2 \sim N(0, 1)$$

$$\mathrm{corr}(u_1, u_2) = \rho$$

When $\rho \neq 0$, standard probit techniques applied to the first equation yield biased results. heckprob provides consistent, asymptotically efficient estimates for all the parameters in such models.

For the model to be well identified, the selection equation should have at least one variable that is not in the probit equation. Otherwise, the model is identified only by functional form, and the coefficients have no structural interpretation.

▷ Example 1

We use the data from Pindyck and Rubinfeld (1998). In this dataset, the variables are whether children attend private school (private), number of years the family has been at the present residence (years), log of property tax (logptax), log of income (loginc), and whether one voted for an increase in property taxes (vote).

In this example, we alter the meaning of the data. Here we assume that we observe whether children attend private school only if the family votes for increasing the property taxes. This assumption is not true in the dataset, and we make it only to illustrate the use of this command.

We observe whether children attend private school only if the head of household voted for an increase in property taxes. We assume that the vote is affected by the number of years in residence, the current property taxes paid, and the household income. We wish to model whether children are sent to private school on the basis of the number of years spent in the current residence and the current property taxes paid.

```
. use http://www.stata-press.com/data/r10/school
. heckprob private years logptax, sel(vote=years loginc logptax)
Fitting probit model:
Iteration 0:   log likelihood = -17.122381
Iteration 1:   log likelihood = -16.407192
  (output omitted )
Iteration 5:   log likelihood = -15.883666
Iteration 6:   log likelihood = -15.883655
Fitting selection model:
Iteration 0:   log likelihood = -63.036914
Iteration 1:   log likelihood = -58.581911
Iteration 2:   log likelihood = -58.497419
Iteration 3:   log likelihood = -58.497288
Comparison:    log likelihood = -74.380943
Fitting starting values:
Iteration 0:   log likelihood = -40.895684
Iteration 1:   log likelihood = -17.920826
  (output omitted )
Iteration 7:   log likelihood = -15.753765
Fitting full model
Iteration 0:   log likelihood = -75.010619  (not concave)
Iteration 1:   log likelihood = -74.287753
Iteration 2:   log likelihood = -74.250148
Iteration 3:   log likelihood = -74.245088
Iteration 4:   log likelihood = -74.244973
Iteration 5:   log likelihood = -74.244973
```

```
Probit model with sample selection          Number of obs      =        95
                                            Censored obs       =        36
                                            Uncensored obs     =        59

                                            Wald chi2(2)       =       1.04
Log likelihood = -74.24497                  Prob > chi2        =     0.5935
```

	Coef.	Std. Err.	z	P>|z|	[95% Conf. Interval]	
private						
years	-.1142597	.1461717	-0.78	0.434	-.400751	.1722317
logptax	.3516098	1.016485	0.35	0.729	-1.640665	2.343884
_cons	-2.780665	6.905838	-0.40	0.687	-16.31586	10.75453
vote						
years	-.0167511	.0147735	-1.13	0.257	-.0457067	.0122045
loginc	.9923024	.4430009	2.24	0.025	.1240366	1.860568
logptax	-1.278783	.5717545	-2.24	0.025	-2.399401	-.1581647
_cons	-.545821	4.070418	-0.13	0.893	-8.523694	7.432052
/athrho	-.8663156	1.450028	-0.60	0.550	-3.708318	1.975686
rho	-.6994973	.7405342			-.9987984	.9622689

```
LR test of indep. eqns. (rho = 0):    chi2(1) =      0.27   Prob > chi2 = 0.6020
```

The output shows several iteration logs. The first iteration log corresponds to running the probit model for those observations in the sample where we have observed the outcome. The second iteration log corresponds to running the selection probit model, which models whether we observe our outcome of interest. If $\rho = 0$, the sum of the log likelihoods from these two models will equal the log likelihood of the probit model with sample selection; this sum is printed in the iteration log as the comparison log likelihood. The third iteration log shows starting values for the iterations.

The final iteration log is for fitting the full probit model with sample selection. A likelihood-ratio test of the log likelihood for this model and the comparison log likelihood is presented at the end of the output. If we had specified the vce(robust) option, this test would be presented as a Wald test instead of as a likelihood-ratio test.

◁

(Continued on next page)

▷ Example 2

In example 1, we could obtain robust standard errors by specifying the vce(robust) option. We also eliminate the iteration logs by using the nolog option.

```
. heckprob private years logptax, sel(vote=years loginc logptax) vce(robust) nolog
Probit model with sample selection          Number of obs    =         95
                                            Censored obs     =         36
                                            Uncensored obs   =         59

                                            Wald chi2(2)     =       2.55
Log pseudolikelihood = -74.24497            Prob > chi2      =     0.2798
```

	Coef.	Robust Std. Err.	z	P>|z|	[95% Conf.	Interval]
private						
years	-.1142597	.1113977	-1.03	0.305	-.3325951	.1040757
logptax	.3516098	.7358264	0.48	0.633	-1.090583	1.793803
_cons	-2.780665	4.786677	-0.58	0.561	-12.16238	6.60105
vote						
years	-.0167511	.0173344	-0.97	0.334	-.0507259	.0172237
loginc	.9923024	.4228044	2.35	0.019	.1636209	1.820984
logptax	-1.278783	.5095156	-2.51	0.012	-2.277415	-.2801508
_cons	-.545821	4.543892	-0.12	0.904	-9.451686	8.360044
/athrho	-.8663156	1.630643	-0.53	0.595	-4.062317	2.329686
rho	-.6994973	.832775			-.9994079	.9812329

```
Wald test of indep. eqns. (rho = 0): chi2(1) =     0.27    Prob > chi2 = 0.6020
```

Regardless of whether we specify the vce(robust) option, the outcome is not significantly different from the outcome obtained by fitting the probit and selection models separately. This result is not surprising since the selection mechanism estimated was invented for the example rather than borne from any economic theory.

◁

Saved Results

heckprob saves the following in e():

Scalars

e(N)	number of observations	e(N_clust)	number of clusters
e(k)	number of parameters	e(chi2)	χ^2
e(k_eq)	number of equations	e(chi2_c)	χ^2 for comparison test
e(k_eq_model)	number of equations in model	e(p_c)	p-value for comparison test
	Wald test	e(p)	significance of comparison test
e(k_aux)	number of auxiliary parameters	e(rho)	ρ
e(k_dv)	number of dependent variables	e(rank)	rank of e(V)
e(N_cens)	number of censored observations	e(rank0)	rank of e(V) for constant-only
e(df_m)	model degrees of freedom		model
e(ll)	log likelihood	e(ic)	number of iterations
e(ll_0)	log likelihood, constant-only model	e(rc)	return code
e(ll_c)	log likelihood, comparison model	e(converged)	1 if converged, 0 otherwise

Macros

e(cmd)	heckprob	e(vce)	vcetype specified in vce()
e(cmdline)	command as typed	e(vcetype)	title used to label Std. Err.
e(depvar)	names of dependent variables	e(diparm#)	display transformed parameter #
e(wtype)	weight type	e(opt)	type of optimization
e(wexp)	weight expression	e(ml_method)	type of ml method
e(title)	title in estimation output	e(user)	name of likelihood-evaluator program
e(clustvar)	name of cluster variable	e(technique)	maximization technique
e(offset1)	offset for regression equation	e(crittype)	optimization criterion
e(offset2)	offset for selection equation	e(properties)	b V
e(chi2type)	Wald or LR; type of model χ^2 test	e(predict)	program used to implement predict
e(chi2_ct)	type of comparison χ^2 test		

Matrices

e(b)	coefficient vector	e(V)	variance–covariance matrix
e(ilog)	iteration log (up to 20 iterations)		of the estimators
e(gradient)	gradient vector		

Functions

e(sample)	marks estimation sample

Methods and Formulas

heckprob is implemented as an ado-file. Van de Ven and Van Pragg (1981) provide an introduction and an explanation of this model.

The probit equation is

$$y_j = (\mathbf{x}_j\boldsymbol{\beta} + u_{1j} > 0)$$

The selection equation is

$$\mathbf{z}_j\boldsymbol{\gamma} + u_{2j} > 0$$

where

$$u_1 \sim N(0,1)$$
$$u_2 \sim N(0,1)$$
$$\mathrm{corr}(u_1, u_2) = \rho$$

The log likelihood is

$$\ln L = \sum_{\substack{j \in S \\ y_j \neq 0}} w_j \ln \left\{ \Phi_2 \left(x_j \beta + \text{offset}_j^\beta, z_j \gamma + \text{offset}_j^\gamma, \rho \right) \right\}$$

$$+ \sum_{\substack{j \in S \\ y_j = 0}} w_j \ln \left\{ \Phi_2 \left(-x_j \beta + \text{offset}_j^\beta, z_j \gamma + \text{offset}_j^\gamma, -\rho \right) \right\}$$

$$+ \sum_{j \notin S} w_j \ln \left\{ 1 - \Phi \left(z_j \gamma + \text{offset}_j^\gamma \right) \right\}$$

where S is the set of observations for which y_j is observed, $\Phi_2()$ is the cumulative bivariate normal distribution function (with mean $\begin{bmatrix} 0 & 0 \end{bmatrix}'$), $\Phi()$ is the standard cumulative normal, and w_j is an optional weight for observation j.

In the maximum likelihood estimation, ρ is not directly estimated. Directly estimated is atanh ρ:

$$\text{atanh } \rho = \frac{1}{2} \ln \left(\frac{1 + \rho}{1 - \rho} \right)$$

From the form of the likelihood, it is clear that if $\rho = 0$, the log likelihood for the probit model with sample selection is equal to the sum of the probit model for the outcome y and the selection model. We can perform a likelihood-ratio test by comparing the likelihood of the full model with the sum of the log likelihoods for the probit and selection models.

References

Baum, C. F. 2006. *An Introduction to Modern Econometrics Using Stata*. College Station, TX: Stata Press.

Greene, W. H. 2003. *Econometric Analysis*. 5th ed. Upper Saddle River, NJ: Prentice Hall.

Heckman, J. 1979. Sample selection bias as a specification error. *Econometrica* 47: 153–161.

Miranda, A., and S. Rabe-Hesketh. 2006. Maximum likelihood estimation of endogenous switching and sample selection models for binary, ordinal, and count variables. *Stata Journal* 6: 285–308.

Pindyck, R. S., and D. L. Rubinfeld. 1998. *Econometric Models and Economic Forecasts*. 4th ed. New York: McGraw–Hill.

Van de Ven, W. P. M. M., and B. M. S. Van Pragg. 1981. The demand for deductibles in private health insurance: A probit model with sample selection. *Journal of Econometrics* 17: 229–252.

Also See

[R] **heckprob postestimation** — Postestimation tools for heckprob

[R] **heckman** — Heckman selection model

[R] **probit** — Probit regression

[R] **treatreg** — Treatment-effects model

[U] **20 Estimation and postestimation commands**

Title

heckprob postestimation — Postestimation tools for heckprob

Description

The following postestimation commands are available for `heckprob`:

command	description
adjust[1]	adjusted predictions of $\mathbf{x}\beta$
estat	AIC, BIC, VCE, and estimation sample summary
estat (svy)	postestimation statistics for survey data
estimates	cataloging estimation results
lincom	point estimates, standard errors, testing, and inference for linear combinations of coefficients
lrtest[2]	likelihood-ratio test
mfx	marginal effects or elasticities
nlcom	point estimates, standard errors, testing, and inference for nonlinear combinations of coefficients
predict	predictions, residuals, influence statistics, and other diagnostic measures
predictnl	point estimates, standard errors, testing, and inference for generalized predictions
suest	seemingly unrelated estimation
test	Wald tests for simple and composite linear hypotheses
testnl	Wald tests of nonlinear hypotheses

[1] `adjust` is not appropriate with time-series operators.

[2] `lrtest` is not appropriate with `svy` estimation results.

See the corresponding entries in the *Stata Base Reference Manual* for details, but see
[SVY] **estat** for details about `estat` (svy).

(Continued on next page)

Syntax for predict

> predict [*type*] *newvar* [*if*] [*in*] [, *statistic* <u>nooff</u>set]

> predict [*type*] { *stub** | *newvar*$_{\text{reg}}$ *newvar*$_{\text{sel}}$ *newvar*$_{\text{athrho}}$ } [*if*] [*in*] , <u>sc</u>ores

statistic	description
Main	
<u>pmargin</u>	$\Phi(\mathbf{x}_j\mathbf{b})$, success probability; the default
p11	$\Phi_2(\mathbf{x}_j\mathbf{b}, \mathbf{z}_j\mathbf{g}, \rho)$, predicted probability $\Pr(y_j^{\text{probit}} = 1, y_j^{\text{select}} = 1)$
p10	$\Phi_2(\mathbf{x}_j\mathbf{b}, -\mathbf{z}_j\mathbf{g}, -\rho)$, predicted probability $\Pr(y_j^{\text{probit}} = 1, y_j^{\text{select}} = 0)$
p01	$\Phi_2(-\mathbf{x}_j\mathbf{b}, \mathbf{z}_j\mathbf{g}, -\rho)$, predicted probability $\Pr(y_j^{\text{probit}} = 0, y_j^{\text{select}} = 1)$
p00	$\Phi_2(-\mathbf{x}_j\mathbf{b}, -\mathbf{z}_j\mathbf{g}, \rho)$, predicted probability $\Pr(y_j^{\text{probit}} = 0, y_j^{\text{select}} = 0)$
<u>psel</u>	$\Phi(\mathbf{z}_j\mathbf{g})$, selection probability
<u>pcond</u>	$\Phi_2(\mathbf{x}_j\mathbf{b}, \mathbf{z}_j\mathbf{g}, \rho)/\Phi(\mathbf{z}_j\mathbf{g})$, probability of success conditional on selection
xb	linear prediction
stdp	standard error of the linear prediction
<u>xbsel</u>	linear prediction for selection equation
<u>stdpsel</u>	standard error of the linear prediction for selection equation

where $\Phi()$ is the standard normal distribution function and $\Phi_2()$ is the bivariate normal distribution function.

These statistics are available both in and out of sample; type predict ... if e(sample) ... if wanted only for the estimation sample.

Options for predict

   Main

pmargin, the default, calculates the univariate (marginal) predicted probability of success $\Pr(y_j^{\text{probit}} = 1)$.

p11 calculates the bivariate predicted probability $\Pr(y_j^{\text{probit}} = 1, y_j^{\text{select}} = 1)$.

p10 calculates the bivariate predicted probability $\Pr(y_j^{\text{probit}} = 1, y_j^{\text{select}} = 0)$.

p01 calculates the bivariate predicted probability $\Pr(y_j^{\text{probit}} = 0, y_j^{\text{select}} = 1)$.

p00 calculates the bivariate predicted probability $\Pr(y_j^{\text{probit}} = 0, y_j^{\text{select}} = 0)$.

psel calculates the univariate (marginal) predicted probability of selection $\Pr(y_j^{\text{select}} = 1)$.

pcond calculates the conditional (on selection) predicted probability of success $\Pr(y_j^{\text{probit}} = 1, y_j^{\text{select}} = 1)/\Pr(y_j^{\text{select}} = 1)$.

xb calculates the probit linear prediction $\mathbf{x}_j\mathbf{b}$.

stdp calculates the standard error of the prediction, which can be thought of as the standard error of the predicted expected value or mean for the observation's covariate pattern. The standard error of the prediction is also referred to as the standard error of the fitted value.

xbsel calculates the linear prediction for the selection equation.

stdpsel calculates the standard error of the linear prediction for the selection equation.

nooffset is relevant only if you specified offset(*varname*) for heckprob. It modifies the calculations made by predict so that they ignore the offset variable; the linear prediction is treated as $\mathbf{x}_j\mathbf{b}$ rather than as $\mathbf{x}_j\mathbf{b} + \text{offset}_j$.

scores calculates equation-level score variables.

The first new variable will contain $\partial \ln L / \partial(\mathbf{x}_j\boldsymbol{\beta})$.

The second new variable will contain $\partial \ln L / \partial(\mathbf{z}_j\boldsymbol{\gamma})$.

The third new variable will contain $\partial \ln L / \partial(\text{atanh}\,\rho)$.

Remarks

▷ Example 1

It is instructive to compare the marginal predicted probabilities with the predicted probabilities that we would obtain by ignoring the selection mechanism. To compare the two approaches, we will synthesize data so that we know the "true" predicted probabilities.

First, we need to generate correlated error terms, which we can do using a standard Cholesky decomposition approach. For our example, we will clear any data from memory and then generate errors that have a correlation of .5 by using the following commands. We set the seed so that interested readers can type in these same commands and obtain the same results.

```
. clear
. set seed 12309
. set obs 5000
obs was 0, now 5000
. gen c1 = invnormal(uniform())
. gen c2 = invnormal(uniform())
. matrix P = (1,.5\.5,1)
. matrix A = cholesky(P)
. local fac1 = A[2,1]
. local fac2 = A[2,2]
. gen u1 = c1
. gen u2 = `fac1'*c1 + `fac2'*c2
```

We can check that the errors have the correct correlation by using the correlate command. We will also normalize the errors so that they have a standard deviation of one, so we can generate a bivariate probit model with known coefficients. We do that with the following commands:

```
. correlate u1 u2
(obs=5000)

             |      u1       u2
-------------+------------------
          u1 |  1.0000
          u2 |  0.5056   1.0000

. summarize u1
  (output omitted )

. replace u1 = u1/r(sd)
(5000 real changes made)
```

```
. summarize u2
(output omitted)

. replace u2 = u2/r(sd)
(5000 real changes made)

. drop c1 c2

. gen x1 = uniform()-.5

. gen x2 = uniform()+1/3

. gen y1s = 0.5 + 4*x1 + u1

. gen y2s = 3 - 3*x2 + .5*x1 + u2

. gen y1 = (y1s>0)

. gen y2 = (y2s>0)
```

We have now created two dependent variables, y1 and y2, which are defined by our specified coefficients. We also included error terms for each equation, and the error terms are correlated. We run heckprob to verify that the data have been correctly generated according to the model

$$y_1 = .5 + 4x_1 + u_1$$
$$y_2 = 3 + .5x_1 - 3x_2 + u_2$$

where we assume that y_1 is observed only if $y_2 = 1$.

```
. heckprob y1 x1, sel(y2 = x1 x2) nolog
```

Probit model with sample selection		Number of obs	=	5000
		Censored obs	=	1790
		Uncensored obs	=	3210
		Wald chi2(1)	=	941.68
Log likelihood = -3600.854		Prob > chi2	=	0.0000

	Coef.	Std. Err.	z	P>|z|	[95% Conf. Interval]
y1						
x1	3.985923	.1298904	30.69	0.000	3.731342	4.240503
_cons	.4852946	.0464037	10.46	0.000	.3943449	.5762442
y2						
x1	.5998148	.0716655	8.37	0.000	.4593531	.7402765
x2	-3.004937	.0829469	-36.23	0.000	-3.16751	-2.842364
_cons	3.011587	.0782817	38.47	0.000	2.858157	3.165016
/athrho	.574063	.0860559	6.67	0.000	.4053964	.7427295
rho	.5183369	.062935			.3845569	.6307914

LR test of indep. eqns. (rho = 0): chi2(1) = 46.58 Prob > chi2 = 0.0000

Now that we have verified that we have generated data according to a known model, we can obtain and then compare predicted probabilities from the probit model with sample selection and a (usual) probit model.

```
. predict pmarg
(option pmargin assumed; Pr(y1=1))

. probit y1 x1 if y2==1

(output omitted)

. predict phat
(option pr assumed; Pr(y1))
```

Using the (marginal) predicted probabilities from the probit model with sample selection (pmarg) and the predicted probabilities from the (usual) probit model (phat), we can also generate the "true" predicted probabilities from the synthesized y1s variable and then compare the predicted probabilities:

```
. gen ptrue = normal(y1s)
. summarize pmarg ptrue phat
```

Variable	Obs	Mean	Std. Dev.	Min	Max
pmarg	5000	.619436	.3209494	.0658356	.9933942
ptrue	5000	.6090161	.3499095	1.02e-06	1
phat	5000	.6723897	.3063207	.096498	.9971064

Here we see that ignoring the selection mechanism (comparing the phat variable with the true ptrue variable) results in predicted probabilities that are much higher than the true values. Looking at the marginal predicted probabilities from the model with sample selection, however, results in more accurate predictions.

◁

Methods and Formulas

All postestimation commands listed above are implemented as ado-files.

Also See

[R] **heckprob** — Probit model with sample selection

[U] **20 Estimation and postestimation commands**

Title

help — Display online help

Syntax

Display help information in Viewer

<u>h</u>elp [*command_or_topic_name*] [, <u>non</u>ew name(*viewername*) <u>mar</u>ker(*markername*)]

Display help information in Results window

<u>ch</u>elp [*command_or_topic_name*]

Description

The help command displays help information about the specified command or topic.

Stata for Windows, Stata for Macintosh, and Stata for Unix(GUI):

help launches a new Viewer to display help for the specified command or topic. If help is not followed by a command or a topic name, Stata launches the Viewer and displays help contents, the table of contents for the online help.

Help may be accessed either by selecting **Help > Stata command** and filling in the desired command name or by typing help followed by a command or topic name.

chelp will display help in the Results window.

Stata for Unix(console):

Typing help followed by a command name or a topic name will display help on the console.

If help is not followed by a command or a topic name, a description of how to use the help system is displayed.

Stata for Unix(both GUI and console):

man is a synonym for chelp.

Options

nonew specifies that a new Viewer window not be opened for the help topic if a Viewer window is already open. The default is for a new Viewer window to be opened each time help is typed so that multiple help files may be viewed at once. nonew causes the help file to be displayed in the topmost open Viewer.

name(*viewername*) specifies that help be displayed in a Viewer window named *viewername*. If the named window already exists, its contents will be replaced. If the named window does not exist, it will be created.

marker(*markername*) specifies that the help file be opened to the position of *markername* within the help file.

Remarks

To obtain help for any Stata command, type `help` *command* or select **Help > Stata Command...** and fill in *command*.

`help` is best explained by examples.

To obtain help for	Type
`regress`	`help regress`
postestimation tools for `regress`	`help regress postestimation` or `help regress post`
graph option `xlabel()`	`help graph xlabel()`
Stata function `strpos()`	`help strpos()`
Mata function `optimize()`	`help mata optimize()`

Tips

- `help` displays a subject table of contents to the online help.

- `help guide` displays a table of contents for basic Stata concepts.

- `help estimation commands` displays an alphabetical listing of all Stata estimation commands.

- `help functions` displays help on Stata functions by category.

- `help mata functions` displays a subject table of contents to Mata's functions.

- `help ts glossary` displays the glossary for the time-series manual, and similarly for the other Stata specialty manuals.

See [U] **4 Stata's online help and search facilities** for a complete description of how to use `help`.

❑ Technical Note

When you type `help` *topic*, Stata first looks along the adopath for *topic*`.sthlp`; see [U] **17.5 Where does Stata look for ado-files?**.   ❑

Also See

[R] **hsearch** — Search help files

[R] **search** — Search Stata documentation

[R] **net search** — Search the Internet for installable packages

[GSM] **6 Getting help**

[GSW] **6 Getting help**

[GSU] **6 Getting help**

[U] **4 Stata's online help and search facilities**

Title

> **hetprob** — Heteroskedastic probit model

Syntax

hetprob *depvar* [*indepvars*] [*if*] [*in*] [*weight*] ,

    het(*varlist* [, <u>off</u>set(*varname*)]) [*options*]

options	description
Model	
*<u>het</u>(*varlist*[ ])*	independent variables to model the variance and possible offset variable
<u>nocons</u>tant	suppress constant term
<u>off</u>set(*varname*)	include *varname* in model with coefficient constrained to 1
asis	retain perfect predictor variables
<u>constraints</u>(*constraints*)	apply specified linear constraints
<u>coll</u>inear	keep collinear variables
SE/Robust	
vce(*vcetype*)	*vcetype* may be oim, <u>r</u>obust, <u>cl</u>uster *clustvar*, opg, <u>boot</u>strap, or <u>jack</u>knife
Reporting	
<u>level</u>(#)	set confidence level; default is level(95)
noskip	perform likelihood-ratio test
<u>nolr</u>test	perform Wald test on variance
Max options	
maximize_options	control the maximization process; seldom used

*het() is required. The full specification is het(*varlist* [, <u>off</u>set(*varname*)])

depvar, *indepvars*, and *varlist* may contain time-series operators; see [U] **11.4.3 Time-series varlists**.

bootstrap, by, jackknife, rolling, statsby, svy, and xi are allowed; see [U] **11.1.10 Prefix commands**.

Weights are not allowed with the bootstrap prefix.

vce(), noskip, and weights are not allowed with the svy prefix.

fweights, iweights, and pweights are allowed; see [U] **11.1.6 weight**.

See [U] **20 Estimation and postestimation commands** for more capabilities of estimation commands.

Description

hetprob fits a maximum-likelihood heteroskedastic probit model.

See [R] **logistic** for a list of related estimation commands.

Options

het(*varlist* [, offset(*varname*)]) specifies the independent variables and the offset variable, if there is one, in the variance function. het() is required.

noconstant, offset(*varname*); see [R] **estimation options**.

asis forces the retention of perfect predictor variables and their associated perfectly predicted observations and may produce instabilities in maximization; see [R] **probit**.

constraints(*constraints*), collinear; see [R] **estimation options**.

vce(*vcetype*) specifies the type of standard error reported, which includes types that are derived from asymptotic theory, that are robust to some kinds of misspecification, that allow for intragroup correlation, and that use bootstrap or jackknife methods; see [R] *vce_option*.

level(*#*); see [R] **estimation options**.

noskip requests fitting of the constant-only model and calculation of the corresponding likelihood-ratio χ^2 statistic for testing significance of the full model. By default, a Wald χ^2 statistic is computed for testing the significance of the full model.

nolrtest specifies that a Wald test of whether lnsigma2 $= 0$ be performed instead of the LR test.

maximize_options: difficult, technique(*algorithm_spec*), iterate(*#*), [no]log, trace, gradient, showstep, hessian, shownrtolerance, tolerance(*#*), ltolerance(*#*), gtolerance(*#*), nrtolerance(*#*), nonrtolerance, from(*init_specs*); see [R] **maximize**. These options are seldom used.

Setting the optimization type to technique(bhhh) resets the default *vcetype* to vce(opg).

Remarks

Remarks are presented under the following headings:

> *Introduction*
> *Robust standard errors*

Introduction

hetprob fits a maximum-likelihood heteroskedastic probit model, which is a generalization of the probit model. Let $y_j, j = 1, \ldots, N$ be a binary outcome variable taking on the value 0 (failure) or 1 (success). In the probit model, the probability that y_j takes on the value 1 is modeled as a nonlinear function of a linear combination of the k independent variables $\mathbf{x}_j = (x_{1j}, x_{2j}, \ldots, x_{kj})$,

$$\Pr(y_j = 1) = \Phi(\mathbf{x}_j\mathbf{b})$$

in which $\Phi()$ is the cumulative distribution function (CDF) of a standard normal random variable, that is, a normally distributed (Gaussian) random variable with mean 0 and variance 1. The linear combination of the independent variables, $\mathbf{x}_j\mathbf{b}$, is commonly called the *index function*, or *index*. Heteroskedastic probit generalizes the probit model by generalizing $\Phi()$ to a normal CDF with a variance that is no longer fixed at 1 but can vary as a function of the independent variables. `hetprob` models the variance as a multiplicative function of these m variables $\mathbf{z}_j = (z_{1j}, z_{2j}, \ldots, z_{mj})$, following Harvey (1976):

$$\sigma_j^2 = \{\exp(\mathbf{z}_j\boldsymbol{\gamma})\}^2$$

Thus the probability of success as a function of all the independent variables is

$$\Pr(y_j = 1) = \Phi\Big\{\mathbf{x}_j\mathbf{b}/\exp(\mathbf{z}_j\boldsymbol{\gamma})\Big\}$$

From this expression, it is clear that, unlike the index $\mathbf{x}_j\mathbf{b}$, no constant term can be present in $\mathbf{z}_j\boldsymbol{\gamma}$ if the model is to be identifiable.

Suppose that the binary outcomes y_j are generated by thresholding an unobserved random variable, w, which is normally distributed with mean $\mathbf{x}_j\mathbf{b}$ and variance 1 such that

$$y_j = \begin{cases} 1 & \text{if } w_j > 0 \\ 0 & \text{if } w_j \leq 0 \end{cases}$$

This process gives the probit model:

$$\Pr(y_j = 1) = \Pr(w_j > 0) = \Phi(\mathbf{x}_j\mathbf{b})$$

Now suppose that the unobserved w_j are heteroskedastic with variance

$$\sigma_j^2 = \{\exp(\mathbf{z}_j\boldsymbol{\gamma})\}^2$$

Relaxing the homoskedastic assumption of the probit model in this manner yields our multiplicative heteroskedastic probit model:

$$\Pr(y_j = 1) = \Phi\Big\{\mathbf{x}_j\mathbf{b}/\exp(\mathbf{z}_j\boldsymbol{\gamma})\Big\}$$

▷ Example 1

For this example, we generate simulated data for a simple heteroskedastic probit model and then estimate the coefficients with `hetprob`:

```
. set obs 1000
obs was 0, now 1000
. set seed 1234567
. gen x = 1-2*uniform()
. gen xhet = uniform()
. gen sigma = exp(1.5*xhet)
. gen p = normal((0.3+2*x)/sigma)
. gen y = cond(uniform()<=p,1,0)
```

```
. hetprob y x, het(xhet)

Fitting probit model:

Iteration 0:   log likelihood = -688.53208
Iteration 1:   log likelihood = -592.23614
Iteration 2:   log likelihood = -591.50687
Iteration 3:   log likelihood = -591.50674

Fitting full model:

Iteration 0:   log likelihood = -591.50674
Iteration 1:   log likelihood = -572.12219
Iteration 2:   log likelihood =  -570.7742
Iteration 3:   log likelihood = -569.48921
Iteration 4:   log likelihood = -569.47828
Iteration 5:   log likelihood = -569.47827
```

Heteroskedastic probit model

	Number of obs	=	1000
	Zero outcomes	=	452
	Nonzero outcomes	=	548
	Wald chi2(1)	=	78.66
Log likelihood = -569.4783	Prob > chi2	=	0.0000

	Coef.	Std. Err.	z	P>|z|	[95% Conf. Interval]	
y						
x	2.228031	.2512073	8.87	0.000	1.735673	2.720388
_cons	.2493822	.0862833	2.89	0.004	.08027	.4184943
lnsigma2						
xhet	1.602537	.2640131	6.07	0.000	1.085081	2.119993

Likelihood-ratio test of lnsigma2=0: chi2(1) =   44.06   Prob > chi2 = 0.0000

Above we created two variables, x and xhet, and then simulated the model

$$\Pr(y = 1) = F\left\{ (\beta_0 + \beta_1 x) / \exp(\gamma_1 xhet) \right\}$$

for $\beta_0 = 0.3$, $\beta_1 = 2$, and $\gamma_1 = 1.5$. According to hetprob's output, all coefficients are significant, and, as we would expect, the Wald test of the full model versus the constant-only model—e.g., the index consisting of $\beta_0 + \beta_1 x$ versus that of just β_0—is significant with $\chi^2(1) = 79$. Likewise, the likelihood-ratio test of heteroskedasticity, which tests the full model with heteroskedasticity against the full model without, is significant with $\chi^2(1) = 44$. See [R] **maximize** for more explanation of the output. For this simple model, hetprob took five iterations to converge. As stated elsewhere (Greene 2003, 681), this is a difficult model to fit, and it is not uncommon for it to require many iterations or for the optimizer to print out warnings and informative messages during the optimization. Slow convergence is especially common for models in which one or more of the independent variables appear in both the index and variance functions.

◁

❑ Technical Note

Stata interprets a value of 0 as a negative outcome (failure) and treats all other values (except missing) as positive outcomes (successes). Thus if your dependent variable takes on the values 0 and 1, 0 is interpreted as failure and 1 as success. If your dependent variable takes on the values 0, 1, and 2, 0 is still interpreted as failure, but both 1 and 2 are treated as successes.

❑

Robust standard errors

If you specify the vce(robust) option, hetprob reports robust standard errors as described in [U] **20.15 Obtaining robust variance estimates**. To illustrate the effect of this option, we will reestimate our coefficients by using the same model and data in our example, this time adding vce(robust) to our hetprob command:

▷ Example 2

```
. hetprob y x, het(xhet) vce(robust) nolog
Heteroskedastic probit model                    Number of obs   =       1000
                                                Zero outcomes   =        452
                                                Nonzero outcomes =       548
                                                Wald chi2(1)    =      65.23
Log pseudolikelihood = -569.4783                Prob > chi2     =     0.0000
```

	Coef.	Robust Std. Err.	z	P>|z|	[95% Conf. Interval]
y					
x	2.228031	.2758598	8.08	0.000	1.687356 2.768706
_cons	.2493822	.0843367	2.96	0.003	.0840853 .414679
lnsigma2					
xhet	1.602537	.2671326	6.00	0.000	1.078967 2.126107

Likelihood-ratio test of lnsigma2=0: chi2(1) =   44.06   Prob > chi2 = 0.0000

The vce(robust) standard errors for two of the three parameters are larger than the previously reported conventional standard errors. This is to be expected, even though (by construction) we have perfect model specification since this option trades off efficient estimation of the coefficient variance–covariance matrix for robustness against misspecification.

◁

Specifying the vce(cluster *clustvar*) option relaxes the usual assumption of independence between observations to the weaker assumption of independence just between clusters; that is, hetprob, vce(cluster *clustvar*) is robust with respect to within-cluster correlation. This option is less efficient than the xtgee population-averaged models since hetprob inefficiently sums within cluster for the standard error calculation rather than attempting to exploit what might be assumed about the within-cluster correlation.

Saved Results

hetprob saves the following in e():

Scalars

e(N)	number of observations	e(N_clust)	number of clusters
e(k)	number of parameters	e(chi2)	χ^2
e(k_eq)	number of equations	e(chi2_c)	χ^2 for heteroskedasticity LR test
e(k_eq_model)	number of equations in model Wald test	e(p_c)	p-value for heteroskedasticity LR test
e(k_dv)	number of dependent variables	e(df_m_c)	degrees of freedom for heteroskedasticity LR test
e(N_f)	number of zero outcomes	e(p)	significance
e(N_s)	number of nonzero outcomes	e(rank)	rank of e(V)
e(df_m)	model degrees of freedom	e(rank0)	rank of e(V) for constant-only model
e(ll)	log likelihood	e(ic)	number of iterations
e(ll_0)	log likelihood, constant-only model	e(rc)	return code
e(ll_c)	log likelihood, comparison model	e(converged)	1 if converged, 0 otherwise

Macros

e(cmd)	hetprob	e(vce)	*vcetype* specified in vce()
e(cmdline)	command as typed	e(vcetype)	title used to label Std. Err.
e(depvar)	name of dependent variable	e(opt)	type of optimization
e(wtype)	weight type	e(method)	requested estimation method
e(wexp)	weight expression	e(ml_method)	type of ml method
e(title)	title in estimation output	e(user)	name of likelihood-evaluator
e(clustvar)	name of cluster variable	e(technique)	maximization technique
e(offset1)	offset for probit equation	e(crittype)	optimization criterion
e(offset2)	offset for variance equation	e(properties)	b V
e(chi2type)	Wald or LR; type of model χ^2 test	e(predict)	program used to implement predict
e(chi2_ct)	Wald or LR; type of model χ^2 test corresponding to e(chi2_c)		

Matrices

e(b)	coefficient vector	e(V)	variance–covariance matrix of the estimators
e(ilog)	iteration log (up to 20 iterations)		
e(gradient)	gradient vector		

Functions

e(sample)	marks estimation sample

Methods and Formulas

hetprob is implemented as an ado-file.

The heteroskedastic probit model is a generalization of the probit model since it allows the scale of the inverse link function to vary from observation to observation as a function of the independent variables.

The log-likelihood function for the heteroskedastic probit model is

$$\ln L = \sum_{j \in S} w_j \ln \Phi\{\mathbf{x}_j \boldsymbol{\beta} / \exp(\mathbf{z}\boldsymbol{\gamma})\} + \sum_{j \notin S} w_j \ln\left[1 - \Phi\{\mathbf{x}_j \boldsymbol{\beta} / \exp(\mathbf{z}\boldsymbol{\gamma})\}\right]$$

where S is the set of all observations j such that $y_j \neq 0$ and w_j denotes the optional weights. $\ln L$ is maximized as described in [R] **maximize**.

References

Greene, W. H. 2003. *Econometric Analysis*. 5th ed. Upper Saddle River, NJ: Prentice Hall.

Harvey, A. C. 1976. Estimating regression models with multiplicative heteroscedasticity. *Econometrica* 44: 461–465.

Also See

[R] **hetprob postestimation** — Postestimation tools for hetprob

[R] **logistic** — Logistic regression, reporting odds ratios

[R] **probit** — Probit regression

[SVY] **svy estimation** — Estimation commands for survey data

[XT] **xtprobit** — Random-effects and population-averaged probit models

[U] **20 Estimation and postestimation commands**

Title

> **hetprob postestimation** — Postestimation tools for hetprob

Description

The following postestimation commands are available for `hetprob`:

command	description
adjust[1]	adjusted predictions of $\mathbf{x}\beta$
estat	AIC, BIC, VCE, and estimation sample summary
estat (svy)	postestimation statistics for survey data
estimates	cataloging estimation results
lincom	point estimates, standard errors, testing, and inference for linear combinations of coefficients
linktest	link test for model specification
lrtest[2]	likelihood-ratio test
mfx	marginal effects or elasticities
nlcom	point estimates, standard errors, testing, and inference for nonlinear combinations of coefficients
predict	predictions, residuals, influence statistics, and other diagnostic measures
predictnl	point estimates, standard errors, testing, and inference for generalized predictions
suest	seemingly unrelated estimation
test	Wald tests for simple and composite linear hypotheses
testnl	Wald tests of nonlinear hypotheses

[1] `adjust` is not appropriate with time-series operators.

[2] `lrtest` is not appropriate with svy estimation results.

See the corresponding entries in the *Stata Base Reference Manual* for details, but see [SVY] **estat** for details about `estat` (svy).

Syntax for predict

> predict [*type*] *newvar* [*if*] [*in*] [, *statistic* <u>nooff</u>set]

> predict [*type*] { *stub** | *newvar*_{reg} *newvar*_{lnsigma2} } [*if*] [*in*] , <u>sc</u>ores

statistic	description
Main	
<u>pr</u>	probability of a positive outcome; the default
xb	linear prediction
<u>sigma</u>	standard deviation of the error term

These statistics are available both in and out of sample; type `predict ... if e(sample) ...` if wanted only for the estimation sample.

Options for predict

⌐ Main ⌐

pr, the default, calculates the probability of a positive outcome.

xb calculates the linear prediction.

sigma calculates the standard deviation of the error term.

nooffset is relevant only if you specified offset(*varname*) for hetprob. It modifies the calculations made by predict so that they ignore the offset variable; the linear prediction is treated as $\mathbf{x}_j\mathbf{b}$ rather than as $\mathbf{x}_j\mathbf{b} + \text{offset}_j$.

scores calculates equation-level score variables.

The first new variable will contain $\partial \ln L / \partial(\mathbf{x}_j\boldsymbol{\beta})$.

The second new variable will contain $\partial \ln L / \partial(\mathbf{z}_j\boldsymbol{\gamma})$.

Remarks

Once you have fitted a model, you can use the predict command to obtain the predicted probabilities for both the estimation sample and other samples; see [U] **20 Estimation and postestimation commands** and [R] **predict**. predict without arguments calculates the predicted probability of a positive outcome. With the xb option, predict calculates the index function combination, $\mathbf{x}_j\mathbf{b}$, where $\mathbf{x}_j$ are the independent variables in the jth observation and $\mathbf{b}$ is the estimated parameter vector. With the sigma option, predict calculates the predicted standard deviation, $\sigma_j = \exp(\mathbf{z}_j\boldsymbol{\gamma})$.

▷ Example 1

We use predict to compute the predicted probabilities and standard deviations based on the model in example 2 in [R] **hetprob** to compare these with the actual values:

```
. predict phat
(option pr assumed; Pr(y))
. gen diff_p = phat - p
. summarize diff_p
```

Variable	Obs	Mean	Std. Dev.	Min	Max
diff_p	1000	-.0107081	.0131869	-.0466331	.010482

```
. predict sigmahat, sigma
. gen diff_s = sigmahat - sigma
. summarize diff_s
```

Variable	Obs	Mean	Std. Dev.	Min	Max
diff_s	1000	.1558881	.1363698	.0000417	.4819107

◁

Methods and Formulas

All postestimation commands listed above are implemented as ado-files.

Also See

[R] **hetprob** — Heteroskedastic probit model

[U] **20 Estimation and postestimation commands**

Title

histogram — Histograms for continuous and categorical variables

Syntax

histogram *varname* [*if*] [*in*] [*weight*] [, [*continuous_opts* | *discrete_opts*] *options*]

continuous_opts	description
Main	
bin(#)	set number of bins to #
width(#)	set width of bins to #
start(#)	set lower limit of first bin to #

discrete_opts	description
Main	
discrete	specify that data are discrete
width(#)	set width of bins to #
start(#)	set theoretical minimum value to #

options	description
Main	
density	draw as density; the default
fraction	draw as fractions
frequency	draw as frequencies
percent	draw as percentages
bar_options	rendition of bars
addlabels	add height labels to bars
addlabopts(*marker_label_options*)	affect rendition of labels
Density plots	
normal	add a normal density to the graph
normopts(*line_options*)	affect rendition of normal density
kdensity	add a kernel density estimate to the graph
kdenopts(*kdensity_options*)	affect rendition of kernel density
Add plots	
addplot(*plot*)	add other plots to the histogram
Y axis, X axis, Titles, Legend, Overall, By	
twoway_options	any options documented in [G] *twoway_options*

See [G] **graph twoway line**, [G] **graph twoway kdensity**, [G] *marker_label_options*,

  [G] **graph twoway bar**, [G] *twoway_options*, and [G] *addplot_option*.

fweights are allowed; see [U] **11.1.6 weight**.

Description

histogram draws histograms of *varname*, which is assumed to be the name of a continuous variable unless the discrete option is specified.

Options for use in the continuous case

Main

bin(#) and width(#) are alternatives. They specify how the data are to be aggregated into bins; bin() by specifying the number of bins (from which the width can be derived) and width() by specifying the bin width (from which the number of bins can be derived).

If neither option is specified, results are the same as if bin(k) were specified, where

$$k = \min\left\{ \mathrm{sqrt}(N), 10\ln(N)/\ln(10) \right\}$$

and where N is the (weighted) number of observations.

start(#) specifies the theoretical minimum of *varname*. The default is start(m), where m is the observed minimum value of *varname*.

Specify start() when you are concerned about sparse data, for instance, if you know that *varname* can have a value of 0, but you are concerned that 0 may not be observed.

start(#), if specified, must be less than or equal to *m*, or else an error will be issued.

Options for use in the discrete case

┌─ Main ┐

discrete specifies that *varname* is discrete and that you want each unique value of *varname* to have its own bin (bar of histogram).

width(#) is rarely specified in the discrete case; it specifies the width of the bins. The default is width(*d*), where *d* is the observed minimum difference between the unique values of *varname*.

Specify width() if you are concerned that your data are sparse. For example, in theory *varname* could take on the values, say, 1, 2, 3, ..., 9, but because of the sparseness, perhaps only the values 2, 4, 7, and 8 are observed. Here the default width calculation would produce width(2), and you would want to specify width(1).

start(#) is also rarely specified in the discrete case; it specifies the theoretical minimum value of *varname*. The default is start(*m*), where *m* is the observed minimum value.

As with width(), specify start(#) if you are concerned that your data are sparse. In the previous example, you might also want to specify start(1). start() does nothing more than add white space to the left side of the graph.

The value of # in start() must be less than or equal to *m*, or an error will be issued.

Options for use in both the continuous and discrete cases

┌─ Main ┐

density, fraction, frequency, and percent specify whether you want the histogram scaled to density units, fractional units, frequencies, or percentages. density is the default.

density scales the height of the bars so that the sum of their areas equals 1.

fraction scales the height of the bars so that the sum of their heights equals 1.

frequency scales the height of the bars so that each bar's height is equal to the number of observations in the category. Thus the sum of the heights is equal to the total number of observations.

percent scales the height of the bars so that the sum of their heights equals 100.

bar_options are any of the labels allowed by graph twoway bar; see [G] **graph twoway bar**.

One of the most useful *bar_options* is barwidth(#), which specifies the width of the bars in *varname* units. By default, histogram draws the bars so that adjacent bars just touch. If you want gaps between the bars, do not specify histogram's width() option—which would change how the histogram is calculated—but specify the *bar_option* barwidth() or histogram option gap, both of which affect only how the bar is rendered.

The *bar_option* horizontal cannot be used with the addlabels option.

addlabels specifies that the top of each bar be labeled with the density, fraction, or frequency, as determined by the density, fraction, and frequency options.

addlabopts(*marker_label_options*) specifies how to render the labels atop the bars. See [G] **_marker_label_options_**. Do not specify the *marker_label_option* mlabel(*varname*), which specifies the variable to be used; this is specified for you by histogram.

addlabopts() will accept more options than those documented in [G] **_marker_label_options_**. All options allowed by twoway scatter are also allowed by addlabopts(). One particularly useful option is yvarformat(); see [G] **_advanced_options_**.

Density plots

normal specifies that the histogram be overlaid with an appropriately scaled normal density. The normal will have the same mean and standard deviation as the data.

normopts(*line_options*) specifies details about the rendition of the normal curve, such as the color and style of line used. See [G] **graph twoway line**.

kdensity specifies that the histogram be overlaid with an appropriately scaled kernel density estimate of the density. By default, the estimate will be produced using the Epanechnikov kernel with an "optimal" half-width. This default corresponds to the default of kdensity; see [R] **kdensity**. How the estimate is produced can be controlled using the kdenopts() option described below.

kdenopts(*kdensity_options*) specifies details about how the kernel density estimate is to be produced along with details about the rendition of the resulting curve, such as the color and style of line used. The kernel density estimate is described in [G] **graph twoway kdensity**. As an example, if you wanted to produce kernel density estimates by using the Gaussian kernel with optimal half-width, you would specify kdenopts(gauss), and if you also wanted a half-width of 5, you would specify kdenopts(gauss width(5)).

Add plots

addplot(*plot*) allows adding more graph twoway plots to the graph; see help [G] **_addplot_option_**.

Y axis, X axis, Titles, Legend, Overall, By

twoway_options are any of the options documented in [G] **_twoway_options_**. This includes, most importantly, options for titling the graph (see [G] **_title_options_**), options for saving the graph to disk (see [G] **_saving_option_**), and the by() option, which will allow you to simultaneously graph histograms for different subsets of the data (see [G] **_by_option_**).

Remarks

Remarks are presented under the following headings:

> *Histograms of continuous variables*
> *Overlaying normal and kernel density estimates*
> *Histograms of discrete variables*
> *Use with by()*

Histograms of continuous variables

histogram assumes that the variable is continuous, so you need only type histogram followed by the variable name:

```
. use http://www.stata-press.com/data/r10/sp500
(S&P 500)
```

```
. histogram volume
(bin=15, start=4103, width=1280.3533)
```

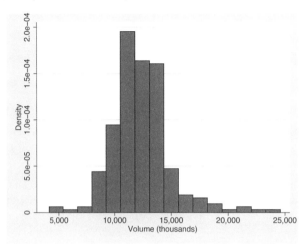

The small values reported for density on the y axis are correct; if you added up the area of the bars, you would get 1. Nevertheless, many people are used to seeing histograms scaled so that the bar heights sum to 1,

```
. histogram volume, fraction
(bin=15, start=4103, width=1280.3533)
```

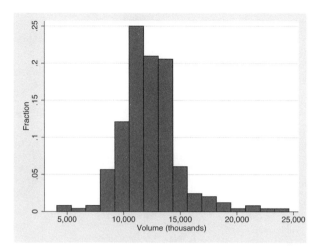

and others are used to seeing histograms so that the bar height reflects the number of observations,

```
. histogram volume, frequency
(bin=15, start=4103, width=1280.3533)
```

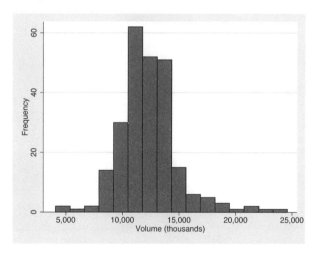

Regardless of the scale you prefer, you can specify other options to make the graph look more impressive:

(*Continued on next page*)

```
. summarize volume
```

Variable	Obs	Mean	Std. Dev.	Min	Max
volume	248	12320.68	2585.929	4103	23308.3

```
. histogram volume, freq
>         xaxis(1 2)
>         ylabel(0(10)60, grid)
>         xlabel(12321 "mean"
>             9735 "-1 s.d."
>            14907 "+1 s.d."
>             7149 "-2 s.d."
>            17493 "+2 s.d."
>            20078 "+3 s.d."
>            22664 "+4 s.d."
>                              , axis(2) grid gmax)
>         xtitle("", axis(2))
>         subtitle("S&P 500, January 2001 - December 2001")
>         note("Source:  Yahoo! Finance and Commodity Systems, Inc.")
(bin=15, start=4103, width=1280.3533)
```

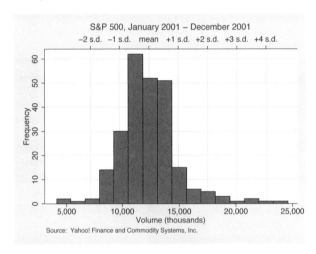

For an explanation of the xaxis() option—it created the upper and lower x axis—see [G] *axis_choice_options*. For an explanation of the ylabel() and xlabel() options, see [G] *axis_label_options*. For an explanation of the subtitle() and note() options, see [G] *title_options*.

Overlaying normal and kernel density estimates

Specifying normal will overlay a normal density over the histogram. It would be enough to type

```
. histogram volume, normal
```

but we will add the option to our more impressive rendition:

```
. summarize volume
```

Variable	Obs	Mean	Std. Dev.	Min	Max
volume	248	12320.68	2585.929	4103	23308.3

```
. histogram volume, freq normal
>        xaxis(1 2)
>        ylabel(0(10)60, grid)
>        xlabel(12321 "mean"
>            9735 "-1 s.d."
>           14907 "+1 s.d."
>            7149 "-2 s.d."
>           17493 "+2 s.d."
>           20078 "+3 s.d."
>           22664 "+4 s.d."
>                              , axis(2) grid gmax)
>        xtitle("", axis(2))
>        subtitle("S&P 500, January 2001 - December 2001")
>        note("Source:  Yahoo! Finance and Commodity Systems, Inc.")
(bin=15, start=4103, width=1280.3533)
```

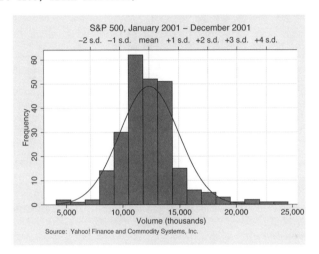

If we instead wanted to overlay a kernel density estimate, we could specify `kdensity` in place of `normal`.

Histograms of discrete variables

Specify `histogram`'s discrete option when you wish to treat the data as discrete—when you wish each unique value of the variable to be assigned its own bin. For instance, in the automobile data, `mpg` is a continuous variable, but the mileage ratings have been measured to integer precision. If we were to type

```
. use http://www.stata-press.com/data/r10/auto
(1978 Automobile Data)

. histogram mpg
(bin=8, start=12, width=3.625)
```

`mpg` would be treated as continuous and categorized into eight bins by the default number-of-bins calculation, which is based on the number of observations, 74.

Adding the discrete option makes a histogram with a bin for each of the 21 unique values.

```
. histogram mpg, discrete
(start=12, width=1)
```

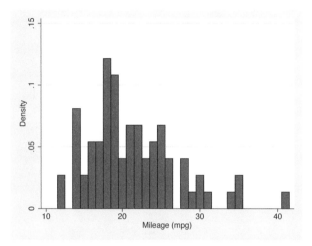

Just as in the continuous case, the y axis was reported in density, and we could specify the fraction or frequency options if we wanted it to be reported differently. Below we specify frequency, we specify addlabels to add a report of frequencies printed above the bars, we specify ylabel(,grid) to add horizontal grid lines, and we specify xlabel(12(2)42) to label the values 12, 14, ..., 42 on the x axis:

```
. histogram mpg, discrete freq addlabels ylabel(,grid) xlabel(12(2)42)
(start=12, width=1)
```

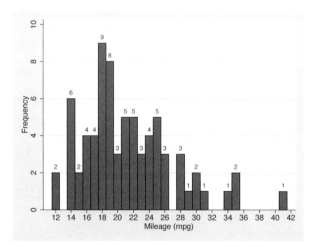

Use with by()

histogram may be used with graph twoway's by(); for example,

```
. use http://www.stata-press.com/data/r10/auto
(1978 Automobile Data)
. histogram mpg, discrete by(foreign)
```

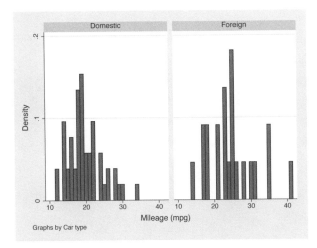

Here results would be easier to compare if the graphs were presented in one column:

```
. histogram mpg, discrete by(foreign, col(1))
```

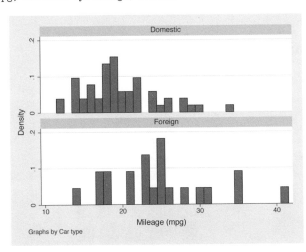

col(1) is a by() suboption—see [G] *by_option*—and there are other useful suboptions, such as total, which will add an overall total histogram. total is a suboption of by(), not an option of histogram, so you type

```
. histogram mpg, discrete by(foreign, total)
```

and not histogram mpg, discrete by(foreign) total.

As another example, Lipset (1993) reprinted data from the *New York Times* (November 5, 1992) collected by the Voter Research and Surveys based on questionnaires completed by 15,490 US presidential voters from 300 polling places on election day 1992.

```
. use http://www.stata-press.com/data/r10/voter, clear
. histogram candi [freq=pop], discrete fraction by(inc, total)
> gap(40) xlabel(2 3 4, valuelabel)
```

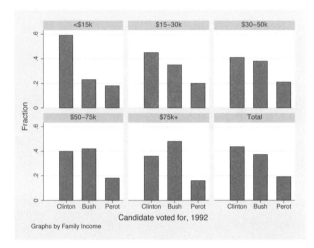

We specified gap(40) to reduce the width of the bars by 40%. We also used xlabel()'s valuelabel suboption, which caused our bars to be labeled "Clinton", "Bush", and "Perot", rather than 2, 3, and 4; see [G] *axis_label_options*.

Methods and Formulas

histogram is implemented as an ado-file.

References

Cox, N. J. 2004. Speaking Stata: Graphing distributions. *Stata Journal* 4: 66–88.

——. 2005. Speaking Stata: Density probability plots. *Stata Journal* 5: 259–273.

Harrison, D. A. 2005. Stata tip 20: Generating histogram bin variables. *Stata Journal* 5: 280–281.

Lipset, S. M. 1993. The significance of the 1992 election. *Political Science and Politics* 26: 7–16.

Also See

[R] **kdensity** — Univariate kernel density estimation

[R] **spikeplot** — Spike plots and rootograms

[G] **graph twoway histogram** — Histogram plots

Title

hsearch — Search help files

Syntax

hsearch *word(s)*

hsearch *word(s)*, build

hsearch, build

Description

hsearch *word(s)* searches the help files for *word(s)* and presents a clickable list in the Viewer.

hsearch *word(s)*, build does the same thing but builds a new index first.

hsearch, build rebuilds the index but performs no search.

Option

build forces the index that hsearch uses to be built or rebuilt.

The index is automatically built the first time you use hsearch, and it is automatically rebuilt if you have recently installed an ado-file update by using update; see [R] **update**. Thus option build is rarely specified.

You should specify build if you have recently installed user-written ado-files by using net install (see [R] **net**) or ssc (see [R] **ssc**) or if you have recently updated any of your own help files.

Remarks

Remarks are presented under the following headings:

> *Using hsearch*
> *Alternatives to hsearch*
> *Recommendations*
> *How hsearch works*

Using hsearch

You use hsearch to find help for commands and features installed on your computer. If you wanted to find commands related to Mills' ratio, you would type

```
. hsearch Mills' ratio
```

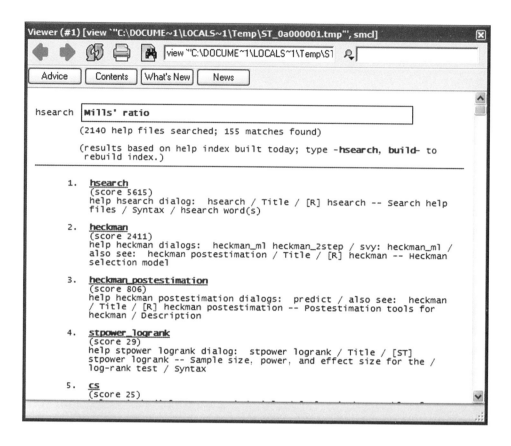

You could just as well type

    . hsearch Mill's ratio

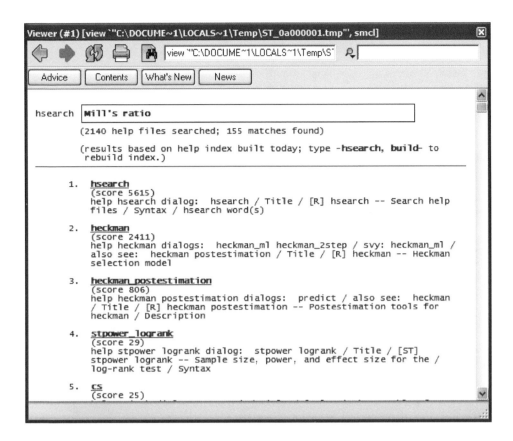

or type any of the following

. hsearch Mills ratio
. hsearch mills ratio

or even

. hsearch ratio mills

because word order, capitalization, and punctuation do not matter.

Alternatives to hsearch

Alternatives to hsearch are search and findit:

. search mills ratio
. findit mills ratio

search, like hsearch, searches commands already installed on your computer. search searches the keywords; hsearch searches the help files themselves. Hence, hsearch usually finds everything that search finds, and more. The fewer things that search finds should be more relevant.

findit searches keywords just as search does, but findit searches the web as well as your computer and so may find commands that you might wish to install.

Recommendations

1. In general, hsearch is better than search. hsearch finds more and better organizes the list of what it finds.

2. When you know that Stata can do what you are looking for but you cannot remember the command name, or when you know you installed a relevant user-written package, use hsearch.

3. When you are unsure whether Stata can do a certain task, use hsearch first and then use findit.

How hsearch works

hsearch searches the .sthlp files.

Finding all those files and then looking through them would take a long time if hsearch did that every time you used it. Instead, hsearch builds an index of the .sthlp files and then searches that.

That file is called sthlpindex.idx and is stored in your PERSONAL directory.

Every so often, hsearch automatically rebuilds the index so that it accurately reflects what is installed on your computer. You can force hsearch to rebuild the index at any time by typing

        . hsearch, build

Methods and Formulas

hsearch is implemented as an ado-file.

Also See

Related: [R] **net search**, [R] **search**